TOWN AND COUNTRY PLANNING
IN THE UK

Twelfth Edition

This is the twelfth edition of a book which has become the leading text on the British planning system. This new edition has been completely revised and extended to cover the whole of the United Kingdom and is an authoritative and indispensable guide for the student, the practitioner and the general reader.

Following an introduction on the nature of planning, a historical account traces the evolution of British planning from the prewar years to the present time. *Town and Country Planning in the UK* examines the institutions and organisations involved in the planning process; and the role of central and local government and the European Union in the planning process. Areas selected for more detailed treatment include:

- land and development policies;
- environmental and countryside planning;
- sustainable development, waste and pollution;
- heritage and transport planning;
- urban policies and regeneration;
- public participation in planning.

J. Barry Cullingworth is a Senior Research Fellow in the Department of Land Economy at the University of Cambridge, and Emeritus Professor of Urban Affairs and Public Policy at the University of Delaware. In addition to the previous editions of *Town and Country Planning in Britain*, he is author of *Planning in the USA*. **Vincent Nadin** is Reader in the Faculty of the Built Environment, University of the West of England, Bristol, joint editor of *The EU Compendium of Planning Systems and Policies* and editor of the journal *Planning Practice and Research*.

TOWN AND COUNTRY PLANNING IN THE UK

Twelfth Edition

J. Barry Cullingworth and Vincent Nadin

London and New York

First published 1964
Twelfth edition published 1997
by Routledge
11 New Fetter Lane, London EC4P 4EE

Simultaneously published in the USA and Canada
by Routledge
29 West 35th Street, New York, NY 10001

Typeset in Garamond by J&L Composition Ltd, Filey, North Yorkshire

Printed and bound in Great Britain by The Bath Press

British Library Cataloguing in Publication Data
A catalogue record for this book is available from the British Library

Library of Congress Cataloguing in Publication Data
Cullingworth, J. B.
Town and country planning in Britain / J. Barry Cullingworth and
Vincent Nadin. – 12th ed.
p. cm.
Includes bibliographical references and index.
1. City planning – Great Britain. 2. Regional planning – Great
Britain. I. Nadin, Vincent. II. Title.
HT169.G7C8 1997
361.6′0941–dc21 96–52768

ISBN 0–415–13912–0
0–415–13913–9 (pbk)

CONTENTS

FIGURES

TABLES

PREFACE

The first edition of this book was drafted in 1963. At that time there was little difficulty in deciding what should be included under the title of *Town and Country Planning*. It was largely defined by a limited number of Acts of Parliament – above all the Town and Country Planning Acts. Such a convenient benchmark no longer exists: planning policies are now far broader, and there is general acceptance of the important interrelationships that exist with other spheres of policy (though catering for these has proved exceptionally problematical). It is therefore difficult to define the boundaries of town and country planning. No longer can it be claimed (as it was in the first edition) that the book provides 'an outline of town and country planning and the problems with which it is faced': that would take several volumes.

However, rather than agonising about definitions, the current edition follows the pragmatic course of updating its predecessors – adding in some parts; deleting in others. The result does not satisfy the authors: too many compromises have had to be made, and too much has had to be omitted. Indeed, it is now an open question as to whether a single-volume text can be adequate. But, like practising planners, authors have to operate within constraints which are externally determined.

A significant change to this edition is the addition of a new chapter on 'the nature of planning'. This is intended to provide an introductory background to the main contents of the book. It is supplemented by suggestions for further reading. The same has also been provided for all the chapters.

These 'Further Reading' sections have been substituted for the endnotes (which, as the book continued to grow, were threatening to get out of hand). They are intended to assist students who wish to follow up the discussion in the text. However, they are only an introductory guide to some of the useful available material: they are in no way comprehensive. Though there may well be a need for an annotated bibliography of planning literature, this is not the place to provide it. The literature is now so vast that the selection of titles for recommendation is inevitably a personal (and, to some extent, an arbitrary) matter. However, it is hopefully not idiosyncratic, though no doubt other teachers may prefer alternatives.

The geographical coverage has been extended to cover the whole of the UK, though it is inevitably very uneven. This is in part because of the unevenness of published material on planning in the four countries of the UK. (Hopefully, the situation is improving, even if slowly.) This is a pity: there is a loss of the insights and lessons that might be learned from the planning initiatives and experience of Northern Ireland, Scotland and Wales. The probability that this is a real loss is evident from the limited amount of planning policy evaluation which is available. When so much interesting study is being made of comparative planning systems among the countries of Europe, it is sad that so little attention is paid to the comparative experience of the UK system within its own boundaries – particularly since this could be expected to be more readily applicable.

Acknowledgement is made with sincere thanks to Mark Southgate who commented on the draft chapter on countryside planning. We also wish to thank Sally Jones and Steve Roddie for the preparation of the figures, and the many people who have provided information, comment and advice on the preparation of the book.

Barry Cullingworth
Vincent Nadin

ACRONYMS AND ABBREVIATIONS

Acronyms and abbreviations are a major growth area in public policy. The following list includes all that are used in the text and others that readers will come across in the planning literature. No claim is made for comprehensiveness.

AAI	Area of archaeological importance	AMA	Association of Metropolitan Authorities
ACC	Association of County Councils		
ACCORD	Assistance for coordinated rural development	ANPA	Association of National Park Authorities
ACO	Association of Conservation Officers	AONB	Area of outstanding natural beauty
		AOSP	Areas of special protection for birds
ACOST	Advisory Council on Science and Technology	APRS	Association for the Protection of Rural Scotland
ACRE	Action with Communities in Rural England (formerly Association of Community Councils in Rural England)		
		ARC	Action Resource Centre
		ASNW	Area of semi-natural woodland
		ASSI	Area of special scientific interest (Northern Ireland)
ADAS	Agricultural Development and Advisory Service		
		ATB	Agricultural Training Board
ADC	Association of District Councils		
AESOP	Association of European Schools of Planning	BACMI	British Aggregate Construction Materials Industries
AGR	Advanced gas-cooled reactor	BATNEEC	Best available techniques not entailing excessive cost (Environmental Protection Act 1990, S.7 (10))
AIS	Agricultural improvement scheme		
ALA	Association of London Authorities (now ALG)		
ALG	Association of London Government	BIC	Business in the Community
		BNFL	British Nuclear Fuels Ltd
ALNI	Association of Local Authorities in Northern Ireland	BPEO	Best practical environmental option
ALURE (1)	Alternative land use and rural economy	BPF	British Property Federation
		bpk	billion passenger kilometres
ALURE (2)	Action with Communities in Rural England	BPM	Best practical means
		BR	British Rail

BRE	Building Research Establishment
BRF	British Road Federation
BSI	British Standards Institution
BUD	British Urban Development
BUFT	British Upland Footpath Trust
BWB	British Waterways Board
CADW	Not an acronym, but the Welsh name for the Welsh Historic Monuments Agency. (The word means to keep, to preserve.)
CAF	Coalfields Area Fund
CAP	Common Agricultural Policy
CAT	City action team
CBI	Confederation of British Industry
CC	Countryside Commission
CCS	Countryside Commission for Scotland (now Scottish Natural Heritage)
CCT	Compulsory competitive tendering
CCW	Countryside Council for Wales
CDA	Comprehensive development area
CDP	Community development project
CEGB	Central Electricity Generating Board
CEMAT	Conférence Européen des Ministres d'Aménagement du Territoire (European Conference of Ministers responsible for Regional Planning)
CEMR	Council of European Municipalities and Regions
CES	Centre for Environmental Studies
CFC	Chlorofluorocarbon
CHAC	Central Housing Advisory Committee
CHP	Combined heat and power
CIA	Commercial improvement area
CIEH	Chartered Institute of Environmental Health
CIPFA	Chartered Institute of Public Finance and Accountancy
CITES	Convention on International Trade in Endangered Species
CLA	Country Landowners' Association

CLES	Centre for Local Economic Strategies
CLEUD	Certificate of lawfulness of existing use or development
CLOPUD	Certificate of lawfulness of proposed use or development
CLRAE	Conference of Local and Regional Authorities of Europe (Council of Europe)
CMS	Countryside management system
CNCC	Council for Nature Conservation and Countryside (Northern Ireland)
COBA	Cost-benefit analysis
COE	Council of Europe
COI	Central Office of Information
COMARE	Committee on Medical Aspects of Radiation in the Environment
COPA	Control of Pollution Act (1974)
COR	Committee of the Regions (EU)
CORINE	Community Information System on the State of the Environment (EU)
CoSIRA	Council for Small Industries in Rural Areas
COSLA	Convention of Scottish Local Authorities
CPO	Compulsory purchase order
CPOS	County Planning Officers' Society
CPRE	Council for the Protection of Rural England
CPRS	Central Policy Review Staff
CPRW	Campaign (formerly Council) for the Protection of Rural Wales
CRE	Commission for Racial Equality
CRRAG	Countryside Recreation Research Advisory Group
CSD (1)	Commission on Sustainable Development (UN)
CSD (2)	Committee on Spatial Development (EU)
CSERGE	Centre for Social and Economic Research on the Global Environment
CSF	Community Support Framework

CSO	Central Statistical Office	ECOSOC	Economic and Social Council (United Nations)
CWI	Controlled Waste Inspectorate	ECS	Economic and Social Committee (EU)
DAFS	Department of Agriculture and Fisheries for Scotland	ECSC	European Coal and Steel Community
DATAR	Délégation à l'aménagement du territoire et à l'action régionale (French national planning agency)	ECTP	European Council of Town Planners
DBFO	Design, build, finance and operate (roads by the private sector)	ECU	European currency unit
DBRW	Development Board for Rural Wales	EEA (1)	European Economic Area (EU plus Iceland, Liechtenstein, Norway, and Switzerland)
DC	Development corporation	EEA (2)	European Environmental Agency
DCC	Docklands Consultative Committee	EEC	European Economic Community
		EFTA	European Free Trade Association
DEA	Department of Economic Affairs	EIA	Environmental impact assessment
DG	Directorate General of the European Commission (DGXVI: Regional Policy and Cohesion; DGXI: Environment and Nuclear Safety; DGVII: Transport)	EIB	European Investment Bank
		EIF	European Investment Fund
		EIP	Examination in public
		EIS	Environmental impact statement
		EMAS	Eco-management and audit scheme
DLG	Derelict land grant	EMP	Environmental management areas
DLGA	Derelict land grant advice note	EN	English Nature
DLT	Development land tax	EP	English Partnerships
DNH	Department of National Heritage	EPA (1)	Educational priority area
DoE	Department of the Environment	EPA (2)	Environmental Protection Act (1990)
DoENI	Department of the Environment Northern Ireland	EPC	Economic planning council
DoT	Department of Transport (formerly DTp)	ERDF	European regional development fund
DPOS	District Planning Officers Society	ERP	Electronic road pricing
DRIVE	Dedicated road infrastructure for vehicle safety in Europe	ES	Environmental statement (UK)
		ESA	Environmentally sensitive area
DTI	Department of Trade and Industry	ESDP	European spatial development perspective
DWI	Drinking Water Inspectorate	ESF	European social fund
EA (1)	Environment Agency	ESRC	Economic and Social Research Council
EA (2)	Environmental assessment		
EAGGF	European Agricultural Guidance and Guarantee Fund	ETB	English Tourist Board
		EU	European Union
EBRD	European Bank for Reconstruction and Development	EUCC	European Union for Coastal Conservation
EC	European Community	EURATOM	European Atomic Energy Community
ECMT	European Conference of Ministers of Transport		

EZ	Enterprise zone	GO-SW	Government Office for the South-West
FA	Forestry Authority	GO-WM	Government Office for the West Midlands
FC	Forestry Commission		
FCGS	Farm and Conservation Grant Scheme	GO-YH	Government Office for Yorkshire and Humberside
FEOGA	Fonds européan d'orientation et de garantie agricole (European Agricultural Guidance and Guarantee Fund)	GPDO	General permitted development order
		HAA	Housing action area
FIG	Financial Institutions Group	HAG	Housing association grant
FMI	Financial management initiative	HAT	Housing action trust
FoE	Friends of the Earth	HBF	House Builders' Federation
FTA	Freight Transport Association	HBMC	Historic Buildings and Monuments Commission
FWAG	Farming and Wildlife Advisory Group	HC	House of Commons
FWGS	Farm woodland grant scheme	HCiS	Housing Corporation in Scotland
FWPS	Farm woodland premium scheme	HCLA	Hill livestock compensatory allowance
GATT	General Agreement on Tariffs and Trade	HIDB	Highlands and Islands Development Board (now HIE)
GCR	Geological Conservation Review	HIE	Highlands and Islands Enterprise
GDO	General development order	HIP	Housing investment programme
GDP	Gross domestic product	HL	House of Lords
GDPO	General development procedure order	HLCA	Hill livestock compensatory allowance
GEAR	Glasgow Eastern Area Renewal	HLW	High-level waste
GIA	General improvement area	HMIP	Her Majesty's Inspectorate of Pollution
GLC	Greater London Council		
GLDP	Greater London development plan	HMIPI	Her Majesty's Industrial Pollution Inspectorate (Scotland)
GOR	Government Offices for the Regions:	HMNII	Her Majesty's Nuclear Installation Inspectorate
GO-EM	Government Office for the East Midlands	HMO	Hedgerow management order
GO-ER	Government Office for Eastern Region	HMSO	Her Majesty's Stationery Office
		HRF	Housing Research Foundation
GO-L	Government Office for London	HSA	Hazardous Substances Authority
GO-M	Government Office for Merseyside	HSE	Health and Safety Executive
		HWI	Hazardous Waste Inspectorate
GO-NE	Government Office for the North-East		
GO-NW	Government Office for the North-West	IACGEC	Inter-Agency Committee on Global Environmental Change
GO-SE	Government Office for the South-East	IAEA	International Atomic Energy Agency
		IAP	Inner area programme

IAPI	Industrial Air Pollution Inspectorate	LEC	Local enterprise company (Scotland)
IAURIF	Institut d'aménagement du territoire et d'urbanisme de la région d'Ile de France	LEG-UP	Local enterprise grants for urban projects (Scotland)
ICE	Institution of Civil Engineers	LETS	Local employment and trading systems
ICOMOS	International Council on Monuments and Sites	LFA	Less-favoured area (agriculture)
		LGA	Local Government Association
IDC	Industrial development certificate	LGC	Local Government Commission for England
IEEP	Institute for European Environmental Policy		
		LGMB	Local Government Management Board
IIA	Industrial improvement area		
ILW	Intermediate-level waste	LLW	Low-level waste
INLOGOV	Institute of Local Government Studies (University of Birmingham)	LNR	Local nature reserve
		LOTS	Living over the shop
		LPA	Local planning authority
IPC	Integrated pollution control	LPAC	London Planning Advisory Committee
IPPC	Intergovernmental Panel on Climate Change		
		LSPU	London Strategic Policy Unit
IRD	Integrated rural development (Peak District)	LWRA	London Waste Registration Authority
ISOCARP	International Society of City and Regional Planners		
		MAFF	Ministry of Agriculture, Fisheries and Food
IWA	Inland Waterways Association		
IWAAC	Inland Waterways Amenity Advisory Committee	MCC	Metropolitan county council
		MEA	*Manual of Environmental Assessment* (for trunk roads)
JNCC	Joint Nature Conservation Committee	MEP	Member of the European Parliament
JPL	*Journal of Planning and Environment Law* (formerly *Journal of Planning and Property Law*)	MHLG	Ministry of Housing and Local Government
		MINIS	Management information system for ministers
LA21	Local Agenda 21 (UNCED)	MLGP	Ministry of Local Government and Planning
LAAPC	Local Authority Air Pollution Control		
		MNR	Marine nature reserve
LAW	Land Authority for Wales	MOD	Ministry of Defence
LAWDC	Local authority waste disposal company	MPA	Mineral planning authority
		MPG	Minerals policy guidance note
LBA	London Boroughs Association (now ALG)	MPOS	Metropolitan Planning Officers Society
LCO	Landscape conservation order	MSC	Manpower Services Commission
LDDC	London Docklands Development Corporation	MTCP	Ministry of Town and Country Planning

NACRT	National Agricultural Centre Rural Trust		territorial units; designates levels of regional subdivision in the EU
NAO	National Audit Office	NVZ	Nitrate vulnerable zone
NARIS	National roads information system		
NCB	National Coal Board	OECD	Organisation for Economic Cooperation and Development
NCBOE	National Coal Board Opencast Executive	OEEC	Organisation for European Economic Cooperation
NCC	Nature Conservancy Council		
NCCI	National Committee for Commonwealth Immigrants	OFLOT	Office of the National Lottery
		OJ	*Official Journal of the European Communities*
NCCS	Nature Conservancy Council for Scotland (now Scottish Natural Heritage)	ONS	Office for National Statistics
		OPCS	Office of Population Censuses and Surveys (now part of ONS)
NCVO	National Council of Voluntary Organisations		
NDPB	Non-departmental public body	PAG (1)	Planning Advisory Group
NEDC	National Economic Development Council	PAG (2)	Property Advisory Group
		PAN	Planning advice note (Scotland)
NEDO	National Economic Development Office	PDO	Permitted development order
		PDR	Permitted development right
NERC	National Environment Research Council	PEP	Political and Economic Planning (now PSI)
NFFO	Non-fossil fuel obligation	PFI	Private Finance Initiative
NFU	National Farmers Union	PI	Planning Inspectorate
NGO	Non-governmental organisation	PIC	Planning Inquiry Commission
NHA	Natural heritage area (Scotland)	PLI	Public local inquiry
NHMF	National Heritage Memorial Fund	PPG	Planning policy guidance note
NII	Nuclear Installations Inspectorate	PPP	Polluter pays principle
NIO	Northern Ireland Office	PRIDE	Programmes for rural initiatives and developments (Scotland)
NIREX	Nuclear Industries Radioactive Waste Executive	PSA	Property Services Agency
NNR	National nature reserve	PSI	Policy Studies Institute
NPA	National park authority	PTA	Passenger transport authority
NPF	National Planning Forum	PTE	Passenger transport executive
NPG	National planning guideline (Scotland)	PTRC	Planning and Transport Research and Computation
NPPG	National planning policy guideline (Scotland)	PVC	Polyvinyl chloride
NRA	National Rivers Authority	PWR	Pressurised water reactor
NRTF	National road traffic forecasts (GB)		
NSA (1)	National scenic area (Scotland)	QUANGO	Quasi-autonomous non-governmental organisation
NSA (2)	Nitrate sensitive area		
NTDC	New town development corporation	RA	Renewal area
NUTS	The nomenclature of statistical	RAC	Royal Automobile Club

RAWP	Regional aggregates working parties	SEEDS	South-East Economic Development Strategy
RCC	Rural community council	SEM	Single European market
RCEP	Royal Commission on Environmental Pollution	SEPA	Scottish Environment Protection Agency
RCHME	Royal Commission on the Historical Monuments of England	SERC	Science and Engineering Research Council
RCI	Radiochemical Inspectorate	SERPLAN	London and South-East Regional Planning Conference
RCU	Road construction unit		
RDA	Rural development area	SHAC	Scottish Housing Advisory Committee
RDC	Rural Development Commission		
RDG	Regional development grant	SI	Statutory instrument
RDP	Rural development programme	SINC	Site of importance for nature conservation
REG	Regional enterprise grant		
RIBA	Royal Institute of British Architects	SLF	Scottish Landowners Federation
		SME	Small and medium-sized enterprises (Europe)
RICS	Royal Institution of Chartered Surveyors		
RIGS	Regionally important geological/ geomorphological sites	SMR	Sites and monuments records (counties)
RPG	Regional planning guidance note	SNAP	Shelter Neighbourhood Action Project
RSA (1)	Regional selective assistance		
RSA (2)	Regional Studies Association	SNH	Scottish Natural Heritage
RSPB	Royal Society for the Protection of Birds	SO	Scottish Office
		SOAEFD	Scottish Office Agriculture, Environment and Fisheries Department
RTB	Right to buy (public sector housing)		
RTPI	Royal Town Planning Institute	SODD	Scottish Office Development Department
RWMAC	Radioactive Waste Management Advisory Committee	SOEnD	Scottish Office Environment Department (now SOAEFD)
SAC	Special area of conservation (habitats)	SOID	Scottish Office Industry Department
SACTRA	Standing Advisory Committee on Trunk Road Assessment	SPA	Special protection area (for birds) (EU)
SAGA	Sand and Gravel Association	SPZ	Simplified planning zone
SCLSERP	Standing Conference on London and South East Regional Planning (see also SERPLAN)	SRB	Single regeneration budget
		SSHA	Scottish Special Housing Association
SDA	Scottish Development Agency (now Scottish Enterprise)	SSSI	Site of special scientific interest
		STB	Scottish Tourist Board
SDD	Scottish Development Department		
SDO	Special development order	TAN	Technical advice notes (Wales)
SEA	Strategic environmental assessment	TCPA	Town and Country Planning Association

TCPSS	Town and Country Planning Summer School
TEC	Training and Enterprise Council
TEN	Trans-European Network(s)
TEST	Transport and Environment Studies
THORP	Thermal oxide reprocessing plant
TPO	Tree preservation order
TPP	Transport policy and programme
TPR	*Town Planning Review*
TRL	Transport Research Laboratory (formerly Transport and Road Research Laboratory)
TSG	Transport supplementary grant
TUC	Trades Union Congress
UCO	Use classes order
UDA	Urban development area
UDC	Urban development corporation
UDG	Urban development grant
UDP	Unitary development plan
UKAEA	United Kingdom Atomic Energy Authority
UNCED	United Nations Conference on Environment and Development ('Earth Summit', Rio, 1992)
UNCSD	United Nations Commission on Sustainable Development
UNCTAD	United Nations Conference on Trade and Development
UNECE	United Nations Economic Commission for Europe
UNEP	United Nations Environment Programme
UNESCO	United Nations Educational, Scientific and Cultural Organisation
UP	Urban programme
URA	Urban Regeneration Agency
URG	Urban regeneration grant
VFM	Value for money
VISEGRAD	Four former communist countries: Poland, Czech Republic, Slovakia and Hungary
WCA	Waste collection authority
WCED	World Commission on Environment and Development
WDA (1)	Welsh Development Agency
WDA (2)	Waste disposal authority
WDP	Waste disposal plan
WMEB	West Midlands Enterprise Board
WMO	World Meteorological Organisation
WO	Welsh Office
WOAD	Welsh Office Agriculture Department
WQO	Water quality objectives
WRA	Waste regulation authority
WRAP	Waste reduction always pays
WTB	Welsh Tourist Board
WTO	World Trade Organisation
WWF	World Wide Fund for Nature (formerly World Wildlife Fund)
YTS	Youth training scheme

Encyclopedia refers to Malcolm Grant's *Encyclopedia of Planning Law and Practice*, London: Sweet and Maxwell, loose-leaf, regularly updated by supplements.

1

THE NATURE OF PLANNING

If planning were judged by results, that is, by whether life followed the dictates of the plan, then planning has failed everywhere it has been tried. No one, it turns out, has the knowledge to predict sequences of actions and reactions across the realm of public policy, and no one has the power to compel obedience.

Wildavsky 1987

The challenge for planning in the 1990s is to 'adapt' not only to new substantive agendas about the environment and how to manage it, but to address new ways of thinking about the relation of state and market and state and citizen, in the field of land use and environmental change.

Healey 1992b

INTRODUCTION

It is the purpose of this chapter to give a general introduction to the character of land use planning. Since this is so much a product of culture, it differs among countries. The understanding of one system is helped by comparing it with others and, for this reason, some international comparisons are introduced. The chapter presents a broad discussion of some basic features of the UK planning system, which is essentially a means for reconciling conflicting interests in land use. Many of the arguments about planning revolve around the relationships between theory and practice. Planning theories (along with related theories on management, government and other facets of human interaction) have often been founded on abstract models based on notions of rationality, defined in normative terms. There are difficulties with the concept of rationality. Some of these stem from the fact that planning operates within an economic system which has a

'market rationality' which can differ from, and conflict with, the rationality which is espoused in some planning theories. But the crucial issue is that the concept of rationality cannot be divorced from objectives, ambitions and interests – as well as place and time. These variables are the very stuff of planning: disputes and conflict arise, not because of irrationalities (though these may be present), but because different interests are rationally seeking different objectives. A brief discussion of these matters leads into issues such as those of incrementalism and implementation, both of which present their own rationales for behaviour and attitudes.

A notable feature of the UK system is the unusual extent to which it embraces discretion. This allows for flexibility in interpreting the public interest. It is in sharp contrast to other systems which, more typically, explicitly aim at reducing such uncertainty. The European and US systems, for example, eschew flexibility, and lay emphasis on the protection of property rights. Flexibility is highly regarded in the UK because it enables the planning system to meet diverse requirements and the constantly changing nature of the problems with which it attempts to deal. Shifts in planning policy have been dramatic, and seem to be accelerating – with a greater concern for market forces, though only within circumscribed limits, and with surges in public and political support for and against development and conservation.

CONFLICT AND DISPUTES

Land use planning is a process concerned with the determination of land uses, the general objectives of which are set out in legislation or in some document of legal or accepted standing. The nature of this process will depend in part on the objectives which it is to serve. The broad objective of the UK system is to 'regulate the development and use of land in the public interest' (PPG 1: 2). Like all such policy statements, this is a very wide formulation of purpose. It is not, however, empty of content: it enshrines the essential character of the UK planning system. Its significance is highlighted when it is compared with possible alternatives. These might be 'to encourage the development and use of land' or 'to facilitate the development of land by private persons and corporations'. Other alternatives include: 'to plan the use of land to ensure that private property rights are protected and that the public interest is served'; and (an example from Indiana) 'to guide the development of a consensus land use and circulation plan'. These scene-setting statements convey the overall philosophy or principles which are to guide the planning system. They are important for that reason, and they are of direct concern in disputes about the validity or appropriateness of policies which are elaborated within their framework. They are called upon in support of arguments about specific policies.

Politics, conflict and dispute are at the centre of land use planning. Conflict arises because of the competing demands for the use of land, because of the externality effects that arise when the use of land

changes, and because of the uneven distribution of costs and benefits which result from development. If there were no conflicts, there would be no need for planning. Indeed, planning might usefully be defined as the process by which government resolves disputes about land uses.

Alternatives arise at every level of planning – from the highest (supranational) level to the lowest (site) level. The planning system is the machinery by which these levels of choice are managed – from plan-making to development control. Though planning systems vary among countries, they can all be analysed in these terms. The processes involved encompass the determination of objectives, policy-making, consultation and participation, formal dispute resolution, development control, implementation, and the evaluation of outcomes.

The explicit function of the processes is to ensure that the wide variety of interests at stake are considered and that outcomes are in the general 'public interest'. In reality there are very many interests that might be served. Four main groups of participants are politicians serving various levels of government, the development industry, landowners, and 'the public'. The latter is a highly diverse group which is achieving an increased role (not always meaningfully) by way of pressure groups and public involvement. Governments usually argue that a reasonable balance is being achieved between the different interests. Critics argue that intervention through land use planning serves to maintain the dominance of particular interests. Evaluations of planning suggest that those with a property interest are more influential and get more out of the planning system, but organised interest groups and even some individuals have had success in individual cases, so the outcomes are by no means certain.

One of the reasons for the increased importance attached to planning processes, and public involvement in them (apart from questions of democracy), lies in the belief that it is effective in reducing the scope for later conflict. The clearer and firmer the policy and the wider its support, the less room there is for arguing about its application and implementation. Thus for the managers of the system efficiency is increased. But there are limits to this: there is no way that conflict can be planned away.

A central problem for the planning system is to devise a means for predicting likely future changes that may impact on the system. In fact, this is extremely difficult, and past attempts have demonstrated that there is a severe limit to prediction. This is one of the reasons why discretion has to be built into the processes: without this, it is difficult to take account of changing circumstances. A second, more immediate reason for discretion is the impossibility of devising a process which can be applied automatically to the enormous variety of circumstances that come to light when action is being taken. Plans and other policy documents provide a reference point for what has been agreed through the planning process, and against which proposals will be measured. Professional research and analysis together with opportunities for consultation, public participation and formal objection and adoption by political representatives give such documents legitimacy. But they cannot be blueprints. The implementation of a plan always differs from what is anticipated.

There are several reasons for this. For the individuals concerned, the actuality of implementation may appear different from the perceived promise of a plan. For the landowners involved, the market implications may prove to be unwelcome (whether or not market conditions have changed). There will generally be those whose objections at the plan-making stage were rejected in favour of alternatives, and who will naturally take advantage of any opportunity to repeat their objections at the stage of implementation: the passage of time and the changes that it has wrought provide that opportunity. Changed conditions may be so great that the plan is outdated or even counterproductive. Where this is the case, there is a clear need for a revised plan, but problems mount in the meantime and (since the process for elaborating a new plan has still to be completed) the areas of dispute multiply. In addition to these general issues, there is always scope for dispute on the detailed application of policy to individual cases: no plan can be so detailed as to be self-implementing. Finally, there

are the cases where there is no policy; or where the policy is simply not relevant to the action that needs to be taken. For these and similar reasons, there has to be machinery for settling disputes concerning implementation.

Adjudication of disputes may be the responsibility of an administrative system (which is theoretically subservient to the political system), the courts, or an *ad hoc* machine. The courts have a major role in countries where planning involves issues which are subject to constitutional safeguards (of property rights or due process), or where plans have the force of law. Where there are no such complications, as is typical in the UK, matters of dispute are more likely to be dealt with by an administrative appeal system. However, there is no hard and fast rule about this, and recent years have witnessed an increasing role for the UK courts. Nevertheless, there remains a huge difference in the role of the courts in the UK compared with the USA and most European countries.

PLANNING, THE MARKET AND THE DEVELOPMENT PROCESS

In the early years of the less sophisticated postwar period, plans were drawn up in a vacuum which blissfully ignored the manner in which the property market and development processes worked. Land was allocated to uses which seemed sensible in planning terms, but with little regard to the market. Indeed, market considerations were often explicitly expressed in terms which cast them as subservient to needs. Given the positive role which was envisaged for public development, this had some semblance of logic, but it rapidly disintegrated in the face of the realities of public finance and the incapacity of public authorities to take on the primary development role.

There is today a better (though very incomplete) understanding of how land and property markets work, and a greater appreciation of the need to take account of market trends (even if they have to be subjected to public control or influence). There is also a greater willingness on the part of both the public and the private sectors to pool their efforts and resources: the word 'partnership' is an important addition to the planning lexicon. Of course, this has not ushered in a new era of sweet harmonious cooperation: there are inherent conflicts of interest between (and within) the two sectors. The planning system provides an important mechanism for mediating among these conflicts.

There have been serious difficulties here (by no means resolved) which stem, in part, from a mutual ignorance between the planning and the development sectors. However, attitudes have changed somewhat as a result of a miscellany of forces, ranging from an increased concern on the part of local authorities to promote economic development, to the changed fortunes of the constituent parts of the development industry. Since there is no prospect of a future period of calm stability, attitudes will continue to change. This hardly promises a good base for traditional type planning (as caricatured by the term 'end state planning'); rather, it promises an even greater role for flexibility and discretion.

A mention of some crucial features of the development process highlights the nature of the conflicts with which planning is concerned:

- Developers are concerned with investment and profits, particularly in a time-frame considerably shorter than is typical in the planning world where the preoccupation is with long-term land use.
- Developers need to act quickly in response to market opportunities and the cost of capital. Planners operate in a different time scale.
- Development is much easier on greenfield sites than on inner-city locations: to developers, projects are risky enough without being burdened with 'extraneous' problems. Problems which are 'extraneous' to developers can be central to the concerns and objectives of planners.
- Markets are very diverse: and one location is not as good as another. They are, moreover, dynamic: timing may be the crucial factor in the feasibility of a development. Markets are frequently simply not understood by planners: their concern is more

generally with the unfolding of a long-term plan. Any pressures they experience are more likely to be political rather than economic in origin.
- Developers are concerned with the particular; for planners, the particular is only one among many which add up to the general policy matters with which they are concerned.

Given such major differences between the two sectors, it is not surprising that their relationships can be difficult. The problem for planning is that full consideration for developers' concerns can quickly lead to *ad hoc* responses which undermine planning policy. The very vagueness of policy statements and the high degree of discretion in the system increases the likelihood of this. The dilemma is inherent: there is no simple solution.

It is not surprising that the comprehensive planning philosophy which was dominant in the early postwar period is now discredited. It relates to a world which does not exist. Thus, planning has moved from a preoccupation with grand plans to a concern for finding ways of reconciling the conflicting interests which are affected by development. This leads away from development control to negotiation and mediation. Paradoxically, this is happening at the same time that the central government is attempting to secure a greater degree of certainty through a plan-led system. Perhaps the circle will be squared if it is found that mediation leads to greater certainty? US experience shows how developers can work more easily under a negotiatory system than within a regulatory framework: they know how to operate it to their benefit.

RATIONALITY AND COMPREHENSIVE PLANNING

Central to planning is the concept of rationality. Since rationality requires all relevant matters to be taken into account, the use of the concept readily leads to a comprehensive conception of planning. This stems from two simple ideas: the first (valid) is that, in the real world, everything is related to everything else; the second (invalid) is that the planning of one sector cannot properly proceed without coordinated planning of others. Rationality also requires the determination of objectives (and therefore, though not always explicitly, of values), the definition of the problems to be solved, the formulation of alternative solutions to these problems, the evaluation of these alternatives, and the choice of the optimum policy.

This application of the idea of rational scientific method to policy-making has been subject to scrutiny from many different perspectives, giving rise to a wide range of ideas about the nature of planning. First, there are those who have criticised the simple notion of rationality noted above but have continued to maintain that the task for planning theorists is to elaborate the notion of planning as a set of procedures by which decisions are made (Davidoff and Reiner 1962; Faludi 1987). Second are those who reject the objectivity implied by the simple rational approach and instead focus on the role that planning plays in the distribution of resources among different interests in society. Part of this criticism has included the development of a body of thought variously described as social, community or equity planning, where planning is promoted as a tool that can redress inequalities and work to benefit minority and disadvantaged groups (Gans 1991). Third, there are more fundamental criticisms from those who have used a neo-Marxist critique to draw attention to more fundamental divisions of power in the political and economic structure of capitalist society (Paris 1982). This view asserts that, irrespective of its explicit intentions, planning will inevitably 'serve those interests it seeks to regulate' (Ambrose 1986).

The persuasiveness of the concept of comprehensive rational planning is seen at every stage of the plan-making process, from the initial production of goals and objectives, to definitions of problems, and proposed solutions. But all this is done in the context of the politics of the place and the time, and against the background of public opinion and the acceptability or otherwise of governmental action. Some important issues may be regarded not as problems capable of solution but as powerful economic trends which cannot be reversed. Others may be of a nature for which possible solutions are conceivable but untried, too costly, too administratively difficult, too uncertain, or even dangerous to the long-term future of the area. And, as will be apparent from later chapters, these acutely difficult problems (of urbanisation, congestion, inner-city decay, for example) have continually proved beyond the powers of governments to solve, at least in the short run; and the long run is unpredictable. Big differences of opinion exist among experts, politicians and electors on these matters. As a result, there are severe constraints operating on the planning process, and there is little resembling a logical calm set of procedures informed by intellectual debate. Certainly, the process is far from scientific or rational.

Practitioners are quick to point out that planning involves deciding between opposing interests and objectives: personal gain versus sectional advantage or public benefit, short-term profit versus long-term gain, efficiency versus cheapness, to name but a few. It entails mediation among different groups, and compromise among the conflicting desires of individual interests. Above all, it necessitates the balancing of a range of individual and community concerns, costs and rights. It is essentially a political as distinct from a technical or legal process, though it embraces important elements of both.

Currently, one of the most difficult planning issues is concerned with reconciling the implications of the growth in traffic with traditional ideas about town centres and urban growth. The policy response has been as confused as the issues are complex and politically daunting. So far, the main focus has been on controlling the number of additional out-of-town shopping centres, and exhorting localities to make their town centres more vital and attractive. Conflicts which arise here include the apparently irrepressible demand for car ownership and use, the traditional view that road space is free and should not incur any type of congestion charge, the desire of town centre businesses to maintain their custom and to avoid the risk of losing it because of tighter parking restrictions, the financial difficulties facing public transport, and so on. Any one of these issues on its own would be difficult enough: all of them together constitutes a planning witches' brew. And, as so often happens, ends and means become intertwined in a hopelessly confusing way: protecting city centres, safeguarding inner-city jobs, conserving the countryside, reducing pollution, facilitating ease of access for the car-less as well as the car traveller, providing greater choice for the shopper – which is the objective and which the solution? These issues are examined in later chapters; they are listed here to underline the essentially political nature of planning. Grand phrases about rational planning to 'coordinate land uses' crumble against the stark reality of the complex real world.

The concept of comprehensive planning in theory may be contrasted with the narrowly focused planning which takes place in practice. Each administrative agency takes its decisions within its particular sphere of interest, understanding, resources and competence. How can it be otherwise? The task of any agency is to undertake the task for which it is established, not to take on the complicating and possibly conflicting responsibilities of others (which, in any case would be resistant to a takeover). Thus, a conservation agency will take decisions of a very different character from an economic development agency: they have separate and potentially conflicting goals. The idea that there is some level of planning which can rise above the narrow sectionalism of individual agencies is not only inconceivable in terms of implementation: it also assumes that an overriding objective can be identified and articulated. This is typically expressed in terms of the public interest; yet there are very many 'publics'. They have conflicting interests which are represented by, or reflected in, different agencies of government.

Change takes place not only in physical terms but also socially, economically, institutionally and indeed in many other ways. The spatial restructuring is the most dramatic visually but, in terms of the quality of everyday life, other dimensions are of greater importance: income and income security, employment, health services and education, and

also matters relating to race, handicap, age and gender. Each of these has its own brand of planning, and it is sometimes suggested that there should be an overarching planning system which coordinates all of these. This is an extreme form of comprehensive planning which, even if inconceivable, frequently arises in discussions of the limitations of purely physical planning.

Even coordination among the various agencies of planning is difficult and, not surprisingly, rare. Planners have made more claims for comprehensiveness than other professionals; indeed, the search for this has been their distinguishing feature. The fact that they have neither the responsibilities nor the resources for such ambitious aims has not, in the past, prevented them from being articulated. A classic example is the Greater London Development Plan which was subject to a searching inquiry in 1970. The plan dealt not only with the land use issues which fell within the remit of the Greater London Council, but also a wide range of policy areas including employment, education, transport, health, and income distribution. There was no doubt about the importance of these and similar issues, but they did not fall within the responsibility of the Council; indeed, it had no way of exerting any influence in these fields (Centre for Environmental Studies 1970).

The experience of the Greater London Development Plan cast a shadow over the hopes of the more ambitious planners of the time, and it affected the attitude of central government to the definition of matters which were relevant to an official development plan. It has taken a long time for central government to feel able to countenance giving official blessing to the need even to take account of 'social needs and problems'. That it has now done so (in PPG 12, quoted later) suggests that the fear of over-ambitious plan-making has receded: local planning authorities are now too experienced in the implementation of plans to seek the impossible. Indeed, the contemporary problem may be the opposite one of believing that too little is possible. The apparent impossibility of tackling problems raised by increased traffic, rising housing demands and such like could lead to a virtual demise of all planning that is not simply regulatory: planning could degenerate into nothing more than a sophisticated form of building control. This is probably an exaggeration – but it is not impossible.

INCREMENTALISM

The obvious failure of comprehensive planning to attain desired goals has led to a number of alternative models of decision-making processes. Many of these revolve around the problem of making planning effective in a world where values, attitudes and aspirations differ, where market and political forces predominate, and where uncertainty prevails. Lindblom (1959) dismissed rational-comprehensive planning as an impractical ideal. In his view, it is necessary to accept the realities of the processes by which planning decisions are taken: for this he outlined a 'science of muddling through'. Essentially, this incrementalist approach replaces grand plans by a modest step-by-step approach which aims at realisable improvements on an existing situation. This is a method of 'successive limited comparisons' of circumscribed problems and actions to deal with them. Lindblom argues that this is what happens in the real world: rather than attempt major change to achieve lofty ends, planners are compelled by reality to limit themselves to acceptable modifications of the status quo. On this argument, it is impossible to take all relevant factors into account or to separate means from ends. Rather than attempting to reform the world, the planner should be concerned with incremental practicable improvements. There has been much debate on Lindblom's ideas (a good selection is given in Faludi 1973); here it is necessary only to make two points.

First, incrementalism is theoretically different from opportunism: it is a rational and realistic approach to dealing with problems. It rejects a comprehensive analysis of all the available options, and concentrates on what appears practicable and sensible given the constraints of time and resources. The classic illustration of the infeasibility of its

opposite is the zero-base budget which, instead of being based on a previous year's figures, rejects history and questions the justification for every individual item. (The term comes from the base line of the new budget – zero.) As Wildavsky (1987) has demonstrated, this is a completely unmanageable approach: it overwhelms, frustrates and finally exhausts those who try it. (Of course, selected items may with benefit be isolated for such treatment; but that is a very different matter.)

Second, incrementalism is more of a practical necessity than a desirable model to be followed. All policies need thorough review at times – particularly those embedded in an established development plan. Without the occasional upheaval (and that is what zero-base budgeting or policy-making implies), policies can continue well after they have served their purpose: they may even have become counterproductive. Indeed, incrementalism can lead to disasters, wars often being dramatic illustrations of the point. (The escalation of the Vietnam war is an horrendous example.) The ease with which incrementalism continues does, in fact, make a break in the continuity difficult; and often both the political and the administrative systems are averse to change. Nevertheless, changes in direction are sometimes essential; and events (particularly unexpected ones) may create the basis for a change, despite any fears about uncertain outcomes. The difficulties are well illustrated by the current heart-searching about transport policy. Here, a reversal of trends is necessary, and is increasingly being recognised; but how change shall be brought about, and what its character shall be, is highly problematic.

IMPLEMENTATION

The rational model of planning embraces the simplistic view that there is a logical progression through successive stages of 'planning', culminating in implementation. The beguiling logic does not translate into reality. On the contrary, it is highly misleading and dangerous to separate policy and implementation matters. In fact, sometimes policy emanates from ideas about implementation rather than the other way round. Thus a policy of slum clearance or redevelopment focuses on the clearly indicated types of action. The implementation becomes the policy, and the underlying purpose is left in doubt. If the objective is to improve the living conditions of those living in slum areas, there might be better ways of doing this, such as rehabilitation, or area improvement through local citizen action. With such an approach, demolition might be merely an incidental element in the local programme. With clearance as a policy, however, there is a danger that quite different objectives might be served, such as commercial development, or provision for roads and car parking.

Unfortunately, the difficulties involved here are even greater than this suggests since focused efforts are clearly not enough. For instance, a policy of improving a low-income area by environmental improvements may be explicitly intended to benefit the existing inhabitants, but the added attraction of the area may become reflected in higher rents and prices which could lead to gentrification, thus benefiting a very different group. Similarly, a policy of providing grants to industrialists to move to an area of unemployment may result in the substitution of capital for labour, or the influx of workers with skills not possessed by the local people. A policy of preserving historic buildings by prohibiting demolition or alteration may lead to accelerated deterioration as owners seek ways of circumventing the regulations (and, in the period before the prohibition comes into effect, a rash of demolitions – as with the Firestone factory on the Great West Road in London which was demolished on a holiday weekend). A policy of reducing urban congestion by controlling growth through the designation of green belts may result in 'leap-frogging' of development, increased commuting, and thus increased urban congestion. Examples could easily be multiplied (Derthick 1972; P. Hall 1980; Kingdon 1984).

To confuse matters further, arguments about such effects are often complicated by differing views on what the objectives of the policy really are. The green belt case is a particularly good illustration of

the point, since defenders (and there are many) can slip from one objective to another with ease. If the green belt does not reduce urban congestion, it provides 'opportunities for access to the open countryside for the urban population' and 'opportunities for outdoor sport and outdoor recreation near urban areas'; and if it does not do this, it does 'retain attractive landscapes and enhance landscapes, near to where people live'. Other objectives (all listed in PPG 2) are 'to improve damaged and derelict land around towns'; 'to secure nature conservation interest'; and 'to retain land in agricultural, forestry and related uses'. There is nothing unique in such a long list of miscellaneous policy objectives (though this one is unusual in the manner in which it is conveniently assembled and articulated). It would be a very sad policy indeed that was unable to meet any objectives in such a list! (To quote Wildavsky's pithy observation: 'objectives are kept vague and multiple to expand the range within which observed behaviour fits'; Wildavsky 1987: 35).

It should be added that sometimes policies have unintended good byproducts. Unfortunately, it is often difficult to relate cause and effect, but one example is the imposition by the US federal government of a 55-miles per hour speed limit. This was introduced to reduce petrol consumption, but a welcome effect was a reduction in road accidents: this, for a time, became the basis of a powerful argument for retaining the speed limit after the fuel crisis had passed.

The points do not need labouring: the certainty which is required for the type of rational planning envisaged in some traditional theories is impossible. The underlying assumptions, relevance and political support can change dramatically; and the outcomes of policy are difficult to predict, are frequently different from expectations, are hard to identify and to separate from all the other forces at work, and rarely clear. Thus, not only is planning a hazardous exercise, with serious likelihood of failure: it is also an exercise whose outcomes are remarkably difficult to evaluate, even when they are felt to be a resounding success. It is perhaps unsurprising that most planners have neither the time nor remit to examine what went wrong with the last plan: they have moved on to the next one! It is, however, a matter of some surprise that there have been so few analyses of the (UK) planning scene to fill the vacuum. The wealth of US studies indicates how valuable this can be. Perhaps it is another cultural characteristic that there is little interest in learning why things go wrong?

THE BRITISH PLANNING SYSTEM IN COMPARATIVE PERSPECTIVE

Since it is easier to understand one planning system by comparing it with another, it is worth exploring a little further the differences between the UK, the USA and other European systems. Three features are of particular interest: first, the extent to which a planning system operates within a framework of constitutionally protected rights; second, the degree to which a system embodies discretion; third, the importance of history and culture.

In many countries, the constitution limits governmental action in relation to land and property. In the USA, the Bill of Rights provides that 'no person shall . . . be deprived of life, liberty or property without due process of law; nor shall private property be taken without just compensation'. These words mean much more than is apparent to the casual (non-American) reader. Since land use regulations affect property rights, they are subject to constitutional challenge. They can be disputed not only on the basis of their effect on a particular property owner (that is, *as applied*), but also in principle: a regulation can be challenged on the argument that, in itself, it violates the constitution (this is described in the legal jargon as being *facially* unconstitutional). Moreover, the constitution protects against arbitrary government actions, and this further limits what can be done in the name of land use planning. No such restraints exist in the UK system. Indeed, the UK does not have a codified constitution of the type common to most other countries (Yardley 1995).

Constitutions can influence the system in more subtle ways. In some European countries, including Italy, the Netherlands and Spain, the constitution provides that all citizens have the right to a decent home. This may limit planning action, but it may also influence policy priorities and provide legitimacy for intervention. In Finland and Portugal, landowners are granted the constitutional right to build on their land. This presents obvious difficulty in pursuing policies of restraining urban growth. Constitutions also often allocate powers to different tiers of government, which effectively ensures a minimum degree of autonomy for regional and local governments. Again, there is no such constitutional safeguard in the UK. As a result, the Thatcher government was able to abolish a whole tier of metropolitan local government in England and, in consequence, that part of the planning system that went with it. Such haughty action would be inconceivable in most countries. In the United States, for example, there is little to compare with the central power which is exercised by the national government in Britain. Plan-making and implementation are essentially local issues, even though the federal government has become active in highways, water and environmental matters and, in recent years, a number of states have become involved in land use planning. So local is the responsibility that even the decision on whether to operate land use controls is a local one; and many US local governments have only minimal systems.

Lack of constitutional constraint allows for a wide degree of discretion in the UK planning system. In determining applications for planning permission, a local authority is guided by the development plan, but is not bound by it: other 'material considerations' are taken into account. In most of the rest of the world, plans become legally binding documents. Indeed, they are part of the law, and the act of giving a permit is no more than a certification that a proposal is in accordance with the plan. In practice, there are mechanisms that allow for variations from the provisions of a plan but, since these are by definition contrary to the law, they may entail lengthy procedures and perhaps an amendment to the plan.

This discretion is further enlarged by the fact that the preparation of a local plan is carried out by the same local authority that implements it. This is so much a part of the tradition of British planning that no one comments on it. The American situation is different, with great emphasis being placed on the separation of powers. (Typically the plan is prepared by the legislative body – the local authority – but administered by a separate board.) The British system has the advantage of relating policy and administration (and easily accommodating policy changes) but, to American eyes, 'this institutional framework blurs the distinction between policy making and policy applying, and so enlarges the role of the administrator who has to decide a specific case' (Mandelker 1962: 4).

Above all, in comparing planning systems, there are fundamental differences in the philosophy that underpins them. Thus, put simply (and therefore rather exaggeratedly), American planning is largely a matter of anticipating trends, while in the UK there is a conscious effort to bend them in publicly desirable directions. In France, *aménagement du territoire* (the term often incorrectly used as a translation for town and country planning) deals with the planning of the activities of different government sectors to meet common social and economic goals, while in the UK town and country planning is about the management of land use, albeit taking into account social and economic concerns.

Planning systems are rooted in the particular historical, legal and physical conditions of individual countries and regions. In the UK, some of the many important factors which have shaped the system are the strong land preservation ethic, epitomised in the work of the Council for the Protection of Rural England (and its Scottish and Welsh counterparts) and, of longer standing, the husbandry of the landowning class. Added (but not unrelated to this) are the popular attitudes to the preservation of the countryside and the containment of urban sprawl which in turn is related to the early industrialisation of the UK; the small size of the country; the long history of parliamentary government; and

the power of the civil service in central government and the professions in local government.

In comparison, land in the USA has historically been a replaceable commodity that could and should be parcelled out for individual control and development; and if one person saw fit to destroy the environment of his valley in pursuit of profit, well, why not? There was always another valley over the next hill. Thus the seller's concept of property rights in land came to include the right of the owner to earn a profit from his land, and indeed to change the very essence of the land, if necessary, to obtain that profit.

In the Mediterranean countries of Greece, Italy, Portugal and Spain there has been a short history of democratic government, and planning regulation has enjoyed little general public support. Controlling land use has been much less a political priority than housing the population. In large parts of these countries rapid urbanisation has proceeded with little regard to regulations or plans. The historic cores of cities, meanwhile, have not until recently felt the scale of pressure for redevelopment which has been the norm in northern Europe.

However, in all countries land for development is becoming more valuable, and the problem of coping with land use conflicts is of increasing importance. In Europe this has led to the growth of a conservationist ethic, with the restraint of urban growth being a top priority. In the USA, this has happened to a limited extent, particularly with environmentally valuable resources, but a major effect was in the opposite direction: to increase the attractiveness of land as a source of profit. Speculation has never been frowned upon in the USA. In many countries, land is regarded as different from commodities: it is something to be preserved and husbanded. In the US, the dominant ethic regards land as a commodity, no different from any other. Though there is much rhetoric to the contrary, actions speak louder than words.

The contrast in the operation of planning in different countries is abundantly clear to anyone who travels.

ACCOMMODATING CHANGE

Having drawn the comparison, it is immediately necessary to qualify it: times and attitudes change, sometimes slowly, sometimes dramatically. The largest postwar change in the UK has been the move from 'positive planning' to a more market-conscious (and sometimes market-led) approach. The elements of this (which range from the abolition of development charges to the embrace of property-led urban regeneration) are discussed later, but it needs to be stressed that the extent of the change in planning attitudes toward market forces has been dramatic. The limits of the possible have been redefined in the light of experience and a recognition of the character of the forces at work in the modern world.

Governments are responsive to shifts in electoral opinion, particularly when changes can be made painlessly. The UK planning system provides a route by which change can be implemented, not only without pain, but also with little effort. Indeed, the ease with which it can accommodate change is quite remarkable. There has, for instance, been a see-saw in the extent to which economic development, social needs and environmental concerns have had a high profile. In the 1980s, economic efficiency rose to prime place in the government's order of priorities. (This was the time when the planning system was attacked for its restrictive character: 'locking away jobs in filing cabinets'.) Environmental concerns later became salient: a result of a fascinating combination of conservative forces, ranging from green-belt voters keen to protect the belt, to a younger generation of protestors who had less to lose but saw more to protect. Social considerations have for long been regarded by central government as being outside the legitimate responsibility of the planning system (a curiously British myopia). Major arguments have raged between the centre and the localities on what is and what is not appropriate for inclusion in a development plan. After many years of pressing local authorities to exclude 'social factors', the central government made a curious about-turn in a planning policy guidance note of 1992: 'authorities will wish to consider the relationship of planning policies and proposals to social needs and problems' (PPG 12, para. 5.48).

This flexibility (another aspect of the discretionary nature of the system) is a built-in feature. The statutory framework is essentially procedural; it is almost devoid of substantive content. Local authorities are given the duty to prepare development plans (a rare case where no discretion is allowed; unlike their US counterparts, a local authority is not free to decide not to have a plan). What goes into the plan, however, is very imprecise: 'general policies in respect of the conservation of the natural beauty and amenity of the land; the improvement of the physical environment; and the management of traffic' (TCP Act 1990, section 13.3A). More detailed requirements are, of course, spelled out in a range of directions and advice from the central government. But that is the point: the content is added separately, and can be changed in line with what 'the Secretary of State may prescribe'.

Yet changes are often not easy to evaluate, even if only because the implementation of planning policy rests with local authorities and, despite much bandying of words, central government powers over them are limited. There are, of course, various control mechanisms and default powers, but these are cumbersome to use, and they carry political risks. Moreover, central government's understanding of how local government works, and its awareness of what happens in practice, is even more circumscribed. These depths of ignorance have had surprisingly little academic light shed on them: few studies have been undertaken of the actual working of the planning machine. (Note the surprise which was expressed when the report on development control in North Cornwall (Lees 1993) revealed that the local councillors gave favourable consideration to the personal circumstances of local applicants for planning permission.)

Given such considerations, it can be difficult to chart (or even to be aware of) important changes. Legislative amendments and new policy statements are more apparent, but they may not be as important as trends which emerge over time. (For example, it *may* be that one of the most significant operational changes has been the way in which local authorities and housebuilders have evolved a system for negotiating housing land allocations; perhaps in time this model might be followed in other development sectors?) Moreover, major political statements and new laws typically follow rather than precede changes in attitudes and perceptions. The picture is also confused by grand claims for new approaches which seldom last far beyond an initial flurry. Much is obscured by political debate and the use of fuzzy jargon. Changes are more easily seen in retrospect than contemporaneously.

PLANNING QUESTIONS

Planning is an imperative: only the form it takes is optional. At a minimum, some system is required to provide infrastructure – preferably in the right place at the right time. Something more than this is generally accepted to be necessary (and general acceptance is the bedrock of any form of effective planning). But there is no way of determining the extent to which a planning system *ought* to go in determining 'how much of what should go where and when'. The decision is a political one, even if it is taken by default (i.e. with no effective opposition to its growth). Hence, as stressed earlier, cultural influences are crucial. However, this does not mean that a planning system is hallowed or immune from review and radical change. It may be that the UK system has reached precisely the stage at which this is required, though this is not the place to elaborate such an argument. It is, nevertheless, appropriate to point to some issues which need addressing if the planning system is to adapt to conditions which are very different from those which existed half a century ago when it was introduced.

First, the UK planning system is highly effective in stopping development: it is much less effective in facilitating it. Comparative research on property markets in Europe (Williams and Wood 1994) underlines the lack of 'positive planning'. There are serious weaknesses in anticipating needs and allocating sufficient land for these to be met; in

the assembling and acquisition of land (especially in inner cities); and integrating the planning of infrastructure with new development. Powers exist for such important planning actions, but they are underused since there is insufficient relationship between the (public) planning process and the (largely private) development process. In the 1947 Act, it was envisaged that 'positive' planning would be undertaken directly by the public sector. This proved infeasible; and alternative mechanisms are underdeveloped.

Second, the most difficult issue facing any policy is defining the right questions. A mechanism is needed to facilitate this. It could be argued that current UK debates are focused on the wrong questions. Too many are concerned with the 'how' of planning policy rather than the 'why'. Why is the countryside to be protected? Why are city centres to be rehabilitated? Why are additional facilities for travel to be provided? There are many such questions, and though they do not have simple (or readily acceptable) answers, debate upon them would provide a firmer base for policy than exists at present. The debate would, however, raise a further level of policy questions. Thus, it might be asked where retail outlets should be located to maximise convenience, service and profitability, (or whatever other criteria are to employed), rather than posing the questions in terms of safeguarding existing patterns of development (particularly existing town centres). Instead of asking where the best locations are for housing an additional *x* million households, argument rages over protecting the countryside from housing development and concentrating new housing in urban areas.

Third, planning deals with a highly complex series of interrelated processes which are imperfectly understood. Though better understanding should be high on the research agenda, these processes will inevitably remain beyond the comprehension needed for fully competent land use planning. It follows that planning must proceed on the basis of either a high degree of ignorance or belief in the efficacy of some overriding political or economic philosophy. In practical terms, this implies debating how far the planning process should ally itself to market forces (or socio-economic trends if that term is preferred).

These issues arise throughout this book. It should be evident from this introductory discussion that they are not easily resolved. Indeed, much 'planning' effort is spent on wrestling with them. There seems no doubt that this will continue.

FURTHER READING

Good starting points on the nature of planning are chapter 2 of Healey *et al.* (1982) *Planning Theory: Prospects for the 1980s*, and Ravetz (1986) *The Government of Space* (which contains a chapter on 'theoretical perspectives'). A useful collection of early articles is contained in Faludi (1973) *A Reader in Planning Theory*. (This includes the paper by Lindblom referred to in the chapter, together with important papers by writers such as Davidoff, Etzioni, Friedmann, and Meyerson). Later collections contain more recent writings: Campbell and Fainstein (1996) *Readings in Planning Theory*, and Fainstein and Campbell (1996) *Readings in Urban Theory*. A helpful analysis of 'Arguments for and against planning' is given by Klosterman (1985). For an insightful, succinct discussion of the constant flood of changes which besets planning, see Batty (1990) 'How can we best respond to changing fashions in urban and regional planning?'. Sillince (1986) *A Theory of Planning* gives useful summaries of rational comprehensive and incremental models of procedural planning theory. A fuller account is provided by Alexander (1992) *Approaches to Planning*.

A particularly useful introduction to the analysis of policy issues is Kingdon (1984) *Agendas, Alternatives and Public Policies*. A clear and succinct account of policy processes is given in Ham and Hill (1993) *The Policy Process in the Modern Capitalist State*. There is a good range of readings in an accompanying volume edited by Hill (1993) *The Policy Process*. There are a number of useful papers in Tewdwr-Jones (1996) *British Planning Policy in Transition*. Hall (1980) *Great Planning Disasters* is required reading for all planners, as well as for non-planners who want to know why planning is so difficult. A more complex, but fascinating, study focused on the operation of US federal policy in one urban area is Pressman and Wildavsky (third edition, 1984) *Implementation*. Very interesting as well as revealing is Derthick (1972) *New Towns In-town: Why a Federal Program Failed*. Such case studies are much more common in the US than in the UK (a reflection of the cultural differences in the openness of government). Among the small number of British studies, see Muchnick (1970) *Urban Renewal in Liverpool*; Levin (1976) *Government and the Planning Process* (which focuses on the new and expanded towns); Barrett and Fudge (1981) *Policy and Action*; Healey (1983) *Local Plans in British Land Use Planning*; and Blowers (1984) *Something in the Air: Corporate Power and the Environment*. Simmie (1994) *Planning at the Crossroads* summarises research findings on the impact of planning in the UK. A radical critique of the role of planning in society is Ambrose (1986) *Whatever Happened to Planning?* See also his *Urban Process and Power* (1994).

For a comparative study of 'certainty and discretion' in planning, see Booth (1996) *Controlling Development*. Discretion is discussed at length (comparing the UK and the US) in Cullingworth (1993) *The Political Culture of Planning*. A broader discussion of the two countries is given by Vogel (1986) *National Styles of Regulation*.

2

THE EVOLUTION OF TOWN AND COUNTRY PLANNING

The first assumption that we have made is that national planning is intended to be a reality and a permanent feature of the internal affairs of this country.

Uthwatt Report 1942

THE PUBLIC HEALTH ORIGINS

Town and country planning as a task of government has developed from public health and housing policies. The nineteenth-century increase in population and, even more significant, the growth of towns led to public health problems which demanded a new role for government. Together with the growth of medical knowledge, the realisation that over-crowded insanitary urban areas resulted in an economic cost (which had to be borne at least in part by the local ratepayers) and the fear of social unrest, this new urban growth eventually resulted in an appreciation of the necessity for interfering with market forces and private property rights in the interest of social well-being. The nineteenth-century public health legislation was directed at the creation of adequate sanitary conditions. Among the measures taken to achieve these were powers for local authorities to make and enforce building bylaws for controlling street widths, and the height, structure and layout of buildings. Limited and defective though these powers proved to be, they represented a marked advance in social control and paved the way for more imaginative measures. The physical impact of bylaw control on British towns is depressingly still very much in evidence; and it did not

escape the attention of contemporary social reformers. In the words of Unwin (1909: 3):

much good work has been done. In the ample supply of pure water, in the drainage and removal of waste matter, in the paving, lighting and cleansing of streets, and in many other such ways, probably our towns are as well served as, or even better than, those elsewhere. Moreover, by means of our much abused bye-laws, the worst excesses of overcrowding have been restrained; a certain minimum standard of air-space, light and ventilation has been secured; while in the more modern parts of towns, a fairly high degree of sanitation, of immunity from fire, and general stability of construction have been maintained, the importance of which can hardly be exaggerated. We have, indeed, in all these matters laid a good foundation and have secured many of the necessary elements for a healthy condition of life; and yet the remarkable fact remains that there are growing up around our big towns vast districts, under these very bye-laws, which for dreariness and sheer ugliness it is difficult to match anywhere, and compared with which many of the old unhealthy slums are, from the point of view of picturesqueness and beauty, infinitely more attractive.

It was on this point that public health and architecture met. The enlightened experiments at Saltaire (1853), Bournville (1878), Port Sunlight (1887) and elsewhere had provided object lessons. Ebenezer Howard and the garden city movement were now exerting considerable influence on contemporary thought. The National Housing Reform Council (later the National Housing and Town Planning Council) was campaigning for the introduction of town planning. Even more significant was a similar demand from local government and professional associations such as the Association of Municipal Corporations, the Royal Institute of British Architects, the Surveyors' Institute and the Association of Municipal and County Engineers. As Ashworth (1954: 180) has pointed out:

> the support of many of these bodies was particularly important because it showed that the demand for town planning was arising not simply out of theoretical preoccupations but out of the everyday practical experience of local administration. The demand was coming in part from those who would be responsible for the execution of town planning if it were introduced.

THE FIRST PLANNING ACT

The movement for the extension of sanitary policy into town planning was uniting diverse interests. These were nicely summarised by John Burns, President of the Local Government Board, when he introduced the first legislation bearing the term 'town planning' – the Housing, Town Planning, Etc. Act 1909:

> The object of the bill is to provide a domestic condition for the people in which their physical health, their morals, their character and their whole social condition can be improved by what we hope to secure in this bill. The bill aims in broad outline at, and hopes to secure, the home healthy, the house beautiful, the town pleasant, the city dignified and the suburb salubrious.

The new powers provided by the Act were for the preparation of 'schemes' by local authorities for controlling the development of new housing areas. Though novel, these powers were logically a simple extension of existing ones. It is significant that this first legislative acceptance of town planning came in an Act dealing with health and housing. The gradual development and the accumulated experience of public health and housing measures facili-

tated a general acceptance of the principles of town planning:

> Housing reform had gradually been conceived in terms of larger and larger units. Torrens' Act (Artizans and Labourers Dwellings Act, 1868) had made a beginning with individual houses; Cross's Act (Artizans and Labourers Dwellings Improvement Act, 1875) had introduced an element of town planning by concerning itself with the reconstruction of insanitary areas; the framing of bylaws in accordance with the Public Health Act of 1875 had accustomed local authorities to the imposition of at least a minimum of regulation on new building, and such a measure as the London Building Act of 1894 brought into the scope of public control the formation and widening of streets, the lines of buildings frontage, the extent of open space around buildings, and the height of buildings. Town planning was therefore not altogether a leap in the dark, but could be represented as a logical extension, in accordance with changing aims and conditions, of earlier legislation concerned with housing and public health.
> (Ashworth 1954: 181)

The 'changing conditions' were predominantly the rapid growth of suburban development: a factor which increased in importance in the following decades:

> In fifteen years 500,000 acres of land have been abstracted from the agricultural domain for houses, factories, workshops and railways . . . If we go on in the next fifteen years abstracting another half a million from the agricultural domain, and we go on rearing in green fields slums, in many respects, considering their situation, more squalid than those which are found in Liverpool, London and Glasgow, posterity will blame us for not taking this matter in hand in a scientific spirit. Every two and a half years there is a County of London converted into urban life from rural conditions and agricultural land. It represents an enormous amount of building land which we have no right to allow to go unregulated.
> (*Parliamentary Debates*, 12 May 1908)

The emphasis was entirely on raising the standards of *new* development. The Act permitted local authorities (after obtaining the permission of the Local Government Board) to prepare town planning schemes with the general object of 'securing proper sanitary conditions, amenity and convenience', but only for land which was being developed or appeared likely to be developed.

Strangely it was not at all clear what town planning involved. Aldridge (1915: 459) noted that it certainly did not include 'the remodelling of the existing town, the replanning of badly planned areas, the driving of new roads through old parts of a town – all these are beyond the scope of the new planning powers'. The Act itself provided no definition: indeed, it merely listed nineteen 'matters to be dealt with by general provisions prescribed by the Local Government Board'. The restricted and vague nature of this first legislation was associated in part with the lack of experience of the problems involved: Nettlefold (1914: 179) even went so far as to suggest that 'when this Act was passed, it was recognised as only a trial trip for the purpose of finding out the weak spots in local government with regard to town and estate development so that effective remedies might be later on devised'.

Nevertheless, the cumbersome administrative procedure devised by the Local Government Board (in order to give all interested parties 'full opportunity of considering all the proposals at all stages') might well have been intended to deter all but the most ardent of local authorities. The land taxes threatened by the 1910 Finance Act, and then the First World War, added to the difficulties. It can be the occasion of no surprise that very few schemes were actually completed under the 1909 Act.

INTERWAR LEGISLATION

The first revision of town planning legislation which took place after the war (the Housing and Town Planning Act of 1919) did little in practice to broaden the basis of town planning. The preparation of schemes was made obligatory on all borough and urban districts with a population of 20,000 or more, but the time limit (January 1926) was first extended (by the Housing Act 1923) and finally abolished (by the Town and Country Planning Act 1932). Some of the procedural difficulties were removed, but no change in concept appeared. Despite lip-service to

the idea of town planning, the major advances made at this time were in the field of housing rather than planning. It was the 1919 Act which began what Marion Bowley (1945: 15) has called 'the series of experiments in State intervention to increase the supply of working-class houses'. The 1919 Act accepted the principle of state subsidies for housing and thus began the nationwide growth of council house estates. Equally significant was the entirely new standard of working-class housing provided: the three-bedroom house with kitchen, bath and garden, built at the density recommended by the Tudor Walters Committee (1918) of not more than twelve houses to the acre. At these new standards, development could generally take place only on virgin land on the periphery of towns, and municipal estates grew alongside the private suburbs: 'the basic social products of the twentieth century', as Asa Briggs (1952 vol. 2: 228) has termed them.

This suburbanisation was greatly accelerated by rapid developments in transportation – developments with which the young planning machine could not keep pace. The ideas of Howard (1898) and the garden city movement, of Geddes (1915) and of those who, like Warren and Davidge (1930), saw town planning not just as a technique for controlling the layout and design of residential areas, but as part of a policy of national economic and social planning, were receiving increasing attention, but in practice town planning typically meant little more than an extension of the old public health and housing controls.

Various attempts were made to deal with the increasing difficulties. Of particular significance were the Town and Country Planning Act of 1932, which extended planning powers to almost any type of land, whether built-up or undeveloped, and the Restriction of Ribbon Development Act 1935, which, as its name suggests, was designed to control the spread of development along major roads. But these and similar measures were inadequate. For instance, under the 1932 Act, planning schemes took about three years to prepare and pass through all their stages. Final approval had to be given by Parliament, and schemes then had the force of law, as a result of which variations or amendments were not possible except by a repetition of the whole procedure. Interim development control operated during the time between the passing of a resolution to prepare a scheme and its date of operation (as approved by Parliament). This enabled, but did not require, developers to apply for planning permission. If they did not obtain planning permission, and the development was not in conformity with the scheme when approved, the planning authority could require the owner (without compensation) to remove or alter the development.

All too often, however, developers preferred to take a chance that no scheme would ever come into force, or that if it did no local authority would face pulling down existing buildings. The damage was therefore done before the planning authorities had a chance to intervene (Wood 1949: 45). Once a planning scheme was approved, on the other hand, the local authority ceased to have any planning control over individual developments. The scheme was in fact a zoning plan: land was zoned for particular uses such as residential or industrial, though provision could be made for such controls as limiting the number of buildings and the space around them. In fact, so long as developers did not try to introduce a non-conforming use they were fairly safe. Furthermore, most schemes did little more than accept and ratify existing trends of development, since any attempt at a more radical solution would have involved the planning authority in compensation which they could not afford to pay. In most cases, the zones were so widely drawn as to place hardly more restriction on the developer than if there had been no scheme at all. Indeed, in the half of the country covered by draft planning schemes in 1937 there was sufficient land zoned for housing to accommodate 291 million people (Barlow Report 1940: para. 241).

ADMINISTRATIVE SHORTCOMINGS

A major weakness was, of course, the administrative structure itself. At the local level, the administrative unit outside the county boroughs was the district council. Such authorities were generally small and weak. This was implicitly recognised as early as 1919, for the Act of that year permitted the establishment of joint planning committees. The 1929 Local Government Act went further, by empowering county councils to take part in planning, either by becoming constituent members of joint planning committees or by undertaking powers relinquished by district councils. A number of regional advisory plans were prepared, but these were generally ineffective and, in fact, were conceived as little more than a series of suggestions for controlling future development, together with proposals for new main roads.

The noteworthy characteristic of a planning scheme was its regulatory nature. It did not secure that development would take place: it merely ensured that if it did take place in any particular part of the area covered by the scheme it could be controlled in certain ways. Furthermore, as the Uthwatt Report (1942: para. 13) stressed, the system was

> essentially one of local planning based on the initiative and financial resources of local bodies responsible to local electorates . . . The local authorities naturally consider questions of planning and development largely with a view to the effect they will have on the authorities' own finances and the trade of the district. Proposals by landowners involving the further development of an existing urban area are not likely in practice to be refused by a local authority if the only reason against the development taking place is that from the national standpoint its proper location is elsewhere, particularly when it is remembered that the prevention of any such development might not only involve the authority in liability to pay heavy compensation but would, in addition, deprive them of substantial increases in rate income.

The central authority, the Ministry of Health, had no effective powers of initiation and no power to grant financial assistance to local authorities. Indeed, their powers were essentially regulatory and seemed to be designed to cast them in the role of a quasi-judicial body to be chiefly concerned with ensuring that local authorities did not treat property owners unfairly.

The difficulties were not, however, solely administrative. Even the most progressive authorities were greatly handicapped by the inadequacies of the law relating to compensation. The compensation paid either for planning restrictions or for compulsory acquisition had to be determined in relation to the most profitable use of the land, even if it was unlikely that the land would be so developed, and without regard to the fact that the prohibition of development on one site usually resulted in the development value (which had been purchased at high cost) shifting to another site. Consequently, in the words of the Uthwatt Committee, 'an examination of the town planning maps of some of our most important built-up areas reveals that in many cases they are little more than photographs of existing uses and existing lay-outs, which, to avoid the necessity of paying compensation, become perpetuated by incorporation in a statutory scheme irrespective of their suitability or desirability'.

These problems increased as the housing boom of the 1930s developed; 2,700,000 houses were built in England and Wales between 1930 and 1940. At the outbreak of the Second World War, one-third of all the houses in England and Wales had been built since 1918. The implications for urbanisation were obvious, particularly in the London area. Between 1919 and 1939 the population of Greater London rose by about two millions, of which three-quarters of a million was natural increase, and over one and a quarter million was migration (Abercrombie 1945). This growth of the metropolis was a force which existing powers were incapable of halting, despite the large body of opinion favouring some degree of control.

THE DEPRESSED AREAS

The crux of the matter was that the problem of London was closely allied to that of the declining areas of the North and of South Wales, and both were part of the much wider problem of industrial location. In the South-East the insured employed population rose by 44 per cent between 1923 and 1934, but in the North-East it fell by 5.5 per cent and in Wales by 26 per cent. In 1934, 8.6 per cent of insured workers in Greater London were unemployed, but in Workington the proportion was 36.3 per cent, in Gateshead 44.2 per cent, and in Jarrow 67.8 per cent. In the early stages of political action these two problems were divorced. For London, various advisory committees were set up and a series of reports issued: the Royal Commission on the Local Government of Greater London (1921–3); the London and Home Counties Traffic Advisory Committee (1924); the Greater London Regional Planning Committee (1927); the Standing Conference on London Regional Planning (1937); as well as *ad hoc* committees and inquiries, for example, on Greater London Drainage (1935) and a Highway Development Plan (Bressey Plan, 1938).

For the depressed areas, attention was first concentrated on encouraging migration, on training and on schemes for establishing the unemployed in smallholdings. Increasing unemployment accompanied by rising public concern necessitated further action. Government 'investigators' were appointed and, following their reports, the Depressed Areas Bill was introduced in November 1934, to pass (after the Lords had amended the title) as the Special Areas (Development and Improvement) Act.

Under the Act, a Special Commissioner for England and Wales and one for Scotland were appointed, with very wide powers for 'the initiation, organisation, prosecution and assistance of measures to facilitate the economic development and social improvement' of the special areas. The areas were defined in the Act and included the North-East coast, West Cumberland, industrial South Wales, and in Scotland, the industrial area around Glasgow. By September 1938, the Commissioners had spent, or approved the spending of, nearly £21 million, of which £15 million was for the improvement of public and social services, £3 million for smallholdings and allotment schemes, and £500,000 on amenity schemes such as the clearance of derelict sites. Physical and social amelioration, however, was intended to be complementary to the Commissioners' main task: the attraction of new industry. Appeals to industrialists proved inadequate; in his second report, Sir Malcolm Stewart, the Commissioner for England and Wales, concluded that 'there is little prospect of the special areas being assisted by the spontaneous action of industrialists now located outside these areas'. On the other hand, the attempt actively to attract new industry by the development of trading estates achieved considerable success, which at least warranted the comment of the Scottish Commissioner that there had been 'sufficient progress to dispel the fallacy that the areas are incapable of expanding their light industries'.

Nevertheless, there were still 300,000 unemployed in the special areas at the end of 1938, and although 123 factories had been opened between 1937 and 1938 in the special areas, 372 had been opened in the London area. Sir Malcolm Stewart concluded, in his third annual report, that 'the further expansion of industry should be controlled to secure a more evenly distributed production'. Such thinking might have been in harmony with the current increasing recognition of the need for national planning, but it called for political action of a character which would have been sensational. Furthermore, as Neville Chamberlain (then Chancellor of the Exchequer) pointed out, even if new factories were excluded from London it did not follow that they would forthwith spring up in South Wales or West Cumberland. The immediate answer of the government was to appoint the Royal Commission on the Distribution of the Industrial Population.

THE BARLOW REPORT

The Barlow Report (Cmd 6153) is of significance not merely because it is an important historical landmark, but also because, for a period of at least a quarter of a century, some of its major recommendations were accepted as a basis for planning policy.

The terms of reference of the Commission were

> to inquire into the causes which have influenced the present geographical distribution of the industrial population of Great Britain and the probable direction of any change in the distribution in the future; to consider what social, economic or strategic disadvantages arise from the concentration of industries or of the industrial population in large towns or in particular areas of the country; and to report what remedial measures if any should be taken in the national interest.

These very wide terms of reference represented, as the Commission pointed out, 'an important step forward' in contemporary thinking. Reviewing the history of town planning it noted that:

> Legislation has not yet proceeded so far as to deal with the problem of planning from a *national* standpoint; there is no duty imposed on any authority or government department to view the country as a whole and to consider the problems of industrial, commercial and urban growth in the light of the needs of the entire population. The appointment, therefore, of the present Commission marks an important step forward. The evils attendant on haphazard and ill-regulated town growth were first brought under observation; then similar dangers when prevalent over wider areas or regions; now the investigation is extended to Great Britain as a whole. The Causes, Probable Direction of Change and Disadvantages mentioned in the terms of reference are clearly not concerned with separate localities or local authorities, but with England, Scotland and Wales collectively: and the Remedial Measures to be considered are expressly required to be in the national interest.
>
> (Ibid.)

After reviewing the evidence, the Commission concluded that 'the disadvantages in many, if not most of the great industrial concentrations, alike on the strategical, the social and the economic side, do constitute serious handicaps and even in some respects dangers to the nation's life and development, and we are of opinion that definite action should be taken by the government towards remedying them'. The advantages of concentration were clear: proximity to markets, reduction of transport costs and availability of a supply of suitable labour. But these, in the Commission's view, were accompanied by serious disadvantages such as heavy charges on account mainly of high site values, loss of time through street traffic congestion and the risk of adverse effects on efficiency due to long and fatiguing journeys to work. The Commission maintained that the development of garden cities, satellite towns and trading estates could make a useful contribution towards the solution of these problems of urban congestion.

The London area, of course, presented the largest problem, not simply because of its huge size, but also because 'the trend of migration to London and the Home Counties is on so large a scale and of so serious a character that it can hardly fail to increase in the future the disadvantages already shown to exist'. The problems of London were thus in part related to the problems of the depressed areas:

> It is not in the national interest, economically, socially or strategically, that a quarter, or even a larger, proportion of the population of Great Britain should be concentrated within 20 to 30 miles or so of Central London. On the other hand, a policy:
>
> (i) of balanced distribution of industry and the industrial population so far as possible throughout the different areas or regions in Great Britain;
> (ii) of appropriate diversification of industries in those areas or regions
>
> would tend to make the best national use of the resources of the country, and at the same time would go far to secure for each region or area, through diversification of industry, and variety of employment, some safeguard against severe and persistent depression, such as attacks an area dependent mainly on one industry when that industry is struck by bad times.

Such policies could not be carried out by the existing administrative machinery: it was no part of statutory planning to check or to encourage a local or regional growth of population. Planning was essentially on a local basis; it did not, and was not intended to, influence the geographical distribution of the population as between one locality or another. The Commission unanimously agreed that the problems were national in character and required a central authority to deal with them. They argued that the activities of this authority ought to be distinct from and extend beyond those

of any existing government department. It should be responsible for formulating a plan for dispersal from congested urban areas – determining in which areas dispersal was desirable; whether and where dispersal could be effected by developing garden cities or garden suburbs, satellite towns, trading estates, or the expansion of existing small towns or regional centres. It should be given the right to inspect town planning schemes and 'to consider, where necessary, in cooperation with the government departments concerned, the modification or correlation of existing or future plans in the national interest'. It should study the location of industry throughout the country with a view to anticipating cases where depression might probably occur in the future and encouraging industrial or public development before a depression actually occurred.

Whatever form this central agency might take (a matter on which the Commission could not agree), it was essential that the government should adopt a much more positive role: control should be exercised over new factory building at least in London and the Home Counties, that dispersal from the larger conurbations should be facilitated, and that measures should be taken to anticipate regional economic depression.

THE IMPACT OF WAR

The Barlow Report was published in January 1940, some four months after the start of the Second World War. The problem which precipitated the decision to set up the Barlow Commission, that of the depressed areas, rapidly disappeared. The unemployed of the depressed areas now became a powerful national asset. A considerable share of the new factories built to provide munitions or to replace bombed factories were located in these areas. By the end of 1940, 'an extraordinary scramble for factory space had developed'; and out of all this 'grew a wartime, an extempore, location of industry policy covering the country as a whole' (Meynell 1959: 13). This emergency wartime policy, paralleled in other fields, such as hospitals, not only provided some 13 million square feet of munitions factory space in the depressed areas which could be adapted for civilian industry after the end of the war; it also provided experience in dispersing industry and in controlling industrial location which showed the practicability (under wartime conditions at least) of such policies. The Board of Trade became a central clearing-house of information on industrial sites. During the debates on the 1945 Distribution of Industry Bill, its spokesman stressed:

> We have collected a great deal of information regarding the relative advantage of different sites in different parts of the country, and of the facilities available there with regard to local supply, housing accommodation, transport facilities, electricity, gas, water, drainage and so on . . . We are now able to offer to industrialists a service of information regarding location which has never been available before.

Hence, though the Barlow Report (to use a phrase of Dame Alix Meynell) 'lay inanimate in the iron lung of war', it seemed that the conditions for the acceptance of its views on the control of industrial location were becoming very propitious: there is nothing better than successful experience for demonstrating the practicability of a policy.

The war thus provided a great stimulus to the extension of town and country planning into the sphere of industrial location. And this was not the only stimulus it provided. The destruction wrought by bombing transformed 'the rebuilding of Britain' from a socially desirable but somewhat visionary and vague ideal into a matter of practical and defined necessity. Nor was this all: the very fact that rebuilding was clearly going to take place on a large scale provided an unprecedented opportunity for comprehensive planning of the bombed areas and a stimulus to overall town planning. In the Exeter Plan, Thomas Sharp (1947: 10) urged that

> to rebuild the city on the old lines . . . would be a dreadful mistake. It would be an exact repetition of what happened in the rebuilding of London after the Fire – and the results, in regret at lost opportunity, will be the same. While, therefore, the arrangements for rebuilding to the new plan should proceed with all possible speed, some patience and discipline will be

necessary if the new-built city is to be a city that is really renewed.

In Hull, Lutyens and Abercrombie (1945: 1) argued that 'there is now both the opportunity and the necessity for an overhaul of the urban structure before undertaking this second refounding of the great Port on the Humber. Due consideration, however urgent the desire to get back to working conditions, must be given to every aspect of town existence.' The note was one of optimism, of being able to tackle problems which were of long standing. In the metropolis (to quote from Forshaw and Abercrombie's *County of London Plan* 1943):

> London was ripe for reconstruction before the war; obsolescence, bad and unsuitable housing, inchoate communities, uncorrelated road systems, industrial congestion, a low level of urban design, inequality in the distribution of open spaces, increasing congestion of dismal journeys to work – all these and more clamoured for improvement before the enemy's efforts to smash us by air attack stiffened our resistance and intensified our zeal for reconstruction.

This was the social climate of the war and early postwar years. There was an enthusiasm and a determination to undertake social reconstruction on a scale hitherto considered utopian. The catalyst was, of course, the war itself. At one and the same time war occasions a mass support for the way of life which is being fought for and a critical appraisal of the inadequacies of that way of life. Modern total warfare demands the unification of national effort and a breaking down of social barriers and differences. As Titmuss (1958: 85) noted, it 'presupposes and imposes a great increase in social discipline; moreover, this discipline is tolerable if, and only if, social inequalities are not intolerable'. On no occasion was this more true than in the Second World War. A new and better Britain was to be built. The feeling was one of intense optimism and confidence. Not only would the war be won: it would be followed by a similar campaign against the forces of want. That there was much that was inadequate, even intolerable, in prewar Britain had been generally accepted. What was new was the belief that the problems could be tackled in the same way as a military operation. What supreme confidence was evidenced by the setting up in 1941 of committees to consider postwar reconstruction problems: the Uthwatt Committee on Compensation and Betterment, the Scott Committee on Land Utilisation in Rural Areas and the Beveridge Committee on Social Insurance and Allied Services. Perhaps it was Beveridge (1942: 170) who most clearly summed up the spirit of the time, and the philosophy which was to underlie postwar social policy:

> The Plan for Social Security is put forward as part of a general programme of social policy. It is one part only of an attack upon five great evils: upon the physical Want with which it is directly concerned, upon Disease which often causes Want and brings many other troubles in its train, upon Ignorance which no democracy can afford among its citizens, upon Squalor which arises mainly through haphazard distribution of industry and population, and upon Idleness which destroys wealth and corrupts men, whether they are well fed or not, when they are idle. In seeking security not merely against physical want, but against all these evils in all their forms, and in showing that security can be combined with freedom and enterprise and responsibility of the individual for his own life, the British community and those who in other lands have inherited the British tradition, have a vital service to render to human progress.

It was within this framework of a newly acquired confidence to tackle long-standing social and economic problems that postwar town and country planning policy was conceived. No longer was this to be restricted to town planning 'schemes' or regulatory measures. There was now to be the same breadth in official thinking as had permeated the Barlow Report. The attack on squalor was conceived as part of a comprehensive series of plans for social amelioration. To quote the 1944 White Paper *The Control of Land Use* (Cmd 6537):

> Provision for the right use of land, in accordance with a considered policy, is an essential requirement of the government's programme of postwar reconstruction. New houses, whether of permanent or emergency construction; the new layout of areas devastated by enemy action or blighted by reason of age or bad living conditions; the new schools which will be required

under the Education Bill now before Parliament; the balanced distribution of industry which the government's recently published proposals for maintaining employment envisage; the requirements of sound nutrition and of a healthy and well-balanced agriculture; the preservation of land for national parks and forests, and the assurance to the people of enjoyment of the sea and countryside in times of leisure; a new and safer highway system better adapted to modern industrial and other needs; the proper provision of airfields – all these related parts of a single reconstruction programme involve the use of land, and it is essential that their various claims on land should be so harmonized as to ensure for the people of this country the greatest possible measure of individual well-being and national prosperity.

THE NEW PLANNING SYSTEM

The prewar system of planning was defective in several ways. It was optional on local authorities; planning powers were essentially regulatory and restrictive; such planning as was achieved was purely local in character; the central government had no effective powers of initiative, or of coordinating local plans; and the 'compensation bogey', with which local authorities had to cope without any Exchequer assistance, bedevilled the efforts of all who attempted to make the cumbersome planning machinery work.

By 1942, 73 per cent of the land in England and 36 per cent of the land in Wales had become subject to interim development control, but only 5 per cent of England and 1 per cent of Wales was actually subject to operative schemes (Uthwatt Report 1942: 9); and there were several important towns and cities as well as some large country districts for which not even the preliminary stages of a planning scheme had been carried out. Administration was highly fragmented and was essentially a matter for the lower-tier authorities: in 1944 there were over 1,400 planning authorities. Some attempts to solve the problems to which this gave rise were made by the (voluntary) grouping of planning authorities in joint committees for formulating schemes over wide areas but, though an improvement, this was not sufficiently effective.

The new conception of town and country planning underlined the inadequacies. It was generally (and uncritically) accepted that the growth of the large cities should be restricted. Regional plans for London, Lancashire, the Clyde Valley and South Wales all stressed the necessity of large-scale overspill to new and expanded towns. Government pronouncements echoed the enthusiasm which permeated these plans. Large cities were no longer to be allowed to continue their unchecked sprawl over the countryside. The explosive forces generated by the desire for better living and working conditions would no longer run riot. Suburban dormitories were a thing of the past. Overspill would be steered into new and expanded towns which could provide the conditions people wanted, without the disadvantages inherent in satellite suburban development. When the problems of reconstructing blitzed areas, redeveloping blighted areas, securing a 'proper distribution' of industry, developing national parks, and so on, are added to the list, there was a clear need for a new and more positive role for the central government, a transfer of powers from the smaller to the larger authorities, a considerable extension of these powers and, most difficult of all, a solution to the compensation–betterment problem.

The necessary machinery was provided in the main by the Town and Country Planning Acts, the Distribution of Industry Acts, the National Parks and Access to the Countryside Act, the New Towns Act and the Town Development Acts.

The 1947 Town and Country Planning Act brought almost all development under control by making it subject to planning permission. Planning was to be no longer merely a regulative function. Development plans were to be prepared for every area in the country. These were to outline the way in which each area was to be developed or, where desirable, preserved. In accordance with the wider concepts of planning, powers were transferred from district councils to county councils. The smallest planning units thereby became the counties and the county boroughs. Coordination of local plans

was to be effected by the new Ministry of Town and Country Planning. Development rights in land and the associated development values were nationalised. All the owners were thus placed in the position of owning only the existing (1947) use rights and values in their land. Compensation for development rights was to be paid 'once and for all' out of a national fund, and developers were to pay a development charge amounting to 100 per cent of the increase in the value of land resulting from the development. The 'compensation bogey' was thus at last to be completely abolished: henceforth development would take place according to 'good planning principles'.

Responsibility for securing a 'proper distribution of industry' was given to the Board of Trade. New industrial projects (above a minimum size) would require the board's certification that the development would be consistent with the proper distribution of industry. More positively, the board was given powers to attract industries to development areas by loans and grants and by the erection of factories.

New towns were to be developed by *ad hoc* development corporations financed by the Treasury. Somewhat later, new powers were provided for the planned expansion of towns by local authorities. The designation of national parks and 'areas of outstanding natural beauty' was entrusted to a new National Parks Commission, and local authorities were given wider powers for securing public access to the countryside. A Nature Conservancy was set up to provide scientific advice on the conservation and control of natural flora and fauna, and to establish and manage nature reserves. New powers were granted for preserving amenity, trees, historic buildings and ancient monuments. Later, controls were introduced over river and air pollution, litter and noise. Indeed, the flow of legislation has been unceasing, partly because of increased experience, partly because of changing political perspectives, but perhaps above all because of the changing social and economic climate within which town and country planning operates.

The ways in which the various parts of this web of

policies operated, and the ways in which both the policies and the machinery have developed since 1947 are summarized in the following chapters. Here a brief overview sets the scene.

THE EARLY YEARS OF THE NEW PLANNING SYSTEM

The early years of the new system were years of austerity. This was a truly regulatory era, with controls operating over an even wider range of matters than during the war. It had not been expected that there would be any surge in pressures for private development, but even if there were, it was envisaged that these would be subject to the new controls. Additionally, private building was regulated by a licensing system which was another brake on the private market. Building resources were channelled to local authorities, and (after an initial uncontrolled spurt of private house building) council house building became the major part of the housing programme.

The sluggish economy made it relatively easy to operate regulatory controls (since there was little to regulate), but it certainly was not favourable to 'positive planning'. The architects of the 1947 system had assumed that most of this positive planning would take the form of public investment, particularly by local authorities and new town development corporations. Housing, town centre renewal and other forms of 'comprehensive development' were seen as essentially public enterprises. This might have been practicable had resources been plentiful, but they were not, and both new building and redevelopment proceeded slowly. Thus, neither the public nor the private sectors made much progress in 'rebuilding Britain' (to use one of the slogans which had been popular at the end of the war).

The architects of the postwar planning system foresaw a modest economic growth, little population increase (except an anticipated short postwar 'baby boom'), little migration either internally or from abroad, a balance in economic activity among the regions, and a generally manageable administrative

task in maintaining controls. Problems of social security and the initiation of a wide range of social services were at the forefront of attention: welfare for all rather than prosperity for a few was the aim. There was little expectation that incomes would rise, that car ownership would spread, and that economic growth would make it politically possible to declare (as Harold Macmillan later did) that 'you have never had it so good'. The plans for the new towns were almost lavish in providing one garage for every four houses.

The making of plans went ahead at a steady pace, frequently in isolation from wider planning considerations, though the regional offices of the planning ministry made a valiant attempt at coordination; but even here progress was much slower than expected, and it soon became clear that comprehensive planning would have to be postponed for the sake of immediate development requirements.

For a time, the early economic and social assumptions seemed to be borne out but, during the 1950s, dramatic changes took place, some of which were the result of the release of pent-up demand which followed the return of the Conservative government in 1951 – a government which was wedded to a 'bonfire of controls'. One of the first acts of this government in the planning sphere was a symbolic one: a change in the name of the planning ministry – from 'local government and planning' to 'housing and local government'. This reflected the political primacy of housing and the lack of support for 'planning' (now viewed, with justification, as restrictive). The regional offices of the planning ministry were abolished: thus saving a small amount of public funds, but also dismantling the machinery for coordination. Though this machinery was modest in scope (and in resources), it was important because there was no other regional organisation to carry out this function.

The first change to the 1947 system came in 1953 when, instead of amending the development charge in the light of experience (as the Labour government had been about to do), it was abolished. At about the same time, all building licensing was scrapped. Private housebuilding boomed; and curiously so did council house building, since the high building targets set by the Conservative government could be met only by an all-out effort by both private and public sectors. The birthrate (which – as expected – dropped steadily from 1948 to the mid-1950s) suddenly started a large and continuing rise.

The new towns programme went ahead at a slow pace, accompanied by a constant battle for resources which, so the Treasury argued, were just as urgently needed in the old towns. (The provision of 'amenities' was a particular focus of the arguments.) By contrast, public housing estates and private suburban developments mushroomed. Indeed, there was soon a concern that prewar patterns of urban growth were to be repeated. The conflict between town and country moved to centre stage. This was a more difficult matter for the Conservative government than the abolition of building controls, development charges and other restrictive measures. New policies were forged, foremost of which was the control over the urban fringes of the conurbations and other large cities where an acrimonious war was waged between conservative counties seeking to safeguard undeveloped land and the urban areas in great need of more land for their expanding house building programmes. On the side of the counties was the high priority attached to maintaining good quality land in agricultural production. On the side of the urban areas was a huge backlog of housing need. The war reached epic proportions in the Liverpool and Manchester areas where Cheshire fought bitterly 'to prevent Cheshire becoming another Lancashire'. Similar arguments were used in the West Midlands, where a campaign for new towns (led by the Midland New Towns Society) was complicated by the government view that Birmingham was a rich area from which to move industry to the depressed areas. London, of course, had its ring of new towns, but these were inevitably slow in providing houses for needy Londoners, particularly since tenants were selected partly on the basis of their suitability for the jobs which had been attracted to the towns. The London County Council therefore, like its provincial counterparts, built houses for 'overspill' in what were then called 'out-county estates'. Similarly,

Glasgow and Edinburgh built their 'peripheral estates'.

The pressures for development grew as households increased even more rapidly than population (a little-understood phenomenon at the time) (Cullingworth 1960a), and as car ownership spread (the number of cars doubled in the 1950s and doubled again in the following decade). Increased mobility and suburban growth reinforced each other, and new road building began to make its own contribution to the centrifugal forces.

Working in the opposite direction was the implacable opposition of the counties. They received a powerful new weapon when Duncan Sandys initiated the Green Belt circular of 1955 (42/55). Green belts had no longer even to be green: their function was to halt urban development. Hope that all interests could be appeased was raised by the Town Development Acts (1952 in England, 1957 in Scotland). These provided a neat mechanism for housing urban 'overspill' and, at the same time, rejuvenating declining small towns and minimising the loss of agricultural land. But though a number of schemes were (slowly) successful, the local government machinery was generally not equal to such a major regional task.

LOCAL GOVERNMENT REORGANISATION

It was this local machinery which was at the root of many of the difficulties. Few politicians wanted to embark on the unpopular task of reforming local government, and even those who appreciated the need for change could not agree on why it was wanted – whether to resolve the urban–rural conflict, to facilitate a more efficient delivery of services, or to provide a system of more effective political units. These and similar issues were grist to the academic mill, but a treacherous area for politicians. Perhaps the biggest surprise here was the decision to go ahead with the reorganisation of London government. The legislation was passed in 1963: this followed (in sequence but not in content) a wide-ranging inquiry. The surprise was not that the recommendations were altered by the political process, but that anything was done at all. One important factor in the politics of the situation was the desire to abolish the socialist London County Council (though ironically the hoped-for guarantee of a permanent Conservative GLC was dashed by the success of the peripheral districts in maintaining their independence).

One effect of the London reorganisation was that further changes elsewhere were taken very seriously. The writing was now on the wall for local government in the rest of the country, and campaigns and counter-campaigns proliferated. Three inquiries (for England, Scotland and Wales) were established by the Labour government which assumed office in 1964. These reported in 1969, but implementation fell to its successor Conservative government. For Scotland, the recommendations were generally accepted (with a two-tier system of regions and districts over most of the country). The city–region recommendations for England, however, were unacceptable, and a slimmer two-tier system was adopted. Wales was treated in the same way. The result south of the border was that the boundaries for the urban–rural strife, though amended in detail, were basically unchanged in character.

POPULATION PROJECTIONS

In the meantime, truly alarming population projections had appeared which transformed the planning horizon. The population at the end of the century had been projected in 1960 at 64 million; by 1965 the projection had increased to 75 million. At the same time, migration and household fission had added to the pressures for development and the need for an alternative to expanding suburbs and 'peripheral estates'. It seemed abundantly clear that a second generation of new towns was required.

Between 1961 and 1971, fourteen additional new towns were designated. Some, like Skelmers-

dale and Redditch, were 'traditional' in the sense that their purpose was to house people from the conurbations. Others, such as Livingston and Irvine, had the additional function of being growth points in a comprehensive regional programme for Central Scotland. One of the most striking characteristics of the last new towns to be designated was their huge size. In comparison with the Reith Committee's optimum of 30,000–50,000, Central Lancashire's 500,000 seemed massive. But size was not the only striking feature. Another was the fact that four of them were based on substantial existing towns: Northampton, Peterborough, Warrington, and Central Lancashire (Preston-Leyland). Of course, town building had been going on for a long time in Britain, and all the best sites may have already been taken by what had become old towns. The time was bound to come when the only places left for new towns were the sites of existing towns. There were, however, other important factors. First, the older towns were in need of rejuvenation and a share in the limited capital investment programme. Second, there was the established argument that nothing succeeds like success; or, to be more precise, a major development with a population base of 80,000–130,000 or more had a flying start over one with a mere 5,000–10,000. A wide range of facilities was already available, and (hopefully) could be readily expanded at the margin.

No sooner had all this been settled than the population projections were drastically revised downwards. It was too late to reverse the new new towns programme, though it was decided not to go ahead with Ipswich (and Stonehouse was killed by the opposition of Strathclyde because of its irrelevance to the problems of the rapidly declining economy of Clydeside). However, the reduced population growth prevented some problems becoming worse, though little respite was apparent at the time. Household formation continued apace, as did car ownership and migration. The resulting pressures on the South-East were severe – and remained so into the 1990s, with little resolution of the difficulties of 'land allocation'.

URBAN POVERTY

While much political energy was spent on dealing with urban growth, even more intractable problems of urban decay forced their attention on government. Every generation, it seems, has to rediscover poverty for itself, and the postwar British realisation came in the late 1960s (Sinfield 1973). As usual, there were several strands: the reaction against inhuman slum clearance and high-density redevelopment, the impact of these and of urban motorways on communities ('get us out of this hell!' cried the families living alongside the elevated M4), fear of racial unrest (inflamed by the speeches of Enoch Powell). These issues went far beyond even the most ambitious definition of 'planning', and they posed perplexing problems of the coordination of policies and programmes. Not surprisingly, the response was anything but coordinated, and programmes proliferated in confusion.

Housing policy was the clearest field of policy development. Slum clearance had been abruptly halted at the beginning of the war, when demolitions were running at the rate of 90,000 a year. It was resumed in the mid-1950s, and steadily rose towards its prewar peak. Both the scale of this clearance and its insensitivity to community concerns, as well as the inadequate character of some redevelopment schemes, led to an increasing demand for a reappraisal of the policy. Added force was given to this by the growing realisation that demolition alone could not possibly cope with the huge amount of inadequate housing – and the continuing deterioration of basically sound housing. Rent control had played a part in this tide of decay, and halting steps were taken to ameliorate its worst effects, though not with much success. More effective was the introduction of policies to improve, rehabilitate and renovate older housing: changing terminology reflected constant refinements of policy. Increasingly, it was realised that *ad hoc* improvements to individual houses were of limited impact: area rehabilitation paid far higher dividends, particularly in encouraging individual improvement efforts. A succession of area programmes have made a significant impact on some older urban neighbourhoods, but a considerable problem remains; and it is debatable whether the overall position is improving or deteriorating.

Housing policies have typically been aimed at the physical fabric of housing and the residential environment. Their impact on people generally, and the poor in particular, has been less than housing reformers had hoped (the lessons of earlier times being ignored). This realisation followed a spate of social inquiries, of which *The Poor and the Poorest* by Brian Abel-Smith and Peter Townsend (published in 1965) was a landmark in raising public concern. A bewildering rush of programmes was promoted by the Home Office (including the urban programme in 1968, community development projects in 1969, and comprehensive community programmes in 1974), the Department of the Environment (DoE) (urban guidelines in 1972, area management trials in 1974, and 'the policy for the inner cities' in 1977), the Department of Education (educational priority areas in 1968) and the DHSS (cycle of deprivation studies in 1973). This list is by no means complete, but it demonstrates the almost frantic search for effective policies in fields which had hitherto largely been left to local effort. Despite all this, the problems of the 'inner cities' (a misnomer since some of the deprived areas were on the periphery of cities) grew apace. The most important factors were the rapid rate of deindustrialisation and the massive movement of people and jobs to outer areas and beyond. Unlike the interwar years, the problems were not restricted to the 'depressed areas': the South-East, previously the source for moving employment to the North, was badly affected. In absolute (rather than percentage) terms, London suffered severely (losing three-quarters of a million manufacturing jobs between 1961 and 1984; P. Hall 1992: 150).

There was initially little difference between the political parties here: both were searching for solutions which continued to evade them. Lessons from America indicated that more money was not necessarily the answer, and academic writers pointed to the need for societal changes, but there were few politically helpful ideas around. Following a period in which the problems were seen in terms of social pathology, attention was increasingly directed to 'structural' issues, particularly of the local economy. In the 1980s, the Conservative government put its faith in releasing enterprise, though it was never clear how this would benefit the poor. New initiatives included urban development corporations (UDCs), modelled on the new town development corporations but with a different private enterprise ethic. The London Docklands UDC seemed almost determined to ignore, if not override, the community in which it was located, but this attitude eventually changed, and both the LDDC and later UDCs became more attuned to local needs and feelings.

LAND VALUES

The issue of land values was addressed by both the two later Labour governments. In the 1964–70 administration, the Land Commission was established to buy development land at a price excluding a part of the development value and to levy a betterment charge on private sales. Its life and promise were cut short by the incoming Conservative government. Exactly the same happened with the community land scheme and the development land tax introduced by the 1974–9 Labour government. Thus, there were three postwar attempts to wrestle with the problem, and none was given an adequate chance to work. With only the happy anomaly of the Land Authority for Wales, land transactions are now a matter for the market, for local government and for land availability studies. The latter, as with speeding planning permissions, have become a time-demanding ritual for planners.

The abandonment of attempts to solve 'the betterment problem' (which may no longer even be perceived as a problem) is more than a matter of land taxation or even equity. The so-called 'financial provisions' of the 1947 Act underpinned the whole system and made positive planning a real possibility. Though it seems unlikely that the issue will return to the political agenda in the foreseeable future, it should not be forgotten that this vital piece of the planning machinery is missing. (Holford once neatly called the half-dismantled legislation 'a set of spare parts'.) Planning is therefore essentially a servant of the market (in the sense that it comes into operation only when market operations are set in motion). This change, made some forty years ago, is far more fundamental than the high-profile changes made under the Thatcher regime.

ENTREPRENEURIAL PLANNING

The theme of the Conservative era which began in 1979 was a commitment to 'releasing enterprise'. This was translated into a miscellany of policies which had little in the way of a coherent underlying philosophy, but which could be characterised in terms of removing particular barriers which were identified as holding back initiative. The identified problems ranged from inner-city landholding by public bodies (dealt with by requiring publicity for the vacant land; which would thereby automatically trigger a market use); to the 'wasteful' and 'unnecessary' tier of metropolitan government in London and the provincial conurbations (simply abolished). Many areas of public activity were privatised, large parts of government were hived off to executive agencies, and compulsory competitive tendering was imposed on local government. The emphasis on 'market orientation' and the concerted attack (regrettably the word is not an exaggeration) on local government had some strange results. More power was vested in central government and its agencies. Public participation was reduced. But, though the planning system was affected in tangible ways (Thornley 1993), in no sense was it dismantled, or even changed in any really significant way. True, it has been bypassed (by urban development corporations); its procedures have been modified (by government circulars and changes in the General Development and Use Classes Orders); development plans were, for a time, downgraded and threatened with severe curtailment; and simplified planning zones were introduced: a system in which 'simplification' meant less planning control, but might involve even more human resources in negotiation.

The list can be extended, but the rhetoric which preceded and accompanied the changes was harsher than the changes themselves. Moreover, the language of confrontation which the politicians employed disguised the fact that previous governments had done similar things, even if more *sotto voce*. The development corporation initiative, for example, was essentially the brainchild of a much earlier period and, indeed, as applied to redevelopment (as distinct from new town development) had for long been proposed by socialists as a means of assisting local authorities. Some of the early days of the UDC flagship – the London Docklands Development Corporation – were characterised by an excess of zeal, a lack of understanding of the way in which the administration of government is different from the administration of business (and an authoritarian style which was widely – and justifiably – criticised). Time, however, mellowed misplaced enthusiasm, and brought about a better understanding of the inherent slowness of democratic government. There was also a keener awareness of the need to pay attention to the 'social' issues of the locality as well as its physical regeneration.

More generally, an old lesson was relearned: it is extremely difficult for one level of government to impose its will on another unless it has some broad and powerful support from outside, as well as willing cooperation inside. (There is, however, the draconian alternative of simply abolishing a wayward layer of local government, as was done with the Greater London Council and the metropolitan county councils.)

An about-turn on structure plans illustrates the pragmatic nature (what some call the flexibility, and others the inconsistency) of the Conservative government's thinking. The initial decision to abolish them was one option for dealing with a problem which dates back to 1947: how to ensure that plans provide (without overwhelming detail) sufficient guidance for the land use planning of an area, while being adaptable to unforeseen changing circumstances. The option actually adopted was a 'streamlining' – not unlike earlier attempts. The 1965 PAG report had highlighted the problem: 'It has proved extremely difficult to keep these plans not only up to date but forward looking and responsive to the demands of change.' Twenty years later, the 1985 White Paper, *Lifting the Burden* (Cmnd 9571), was in a similar key: 'There is cause for concern that this process of plan review and up-dating is becoming too slow and cumbersome.'

More effective structure plans require a framework of regional policy. It is debatable whether, in a country like Britain, there is any alternative to this being a function of central government; but debate there will always be. The increased role of central government in providing regional guidance (most apparent in the introduction of regional policy guidance statements) can be represented as undue central interference with local planning, even though it has been considerably influenced (probably too much so) by local input. This, again, is a part of a very lengthy debate on regional planning. Various stratagems have been tried: regional economic planning councils (which despite their name, found themselves absorbed in land use matters); proposed constitutional devolution; the establishment (by a Conservative government) of metropolitan counties – and their later abolition (also by a Conservative government). It is, however, unclear whether the current regional planning guidance system will work – not because of too much central government control, but because of too little. The guidance for the South-East is proving highly problematic and, though there do not appear to be similar difficulties in the provincial regions, it is at least arguable that this is because the guidance is bland and inadequate: it hides the problems instead of facing them (Minay 1992). Nevertheless, further permutations on regional planning continue to be discussed on the political and academic sidelines.

A final point, in this somewhat tendentious overview, is that there is now probably more regional planning than ever before. This is undertaken by a wide range of public and private agencies. The regional offices of government departments such as Environment, Trade and Industry, and Transport have now been 'integrated' and given a higher profile as *Government Offices for the Regions*. There are regional organisations of the Environment Agency, English Partnerships (the Urban Regeneration Agency) and British Waterways, regional advisory councils for education and training, regional boards of the Arts Council, and such like. Regionalism in various forms is thus a fact of contemporary administration though much of it, of course, is not the same as 'regional planning'. Even where activities justify the term, the crucial element of coordination is missing. The deficiency becomes of increasing importance as the number of separated planning activities grows: urban development corporations, enterprise zones, groundwork trusts, development agencies and enterprise boards, and new bodies for environmental protection and the administration of the national parks.

THE ENVIRONMENT

All governments operate with some degree of pragmatism: electoral politics force this upon them. So it was with the Conservative government. After many years of relegating environmental issues to a low level of concern, there was a sudden conversion to environmentalism in 1988. This was heralded in a remarkable speech by Mrs Thatcher in which she declared that Conservatives were the guardians and trustees of the earth. At base, this reflected a heightening of public concern for the environment which is partly local and partly global.

The action which followed looks impressive (though critics have been less impressed by the results). A 1990 White Paper *This Common Inheritance* (Cm 1200) spelled out the government's environmental strategy over a comprehensive range of policy areas (untypically this covered the whole of the UK). An update is published each year reporting progress and consolidating policy advances. Environmental protection legislation was passed, 'integrated pollution control' is being implemented, 'green ministers' have been appointed to oversee

the environmental implications of their departmental functions, and new environmental regulation agencies have been established. The latter follow a spate of organisational changes which remind one of the old saying: 'when in doubt reorganise'. But there are difficult issues here which, though including organisational matters, go much deeper. Questions about the protection of the environment underline a perhaps (to the layperson) surprising ignorance of the workings of ecosystems at the local, national and global levels. Additionally, new questions of ethics have come to the fore. Difficult problems of deciding among alternative courses of action are rendered ever more complex. Cost benefit analysis is of little help: indeed, all forms of economic reasoning are being challenged. International pressures have played a role here as, of course, has the coming of age of the EC. This has added a new dimension to the politics of the environment (and much else as well).

Concern for historic preservation (now embraced in the term 'heritage') is of much longer standing. Though many historic buildings were destroyed during the war, the more effective stimulus to preservation came from the clearance, redevelopment, renewal and road building policies which got under way in the 1950s and accelerated rapidly. As with housing, the emphasis has been mainly on individual historic structures, but a conservation area policy was ushered in by the Civic Amenities Act 1967, sponsored by a private member (Duncan Sandys), though with wide support. This proved a popular measure, and there are now over 9,000 of them. Indeed, there has been mounting concern that too many areas are designated and too few resources applied to their upkeep and management. Later legislation, the Ancient Monuments and Archaeological Areas Act 1979, promised more in its title than it could provide, and only five *areas of archaeological importance* were designated. But the Act did more: it consolidated legislation dating back to 1882, and strengthened controls safeguarding ancient monuments. It was substantially amended by the National Heritage Act 1983 whose modern name signified a new and wider appreciation of the historical legacy. A new executive agency, English Heritage (formally called the Historic Buildings and Monuments Commission for England) was established and took over many of the functions previously housed within the DoE. In Northern Ireland, Scotland and Wales, rather different administrative solutions were devised, as befits the distinctive characters of these parts of the UK.

Surprisingly, the new environmental and historical awareness was too late in raising sufficient concern about transport to bring about any significant change from a preoccupation with catering for the car.

ROAD BUILDING POLICIES

Transport policy has for long been largely equated with road building policy, and protests that alternatives need to be considered have been unavailing until recently. On a number of issues, however, the protests could not be ignored. One has already been mentioned: the brutal impact of urban motorways on the communities through which they pass. The outcry against this led to a reassessment of both the location of urban roads and their necessity. Compensation for 'distress' caused by new roads was increased as part of a policy labelled (in a 1972 White Paper) *Putting People First* (Cmnd 5124). Closely related was a growing concern about the inadequacy of the road inquiry process, which resulted in a significant improvement of the provisions for public participation. These and other changes curbed but did not allay the concerns: indeed, they are still vocal, though it seems that further changes can be expected. The turning point came in 1989 when new forecasts of huge increases in car ownership and use were published. It was widely considered to be impossible to accommodate satisfactorily the forecast amount of traffic. The results of a change in attitude are working their way through the political system. Traffic calming has become part of the contemporary vocabulary (and is now statutorily enshrined); road pricing is on the agenda for serious discussion (but little

action); and road building has been hived off from the Department of Transport to an executive agency which, it is believed, will reduce the preoccupation of the Department with road building as the foremost means of meeting transport needs. The latest episode in the story is the conversion of the Department of Transport to a policy of *not* building roads. This reflects the government's interpretation of public attitudes to road building which nicely attune with the political objective of reducing tax-related expenditure. In this area at least, a bankruptcy of political ideas (for which persuasive alternatives are sadly in short supply) have led firmly into the doldrums.

THE COUNTRYSIDE

The countryside has always been dear to hearts of Conservatives, though support for the protection and enjoyment of the countryside has traditionally cut across party and class lines. Increasing concern for the rural landscape, growing use of the countryside for recreation (and investment), and huge changes in the fortunes of the agricultural industry have transformed the arena of debate of rural land use. At the end of the war, and for many years afterwards, the greatest importance was attached to the promotion of agriculture. There were, however, established movements for countryside conservation and recreation, some of which came together with the National Parks and Access to the Countryside Act of 1949 (but a separate Nature Conservancy Council was also established, thus dividing the conservancy function). The pressures for conservation and for recreation have varied over time, and the balance between them is inevitably an ongoing problem, particularly in areas of easy access (which now includes most of the country). Limited budgets held back incipient pressures in the early postwar years, but increasing real incomes and mobility led to mounting pressures which were acknowledged in the 1966 White Paper *Leisure in the Countryside* (Cmnd 2928) and the 1968 Countryside Act. This replaced the National Parks Commission with a

Countryside Commission which was given wider powers and improved finance. At the same time, the powers of local authorities were expanded to include, for instance, the provision of country parks. Unlike national parks, these were not necessarily places of beauty, but were intended primarily for enjoyment. They were also seen as having the added advantage of taking some of the pressure off the national parks and similar areas where added protection was needed. The 1974 reorganisation of local government was accompanied by a requirement that local authorities which were responsible for national parks should establish a separate committee and appoint a park planning officer. The modesty of this provision was clearly a compromise between concerns for local government and for the planning of national parks. It was a step forward, but an enduring case for *ad hoc* park authorities continued. Local authorities had too many local interests to satisfy to give adequate resources for national parks – whose very name indicated their much wider role. The growth of pressures on the parks continued, and the administrative knot was finally cut in 1992 when it was decided to establish national park authorities for all the parks. (These were established in 1996–97.)

More widely, a longstanding debate continued on the divided organisational arrangements for nature conservation and amenity, and for scientific conservation and wildlife. In England, that separation continues (on the basis of arguments which are not easy to follow), but in Scotland and Wales the responsibilities are now vested in a single body: Scottish Natural Heritage and the Countryside Council for Wales. Of particular note is the first outcome of Scottish thinking on integrated countryside planning, which builds upon the simple (but rarely used) notion that all countryside activities 'are based on use, in one way or another, of the natural heritage'. This thought has passed into the realm of what Donald Schon, in *Beyond the Stable State* (1971) has termed 'ideas in good currency', and it is echoed in the three highly coloured White Papers (Cm 3016, Cm 3041 and Cm 3180) on the countryside, issued in 1995–6.

There are many underlying concerns in the changes which are taking place in countryside policy. The most salient is the remarkable change in the fortunes of the agricultural industry. With huge increases in productivity, competition from abroad and the impact of the Common Market, the need now is to reduce agricultural production. It is never easy to change the course of a well-established policy, and still more difficult is it to reverse a policy, particularly one which has had such widespread support as agriculture. Nevertheless, changes are being made, and the thrust of these in reducing agricultural production is clear, though the policy instruments are less so and are in a state of flux. Given the dramatic changes involved, this is hardly surprising.

WHITHER PLANNING?

It is now half a century since the postwar planning system was put into place. Major changes have taken place during this time in society, the economy and the political scene – some of which have been touched upon in this rapid review. In these shifting sands, 'town and country planning' has grown into (or been submerged by) a series of different policy areas which defy description, let alone coordination. Yet 'planning' is nothing if not a coordinative function, and the frenetic activity in reorganising machinery which has absorbed so much energy in the last fifty years must, at some point, give way to substantive progress. The difficulty lies in determining the direction in which this lies.

One thing is clear: some of the most important underlying problems are well beyond any conceivable scope of 'planning': for example, much urban change has been due to global forces which are currently beyond *any* political control. Multinationals and international finance were not in the standard vocabulary in the early postwar years. Planners find it easier to think in terms of 'need'. In recent years, they have been forced to recast some of their thinking in 'market' terms. But could they ever come to terms with the workings of the property investment market? As many studies have shown, 'the channelling of money to promote new urban development is determined not by need or demand, but by the relative profitability of alternative investments' (Bateman 1985: 32) – which may be in different sectors, such as industrial equities, or in quite different geographical locations. Much private sector development is now 'driven more by investment demand and suppliers' decisions than by final user demand – and even less by any sort of final user needs' (Edwards 1990: 175). This widening gap between land use development and 'needs' throws considerable doubt on the adequacy of a planning system which is based on the assumption that land uses can be predicted and appropriate amounts of land 'allocated' for specific types of use.

Overriding all other pressing considerations, of course, is the state of the economy. (It is little comfort that so many other countries share the same problem.) One result has been a strengthening of the 'partnership' philosophy which has gradually grown over the last two decades. The term now means more than coordination of the efforts of different agencies: it implies that planning has to embrace the agents of the market and adapt a regulatory system of planning to the need for negotiation (a style which, as shown in Chapter 7, has traditionally characterised British pollution control). At the least, risks are shared.

The implications of all this are not clear. Though an obvious response may be to try harder to identify emerging trends, this is more difficult to do than ever before. Economic and social trends seem as unpredictable as the weather or the course of scientific inquiry. It is perhaps not surprising that one planning analyst has commented that 'it appears now that almost everybody is working on chaos theory' (Batty 1990: 6). This, of course, is an intended exaggeration, but the point is valid: comprehensive planning based on firm predictions of the future course of events is now clearly impossible. Incrementalism is the order of the day, and Burnham's famous aphorism has now been turned on its head: 'make no big plans'.

But planners have always strained for unattainable goals, whether they be frankly utopian or simply overenthusiastic. Contemporary plans are more practicable in this regard than many earlier ones. The plans prepared at the end of the Second World War were often quite unrealistic in the assumptions that were made about the availability (and control) of resources – though that did not prevent them being very influential in moulding planning ideas. As noted in the previous chapter, the Greater London Development Plan was replete with policies over which the GLC had no control (Centre for Environmental Studies 1970). It remains to be seen whether the lesson has been learned – or whether some currently unpredictable change will transform the future. Be that as it may, there seems little doubt that in the perpetual planning conflict between flexibility and certainty, the former is the clear winner.

FURTHER READING

Though the Barlow and Uthwatt reports are seldom read these days, they are well worth at least a perusal. Like other reports of the period (particularly Beveridge) they give an insight into the spirit of the times which produced the planning system. Hennessy (1992) narrates this wonderfully in *Never Again: Britain 1945–1951*. A little-known but insightful essay is Titmuss (1958) 'War and social policy'. An excellent account of over a longer period (1890–1994) is given by Ward (1994) *Planning and Urban Change*. Two of Hall's books are also essential reading: *Cities of Tomorrow* (1988) and *Urban and Regional Planning* (1992). Ashworth (1954) *The Genesis of Modern British Town Planning* is a thorough account up to the passing of the 1947 Act. A clear exposition of the (original) 1947 Act is given by a civil servant who was heavily involved in drafting the legislation: Wood (1949). Cherry (1996) *Town Planning in Britain since 1900: The Rise and Fall of the Planning Ideal* carries the story up to date. See also his *The Evolution of British Town Planning* (1974) which incorporates a history of the planning profession and its Institute.

A number of earlier writers are quoted in the text, as are several of the wartime and postwar plans.

Analyses and commentaries on the operation of the planning system rapidly become out of date. Among the books published in the last decade are Ambrose (1986) *Whatever Happened to Planning?*; Reade (1987) *British Town and Country Planning*; Healey *et al.* (1988) *Land Use Planning and the Mediation of Change*; Thornley (second edition 1993) *Urban Planning under Thatcherism: The Challenge of the Market*; Simmie (1994) *Planning London*; Adams (1994) *Urban Planning and the Development Process*; Ambrose (1994) *Urban Process and Power*.

3

THE AGENCIES OF PLANNING

(A) CENTRAL GOVERNMENT

Most entrepreneurial governments promote competition between service providers. They empower citizens by pushing control out of the bureaucracy, into the community. They measure the performance of their agencies, focusing not on inputs but on outcomes. They are driven by their goals – their missions – not by their rules and regulations. They define their clients as customers and offer them choices . . . They put their energy into earning money, not simply spending it. They prefer market mechanisms to bureaucratic mechanisms . . .

Osborne and Gaebler, *Reinventing Governments*, 1992

THE CHANGING GOVERNMENTAL SYSTEM

The quotation is from an account of ways in which the American governmental system is changing. Though there are differences, the resemblance to the changes which have taken place in Britain since 1980 is striking. But change is not of recent origin: on the contrary, inquiries into the management of government, reorganisations of departmental and ministerial responsibilities, and innovations in the control of public expenditure and in management have been a long-standing feature of British government. What has distinguished the recent past is the extent and pace of change, though whether this has increased the effectiveness of government is a debatable question. Much depends, of course, on what is meant by 'effectiveness'. To the Thatcher and Major governments, it meant increasing the responsiveness to consumer demands, and improving efficiency and service delivery.

All arms of the state have been affected by these changes: central government, local government, and a multitude of *ad hoc* agencies – including those which have statutory responsibilities for aspects of land use planning, development and environmental protection. Moreover, the experience of private sector and voluntary agencies has had some similarities, indicating that the forces for change are wider than those emanating from the ideological stance of the Conservative government to which they are often ascribed. Of particular relevance to this book is the establishment of executive agencies both within and outside the governmental machine, and the promotion of local authorities as 'enablers' rather than direct service providers. The objective of these changes is to institute 'a quite different way of conducting the business of government'.

A major landmark in this development was the Cabinet Office Efficiency Unit's 1988 report, *Improving Management in Government: The Next Steps* and its acceptance by Mrs Thatcher. The phrase 'the next steps' indicates that there had been previous attempts in this field. One of these was the 1968 Fulton Report on the civil service – a faltering one because of the resistance of Whitehall to its upsetting recommendations. There was no Mrs Thatcher around in those days to do battle with entrenched interests. Mrs Thatcher made various attempts, with the aid of advisers who had proved their competencies in the private sector, including Sir John Hoskyns who regarded the senior civil service as hopelessly out of touch with the changing times: he recommended that a high proportion of senior officials over the age of fifty should be replaced by outsiders. As Birch (1993: 151) comments, this was too radical even for Mrs Thatcher. She did, however, take personal charge of senior promotions to ensure that dynamism and even ruthlessness would prevail over traditional policy attitudes of consensus and continuity. During her term of office she made appointments to the great majority of the 175 top civil service posts.

Other approaches included the 'scrutinies' initiated by Mrs Thatcher's efficiency adviser, Sir Derek Rayner (of Marks & Spencer), which involved examining policies and programmes to find 'savings or increased effectiveness'. Though this became a permanent feature of the Whitehall scene, its immediate impact was disappointing, and a more radical initiative was put forward by Rayner's successor Sir Robin Ibbs (of ICI). His 'next steps' involved the break-up of the monolithic Whitehall machine: this was characterised as consisting of a small top group of senior civil servants concerned with policy, and a huge body (95 per cent) of executives who operated the front-line services. The elite were concerned with policy while the main body dealt with execution; operational efficiency fell between the cracks. The *Next Steps* report proposed that the vast machine should be broken down into its constituent parts, each with clear responsibilities for managing its specified functions.

A very clear statement of the new philosophy is set out in the 1994 edition of *The Civil Service Year Book* which (following four pages of advertisements) notes with obvious pride that the civil

service had fallen in size as a result of efficiency savings and the transfer of activities to the private sector. It continued:

> Since 1988 the functions of central government have been reviewed systematically. The first consideration is whether the work needs to be done at all. If not, the function is abolished. Second, does the work need to be carried out by Government? If not, the activity can be privatised. Third, where Government must retain overall responsibility for the function, does it need to be provided by civil servants, or would the private sector have expertise to offer? Here management of the function might be contracted out to the private sector, or exposed to competition under the Government's market testing programme. Finally, if the service is to remain within Government, and provided at least in part by civil servants, should it be an executive agency?

The *Next Steps* programme is continuing. In 1995, there were over a hundred agencies employing more than a third of a million civil servants. There were a further fifty-seven 'candidates for agency status' at the beginning of 1996 (*Next Steps Agencies in Government, Review* 1995 (Cm 3164)).

Closely linked with the *Next Steps* programme is the Citizens' Charter introduced by John Major in 1991 (supposedly arising from his experience as chairman of the Lambeth housing committee in the 1970s). This is intended to set out the standards and quality of service which users ('consumers') of public services are entitled to expect. The Planning Charter was published in 1995 by the DoE, in cooperation with the local authority associations. This defines the standards of service which the public can expect from planning departments. The DoE has elaborated this with *performance checklists* which result in the regular production of league tables of the speed of local decisions on planning applications.

A further injection of private sector discipline was the introduction of competitive tendering in the National Health Service, local government and finally in central government itself. The basic philosophy here is that the roles of purchaser and provider should be separated: this avoids conflict of interest, gives public purchasers access to private sector resources, and 'forces both sides to redefine the nature of the services and the standards of quality to be provided'.

These various initiatives are examples of the ways in which the operations of government are changing. Additionally, of course, there is the privatisation of public sector activities which removes control and risk to the private sector. More experimental is the Private Finance Initiative (PFI), where the object is increased efficiency by the 'purchase' of services from an 'operator' using private finance. A major factor in the philosophy underlying the PFI is that government is concerned with the provision of services, not the buildings in which the services are provided. Yet buildings are expensive to build, they involve continuing costs which are related to their original design, and they carry risks which are unconnected with the service being provided.

The objective of the PFI is to transfer these risks to a private operator. The concept is a highly elastic one: it is claimed that the only practical limit to its scope is political acceptability. Currently, however, the emphasis is on services involving high capital expenditure (on the lines of the Dartmouth Crossing and the second M4 crossing of the Severn). A claimed advantage of the PFI is that it taps into the superior entrepreneurial skills and management disciplines of the private sector. The PFI also is intended to boost public–private joint ventures and partnerships of various types – through local authorities, by means of agencies such as English Partnerships, and by way of the Single Regeneration Budget. Though based on the established grant-aid systems, these arrangements are more flexible and longer term. It is held that the PFI provides a clearer framework within which these arrangements can flourish (RICS 1995a).

Though the PFI has had a slow start, 'it implies a profound change in the relationship between public and private sectors in the provision of services, and the buildings and infrastructure to support them'. It is envisaged that it will 'in due course, pervade the entire public sector . . . it is hard to exaggerate its importance' (RICS 1995b).

ORGANISATIONAL RESPONSIBILITIES

A large number of governmental departments and agencies are involved in town and country planning. Those having the main responsibility for the planning Acts are the Department of the Environment (DoE) in England, the Scottish Office Development Department (SODD), the Welsh Office (WO) and the Planning Service Executive Agency of the Department of the Environment for Northern Ireland (DoENI). But, of course, there are many planning functions which fall to departments responsible for agriculture, the countryside, the human heritage, nature conservation, trade and industry, transport; and an increasing number of functions have been transferred from government departments to agencies and public bodies. Figure 3.1 shows the main institutional arrangements, and gives a flavour of their complexity.

Planning responsibilities have evolved over time, and though there have been numerous reorganisations, the machinery inevitably has a patchwork appearance. (As an example of the problems involved: in which department should questions of the rural economy be placed – the one concerned with agriculture, economic development, employment, or natural resources?) It is also unstable: changing conditions, problems and objectives demand new policy responses which in turn can lead to organisational changes. For example, increased concern for the devolution of responsibility from London has led to the transfer of functions to the geographical offices – most recently to regional offices in the English regions. Emphasis on co-ordination has resulted in many environmental functions being combined within new environment agencies: one for England and Wales and one for Scotland. Sometimes different patterns emerge in different parts of the UK. Nature conservation and access to the countryside are the responsibility of one agency in Scotland and Wales, but are divided between two in England.

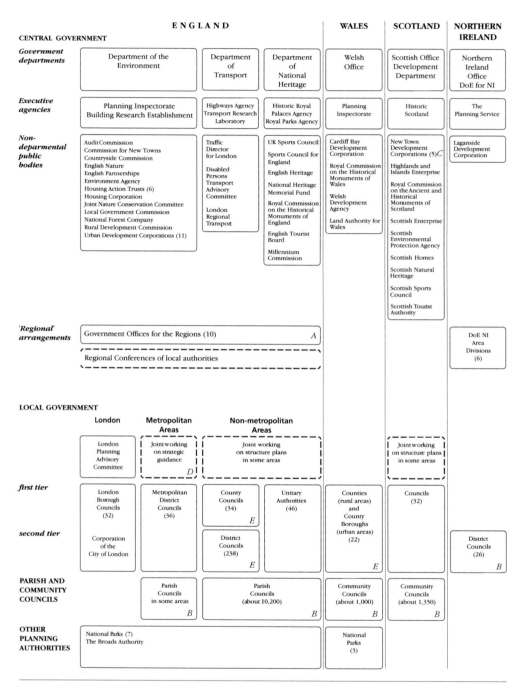

A. The Government Offices for the Regions also include the Department of Trade and Industry, Transport and Education and Employment.
B. Have a consultative role only for planning.
C. The Scottish NTDCs will all be wound up by 31 March 1997.
D. Strategic guidance is now being incorporated in regional guidance.
E. These figures apply from completion of re-organisation on 1 April 1998.

Figure 3.1 The Institutional Arrangements for Town and Country Planning in the UK

THE DEPARTMENT OF THE ENVIRONMENT

The Department of the Environment was established by the newly elected Conservative Government in 1971. This centralised housing, construction, transport, planning, local government and a number of other environmental functions under a Secretary of State for the Environment (Draper 1977). Except in Scotland and Wales, where the Scottish and Welsh Offices (together with their own Secretaries of State) have major responsibilities, the DoE became responsible for 'the whole range of functions which affect people's living environment'. These now include planning, environmental protection, countryside and wildlife, urban policy, local government, housing, construction, and the property holdings of government. The Secretary of State has final responsibility for all the functions of the Department. He is, however, concerned primarily with strategic issues of policy and priority, including public expenditure, which determine the operations of the Department as a whole. In 1996–7, the planned expenditure of the Department was £39,530 million, most of which was support to local government and other bodies such as the Housing Corporation and the urban development corporations. The New Towns Commission has been a net 'earner' of income for the DoE because of the sale of new town assets. The DoE is a large department: in 1996, there were 3,171 staff at headquarters in London, 959 in the ten regional offices, and a further 3,290 in agencies, property holdings and PSA services: a total of 7,520 (1,280 less than in 1994).

Originally, the DoE was divided into three main parts, each under a separate minister: housing and construction; transport industries; and local government and development. Several changes have taken place since 1970, including the re-establishment of transport as an independent department with its own Secretary of State in 1976 and, more recently, following the 1992 general election, the creation of a separate Department for National Heritage. Other recent changes are the gathering together of the pollution regulation functions within Her Majesty's Inspectorate of Pollution (HMIP), and later the establishment of Environmental Agencies for England and Wales and for Scotland; these have taken over the functions of the HMIP, the National Rivers Authority and the waste regulation functions of

---

DoE Objectives for the Planning System, 1996

- To pursue better implementation of sustainable development in the planning system taking especial account of urban forms and transport;
- to monitor and encourage progress on nationwide coverage of up-to-date local plans;
- to promote quality, efficiency and consistency in development plans combined with promptness in individual development control decisions . . . ;
- to maintain and develop a policy framework which enables society's need for an adequate and steady supply of minerals to be met in a way which contributes to economic growth whilst protecting the environment and amenity;
- to develop countryside planning policy, including implementing planning proposals in the Rural White Paper;
- to pursue planning policies for housing, especially considering the implications for planning of the latest household projections; and
- to seek to embed the aims and objectives of the *Quality in Town and Country Initiative* in the work of Government and outside organisations.

(*DoE Annual Report* 1996)

local government. It has been a restless time in Whitehall.

The Department's policy aims are set out in its annual report and are under constant review. Sustainability now figures significantly, and in the latest report has become a separate cross-cutting strategic aim along with good government, the quality of the built and natural environment, and economic competitiveness. The lists of general priority aims and programme aims and objectives include improving the quality of life in urban and rural areas through regeneration. The specific aims of the Department in respect of planning are shown in the box on p. 39. These are, of course, very broad statements of intent. More operational aims and objectives are set for each programme area. These are discussed at appropriate points in later chapters, but it is useful to cite one for the insight it gives into what the DoE actually does. The source here is the annual *Management Information System for Ministers* (MINIS), prepared within the DoE. The land use planning objectives are 'to provide through a plan-led system, a secure framework for individual planning decisions'. This translates into responsibilities for the legislative framework, issuing planning guidance, arranging appeals and call-ins, sponsoring research, providing a statistical service, promoting initiatives such as the one for *Quality in Town and Country*, and assist-ing in the implementation for the *Planning Framework for Thames Gateway*. As this clearly indicates, the DoE has few responsibilities involving the implementation of planning. It provides national and regional planning guidance, and it performs important supportive, promotional, advisory, supervisory and semi-judicial functions, but the operational work of planning lies with local government and other implementing agencies.

THE DEPARTMENT OF TRANSPORT

As already indicated, transport was merged in the DoE in 1970, but was moved back into a separate department in 1976. The Department of Transport is responsible for land, sea and air transport. With the privatisation programme, some railway functions are the responsibility of the Office of Passenger Rail Franchising and the Office of the Rail Regulator. Technically, these are *non-ministerial government departments* (as with the regulatory bodies for telecommunications, electricity, water and gas). *Executive agencies* were established in 1994 for a number of functions, including highways. The Highways Agency is responsible in England for the road programme and the management and maintenance of the trunk road network. These functions are carried out in Wales by the Welsh Office, in Scotland by

---

DoE Planning Directorate

The Planning Directorate comprises seven division which advise Ministers on the land use planning system, including policy on minerals. It operates through instruments of statute, secondary legislation, published planning guidance advice on good practice, individual casework decisions, promotion and persuasion. It has no grant resources to deploy in support of policy responsibilities. Its inputs are primarily concerned with shaping development plans and development control decisions through guidance and through key decisions on appeal and call-in and direct contact with those who use or are influenced by the system. It also has extensive direct casework responsibilities for specialist planning work including mineral and advertisement appeals, for the preparation of sub-regional planning guidance for the Thames Gateway and the planning aspects of the Channel Tunnel Rail Link.

(DoE MINIS 16, Part 2: Local Government and Planning Group, DoE 1995)

the Scottish Office Industry Department, and in Northern Ireland by the NIDoE.

As stated in its 1996 *Annual Report*, the principal aims of the DoT are

- to promote an efficient market, with prices reflecting the true costs of transport;
- to enable the market to provide greater choice by substantially increasing opportunities for private sector investment;
- by means of the above, to assist economic growth and to provide greater choice in good quality, safe and accessible services for all transport users;
- to promote safety, security and mobility across transport systems;
- to reduce the adverse impact of transport on the environment;
- to promote UK transport policies and interests internationally;
- to promote more efficient, effective and responsive use of departmental resources;
- to conduct licensing, regulatory and grant-payment services effectively and economically, maintaining a high quality of service to industry and the general public.

The function of the Transport Departments in building and maintaining trunk roads gives them an operational character; and, since roads account for about a half of the expenditure, this is a very significant role. Indeed, the DoT has frequently been criticised for being a 'ministry for roads'. With the privatisation of the railway system this may appear even more appropriate. However, opposition to road construction has reached unprecedented levels and this, coupled no doubt with the government's desire to cut public expenditure, has led to a major slowing down of the programme. At the same time, pressures for a more 'balanced' transport investment programme have grown. Demands for a greater investment in public transport have come from a wide variety of sources, ranging from pressure groups such as Transport 2000 and the CPRE to the Royal Commission on Environmental Pollution. The motoring interests and bodies such as the CBI seek a different version of 'balance' – one in which new road building is required for industrial efficiency and economic growth. It is unlikely that this heated debate will be readily settled: there are too many conflicting issues at stake.

Meanwhile, the DoT is heavily engaged in

Department of Transport Priorities for 1997

- Securing the flotation of Railtrack;
- the sale of the remaining BR freight business;
- supporting the Franchising Director in selling further passenger rail franchises;
- working with London Transport and the Civil Aviation Authority to let contracts to private sector consortia to build and operate Croydon Tramlink and the new Scottish Centre for air traffic control as PFI projects;
- letting contracts for a further tranche of DBFO road schemes, in addition to continued public funding of priority road schemes and maintenance;
- progressing the privatisation of the trust ports of Tyne and Ipswich;

- amending or repealing the remaining unnecessary regulation identified in 1994, insofar as this can be done without new primary legislation;
- completing the reviews of merchant shipping regulations and some railway regulations;
- ensuring that EC directives are business friendly and that the risks of new regulations are properly assessed, giving priority to the needs of small businesses;
- publishing a response to the Transport Debate [published as *Transport: The Way Forward* in 1996].

(Transport Report 1996)

recruiting private sector participation in transport infrastructure. Some success has attended this with, for example, the privately financed Dartford Bridge and the second Severn Bridge. Private financial participation has also been negotiated for the Heathrow Express, the extension of the Jubilee Line, and (as part of this line) the building of North Greenwich Station by the local property developer British Gas. With railways, the objective is total transfer to the private sector. The fondness for market mechanisms extends over much of the transport field, with one notable exception: it does not extend to motorists. Studies of road and congestion charging have increased, but the political sensitivity of the issue has guaranteed a consistency of inaction.

THE DEPARTMENT OF NATIONAL HERITAGE

In May 1992, the Department of National Heritage (DNH) was created, taking over important planning functions from the DoE. The DNH now has responsibility for heritage and tourism, historic and royal palaces, broadcasting and films, sport and recreation, and the arts. Its general aim was explained in its first annual report as 'to conserve, nurture and enhance, and make more widely accessible the rich and varied cultural heritage of the countries of the United Kingdom'. The 1996 report trims this to 'our aim is to enrich people's lives'. There are now greatly enhanced resources for these worthy objectives by way of the National Lottery, which raised £1.3 billion for distribution in the year November 1994–December 1995.

In the planning area, the Department has a specific remit 'to preserve ancient sites, monuments, and historic buildings, and increase their accessibility for study and enjoyment now and in the future', and 'to encourage inward and domestic tourism so that the industry can make its full contribution to the economy and increase opportunities for access to our culture and heritage'. Thus the DNH has taken over the heritage functions of the DoE and also tourism from the former Department of Employment.

The DNH has responsibilities in relation to scheduled ancient monuments and listed buildings. A review of the estimated 600,000 archaeological sites in England is being undertaken by the Royal Commission on the Historical Monuments of England, and it is anticipated that more than 2,000 scheduling recommendations will be made in each year up to 2003. The Royal Commission received £12 million in 1995–6 from the DNH. The most significant of the sponsored bodies in the heritage area is the Historic Buildings and Monuments Commission, better known as English Heritage, which receives over £100 million to support its properties in care and its grant-aid programme. Also now within the responsibility of the DNH are the Royal Fine Art Commission, which provides advice on major development proposals (£739,000); the Churches Conservation Trust (formerly the Redundant Churches Fund) (£2.5 million); the National Heritage Memorial Fund (£8.8 million); and the Historic Royal Palaces Agency (£4.7 million).

In the tourism field, the DNH provides £44 million of support to the British Tourist Authority, which encourages the development of overseas tourism to the UK, and the English Tourist Board. As in other areas of government spending, there is an increasing emphasis in heritage and tourism on levering partnership funding from the private sector and on focusing spending on economic regeneration.

THE MINISTRY OF AGRICULTURE, FISHERIES AND FOOD

Other departments of government have special status in respect of town and country planning, notably the Ministry of Agriculture, Fisheries and Food (MAFF). An overriding concern of government after the war was the protection of agriculturally productive land. This secured a central place for the MAFF in land use decisions. It has had to be consulted on important proposals, and the MAFF classification of agricultural land quality remains a potentially important consideration in development control. The influence of the MAFF has waned somewhat in parallel with the decline of agriculture in the British economy, but it still retains a special status. For example, if the MAFF has an unresolved objection to a development plan, the LPA must refer the dispute to DoE. The MAFF also has to be consulted on any planning proposal which involves a significant loss of high-quality agricultural land. Objections by the MAFF have fallen considerably over recent years. At the same time, it has assumed an increasing responsibility for countryside protection functions such as environmentally sensitive areas. In the words of the Ministry's annual report for 1995, it has

> an underlying aim of enhancing and improving the rural and marine environment, and to that end operates a number of schemes, at an estimated cost of £100 million in 1994–95, which have an explicit environmental focus. The principles laid down in the UK's Sustainable Development Strategy, of pursuing a balance between economic development and environmental protection, are integrated into the development of MAFF's policies.

THE SCOTTISH OFFICE

Scotland has had a special position in the machinery of government since the Act of Union, 1707. It has maintained its independent legal and judicial systems, its Bar, its established Church (Presbyterian) and its heraldic authority (Lord Lyon King-at-Arms). There has been a Scottish Grand Committee of the House of Commons since 1970 (although it has few powers). There is also a HC Committee on Scottish Affairs. The Scottish Office has a long history and, in its present form, dates from 1939. It now has five main departments: agriculture, environment and fisheries; development; education and industry; health; and home.

Each of these has its own permanent secretary (in addition to the Scottish Office Secretary – termed the Permanent Under-secretary of State). The Secretary of State for Scotland also has ministerial responsibility for the Forestry Commission whose functions extend throughout Britain.

The Scottish Office has a degree of independence

from Whitehall which is more easily felt than described. Many years of responsibility for Scottish services, the relative geographical remoteness of Edinburgh (perhaps now essentially psychological), the nature of the distribution of people and economic activity, the vast areas of open land, the close relationship between central and local administrators and politicians – such are the factors which give the Scottish Office a distinctive character. (This is not to mention the additional party political factors, such as the very small number of Conservative MPs: eleven out of seventy-two seats in 1996).

The reorganisation of governmental functions in England, which gave rise to the establishment of the DoE, were foreshadowed in Scotland in 1962. In that year, town and country planning and environmental services were transferred from the Department of Health to the Scottish Development Department, which at the same time took over all the local government, electricity, roads and industry functions of the Scottish Home Department. A later reorganisation divided functions between the Scottish Office Environment Department and the Scottish Office Industry Department. A further reorganisation in October 1995 recreated a separate Scottish Office Development Department with responsibility for land use planning. Historic buildings are dealt with by Historic Scotland, which is an executive agency. Non-departmental public bodies include Highland and Islands Enterprise (economic and social development in the Highlands); Scottish Enterprise (for the remainder of the country); Scottish Natural Heritage; and the new town development corporations.

THE WELSH OFFICE

In Wales, increasing responsibilities over a wide field have been transferred from Whitehall to the Welsh Office. This transfer has taken many years to achieve. Welsh affairs were dealt with by the Home Secretary until 1960, many services being administered directly by the departments which served England. There has been a minister responsible for

Wales since 1951 when Sir David Maxwell-Fyfe became the first Minister for Welsh Affairs, but it was not until 1964 that the (Labour) government established the Welsh Office and a Secretary of State for Wales who has a seat in the Cabinet.

The Welsh Office now has responsibility for health, community care, education, agriculture and fisheries, forestry, local government, housing, water and sewerage, environmental protection, town and country planning, nature conservation and the countryside, and (through the executive agency of Cadw) ancient monuments and historic buildings. In short, the Welsh Office is responsible for most government services in Wales except a small number which are run on a national basis, such as social security.

The Office is a relatively small one, with only 2,200 staff. It is organised in twelve groups: planning is administered in a Transport, Planning, and Environment Group, which is responsible for transport policy, planning, estates, environment, and the Welsh Historic Monuments Agency (Cadw). The Welsh Office is also responsible for the Land Authority of Wales, the Development Board for Wales, the Countryside Council for Wales, the Welsh Development Agency, the Welsh Tourist Board and the Cardiff Bay Development Corporation. Some of these are discussed in later chapters.

THE NORTHERN IRELAND OFFICE

Government in Northern Ireland has a unique character and structure. National government performs, either directly or through agencies, virtually all governmental functions: local government has few responsibilities. Though there are twenty-six elected district councils, their powers are limited to matters such as building regulations, consumer protection, litter prevention, refuse collection and disposal, and street cleansing. The councils also nominate representatives on the various statutory bodies responsible for regional services such as education, health and personal social services, and the fire service. They also have a consultative status in relation to a number of services including plan-

ning. All the major services, including countryside policies, heritage, pollution control, urban regeneration, transport, roads, and town and country planning are administered directly by the Northern Ireland Office. The NIDoE is the responsible department for these. Housing is administered by the Northern Ireland Housing Executive which was formed in 1971 to take control of the local authority housing stock.

Given the tragic history of Northern Ireland, the Office's priority aims are significantly different from those of other parts of the UK: 'to create the conditions for a peaceful, stable and prosperous society in which the people of Northern Ireland may have the opportunity of exercising greater control over their own affairs'. Planning has an important role in this which is undertaken through an executive agency: the Planning Service of the Northern Ireland Office. The general status of executive agencies is discussed below. The Agency's aim is 'to plan and manage development in ways which will contribute to a quality environment and seek to meet the economic and social aspirations of present and future generations', and has an annual budget of £7.3 million (1995–6).

NEW EXECUTIVE AGENCIES

The proliferation of new government agencies is confusing, Essentially it has taken two main forms: the *executive agencies* (of which the highly successful Driver and Vehicle Licensing Executive was the forerunner); and the well-established quango type exemplified by the Rural Development Commission and the new town development corporations (which technically are known as *non-departmental public bodies*).

Executive agencies, established under the *Next Steps* programme, remain part of their Department, and their staffs are civil servants; but they have a wide degree of managerial freedom (set out in their individual 'framework' documents). They enjoy delegated responsibilities for financial, pay and personnel matters. They work within a framework of objectives, targets and resources agreed by ministers. They are accountable to ministers, but their chief executives are personally responsible for the day-to-day business of the agency. Ministers remain accountable to Parliament. If this sounds somewhat confusing, that is because it is. However, in principle, the stated intention is to increase accountability. In Waldegrave's words, 'what we have done is to make clear the distinction between responsibility, which can be delegated, and accountability, which remains firmly with the minister' – a contention which is the subject of considerable controversy. Examples of executive agencies are the Building Research Establishment and the Planning Inspectorate.

By contrast, *non-departmental public bodies* (NDPBs) are a type of quango (quasi-autonomous non-governmental organisation). These bodies (of which there are over 1,300) range enormously in function, size and importance. They all play a role in the process of national government, but are not government departments or parts of a department. There are three types of NDPB: executive bodies such as the Countryside Commission, Scottish Natural Heritage, the Land Authority for Wales, and the new town and urban development corporations; advisory bodies such as the Advisory Committee on Business and the Environment, the Radioactive Waste Management Committee, and the Royal Commission on the Environment; and tribunals such as the lands tribunals, rent assessment committees, and the Agricultural Land Tribunal.

THE PLANNING INSPECTORATE

The way in which the aims and objectives of executive agencies are cast is illustrated by the case of the Planning Inspectorate. The Planning Inspectorate serves both the DoE and the Welsh Office. English work is run from an office in Bristol, while Welsh operations are handled by a separate Welsh agency group located within the Welsh Office in Cardiff. (Similar functions are performed by the Scottish Office Inquiry Reporters Unit and the Northern

Ireland Planning Appeals Commission, which is a wholly independent body.) The major areas of work are the holding of local plan inquiries (sixty-six opened in 1994–5 and a hundred forecast for 1995–6); determining planning appeals (12,236 in 1994–5); and determining enforcement appeals (2,953 in 1994–5). Now that the development plan has assumed a higher profile in the 'plan-led' system, local plan inquiries are placing heavy demands on the Inspectorate. Plans are currently attracting increasing numbers of objections, and reporting times are inevitably lengthened. Whereas in 1993–4 the duration from the opening of an average inquiry to the submission of the inspector's report was 40 weeks, by 1995–6 this had increased to 49 weeks.

Other responsibilities of the Planning Inspectorate include highway inquiries for the Department of Transport, and footpath orders under the Highways, Town and Country Planning, and Wildlife & Countryside Acts. An increasing amount of work is devoted to environmental matters such as inquiries under the Environmental Protection Act. The Inspectorate has an annual budget of £26 million and a total staff of around 600. The *Business and Corporate Plan 1995/96–1998/99* shows, at the beginning of this plan period, 210 full-time salaried inspectors and 111 fee-paid inspectors (in terms of person-years). In line with current ideas about governmental administration, it has performance targets which include deciding 80 per cent of written representation appeals within eighteen weeks; providing an inspector for local plan inquiries within twenty-six weeks of the end of the objection period; and holding the unit cost of planning appeals decided by written representation at or below £690. The Inspectorate now publishes its own journal.

CENTRAL GOVERNMENT PLANNING FUNCTIONS

Relationships between central and local government vary significantly among various policy areas, 'reflecting, in part, the difference in weight and concern which the centre gives to items on its political agenda, and, in part, differences in the sets of actors involved in particular issue areas' (Goldsmith and Newton 1986: 103).

Under the 1943 Town and Country Planning Act (which preceded the legislation on the scope of the planning system), there was a duty of 'securing consistency and continuity in the framing of a national policy with respect to the use and development of land'. Though this is no longer an explicit statutory duty, the spirit lives on, and the Secretary of State has extensive formal powers. These, in effect, give the Department the final say in all policy matters (subject, of course, to parliamentary control – though this is in practice very limited). For many matters, the Secretary of State is required or empowered to make regulations or orders. Though these are subject to varying levels of parliamentary control, many come into effect automatically. This delegated legislation covers a wide field, including the Use Classes Orders and the General Development Orders. These enable the Secretary of State to change the categories of development which require planning permission.

The formal powers over local authorities are wideranging. If a local authority fails to produce a plan, or a 'satisfactory' plan, default powers can be used. The Secretary of State can require a local authority to make 'modifications' to a plan, or can 'call in' a plan for 'determination'. Decisions of a local planning authority on applications for planning permission can, on appeal, be modified or revoked. Development proposals which the Secretary of State regards as being sufficiently important can be 'called in' for decision.

Though these powers are infrequently used, they are available if there is deadlock between local and central government. This can amount almost to a game of bluff as, for instance, when a local authority wants to make a political protest or to demonstrate to its electors that it is being forced by central government to follow a policy which is unpopular. Thus, opposition in Surrey to the M25 was so strong that the county omitted it from its structure plan. The Secretary of State made a direction requiring it to be included. Another battle arose over the Islington unitary development plan, where the Secretary of State took strong objection to the stringent controls which the borough proposed (*inter alia*) for its thirty-four conservation areas. The Secretary of State issued a direction requiring thirty-two of these to be changed. The borough took the matter to court, which held that it had no power to intervene on the planning aspects of the case; and since the Secretary of State had not acted perversely or in conflict with his own policies, his action was quite legal. Judicial review cannot be used as an oblique appeal. It was therefore the responsibility of the borough and the Secretary of State to resolve their differences to the satisfaction of the Secretary of State (*Journal of Planning and Environment Law* 1995: 121–5). A more recent case was a direction to Berkshire County Council to modify their proposed structure plan to increase the provision for new housing in the county (from 37,500 to 40,000) by the year 2006. As discussed in chapter 6, this is a common issue of friction between central government and a number of county councils, particularly in the South-East.

Perhaps the classic case of open political conflict was the North Southwark Local Plan which was formally called in by the Secretary of State. A DoE press release explained the situation:

> We do not lightly call-in local plans, let alone reject them; indeed this is the first to be rejected and only the second to be called in. However, we will not hesitate to take action where such plans fly in the face of national policies for economic regeneration, especially of the inner city. The Southwark Local Plan is opposed to private investment and is hostile to the London Docklands Development Corporation. The Council has rejected the helpful comments of the inspector following the public local inquiry. The plan is quite unacceptable.
>
> (Read and Wood 1994: 11)

The interest of cases such as these lies in their exceptionality. It is very rare for a local authority to engage in a pitched battle with central government. Equally, it is seldom that central government will feel compelled to use its reserve powers. It is perhaps noteworthy that these two cases arose in the politically charged areas of inner London between radical Labour authorities and a Conservative government that had become openly hostile to local government.

The North Cornwall case was handled in a way more consistent with tradition. Here the local authority was giving planning permissions for development in the open countryside contrary to national policies and the approved county structure plan. Pressure was brought to bear upon the district council by way of a special enquiry carried out by an independent professional planner (Lees 1993). Normally, informal pressures are sufficient: the threat of strong action by the Secretary of State is typically as good as – if not better than – the action itself. With the enhanced position of development plans in the so-called plan-led system, attention now focuses on the provisions of draft plans. Departmental officials pore over the wording of local policies to ensure that they accurately reflect those established at the national level. To the outsider, this has developed into a game of words, sometimes taking on the character of academic hair-splitting. For instance, the word 'normally' is taboo, as is 'a presumption against' – unless this accords with specific national policy (Read and Wood 1994: 11).

In spite of all this, it is not the function of the Secretary of State to decide detailed planning matters. This is the business of local planning authorities. The Secretary of State's function is to coordinate the work of individual local authorities and to ensure that their development plans and development control procedures are in harmony with broad planning policies. That this often involves rather closer relationships than might *prima facie* be supposed follows from the nature of the governmental processes. The line dividing policy from day-to-day administration is a fine one. Policy has to be translated into decisions on specific issues, and a series of decisions can amount to a change in policy. This is particularly important in the British planning system, where a large measure of administrative discretion is given to central and local government bodies. This is a distinctive feature of the planning system. There is little provision for

external judicial review of local planning decisions: instead, there is the system of appeals to the Secretary of State. The Department in effect operates both in a quasi-judicial capacity and as a developer of policy.

The Department's quasi-judicial role stems in part from the vagueness of planning policies. Even if policies are precisely worded, their application can raise problems. Since local authorities have such a wide area of discretion, and since the courts have only very limited powers of action, the Department has to act as arbiter over what is fair and reasonable. This is not, however, simply a judicial process. A decision is not taken on the basis of legal rules as in a court of law: it involves the exercise of a wide discretion in the balance of public and private interest within the framework of planning policies.

Appeals to the Secretary of State against (for example) the refusal of planning permission are normally decided by the Planning Inspectorate. Such decisions are the formal responsibility of the Secretary of State; and there is no right of appeal except on a question of law. Inspectors also consider objections made to local development plans, but in this case the recommendations are made to the local authority rather than the Secretary of State. The local authority is responsible for deciding how to act on the recommendations, but there are additional public safeguards allowing further objections and call-in by the Secretary of State.

Planning authorities, inspectors and others are guided in their decisions and recommendations by government policy. Central government guidance on planning matters is issued by way of circulars and, since 1988, in *planning policy guidance notes* (PPGs), *minerals planning guidance notes* (MPGs) and *derelict land grant advice notes* (DLGAs). *Regional planning guidance notes* (RPGs) are published in a separate series. In Scotland, Wales and Northern Ireland there are similar 'national' policy statements, which are explained in Chapter 4.

Since the introduction of planning guidance documents, circulars have been concerned mainly with the explanation and elaboration of statutory procedures. PPGs deal with government policy in substantive areas, ranging from green belts to outdoor advertising. A full review of DoE policy guidance was completed early in 1993 and resulted in the revision of many PPGs. Circulars and guidance notes are generally subject to some consultation with local authorities and other organisations prior to final publication, but the Secretary of State has the final word.

Circulars and guidance notes are recognised as important sources of government policy and interpretation of the law, although they are not the authoritative interpretation of law (this is the role of the courts), nor are they generally legally binding. Indeed, advice can be conflicting, perhaps as a result of piecemeal revision at different times. Nevertheless, circulars and guidance notes command a great deal of respect and form an important framework for development planning and development control.

Policy, of course, has to be translated into action. This presents inevitable problems: policy is general, action is specific. In applying policy to particular cases, interpretation is required; and often there has to be a balancing of conflicting considerations – of which many examples are given throughout this book. Policies can never be formulated in terms which allow clear application in all cases, since more than one 'policy' is frequently at issue. And even the most hallowed of policies has to be flouted on occasion: as witness developments in the green belts, in protected sites of natural or historic importance and in national parks. Such developments may be unusual (if only because they attract great opposition – of an increasingly strident nature); but they represent only the most obvious and the most public of the conflicts over land use.

Given the realities of land use controls, policies are usually couched in very general terms such as 'preserving amenity', 'sustaining the rural economy', 'enhancing the vitality of town centres', 'restraining urban sprawl', and such like. This is a very different world from that of a zoning ordinance, which is the principal instrument of development regulation in many countries. Such an ordinance may provide (for example) that a building shall be set back at least 15 feet from the road, have a rear yard of 20 feet or more, and side yards of at least 8 feet. Zoning was intended to be clear and precise, and subject to virtually no 'interpretation'. Indeed, it was hoped that it would be virtually self-executing. Though these hopes failed to materialise, it is fundamentally different in approach from the British planning system. Above all, the British system embraces discretion and general planning principles rather than certainty for the landowner and developer.

It is important to recognise that discretion means much more than 'making exceptions' in particular cases. The system requires that *all* cases be considered on their merits within the framework of relevant policies. Local authorities cannot simply follow the letter of the policy: they must consider the character of a particular proposal and decide how policies should apply to it. But they cannot depart from a policy unless there are good and justifiable planning reasons for so doing. The same applies to the Secretary of State, who is equally bound both by the formulated policies and the merits of particular cases. The courts will look into this carefully in cases which come before them and, though they will not question the merits of a policy, they will ensure that the Secretary of State abides by it. Thus, in a curious way, discretion is limited. All material considerations must be taken into account and justified. Arbitrary action is unacceptable, as it is in the USA which has written constitutional safeguards.

(B) LOCAL GOVERNMENT

Local authorities are undergoing a fundamental transformation from being the main providers of services to having responsibility for securing their provision. The task of setting standards, specifying the work to be done and monitoring performance is done better if it is fully separated from the job of providing the services . . . As enablers, local authorities have greater opportunities to choose the best source of service and thus to provide local people with a greater choice. The aim should be to secure the best services at least cost. The private and voluntary sectors should be used to provide services where this is more cost-effective than direct provision by the authority.

DoE (1991) *The Structure of Local Government in England*

The map of local government areas in the UK has been radically redrawn twice in the last thirty years. A uniform two-tier system of local government was established across Great Britain by Acts in 1963 for London, 1972 for England and Wales outside London and 1973 for Scotland. The two tiers were replaced by a unitary system of local government in London and the metropolitan areas of England in 1986 under the banner of 'streamlining the cities'. Further reorganisation into unitary authorities took place in Scotland and Wales in 1996, and a number of unitary authorities were introduced in parts of non-metropolitan England from 1995 to 1997 (although much of the two-tier system remains). In Northern Ireland, a unitary system of local government was set up in 1973. Amidst the various reorganisations of administrative areas, significant changes have taken place in the functions that local authorities perform and the style in which they operate.

Thus the UK has a complex structure of local government with variations in functions across the four constituent countries, as illustrated in Figure 3.1. There is further complexity in England where the latest local government reorganisation was not undertaken on a uniform basis. As a result some areas have a unitary system while others retain a two-tier structure. Figures 3.2 and 3.3 show the principal planning authorities in the UK. (For Northern Ireland the divisional offices of the Planning Service are shown since these are the planning authority.) The joint arrangements for structure planning are explained in Chapter 4.

METROPOLITAN GOVERNMENT IN ENGLAND

In the six English metropolitan areas (Greater Manchester, Merseyside, South Yorkshire, Tyne & Wear, West Midlands and West Yorkshire) metropolitan county councils were established in 1974. There was no difference between the planning powers of the metropolitan and non-metropolitan counties. Their responsibilities included structure planning, agreeing a framework for local plan preparation with the district councils, consulting on matters of common concern, and determining major development control issues.

In fields other than planning, the metropolitan districts had a much wider range of functions (and the counties less) than their non-metropolitan neighbours. Districts also continued to operate some services transferred to the counties through agency arrangements. The overall picture was confusing, with a blurred division of responsibilities. The result, in some cases, was conflict between the tiers – nowhere more so than with planning.

In 1986, under the banner of 'streamlining the cities', the Thatcher government abolished the metropolitan county councils (MCCs) along with the Greater London Council (GLC). This dramatic step, untraditionally, was preceded by no royal commission, committee of inquiry or study.

The arguments put forward by the government were decidedly thin. Efficiency was cited as the main objective, and it was held that this was where the

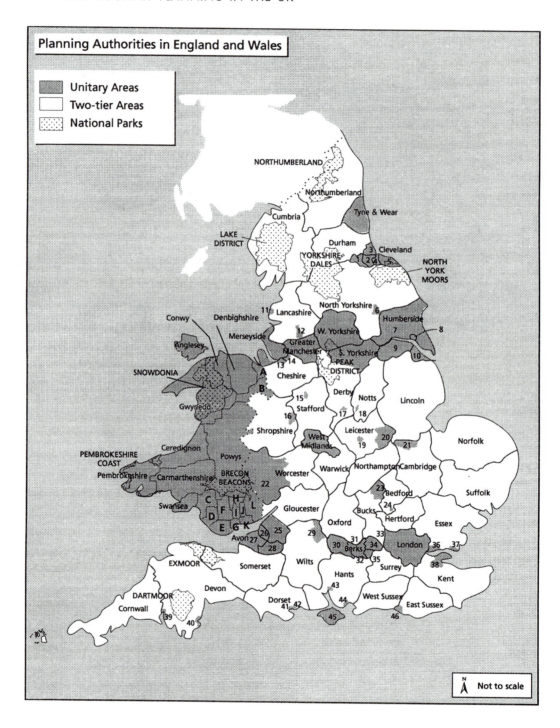

Figure 3.2 Planning Authorities in England and Wales

Non-metropolitan unitary councils

1	Darlington	17	Derby	32	Wokingham
2	Stockton-on-Tees	18	Nottingham	33	Slough
3	Hartlepool	19	Leicester	34	Windsor and Maidenhead
4	Middlesbrough	20	Rutland	35	Bracknell Forest
5	Redcar and Cleveland	21	Peterborough	36	Thurrock
6	York	22	Herefordshire	37	Southend
7	East Riding of Yorkshire	23	Milton Keynes	38	Gillingham and Rochester
8	Kingston upon Hull	24	Luton		upon Medway
9	North Lincolnshire	25	South Gloucestershire	39	Plymouth
10	North-East Lincolnshire	26	Bristol	40	Torbay
11	Blackpool	27	North-West Somerset	41	Poole
12	Blackburn	28	Bath and North-East	42	Bournemouth
13	Halton		Somerset	43	Southampton
14	Warrington	29	Thamesdown	44	Portsmouth
15	Stoke-on-Trent	30	Newbury	45	Isle of Wight
16	The Wrekin	31	Reading	46	Brighton and Hove

Metropolitan unitary district councils

Greater Manchester
  Bolton
  Bury
  Manchester
  Oldham
  Rochdale
  Salford
  Stockport
  Tameside
  Trafford
  Wigan

Merseyside
  Knowsley
  Liverpool
  St Helens
  Sefton
  Wirral

South Yorkshire
  Barnsley
  Doncaster
  Rotherham
  Sheffield

Tyne & Wear
  Gateshead
  Newcastle upon Tyne
  North Tyneside
  South Tyneside
  Sunderland

West Midlands
  Birmingham
  Coventry
  Dudley
  Sandwell
  Solihull
  Walsall
  Wolverhampton

West Yorkshire
  Bradford
  Calderdale
  Kirklees
  Leeds
  Wakefield

London borough councils

Barking and Dagenham
Barnet
Bexley
Brent
Bromley
Camden
City of Westminster
Croydon
Ealing
Enfield
Greenwich
Hackney
Hammersmith and Fulham
Haringey
Harrow
Havering
Hillingdon
Hounslow
Islington
Kensington and Chelsea
Kingston upon Thames
Lambeth
Lewisham
Merton
Newham
Redbridge
Richmond upon Thames
Southwark
Sutton
Tower Hamlets
Waltham Forest
Wandsworth

Wales

A	Flintshire	E	Vale of Glamorgan	I	Caerphilly
B	Wrexham	F	Rhondda, Cynon, Taff	J	Newport
C	Neath and Port Talbot	G	Cardiff	K	Torfaen
D	Bridgend	H	Merthyr Tydfil/Blanau Gwent	L	Monmouthshire

Figure 3.2 continued

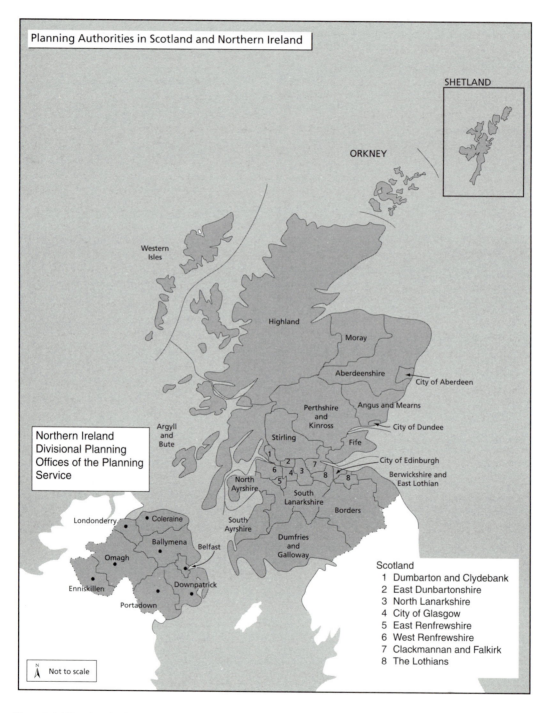

Figure 3.3 Planning Authorities in Scotland and Northern Ireland

GLC and MCCs were particularly deficient. They had few functions, but they spent a great deal of money; they were inefficient and had poor and counterproductive relations with their constituent districts (which were often controlled by a different political party); and their autonomy was frequently annoying and embarrassing to the Thatcher government, particularly after 1981 when all six MCCs returned Labour councils with huge majorities. Finally, in this abbreviated list, it was argued that efficiency could be improved and public resources saved by abolition. The details are too complex to spell out here, but the question arises: what really went wrong?

It is clear that the metropolitan counties 'gained few friends since their establishment' (Flynn *et al.* 1985: 105). Partly, this was a matter of the previous history of local government in the regions, and the acrimonious town–county battle which had for so long bedevilled English local government. Partly, it was a matter of the lack of functional and political power of the regions. The matter of power is crucial. The new counties had the difficulty of establishing themselves between the entrenched 'political arenas' of the district councils and central government; and (at least in the time available to them) they did not succeed (Healey *et al.* 1988: 209). In Young's words, 'a metropolitan authority tends to have too little power to be effective, and too much to be acceptable' (Young 1984: 5). The White Paper (*Streamlining the Cities*: Cmnd 9063) took up this point: whereas the shire counties were major providers of services in their areas, this was not the case with the metropolitan counties.

Given the predispositions of the government, no argument about strategic planning was persuasive: indeed, since strategy in the two-tier system required the agreement of both tiers, this was 'a recipe for conflict and uncertainty'. Moreover, their 'search for a role' brought the GLC and the MCCs into conflict not only with the district councils but also with central government, both in terms of expenditure and policy. The conclusion was clear: the structure was 'fundamentally unsound' and had 'imposed heavy and unnecessary burdens on ratepayers'. The proposed reorganisation involved transferring as many services as possible to the districts; others were to be moved to joint authorities or other statutory bodies.

Thus in 1986, all the metropolitan counties and the GLC were abolished, and the thirty-six metropolitan boroughs, the thirty-two London boroughs and the City of London became single-tier 'unitary' authorities. Functions transferred to these unitary districts included highways and traffic management; minerals and derelict land reclamation; waste regulation and disposal; sport; historic buildings; recreation, parks and green belt land; and gypsy sites. Additionally, and of prime importance, is the planning function. Here major difficulties are those of strategic planning and of establishing a substitute for the metropolitan county structure plans to provide a basis for coordinating the unitary plans of the individual districts. The metropolitan county areas have nothing to compare with the London Planning Advisory Committee. The Manchester experience is probably typical: it is clear that there is little approaching a plan for the conurbation as a whole. Instead, there are 'ten largely isolationist visions born out of separate processes of plan development' (G. Williams *et al.* 1992).

Some coordination is provided by non-statutory strategic guidance, prepared cooperatively by the districts. When agreed by the Secretary of State, this is issued as a regional planning guidance note. The products to date generally have been poor substitutes for the former metropolitan county structure plans. They are discussed in the following chapter.

LONDON

The London Government Act of 1963 came into operation in 1965. In brief, it established a Greater London Council covering an area of about 1,600 square kilometres with a population, at that time, of nearly eight million; and thirty-two new London boroughs, plus the unmolested Corporation of the City of London (which, for the purposes of this book, can be regarded as the thirty-third borough).

These replaced the London County Council, twenty-eight metropolitan borough councils, the county council of Middlesex and the county boroughs of Croydon, East Ham and West Ham. Considerable parts of Essex, Hertfordshire, Kent and Surrey were transferred to the new Greater London area.

The GLC had the responsibility for preparing the strategic development plan for the whole of the Greater London area. (This was published in 1969.) It set out the policies relating to population, housing, employment, transport and, indeed, all major issues which came within the compass of strategic planning. Within this strategic framework, the London borough councils were responsible for preparing their own local development plans. Originally, it was intended that these would be structure plans, but this was changed in 1972 owing to the great length of time required to process the Greater London Development Plan. The boroughs, therefore, were required only to prepare local plans within the framework of the GLC structure plan. With the abolition of the GLC in 1986, the London boroughs became unitary authorities with responsibility for most planning functions. Each borough now has to prepare a unitary development plan, which is described in the following chapter. There is, in addition, a statutory joint committee which has the function of advising on matters of common interest relating to the planning and development of Greater London. This is the London Planning Advisory Committee (LPAC). Each borough is also the mineral planning authority and highway authority for its area. Public transport is operated by London Regional Transport, which is responsible directly to the Secretary of State.

NON-METROPOLITAN GOVERNMENT IN ENGLAND

Outside the six metropolitan areas and London, the 1972 Act divided England into thirty-nine counties and 196 districts. In the shire counties, the planning functions of local government were split between the tiers, with the county preparing structure plans and many local subject plans; and the district preparing local plans. Proposals came forward in 1979 to review the distribution of functions between counties and districts. Although the proposals were described as 'organic change' (the title given to the White Paper: Cmnd 7457) they were seen by some as not going far enough and by others as a radical and important change. However, the proposals for development control, which reduced the 'county matters' category of applications, had wide agreement and were enacted by the Local Government, Planning and Land Act of 1980. This left responsibility for development control almost entirely with districts, whilst counties retained responsibility for only a limited number of special categories such as applications concerning mineral working.

The development planning functions continued to be split between the tiers but, in practice, the counties dominated development plan work until well into the 1990s. Structure plans provided the only guidance for control in the greater part of the shire counties where local plans had not been prepared.

A major function of counties has been in the planning and control of minerals and waste disposal. This has been strengthened under the Planning and Compensation Act 1991, with the mandatory requirement for minerals and waste plans. The counties also have additional statutory functions in archaeology and coast protection. Tourism, economic development and off-street parking were split between the tiers.

The districts were given responsibility for local development plans and most of development control, as well as statutory functions in respect of building control, housing, parks and open spaces and playing fields.

The two-tier structure has been subject to continued dispute which, in some cases has developed into outright conflict. One of the key questions concerned the appropriateness of the two-tier structure for some of the large provincial cities, such as Nottingham, Exeter and Leicester. These and other large towns sought greater standing and a measure of independence from the county, if not full independence. Their opportunity came with the further reorganisation of local government in the non-metropolitan areas. The driving force behind this was Michael Heseltine: hence the title of the INLOGOV report: *The Heseltine Review of Local Government: A New Vision or Opportunities Missed?* (Leach *et al.* 1992). The process of reorganisation in England, handled by the DoE, was conducted in parallel with reorganisation in Scotland and Wales, undertaken by the Scottish and Welsh Offices. However, the process adopted in England was quite different and the outcome far more complex. Whatever Heseltine's original vision for English local government, it disintegrated in the political storms through which his ideas had to pass.

In England a Local Government Commission was established to work within policy and procedural guidance published by the DoE. This 'guidance' proved to have greater power than the government expected: it had the effect of limiting the changes which could be made to the Commission's proposals. Moreover, because of the consultative way in which the Commission operated, these proposals were significantly influenced by the views of articulate local people and organisations.

The government's initial proposals for change were set out prior to the establishment of the Commission in a Consultation Paper on *The Structure of Local Government in England* (DoE 1991). This made the argument, widely accepted across the political spectrum, for a unitary structure of local government in the shires (the pattern which had been put in place in the metropolitan counties). It was argued that a single tier would reduce bureaucracy and costs and improve coordination. It would clarify responsibility for services and, since tax-payers would be able to relate their local tax bills more clearly to local services, would provide for greater accountability. In the early stages of the Commission's review, the Environment Secretary, John Gummer, had stated unequivocally that the aim was to produce a unitary structure in England, with the two-tier system remaining in only excep-

tional circumstances. The way this was to be achieved was left open, with several possibilities: existing districts might become unitary authorities; two or more authorities could be merged into larger ones; and wholly new authorities might be created. The main criteria for judging the need for change were responsiveness to local needs and 'sense of identity' as well as the ubiquitous 'cost-effectiveness'.

During the two years of the Commission's review, district and county authorities sought to justify their existence through an expensive and sometimes bitter propaganda war. In fourteen of the worst cases this led to challenge to the Commission's recommendations in the courts. There was also a legal challenge by the Association of County Councils which successfully prevented the Secretary of State from modifying the guidance he had previously provided to the Commission in an attempt to strengthen the case for unitary authorities. Given the government's wish to see a unitary structure, the eventual undoubted winners were the counties. The Commission found little evidence that change would improve service provision. In the main, changes were limited to renewing unitary status for former county boroughs, and abolishing new and contrived counties created in the 1974 reorganisation.

After much debate, the Commission recommended only fifty new unitary authorities. Most of these were former county boroughs (unitary authorities before the 1974 reforms), although a significant number of 'special cases' were included on the basis of 'substantial local support for change'. The Commission explained the modest extent of their recommendations as due to the 'weight of evidence from national organisations pointing to the problems and risks associated with a breaking up of county wide services' – a view that was strongly supported by local opinion. However, these arguments did not satisfy the many districts which were not proposed for unitary status and which had campaigned hard for this. More significantly, it did not satisfy the government, which was concerned to further increase the number of unitary authorities.

As a result of these disagreements, the chairman of the Commission resigned, and the new chairman was given the remit to review again the cases of twenty-one districts where the government believed there was a strong case for unitary status. Further guidance was issued for this mini-review, stressing the potential benefits of unitary status, particularly for areas in need of economic regeneration (as in the Thames Gateway). It was argued that the 'single focus' of unitary local government would be more effective in promoting multi-agency programmes in these areas. This final review initially recommended unitary status for ten of the twenty-one districts, but this was reduced to eight after consultation.

The process of reorganisation in the shires has been the subject of considerable criticism and, after three years of work, many commentators are asking whether it was worth while. Certainly, reorganisation seems to have been handled much more expeditiously in Wales and Scotland.

WALES

As in England, the Local Government Act of 1972 created for Wales a two-tier local government system: this consisted of eight counties and thirty-seven districts. The separation of functions of the two tiers was also largely the same as in England.

Reorganisation in Wales proceeded more quickly than in England. The review was carried out by the Welsh Office (rather than by an independent commission) and the country was considered as a whole (rather than by separate areas). After a two-year period of consultation, a White Paper *Local Government in Wales: A Charter for the Future* (Cm 2155) was published in 1993, setting out detailed proposals. There was widespread agreement that the new structure should be unitary in character: the debate was focused on the number and boundaries of the new local authorities.

The underlying thinking included a restoration of authorities which had been swept away in the earlier reorganisation – Cardiff, Swansea, Newport, and some of the traditional counties such as Pembrokeshire and Anglesey. However, to fit into a unitary structure, the boundaries had to be stretched somewhat and some counties had to be amalgamated. After consideration of proposals for thirteen, twenty and twenty-four unitary councils, the final outcome was twenty-one authorities, known as counties in rural areas and county boroughs in predominantly urban areas. They range in population from 60,000 in Merthyr Tydfil to 295,000 in Cardiff.

In the White Paper, the unitary system was commended for its administrative simplicity, its roots in history, its familiarity and the relative ease with which residents could identify 'with their own communities and localities'. The intention was to create 'good local government which is close to the communities it serves'. The White Paper continued:

> Its aims are to establish authorities which, so far as possible, are based on that strong sense of community identity that is such an important feature of Welsh life; which are clearly accessible to local people; which can, by taking full advantage of the 'enabling' role of local government, operate in an efficient and responsive way; and which will work with each other, and with other agencies, to promote the well-being of those they serve.

These desirable objectives do not all work in the same direction, of course, and some compromise was inevitable. A number of the areas are very large. Powys, for example, has over 500,000 hectares: this is a very large area for *local* government. There is potential for the community councils to take on an increased role, but the Welsh Office has stressed that there is no intention of forming a second tier of local government.

SCOTLAND

The cultural history and physical conditions of Scotland dictate that, to a degree, the administration of planning will be different. Changes to the law in Scotland require specific legislation, and the Scottish Office has administrative discretion within which it will take account of the very special circumstances which exist in parts of the country. The pursuit of planning objectives, therefore, is sometimes distinctive. Nevertheless, the broad thrust and impact of government policy are much the same (Carmichael 1992). Indeed, some of the government's most fundamental attempts at reform have been tried and tested in Scotland first, including the community charge, rate capping and 'compulsory competitive tendering'.

The Local Government (Scotland) Act 1973 provided for a two-tier system of local government except in the three island areas of Orkney, Shetland and the Western Islands (which became 'most-purpose' authorities). This created nine regional and fifty-three district councils. Together with the three island authorities, there were sixty-five local authorities, of which forty-nine had planning powers. By selectively allocating planning powers in these areas of scattered population to the regional authority, it was possible to increase the number of districts (and thereby also reduce their enormous geographical size). There were thus three different types of area for planning purposes: two-tier areas where planning powers were divided between regional and district authorities, two-tier areas where only regions (termed general planning authorities) had planning powers, and the three unitary island areas where the island councils were general planning authorities. In effect, there was a two-tier planning system in six regions (covering nine-tenths of the population) and a 'general' planning authority system elsewhere.

The regions varied greatly in size: Strathclyde (with a population of over two million) had nearly half the country's population. Districts ranged in population from 10,630 in Nairn to 439,000 in Edinburgh and 688,000 in Glasgow (1991 figures). Despite the variation in size, there was a clearer distinction, at least in concept, between the functions of the regions and the districts than was the case south of the border. The regions were conceived as strategic authorities responsible for 'regional planning functions', all highways, public transport, education, social work and water. The districts were responsible for local planning, housing, refuse collection and disposal and a range of other local services.

In setting out to reorganise Scottish local government, the government was firmly committed to a single-tier structure, and the 1992 Scottish consultation paper provided options only on the number that were to be established.

There were, of course, some political factors involved in this decision: the problem of conflicting interests within the Conservative Party was much less in Scotland since only a handful of the sixty-five Scottish local authorities were in Conservative control. The consultation document in Scotland was also more forthright about the role of local government reform in direct service provision. While the government confirmed its commitment to 'a strong and effective local authority sector' it also argued that local authorities no longer needed to 'maintain a comprehensive range of expertise within their own organisation', since much could be done by outside contractors'.

In reviewing the possible number and size of the proposed unitary authorities, the 1992 consultation paper provided four illustrations showing structures of fifteen, twenty-four, thirty-five and fifty-one authorities, but it also noted that there were other options. The choice between mainly small or mainly large authorities has important implications for the planning function, especially structure plans. Only the fifteen-authority option would allow for unitary authorities to prepare their own structure plans. Even then, special arrangements would be needed for Glasgow to ensure effective strategic planning. The outcome of reorganisation in Scotland was thirty-two unitary councils, each of which has full planning powers for its area. The fragmentation of the strategic planning function across a larger number of authorities threatens a recognised strength of the Scottish system, and the need for special arrangements for strategic planning was acknowledged by the Scottish Office during the review. The country has been divided into seventeen structure planning areas, six of which require joint working between authorities. The plan framework is discussed further in Chapter 4.

NORTHERN IRELAND

Local government in Northern Ireland was last reorganised in 1973, when thirty-eight authorities, made up of counties, county and municipal boroughs and urban districts, were replaced by a single tier of twenty-six district councils. Although this reduced the enormous variation in the size of districts (previously ranging from 2,000 to over 400,000) there is still a wide variation, from Moyre with a 1991 population of 14,700 to Belfast City with a population of 288,700. Planning powers were centralised under the then Northern Ireland Ministry of Development. Since the demise of the power-sharing Northern Ireland Assembly in 1974, planning, like all public services, has been subject to 'direct rule' under the supervision of the Secretary of State for Northern Ireland. The preparation of plans and the control of development are functions of the DoENI, which it exercises through the Planning Service (an executive agency). Local government is only consulted on the preparation of plans and development control matters.

The lack of accountability through local government (described as the 'democratic deficit') obviously needs to be seen in the light of the very special circumstances, though it has been judged to have operated with a 'considerable measure of success' (Hendry 1992: 84). Nevertheless, local councillors have been able to attack planning and to 'represent themselves as the champions of the local electorate against the imposed rule of central government' (Hendry 1989: 121). Even when the central bureaucracy has made determined efforts to open decision-making and involve local people, it has been accused of having ulterior motives (Blackman 1991b).

The promise of a 'lasting peace' in Northern Ireland during the ill-fated cease-fire brought with it ideas for reform, which are still on the agenda. Several possible scenarios have been suggested, including the continuation of a central planning authority accountable to an elected assembly, devolution to joint regional boards, and complete delegation of powers to the districts (RICS 1994). The relatively weak position of local government over many years and the dearth of skills and experience will not be put right quickly. It is likely that any reform will be introduced incrementally. The volatile political conditions in the Province make prediction impossible. In the meantime, the Northern Ireland Office established the Planning Service as an executive agency, described earlier in this chapter.

PARISH, COMMUNITY AND TOWN COUNCILS

There are over 12,000 parishes (or their equivalent) in Britain, many of which are represented by a parish, community or town council. They do not form part of the local government structure formally, but are sometimes described as the 'lowest tier of local democracy'. They can play a significant part in the democratic process in rural areas and small towns by providing an effective voice for local interests. Where they are active, planning issues will make up a significant part of their agendas.

In England, there are about 10,200 'parishes' ranging in population from nil to 48,700 (Bracknell). They are represented by 70,000 parish councillors. About 8,200 have parish or town councils; in the remainder an annual parish meeting should be called. The 1972 Act allowed for additional parishes to be created where they do not exist (especially in urban areas), but little use is made of this power. Several non-statutory neighbourhood councils have been established in the larger urban areas of England.

The statutory functions of parishes are very restricted, and are mostly shared with the district council. They include the management of allotment gardens and some local open spaces, and footpath lighting and maintenance. Of particular importance (and widely used) is the right of parish councils to be consulted on planning applications in their areas. Parish councils can require the LPA to notify them of all, or specific classes of, local applications. In practice, where there is an active parish council, it will usually be consulted by the LPA on all applications, and can play an important role in the preparation of development plans.

The 1972 Act reorganisation of local government in Wales abolished parish councils and provided for statutory 'community councils'. The role of these was considered during the latest reorganisation but no change was made since it was felt that any further responsibility at this level would merely recreate a two-tier system of local government. Instead, the Welsh Office proposed steps to enhance their representational role, in effect ensuring effective consultation with the unitary councils.

Scottish legislation provides for the establishment of community councils where there is a demand for them, under schemes prepared by local authorities. As in England and Wales, their purpose is to represent the local community and 'to take such action in the interests of the community as appears to its members to be desirable and practicable'. However, community councils do not have a statutory right to be consulted. Indeed, the review of local government in Scotland concluded that no functions should be specified by statute. Instead, local authorities are urged to delegate such responsibilities as community councils are capable of fulfilling. A 1996 Scottish Office PAN, *Community Councils and Planning*, sets out a code for consultation arrangements.

LOCAL GOVERNMENT REORGANISATION AND PLANNING

There is general consensus that the new local government structure lacks coherence, especially in England, but a proper evaluation of the consequences of reorganisation for planning will be possible only when the new arrangements have had time to settle down. Nevertheless, there has been no shortage of speculation, most of it pessimistic. The biggest concern is over the difficulties which will face the new local authorities in balancing strategic and local considerations. This is particularly acute in planning since local government must cope with the need to make decisions in the interests

of its area as a whole, whilst managing the effect of new development on local communities (Hollis *et al.* 1992). A specific concern is the need for joint cooperative arrangements for the production of structure plans (an issue which is taken up in the next chapter).

A report on the process of local government change in Scotland suggests that a three-tier system of local government will emerge. Many functions, including strategic planning in many areas, will be performed by groups of authorities acting together, expanding the number of joint boards which are only indirectly accountable. This has raised concerns about the effectiveness of strategic planning in situations where planning officers will be giving advice both to their own authority and to a joint board. Other functions, including development control in some of the large authorities, will need to be provided on an area basis and they may be run by area committees (Boyne *et al.* 1995).

On the positive side, there is potential for closer liaison between functions brought under one roof which were previously the province of separate tiers. This is particularly so for planning and transport policy. More speculative are the claims that the demise of two-tier government paves the way for the introduction of an additional tier of government at the regional level (Leach 1994).

Under the new local authority structure, joint arrangements for structure planning will be widespread. The challenge to raise awareness and engage the public in this task is daunting. The public will relate to their local authority area rather than any 'structure planning area' (in the same way that many planners did in the debate over reorganisation). The defence of territory which has characterised the process has without question diverted effort from strategic planning and inter-authority cooperation. It is interesting to note that whilst the counties and districts in England have been arguing over who should survive, central government has integrated its regional offices and thus has begun to fill the most significant void in the British governmental structure.

CENTRAL–LOCAL RELATIONS

It used to be common to talk of central–local government relationships as constituting a 'partnership' but, for several reasons, this is no longer an appropriate description. First, there has been a fundamental change in the role of local government within the British system of government. In its heyday, local government was largely concerned with the provision of 'public goods' rather than redistributive services. (Public goods are those that benefit all the public – streets, sewers, parks, lighting, etc. – while redistributive services confer benefits on some at the cost of others.) This is no longer the case:

> The prominence of income redistribution in modern economic policy has not only shifted attention from local government to central, but also altered the balance of local authority activities. The slum clearance and community health services that originally were seen as benefiting the community as a whole have been superseded by housing and social services that are designed to be redistributive. In many cases local authorities are trying to fill the gaps or general inadequacies of national cash redistribution programmes. This switch of emphasis from public goods to redistribution has played a part in the postwar tendency to treat local authorities as agents of central government that 'should' act in accordance with the political will of the centre.
>
> (Dawson 1985: 31)

The replacement of domestic rates by the flat-rate community charge (itself now replaced by the council tax) represented a major change in the direction of local government finance back to the notion of *charging* for local authority services.

Second, many functions have been removed from local government: in the early postwar reorganisations, electricity and gas, trunk roads, 'poor relief' and hospitals were all removed. Despite strong opposition from local government, the building of new towns was entrusted to *ad hoc* government-appointed development corporations. Other local health services, water supply and sewage disposal functions were removed in 1974–5. In more recent years further responsibilities have been removed, reduced or transformed, such as public transport, council housing and education.

Most important, however, has been the dramatic challenge to local government by the Thatcher administration (though the trends can certainly be seen earlier). This started as a reaction to 'overspending' by local authorities: central government wanted to divert expenditure from areas such as housing and education to defence and (in response to rising unemployment) social security. But this was not under their control: local authorities still had their own sources of revenue which, though limited, were significant. The result was a tightening of controls, 'caps' on local expenditure, and eventually drastic changes in the system of local taxation, including the notorious 'poll tax' (officially the *community charge*). These controls resulted in a major transfer of power from the localities to the centre. They also led, first, to strained relationships between local and central government, and later, to strife and virtual open warfare. The original intention to restrain public expenditure developed into a drive to make local authorities more accountable.

It was to achieve this that the community charge (payable by all adults) was introduced. The uncooperative and sometimes belligerent stance of some local authorities, particularly in the metropolitan areas, led to further draconian measures based on a firm belief that central government policies had to prevail over local ones. Local authorities were forced to sell council houses (tenants were given the 'right to buy' at heavily discounted prices), and local authority house building programmes were slashed: these and similar measures amounted to the 'nationalisation' of housing policy. Such measures (including the deregulation of passenger transport) significantly reduced the role of local authorities and diminished their position in the overall machinery of government. Not least, a whole tier of local government disappeared with the abolition of the GLC and the metropolitan counties in 1985.

These forces, however significant for the role of local government, were not the product of a clear and coherent strategy from the centre. Rather, they were a series of *ad hoc* reactions to individual problems: first, of controlling expenditure; later, of reducing the discretion of local government; and then of transforming local government into a means of enabling other agencies to provide services to a standard which could not be achieved by traditional means.

THE ENABLING LOCAL AUTHORITY

Towards the end of the 1980s, more far-reaching changes began to take place across virtually all local government functions. The 1988 Education Reform Act introduced a centralised curriculum and the local management of schools. The 1988 Housing Act introduced *housing action trusts* to take over the management of large council housing estates. It also gave groups of council tenants who were not exercising their right to buy the opportunity to opt out of council control and into the tenancy of a housing association or other landlord. The 1988 Local Government Act obliged councils and other specified public bodies to contract-out manual services through compulsory competitive tendering (CCT). This was later extended to construction-related and corporate services, including architecture and property management but, as yet, it does not include the town and country planning function.

In sum, the direct services, responsibilities and powers of local government were systematically and substantially reduced. Local government is now cast in a new 'enabling' role, although exactly what this means is a matter of continuing debate. Leach and his colleagues (1994) argue that, during the 1980s, the dominant central government vision was of local authorities playing a residual role as enablers of a limited range of services that the market could not provide. During the 1990s, this view has given way to one which sees local authorities involved in a wider range of activities but still very much as an enabling broker between central policy, local demands and a wide range of agencies and contractors who provide the service (Audit Commission 1989). Where the market is unable or unwilling to provide a service, the local authority should use the market as a model of how to deliver the service itself through 'quasi-markets'. This mode of operation has been encouraged by the requirement for compulsory competitive tendering, the separation of client and contractor roles, the need to demonstrate value for money (VFM), and performance review.

It is here that the contrast with trends in European local government are most interesting. The notion of 'enabling' is already the norm in much of western Europe, with local government implementing its functions through a diverse range of agencies, often in partnership, but paying more attention to its role in prioritising community needs and acting as the focus for local political activity. By contrast, the debate about local government in the UK has been dominated by disputes over service delivery rather than government.

The attack on local government has been much less successful than either camp would have us believe. Cochrane (1991: 282) argues that 'the market model of local government looks less like an assessment of what has happened and more like a picture of what some people would like to happen (or like to believe has happened)'. Certainly, the government's privatisation ideology has yet to bite deeply into the planning service. Instead, there has been the 'rise of partnership', where 'economic power, social status and the political control of local government' are combined – 'a new closeness between private and public sectors at the local level' (Cochrane 1993: 99, 103).

LOCAL GOVERNMENT CHANGE AND PLANNING

For town and country planning, the apparent and seemingly paradoxical outcome of change in the 1980s and 1990s has been a larger and stronger body of planners with widened statutory functions. The indirect effect of market deregulation, the increasing complexity of development issues and the growing emphasis on environmental protection were bound to lead to a greater demand for planning skills (Healey 1989). The concept of an enabling local government also increases the need for strategic thinking and focuses attention on the corporate planning function (Carter *et al.* 1991). The direct impact on the way in which the planning service is delivered is less significant. So far, planning has been subjected to only minimal change in comparison to other services, and the concept of the local authority as enabler requires more attention to strategy and in-house planning, rather than less.

The conclusion has to be that, in contrast to most local services, planning as a statutory and regulatory function has been somewhat protected from the pressure for change. Nevertheless, there are significant implications for the planning service, not least the need to demonstrate 'value for money' (VFM). This is not an easy concept to define for planning because of the difficulty of assessing quality in plans and planning decisions. The Audit Commission provides guidance for local authorities and district auditors on performance indicators for all services, including planning, but these have been criticised for their reliance on quantitative measures, the clas-

sic example being the proportion of applications decided within eight weeks.

The Audit Commission's 1992 report *Building in Quality* addressed these criticisms and made a real attempt to introduce a wider assessment of performance, recognising that there were many ancillary tasks in providing advice and negotiating with applicants and making 'complex professional and political judgments'. After consultation, the Audit Commission settled on six key indicators. As with the earlier version, these concentrate on matters of efficiency rather than on the effectiveness of the system, though the added breadth to performance review will be a significant improvement on previous practice.

The emphasis on VFM is being reinforced in two ways. First, there is the requirement for all local authorities to establish a 'transparent' internal accounting system. This will ensure that the real costs of services can be identified, will help to bring about internal markets in local authorities, and will enable much closer investigation of 'value for money' and cost effectiveness. The second is the Citizen's Charter: this is discussed in Chapter 12.

(C) EUROPEAN GOVERNMENT

What the Community is doing, and the ways in which it is developing, will affect planning in Britain, as it will affect planning in every other Member State of the Community; through its direct and indirect impact on planning policies and legislation . . . and in the constraints which it may impose, but, more significantly, in the opportunities which it will open up for the development of planning practice.

Davies and Gosling 1994

The impact of the European Union (EU) on town and country planning in the UK is steadily becoming more important, although as yet this has been felt predominantly in the field of environmental controls rather than in mainstream planning practice. The introduction of environmental impact assessment is the most striking example of European influence. Later chapters identify a range of agricul-

tural, environmental, economic and regional policies of the EU which are having an effect on parts of the British planning system. Chapter 4 includes a note on supranational and cross-border planning instruments and policies that are being introduced at the European level. Here, a brief and more general account is given of important EU institutions and actions, and their effect on planning in the UK.

Note is also taken of the increasing interest in the methods, policies and experiences of planning systems in other member states.

BRITAIN IN THE EUROPEAN UNION

The UK was not an enthusiastic supporter of the postwar moves towards a federal Europe. Though it favoured intergovernmental cooperation through such bodies as the Organisation for European Economic Cooperation (1948) and the Council of Europe (1949), it was opposed to the establishment of organisations which would facilitate functional cooperation alongside nation-states. It therefore did not join the European Coal and Steel Community (1952), nor was it a signatory to the 1955 Treaty of Rome which established the European Economic Community and the European Atomic Energy Community. However, along with the other members of the Organisation for European Economic Cooperation, it formed the European Free Trade Association (EFTA) in 1960. Britain envisaged that EFTA would form the base for the development of stronger links with Europe. When it became clear that this was not viable, Britain applied for membership of the European Community. This was opposed by France and, since membership requires the unanimous approval of existing members, negotiations broke down. The opposition continued until a political change took place in France in 1969. Renewed negotiations led finally to membership at the beginning of 1973.

The Treaty of Accession provided for transitional arrangements for the implementation of the Treaty of Rome, which Britain agreed to accept in its entirety. The objectives include the elimination of customs duties between member states and restrictions on the free movement of goods; the free movement of people, services and capital between member states; the adoption of common agricultural and transport policies; and the approximation of the laws of member states to the extent required for the proper functioning of the common market. These objectives are often referred to as the 'four freedoms': the free movement of goods, people, services and capital.

There are currently fifteen members of the EU: Austria, Belgium, Denmark, Finland, France, Germany, Greece, the Republic of Ireland, Italy, Luxembourg, the Netherlands, Portugal, Spain, Sweden and the United Kingdom. The EU has a population of 363 million and a land area of over 3.2 million square kilometres. In comparison with the US, this is about 50 per cent more people living in just over one-third of the space. The EU is also easily the largest trading bloc, having a share of exports more than three times its nearest rivals of the US and Japan.

The organisational and political structure of the EU is complex and, like all such bodies, its actual workings are somewhat different from the formal organisation chart. The main institutions of the EU and their elements which are of particular interest to planning are shown in Figure 3.4. In brief, there is an elected Parliament which operates as an advisory body, and a Council of Ministers which is the legislature and makes policy largely on the basis of proposals made by the European Commission, which is the executive. There is also a Court of Justice which adjudicates matters of legal interpretation and alleged violations of Community law. There is a large number of other organisations within or related to the EU.

THE COMMISSION

The major work of the EU is undertaken by the Commission. This executive body has responsibilities for preparing proposals for decision by the Council and for overseeing their implementation. The Commission has the sole right to initiate legislation. (Only rarely can the Council make a policy decision without a proposal from the Commission.) The Community's decision-making process 'is dominated by the search for consensus among the member states' and this gives the Commission a crucially important role in mediation and conciliation. Of the same nature is the ethos of achieving

THE EUROPEAN COUNCIL

Meeting of Heads of State
Gives broad guidance and impetus to action

COREPER

Committee of Permanent
Representatives.
Manages work of Council

THE COUNCIL

Presidency rotates every 6 months
1996	– Italy	– Ireland	1999	– Germany	– Finland
1997	– Netherlands	– Luxembourg	2000	– Portugal	– France
1998	– UK	– Austria	2001	– Sweden	– Belgium

Informal Meeting (Council) of Ministers of Spatial Planning

COMMITTEE ON SPATIAL DEVELOPMENT

Meeting of civil servants responsible for spatial planning

EUROPEAN COMMISSION
20 Commissioners
30 Directorates, including :

DG VII : TRANSPORT

DG XI : ENVIRONMENT & NUCLEAR SAFETY

DG XVI : REGIONAL POLICY AND COHESION
(includes Spatial Planning and ERDF)

Applies the Treaties by initiating
legislation and issuing notes as
Executive body.

EUROPEAN PARLIAMENT
626 elected members (87 from UK)
20 standing committees, including :
    Agriculture and rural development
    Regional policy
    Transport and tourism
    Environment, public health and consumer protection

Political driving force, supervis-
ing and questioning Council and
Commission. Joint power to
adopt legislation with Council.
Supervises appointment of
Commission.

ECONOMIC AND SOCIAL COMMITTEE
222 nominated members from employers, workers and other interests.
9 sections, including :
    Environment, public health and consumer protection
    Regional policy and spatial planning
    Transport and communications

Is consulted and delivers opinions
on proposed legislation.

COMMITTEE OF THE REGIONS
222 members
8 Commissions, including :
    Commission 1 : REGIONAL POLICY
    Commission 2 : RURAL PLANNING
    Commission 3 : TRANSPORT
    Commission 4 : URBAN POLICIES
    Commission 5 : LAND USE PLANNING, ENVIRONMENT AND ENERGY

Is consulted and delivers opinions
where regional interests involved.

THE COURT OF JUSTICE & COURT OF FIRST INSTANCE
15 Judges in each court

Interpret the Treaties and apply judgments and penalties in cases of non-
compliance.

Figure 3.4 Institutions of the European Union and Spatial Planning

compromise and of progressing in an incremental way. There is thus no challenging bold vision to which the participants aspire: that would inevitably lead to conflict.

Among the Commission's powers is that of dealing with infringements of Community law. If it finds that an infringement has occurred, it serves a formal notice on the state concerned requiring discontinuance or comments within a specified period (usually two months). If the matter is not resolved in this way, the Commission issues a *reasoned opinion*, requiring the state to comply by a given deadline. As a last resort, the Commission can refer a matter to the Court of Justice, whose judgment is legally binding. Most matters are dealt with informally, but Britain has been subject to reasoned opinions on environmental matters, some of which are noted in Chapter 7 (Haigh 1990: 153, 160).

The Commission is organised in thirty Directorates-General (DGs) and other main departments which, in good bureaucratic style, are referred to by their numbers rather than their names. Some 15,000 officials are employed in them. The DGs are further divided into directorates and divisions, but the complexity varies according to the importance of their function, and hence their size. The Directorates are each headed by a director-general but considerable influence over the work of the Directorate is exercised by the personal 'cabinet' of the Commissioner, and in particular the chair, known as the 'chef du cabinet'.

THE EUROPEAN PARLIAMENT

The European Parliament is a directly elected body consisting of 626 members who are elected every five years. Britain has eighty-seven representatives, known as MEPs: Members of the European Parliament. The Single European Act of 1986 increased the legislative powers of the Parliament, and they were further extended by the Treaty of Union (sometimes called the Maastricht Treaty), but it is important to note that the Parliament is essentially an advisory and supervisory body, whilst the Council of Ministers is the legislature. However, the Parliament is consulted on all major Community decisions, and it has powers in relation to the budget which it shares with the Council, and in approving the appointment of the Commission. The assent of Parliament is needed for accession of new members and international agreements. It is organised along party political (not national) lines. The political groups have their own secretariats and are the 'prime determiners of tactics and voting patterns' (Nugent 1989: 130). Much of their work is carried out by standing committees and through questions (3,900 in 1994) to the Commission and Council.

THE COUNCIL OF MINISTERS

The policy-making body of the EU is the Council of Ministers. This is composed of the senior ministers of the fifteen countries – usually the foreign minister and the minister most concerned with the matters to be discussed.

The Council has numerous committees and working groups with various functions and membership. Thus 'while a single institution under the treaties, the Council has many forms, as it is composed of the appropriate ministers for the policy area under discussion' (Hadjilambrinos 1993: 289), such as the Council of Ministers for the Environment or for Culture. Heads of State of each country and the President of the Commission meet at least twice each year when very broad policy directions are discussed. The Presidency of the Council rotates every six months. Voting is by qualified majority on many matters (normally sixty-two votes out of eighty-seven).

There is no formal Council of Ministers responsible for planning, but since 1991 there have been biannual informal meetings of ministers responsible for spatial planning. A sub-committee of this informal council is the Committee on Spatial Development (CSD) which is made up of officials from the member state governments. The UK is represented by the DoE, the DTI and the Scottish Office.

It is the Council which makes Community laws (termed *regulations*) which are binding on member states. These are 'directly binding': they require no additional implementing legislation. They are used mostly for detailed matters of a financial nature or for the technical aspects of (for example) administering the Common Agricultural Policy (CAP). By contrast, *directives* are broad statements of policy which, though equally binding, are implemented by national legislation. This leaves the method of implementation to the member states. Environmental matters are typically dealt with in this way. The Council can also issue *decisions* which are binding on the member state, organisation, firm or individual to whom they are addressed. Finally, there are *recommendations* and *opinions*, which have no binding force.

THE COMMITTEE OF THE REGIONS

The Committee of the Regions (CoR) is the youngest European institution; set up following the Treaty of European Union, it held its first session in March 1994. It is intended to give a voice to the regions and local authorities in European Union debates and decision making. It has 222 members representing the regions, including twenty-four UK members. (The UK representation comprises fourteen from local authorities in England, five from Scotland, three from Wales and two from Northern Ireland.) The CoR is particularly concerned with matters that directly affect the regions. It has taken a particular interest in regional planning and in advocating wider use of the principle of subsidiarity, so as to strengthen the role of regional and local authorities.

The Treaty identifies particular areas where the CoR has to be consulted by the Commission, including trans-European networks, economic and social cohesion, and structural fund regulations. It can also offer opinions in other areas that it thinks appropriate, typically when an issue has a specifically regional dimension. Up to the end of 1995, it had issued about fifty opinions, including important comments on Europe 2000+ and other planning, urban and environmental issues.

COURT OF JUSTICE

The European Court of Justice has fifteen judges, at least one from each member state. It decides on the legality of decisions of the Council and the Commission and it determines violations of the Treaties. Cases can be brought before it by member states, organisations of the Community and private firms and individuals.

THE COUNCIL OF EUROPE

The Council of Europe is a broad organisation and should not be confused with the EU. It was set up in 1949 to promote awareness of a common European identity, to protect human rights and to standardise legal practices across Europe to help achieve these aims. Since 1989, its main role has been to monitor human rights in the post-communist democracies, and to assist them to carry out political, constitutional and legal reform. It has thirty-six member countries (including twelve former communist states) and, at the end of 1995, had five membership applications outstanding, including Russia.

It has a three-tier structure, with a Council and Ministers, a Parliamentary Assembly and a Congress of Local and Regional Authorities. With an annual budget of less than £100 million, it is much less powerful than the EU (which has an annual budget of over £50 billion), but nevertheless it has played an important part in maintaining and establishing democracy on the continent. It is best-known for its Convention on European Human Rights. Anyone who feels that their rights under the Convention have been breached may take a case to the European Court on Human Rights for a decision which will be binding on those states that have signed up to the Convention.

The Council has been active for many years in the field of regional planning and environment, especially through the Congress. It has published conference and other reports on the implications of sustainability for regional planning, the representation of women in urban and regional planning, and many other topics. Perhaps its most important contribution to spatial planning has been the European regional/spatial planning charter, known as the Torremolinos Charter, adopted in 1983: this committed the Council to producing a European regional planning concept. The spatial planning work of the Council is taken forward by the Conference of European Ministers of *Aménagement du territoire* (CEMAT) responsible for regional planning which cooperates closely with the European Commission.

The Council was also responsible for the European Campaign for Urban Renaissance (1980–2). This led to a programme of *ad hoc* conferences, various reports and 'resolutions' on such matters as health in towns, the regeneration of industrial towns, and community development. In 1992, the Conference adopted *The European Urban Charter*. This 'draws together into a single composite text, a series of principles on good urban management at local level'. The 'principles' relate to a wide range of issues, including transport and mobility, environment and nature in towns, the physical form of cities, and urban security and crime prevention.

FURTHER READING

Central Government

Whitehall by Hennessy (1989) is an insightful analysis of the culture of the British civil service and the changes which have been imposed upon it. See also Drewry and Butcher (1991) *The Civil Service Today*. Osborne and Gaebler (1992) *Reinventing Government* give a coloured account of changes which they perceive in the American systems of government: resemblances to the British scene are striking, even if inconclusive. An invaluable overview is given by Dynes and Walker (1995) *The Times Guide to the New British State: The Government Machine in the 1990s* (which is heavily used in the text).

An up-to-date summary description of government departments and their functions is given in the annual *Whitaker's Almanack*.

A principal source of information on the workings of government departments is the *Departmental Annual Reports*. These are the *Government's Expenditure Plans* for the forthcoming three years and are sometimes referenced in this way.

Local Government

The principal textbooks which give a general introduction to local government structure and organisation are Elcock (1994) *Local Government*, and Wilson and Game (1994) *Local Government in the United Kingdom*. For an overview of the politics of local government, including the roles of and relationships between councillors, officers and political parties, see Stoker (1991) *The Politics of Local Government*. At the time of writing, the most up-to-date overall review is Chandler (1996) *Local Government Today*. For European comparisons, see Hirsch (1994) *A Positive Role for Local Government: Lessons for Britain from Other Countries*, and Batley and Stoker (1991) *Local Government in Europe*.

On Scotland and Wales, see Carmichael (1992) 'Is Scotland different?'; Boyne, Jordan and McVicar (1995) *Local Government Reform: A Review of the Process in Scotland and Wales*; Midwinter (1995) *Local Government in Scotland*; On Northern Ireland, see Bannon *et al.* (1989) *Planning: The Irish Experience 1920–1988*, and Hendry (1992) 'Plans and planning policy for Belfast'.

Parish councils are the subject of a survey by the Public Sector Management Research Centre (1992) *Parish Councils in England*. A Welsh Office consultation paper was issued in 1992: *The Role of Community and Town Councils in Wales*.

On central–local government relations, see Cochrane (1993) *Whatever Happened to Local Government?* A case for improving relations is made in Carter and John (1992) *A New Accord: Promoting Constructive Relations between Central and Local Government*. *Local Government in the 1990s* is a collection of essays edited by Stewart and Stoker (1995).

European Union

The institutions and policies of the EU are summarised in a series of booklets *Europe on the Move* which are updated on at least an annual basis (available from the UK Office of the European Commission, Jean Monnet House, 8 Storey's Gate, London, SW1P 3AT). For a summary of the history of the EU, see Borchardt (1995) *European Integration: The Origins and Growth of the European Community*. A concise summary of the UK in the EU is given in the COI booklets (in the *Aspects of Britain* series) *Britain in the European Community* (1992), and *European Union* (1994). The EU publishes information booklets, including *Serving the European Union: A Citizen's Guide to the Institutions of the European Union*.

On European comparative planning, the European Commission's forthcoming *Compendium of EU Spatial Planning Systems and Policies* comprises a volume describing the systems and policies of spatial planning in each member state, volumes of case studies illustrating the operation and contribution of planning practice to the implementation of particular projects, and a comparative review. Williams (1996) *European Union Spatial Policy and Planning* provides the context for understanding planning in Europe, including a full description of the institutions and policies of the EU of relevance to planning. See also Davies, Hooper and Edwards (1989) *Planning Control in Western Europe*. A useful journal is *European Planning Studies*.

For a comparison of the discretionary nature of British planning as compared with other systems, see Booth (1996) *Controlling Development: Certainty and Discretion in Europe, the USA and Hong Kong*, and Cullingworth (1993) *The Political Culture of Planning*.

4

THE PLANNING POLICY FRAMEWORK

We are, I think, entering a new planning era . . . After a period of some uncertainty, we see planning emerging in a new light, and with a subtly changed role.

Sir George Young 1992

My own view is that the 'new system' of plans is essentially a side show, a new and ill-thought out set of procedures derived from a way of thinking which wishes to devolve as much of the work of plan-making to the lowest tier of government, while keeping hold of the possibility of central control of content and procedures.

Healey (1990)

INTRODUCTION

The British planning system is, in one sense, embodied in a huge library of statutes, rules, regulations, directions, policy statements, circulars and other official documents. Much of this chapter discusses the formal system which is mirrored in these documents. However, it is important to appreciate that the formal system is one thing: the way in which matters work in practice may be very different. The informal planning system operates within the formal structure. It may continue with little modification even when major legislative changes are made; alternatively, there may be significant changes in practice within a stable formal system. Political forces, professional attitudes and management styles will all affect the ways in which the system operates.

It is also necessary to note that much development (in the everyday, rather than the legal, sense of that word) takes place without any help or hindrance from the planning system. Even where the development is clearly related to some action within the statutory framework for planning, typically through the granting of permission to undertake the development, the actual outcome is affected by 'extraneous' factors, and it may not be at all clear what effect planning has had on the outcome. The question of whether the same development would have taken place in the same way is equally problematic.

Nevertheless, the framework for planning policy, and the procedures which are followed in its design and operation, provide an important starting point for an understanding of the way in which the system works. It is current government policy to bring much more of the informal operation of planning and development within the statutory planning framework. The emphasis is on plan-led development control. This is intended to reduce the amount of *ad hoc* planning control, and thus provide a firmer foundation for the resolution of conflicts and investment decisions. Plans are seen as providing a more efficient means of conflict mediation than decision-making on a project-by-project basis, as well as a measure of certainty and coordination for the pro-motion of investment (Healey 1990). There had been considerable lobbying for this from a range of interests, including developers, conservationists, investors and Conservative Party supporters (Nadin and Doak 1991).

The plan-led system is intended to produce a comprehensive and systematic hierarchy of national and regional guidance, and development plans. It is envisaged that, once these plans are in place, they will be a more important factor than their predecessors in land use decision-making.

The first of the quotations at the head of this chapter suggests that expectations are high. The second reflects some scepticism about any major shift in emphasis or impact. As we shall see, the fundamental principles of the system remain intact, and recent changes, whilst superficially significant, may have little effect on the patterns of land use and protection, or on who gains and loses from it. Plans may be more significant – and there is certainly a high level of activity in plan-making – but it is too soon to assess what changes, if any, may come about. For planners who experienced the introduction of the 1947 system or the changes made in 1968, there must be a sense of *déjà vu*. Many of the questions of those times bear a marked resemblance to the current debate:

- What framework will ensure the accountability of decision makers and safeguard the interests of those affected by planning, yet be expeditious and efficient in operation?
- How can the framework provide a measure of certainty and commitment, yet allow for flexibility to cope with changing circumstances, local conditions, and new opportunities?
- What objectives should plans pursue, and how will these shape their form and content?
- Who should have influence in the planning process, and what should be the respective roles of central and local government and of local communities?

These perplexing questions have no easy answers: indeed, by their nature, they have to be readdressed constantly. Acceptable answers rarely have stability since conditions and attitudes change over time. The biggest change in recent years has been the gradual growth of supranational planning, led by the European Union. It is appropriate that a discussion of the UK planning framework should begin with this.

(A) SUPRANATIONAL PLANNING

THE RATIONALE FOR PLANNING AT THE EUROPEAN SCALE

The EU is driven by the goals of economic and political union, and the balanced development of economic activities amongst its member states. These objectives have an obvious spatial dimension. The main obstacles to meeting them are the great disparities in wealth, jobs, investment and access to services across different regions. Indeed, recent evidence suggests that, despite the actions of the EU, some disparities (especially between the north and the south) are widening, and that economic and political forces will ensure that they continue to do so. Investment trends identified by the Commission also support this analysis, with mobile economic activities concentrating in the core cities.

The EU has increased regional funding through the European Regional Development Fund (ERDF) and the Cohesion Fund with the objective of balancing economic growth: it now accounts for more than a third of the total budget, and will grow to more than ECU 200 million in the ten years to 1999. This structural funding, linked to member states' own resources for regional aid, is beginning to have an effect on economic growth and development patterns across Europe. The Commission argues that this is *de facto* spatial planning. However, coordination of this investment through explicit spatial strategy is another matter, and it is only recently that the EU has taken a direct interest in such planning. This may be surprising, given the dynamic nature of spatial development patterns across the Union brought about by investment in infrastructure, changing locational decisions and other forces (Williams 1996).

Allocation of the funds is divided according to six 'objectives', three of which have a spatial dimension, being focused on particular regions in economic difficulty. Objective 1 is to assist regions lagging behind in development (where the GDP is normally 75 per cent of the Community average), Objective 2 is targeted at declining industrial areas, and Objective 5b is to develop and adjust the structure of vulnerable rural areas. The areas of the UK which fall under these objectives are identified in Figure 4.1. The designations are effective until 1999. The UK government also has its own measures to promote the development of economically disadvantaged regions, although the locational controls and grant assistance have been reduced since the 1960s. The main instrument remaining is regional selective assistance (RSA) which supports projects to create jobs in areas of high unemployment. The two types of area are illustrated in Figure 4.1 for comparison with the EU designations. Development areas are eligible for higher rates of assistance than intermediate areas. The areas were designated in 1993 and are currently being reviewed. Both the structural funds and RSA are administered by the DTI.

Over two-thirds of the structural funds go to Objective 1 areas, with a further 11 per cent to Objective 2 and 5 per cent to Objective 5b. Thus some 84 per cent of structural funds are targeted on specific regions, which is perhaps not surprising since the main intention is to produce a better economic and social balance across the community. The main beneficiaries in the UK are the Objective 1 regions, covering Merseyside, the Highlands and Islands and Northern Ireland.

The Commission believes that more attention to positive spatial planning at the European level will make the policies of the EU and the member states more effective. However, the Union has no competence in spatial planning, and progress is made through intergovernmental cooperation, with member states rather than the Commission formally taking the lead. The Treaty of Union does empower the EU to 'adopt measures concerning town and country planning, land use with the exception of waste management . . . and management of natural resources', but this is only within the field of environmental protection. Nevertheless, ministers responsible for spatial planning in the member states meet on a six-monthly cycle and have sanctioned increasing attention to supranational planning. The work is formally undertaken by the Committee on Spatial Development (CSD) backed by the involvement of DGXVI (the Regional Policy and Cohesion Directorate).

THE EMERGENCE OF SUPRANATIONAL PLANS

The first initiative on systematic planning at the European scale came from the German government, which prompted the establishment of a permanent Conference of European Ministers of *Aménagement du territoire* (CEMAT) through the Council of Europe. The early regional policy of the EU had little spatial content and instead focused on the need to support particular industrial and commercial sectors. Despite a resolution in the European Parliament to draw up a European scheme for spatial planning, progress has been slow. However, the French, German, Danish and Dutch governments continue to promote supranational planning studies. These have introduced memorable and powerful spatial concepts at the European level such as the now infamous 'blue banana' (a concept used by the French national planning agency DATAR in relation to the area between south-east England and northern Italy where growth and investment has been – and continues to be – concentrated).

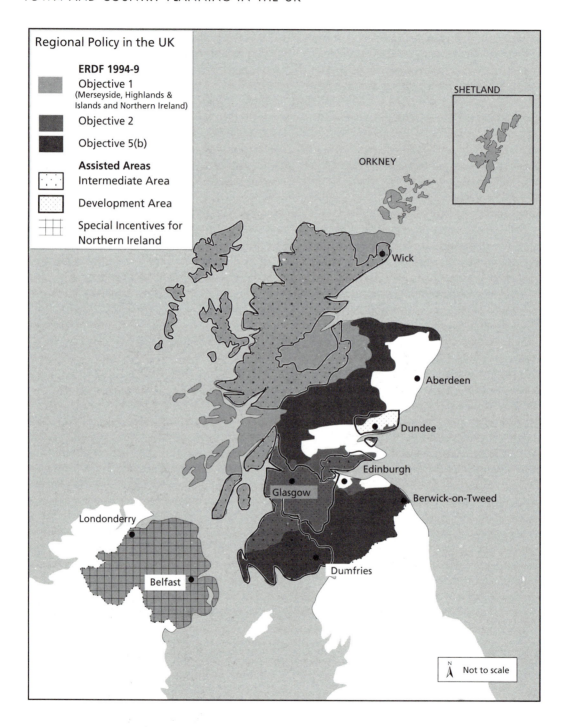

Figure 4.1 Areas Eligible for EU Structural Funds and Selective Regional Assistance

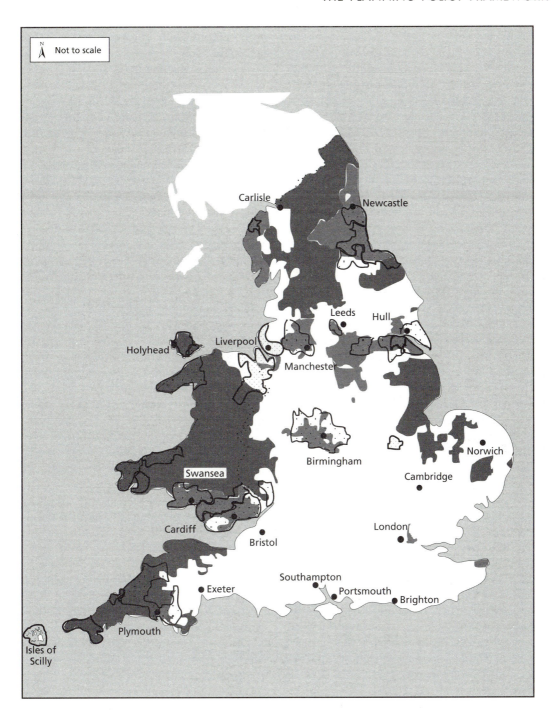

Figure 4.1 continued

The Commission's first major contribution to the development of European supranational planning came with the publication of *Europe 2000: Outlook for the Development of the Community's Territory* (CEC 1991). This document was intended to provide a European reference for planners working on national or regional planning policies. It was effectively a geography text, raising awareness of Europe-wide spatial development issues. It adopted an approach which cut across country borders to identify seven transnational study areas with shared characteristics. (An eighth area was added with the inclusion of the eastern German Länder.) These eight regions, together with adjacent 'external impact areas', became the subject of extensive research studies.

The initial findings for these study areas are reported in *Europe 2000+: Cooperation for European Territorial Development* (CEC 1994). As its title implies, this publication signalled a change of gear on supranational planning, with a clear call for more cooperation between member states. It charts the trends in the physical development of the European territory and crucially makes strong assertions about the preferred development patterns for the future.

Of overriding importance to the EU is the control of unrestricted urban growth. This has been a feature of development, much of it unregulated, in the southern European states. In its place, the Commission wishes to see the strengthening of small and medium size towns, especially where this can help in focusing the provision of services in rural areas. Underlying this policy is the concept of a polycentric urban system. This is much like the existing pattern in Germany, with a spread of urban services among a range of cities; it is quite different from the pattern of development in the UK or France. Its rationale is that it will avoid both the congestion problems of very large conurbations and the decline in service provision in rural areas. It is envisaged that much of this development will take place in the main corridors between existing centres.

Elsewhere, more rigorous protection of areas of environmental importance is promoted together with further policies aimed at containing the problems of land abandonment in rural areas where traditional agricultural practices are increasingly uneconomic. Urban regeneration and the revitalisation of poor neighbourhoods also has a high priority, but a firm line is being taken on the retention and restoration of the built heritage. Additionally, *Europe 2000+* reports on the Trans-European Networks (TEN), and notes the importance of linking spatial development and transport policy. The transport element in this concentrates on eleven priority transport projects which are intended to address missing links in the transport routes in the core area and out to the peripheral regions.

Whilst the Commission has concentrated on providing a more coherent analysis of European regional geography, the Committee on Spatial Development has taken the lead in producing the European Spatial Development Perspective. This is the most important initiative on spatial planning at this level, and perhaps a unique experiment in supranational planning. It is intended (if only at the broadest scale) to be a physical development plan for Europe. It aims to promote 'coherence and complementarity' of the development strategies of the member states' by coordinating the spatial aspects of EC sectoral policies. The task of achieving the necessary consensus among fifteen member states (and many regions within nations) is obviously a difficult one and, needless to say, the statements made are at a very general level, with maps of development patterns in Europe being expressed on A3 paper!

Increasing attention to spatial planning policy at the European level will inevitably affect the planning mechanisms in member states. There are large differences in the way planning operates across Europe: each member state (and in some cases an individual region) has developed its system in response to local economic and physical development problems arising within particular cultural, legal and social contexts. There is likely to be a trend towards some degree of convergence of these systems as countries work together.

In order to facilitate understanding about the way that spatial planning operates in different member states, and thus to promote more effective cross-border and transnational planning, DGXVI commissioned a *Compendium of EU Spatial Planning Systems and Policies*. This demonstrates the diversity in planning systems and policies (especially in their operation) but also notes similar trends as the different countries respond to the same macroeconomic forces. There is a distinct trend in much of Europe towards greater flexibility in the operation of regulation. New mechanisms are being introduced to establish more strategic planning frameworks and to allow for decisions which are contrary to the characteristic binding zoning plans. Another emphasis is on the integration of spatial plans and sectoral spending programmes. The spatial plan is more widely recognised as the coordinative mechanism for sectoral policy and spending. New instruments are being introduced to tackle cross-border issues, and there is increased transnational cooperation between planners dealing with similar issues in different regions. In the UK this is most pronounced in Kent where the County Council has worked cooperatively with four other 'regional authorities' of France and Belgium in the production of *A Vision for Euroregion* (Kent County Council 1995).

These impressive developments in supranational planning have not been made without some resistance. The UK government in particular has been less than enthusiastic about 'universal spatial planning policies' and has instead emphasised the usefulness of exchanges of experience. There is certainly room for debate on the implications of seeking dispersed but concentrated development, urban containment and a focus on corridor development. There will be other important issues that have not yet been considered, such as the availability and price of land for development. The proposals may not fully address powerful market forces, especially at a time when the private sector is taking a greater share of investment in virtually all member states. The limitations of the spatial planning systems in bringing about desired objectives of sustainable and balanced development across the community are recognised, and this may eventually lead the Commission to put more emphasis on other policy options for regulating and promoting development such as taxation measures or development incentives. But there is a general assumption in Brussels (and most member states) that more supranational planning is inevitable and that it is to be welcomed. This sentiment is shared by the former communist countries which, whilst not wishing to reinvent centralised state planning, do want to deal in a coordinated way with their massive problems of environmental degradation and economic decline.

(B) NATIONAL AND REGIONAL PLANNING

NATIONAL PLANNING GUIDANCE

There is no national land use planning in the UK in the sense that policies or plans are prepared for the whole country. There is, however, a growing amount of national land use *guidance*. This developed first in Scotland, where it was prompted by the need for a strategy to deal with the unprecedented problems posed by North Sea oil and gas. Since these were considered to be of national importance, the Scottish Office decided to issue guidelines for use by local authorities, especially in relation to coastal development (Gillett 1983). *North Sea Oil and Gas: Coastal Planning Guidelines* was published in 1974, and national planning guidelines (NPGs) on other topics soon followed. These guidelines were intended to fill a gap between relatively inflexible policy expressed in circulars and general advice that could be ignored (Raemaekers 1995). They had formal status – local authorities were required to notify the Scottish Office if they proposed to grant permission for development which was contrary to guidelines – but they did not tie ministers or local authorities to particular solutions. They did not go so far as to constitute a 'national plan', and they were not intended to be comprehensive, but they were locationally specific.

The benefits of this system soon became apparent (Diamond 1979). First, it enabled local authorities to explain the way in which their plans took account of national policies. Second, a higher degree of coordination was possible among the various branches of central government. Third, national interests in which the Secretary of State needed to be involved could be readily separated from local matters. Later assessments continued to recognise the strengths of the NPGs (Nuffield Report 1986; Rowan-Robinson and Lloyd 1991). Nevertheless, as the series expanded to cover more topics and non-locationally specific guidance, questions arose about the precise status of NPGs and their overlap with other policy statements (Planning Exchange 1989).

In response to these concerns, the Scottish Office introduced in 1993 a rationalised structure for national policy and advice through a series of national planning policy guidelines (NPPGs) together with continued use of planning advice notes (PANs) and circulars. The role of NPPGs is 'to provide statements of government policy on nationally important land use and other planning matters, supported where appropriate by a locational framework.' NPPGs are broader in scope than their predecessor NPGs, and are intended to provide more comprehensive coverage of topics of national concern. The role of PANs is 'to identify and disseminate good practice and to provide advice and other information' (NPPG1: para. 13).

Because of the recognised success of the NPG series, the shift to NPPGs has been watched closely. A review of early experience (Raemaekers *et al.* 1994) assessed NPPGs as 'a convincing effort to produce a successor to the pioneering series of the 1970s, fit for changed and complex circumstances'. It also called for greater breadth in topic coverage, and identified particular omissions in the proposed list, especially on major transport infrastructure. In the light of the impacts of local government reorganisation, it also recommended a tier of regional planning guidance for Scotland, and a return to

regional reports for the city regions in the Central Belt to provide greater coordination in joint structure plan preparation.

In England and Wales, national guidance did not arrive until 1988, although its form and content has provided a model for more recent changes to the Scottish system. National guidance in England is made up of planning policy guidance notes (PPGs) and minerals planning guidance notes (MPGs). As in Scotland, PPGs have taken on the role of expressing national land use and development policy, leaving circulars to be used mainly for the elaboration of procedural matters. They have certainly clarified and extended the national policy framework, but they have tended to be more general than the NPPGs in Scotland (and certainly NPGs), broader in scope, and not at all location-specific.

PPGs have had a considerable impact on planning practice. An evaluation of their effectiveness concluded that they had 'assisted greatly in ensuring a more consistent approach to the formulation of development plan policies and the determination of planning applications and appeals' (Land Use Consultants 1995a: 47). This is because PPGs are important material considerations in development control and have a determining influence on the content of development plans. Conformity between national guidance and plans is ensured through regional office scrutiny of development plans, but the study also found that most professional planners have a high regard for national guidance and welcome the order and consistency in policy that it brings. Councillors are generally more sceptical; this is not surprising because national guidance limits their ability to respond according to their interpretation of local needs.

Having successfully introduced more systematic planning policy at national level, the government's problem is now to maintain consistency and clarity in the series. There have been many calls for more PPGs, whilst at the same time some concerns have been voiced over perceived contradictions between one PPG and another, and between the series and other government policy statements. One example (from the Land Use Consultants 1995 study) is the concern over the different explanations of the term 'sustainable development' in government guidance.

In Wales, the publication of national guidance was, until recently, shared for the most part with England through joint publications of the DoE and the Welsh Office. On occasion, separate advice was thought necessary to reflect distinctive Welsh conditions (for example, on land for housing and on strategic planning), but these followed the English version quite closely. By 1995, the Welsh Office had decided to go its own way, and published two draft planning policy guidance notes intended to replace PPGs that originally were shared with the DoE; final versions were published in 1996. The first summarises policy guidance for a long list of topics covered previously in separate notes. The second deals solely with the new unitary development plan structure following local government reorganisation. The effect is to reduce the amount of guidance considerably, to the great concern of many people in Wales. The RTPI response to the draft guidance was that it 'significantly dilutes essential policy advice, lacks coherence and comprehensiveness, and will provide Wales with inferior guidance to that available to local planning authorities in England'. Technical Advice Notes (TANs) have been promised to fill in much of the lost detail, but concerns centre on the problem of lack of strategic direction, especially in the light of the reforms to the development plan system in response to the reorganisation of local government. However, one interesting innovation is the inclusion of a strategic diagram for the whole of Wales, illustrating areas of growth and protection and main communication routes (although it effectively states little more than the obvious).

In Northern Ireland, guidance takes the form of *development control advice notes* and *planning policy statements*. This guidance is much less comprehensive than in the rest of the UK, and it has been suggested that planning officers in the Province make use of PPGs and NPPGs to keep in touch with policy developments (RSPB, *Planscan: Northern Ireland*, 1993).

In whatever form, guidance carries considerable weight as important statements of government policy. But though local planning authorities are required to have regard to guidance, they are not bound by it. Indeed, other material considerations may be of greater importance in particular cases. Moreover, the advice in one guidance note may contradict another, perhaps as a result of piecemeal revisions at different times. Nevertheless, guidance notes command a great deal of respect and are closely followed in development control and development planning. They exert significant influence in planning policy and decisions, and are quoted profusely in decision making, especially at inquiries.

Some government policy is still to be found in circulars and also, from time to time, in ministerial statements. Major changes in policy are often published in White Papers. All these documents can be regarded as material considerations in planning, and thus central government has an array of instruments through which national policy can be expressed. Indeed, the result can sometimes be confusing, if not actually contradictory.

REGIONAL PLANNING GUIDANCE

In addition to the strengthening of national guidance, the government has taken steps towards expanding regional guidance in England. The 1986 Consultation Paper on *The Future of Development Plans* (Cm 569) noted the progress that had been made in some regions, such as East Anglia and the West Midlands, by local authorities cooperating voluntarily to produce regional strategies. Most prominent amongst these was the London and South East Regional Planning Conference (SERPLAN), which started life in 1962 as the Standing Conference on London Regional Planning.

Official encouragement was given in the 1986 Consultation Paper to the formation of other regional groupings. It was proposed that, as in the South-East, the Secretary of State would issue guidance after issuing a draft for comment. No precise procedures were suggested, and a warning was given that 'such arrangements would not represent a formalised regional structure, nor would they be a return to the type of large-scale regional planning which was attempted in the 1960s and 70s'. The 1989 White Paper (Cm 569) made similar comments and added the recommendation that business organisations and other bodies should be involved as well as local authorities. Conservation and agricultural interests were later added to this list. The need to merge the strategic guidance produced by the metropolitan counties as part of the unitary development planning process (completed by December 1989) with the new regional guidance was also recognised.

In 1990, the government stated that 'the aim should be to have guidance in place for most regions during the early 1990s'. PPG 12 (1992) specified the end of 1993 as the target date for completion. Guidance notes were to look ahead twenty or more years rather than fifteen. Flexibility in the process of production of advice has been retained, but local authorities are now expected to detail the results of consultations they have undertaken and the response. The conversion of the government to a modicum of regional guidance, if not regional planning, is thus complete. Under the present arrangements, 'conferences' of constituent local authorities produce advice to the Secretary of State in the form of draft guidance. The Secretary of State then takes this into account in publishing regional planning guidance. (Initially regional guidance was published in the PPG series, but since 1989 there has been a separate RPG series.) Local authorities thus do most of the preparatory work, but the important decisions are made by central government, and where significant changes are made this is seldom to the liking of the local authorities.

The scale of the task of producing regional guidance should not be underestimated. Even where there are long-standing cooperative arrangements, political conflict within the regions can be intense, and it is questionable whether the effort is always worthwhile. An indication of the varying style and quality is given in a symposium in the *Town Planning Review* (Minay 1992). From this, it appears that much of the guidance is no more than a further detailing of national guidance and restatement of current policies. The Regional Strategic Guidance for East Anglia (RPG 6) is described as:

> bland and incremental . . . it mostly describes existing situations and trends. It is very largely written as a substantive account of what has been happening in the region in recent years, informed mainly by topic PPGs (in other words, by national government policy) and specific programmes of road building and targets for house building.

Most of the others are described in a similar tone, though Roberts (1996) gives a more positive appraisal.

Until the 1980s, Scotland had a tier of regional reports which provided a corporate policy statement for the regions as well as a framework for the preparation of structure plans. There were few formal procedures governing their preparation and they did not require central government approval, but were simply published with the Secretary of State's observations. They were much admired but, as the statutory development plan framework was put into place, they became regarded as redundant.

A series of guideline documents was prepared as Strategic Planning Guidance in Wales by the Welsh Office. The documents were intended to 'consolidate and re-present the wide range of available strategic guidance material in a consistent and accessible form' and particularly to provide a framework for the preparation of structure plans. Given the changes to a unitary structure in Wales, strategic guidance is perhaps more important than ever, but neither of the two Welsh planning guidance documents refers to the strategic guidance.

The first regional plan in Northern Ireland was the Belfast Regional Plan published in 1964 (the Matthew Plan). This proposed the stopline, a system of radial motorways, and a major new town, Craigavon, modelled on the English experience (Hendry 1989). Some of the principal recommendations have been implemented but, like its counterparts in England, the plan was overtaken by the effects of dramatic economic recession. As a result, projects were undertaken without reference to their consequences for the plan. The subsequent *Regional Physical Development Strategy 1975–85* sought to concentrate growth in the Province to twenty-six key centres, but the depressing effects on other areas were widely challenged (Blackman 1985). A new rural planning policy published in 1978 took a much more relaxed approach to development in three-quarters of the rural territory, which led to extensive development of single houses in the countryside and ribbon development. Indeed, 25,000 houses were built in the countryside over ten years, many close to the urban centres, and the annual figure of 2,500–3,000 for Northern Ireland is more than the rest of Great Britain put together. This led to a reappraisal of the need for regional planning and the publication of *A Planning Strategy for Rural Northern Ireland* in 1993. This is unlike any other UK regional planning document in that it includes both strategic objectives for the overall development of the territory and detailed development control policies. This could only be a product of a system where central government sets the strategy, makes local plans and undertakes development control. The strategy introduces new restrictions on development in the countryside whilst introducing the novel designation of 'dispersed rural communities'. These are not villages in the normal sense of nuclear settlements: they are areas of dispersed single houses or clusters of houses in the countryside, but with a focal point (church, school or shop) and a strong community identity. In such communities, the strategy envisages new housing development.

The cease-fire gave added impetus to the increasing activity on regional planning in Ulster. During 1995, new Northern Ireland planning policy statements were published on transport, retailing, town centres and industrial development, reflecting the general content of guidance available to local authorities in England. At the end of that year the Northern Ireland Minister released a consultation document on planning in the Belfast City region. Progress in Northern Ireland reflects a renewed interest in regional planning throughout the UK. Breheny (1991) explains the changes in attitude:

> Strategic planning is back on the professional, political, and, arguably, the popular agendas. Probably the single most important reason for this is the growing concern over the consequences of rapid, and apparently random, development, particularly in the south of England. This concern coincides with, and fuels, the dramatic increase in interest – at local to global scales – in environmental issues. The government has found that groups of its own supporters, who earlier voted for its radical deregulatory polices, are now in the vanguard of moves to strengthen the planning system generally, and strategic planning in particular.

Despite the pressure from many sources, regional planning remains relatively weak in the UK, especially compared with similar countries in the rest of Europe. This is well illustrated by Figures 4.2 and 4.3, which set out the framework of instruments of planning in the UK. Many commentators have called for more planning at the regional level. The RTPI has published a policy statement and guidance note on the regional planning process in England. As well as giving advice on the operation of the current system, it reiterates the RTPI's position on the need for a statutory tier of regional plans, as opposed to guidance, to be prepared by a secretariat and technical staff independent of the constituent planning authorities.

These views are unlikely to prevail in the immediate future, and though central government is paying more attention to regional planning questions in the light of local government reform, achievements to date have not amounted to much.

UK PARLIAMENT

PRIMARY LEGISLATION
Town & Country Planning Act 1990; Planning (Listed Buildings and Conservation Areas) Act 1990

Secretary of State for the Environment	County Councils	District Councils	Unitary Authorities	Metropolitan district councils; London boroughs; Isle of Wight and Herefordshire unitaries
SECONDARY LEGISLATION Statutory Instruments eg: T&CP (General Development Procedure) Order 1995; T&CP (General Permitted Development) Order 1995; T&CP (Use Classes) Order 1987	STRUCTURE PLAN Authority-wide; mandatory; broad framework; 15 year horizon, but longer for some policies, eg green belt; cover complete; prepared jointly with unitaries in some cases		STRUCTURE PLAN joint with adjacent county	UNITARY DEVELOPMENT PLAN Authority-wide; mandatory PART I : Framework of general policies
PLANNING POLICY GUIDANCE NOTES (Listed at the end of the book)	*The Department of the Environment consults all planning authorities and relevant organisations on draft guidance*			PART II : Detailed policies and proposals to guide development control; 10 year horizon but longer for some policies, e.g. green belt
MINERALS PLANNING GUIDANCE NOTES (Listed at the end of the book)	*Local authorities prepare 'advice' to the SoS on regional guidance, through a conference of constituent authorities*	LOCAL PLANS Authority-wide; mandatory; detailed policies and proposals to guide development control; 10 year horizon, but longer for conservation and 'phased development' policies; 43% cover by end 1996		
REGIONAL PLANNING GUIDANCE Provides a framework for structure plans and context for UDPs and local plans for 20 year period or longer	MINERALS PLAN Authority-wide; mandatory; safeguard sites and ensure environmental protection			
Circulars Elaboration of procedural matters	WASTE PLAN Authority-wide; mandatory; policies for treatment and disposal of waste and land use implications	*may be combined*		
Overall power to call-in		SIMPLIFIED PLANNING ZONE Small area; discretionary; gives planning permission for designated uses subject to conditions. Seldom used		
	SUPPLEMENTARY PLANNING GUIDANCE Discretionary; limited to supplements to statutory plan policy and to be clearly cross-referenced to it			

A. Two National Parks are also responsible for development plans (from April 1997, all will be).

B. The Broads Authority is also responsible for preparing a local plan for its area.

C. 'Strategic guidance' has been prepared by joint conferences of local authorities for the metropolitan counties and is being integrated with regional guidance.

Figure 4.2 The Planning Policy Framework in England

SCOTLAND

UK Parliament	Secretary of State for Scotland	Unitary Councils
PRIMARY LEGISLATION Town & Country Planning (Scotland) Act, 1972	SECONDARY LEGISLATION eg: Town & Country Planning (Use Classes) (Scotland) Order, 1989; Town & Country Planning (General Permitted Development) (Scotland) Order, 1992	STRUCTURE PLAN Provides a framework for local plan production and general guidance for development control 6 joint plans 11 single authority plans new structure plan arrangements introduced 1996 - old plans will remain in force until superseded
	CIRCULARS Elaboration of procedural matters	LOCAL PLAN Complete coverage mandatory but many small area plans; provide detailed guidance for development control
	NATIONAL PLANNING POLICY GUIDELINES Statements of government policy on nationally important land use issues	
	PLANNING ADVICE NOTES Identify and disseminate good practice	SIMPLIFIED PLANNING ZONE Grants planning permission in advance. Seldom used
		SUPPLEMENTARY PLANNING GUIDANCE

NORTHERN IRELAND

UK Parliament	Secretary of State for Northern Ireland	The Planning Service Executive Agency
LEGISLATION The Planning (Northern Ireland) Order, 1991 under the Northern Ireland Act 1974	SECONDARY LEGISLATION eg: Planning (Use Classes) Order (Northern Ireland), 1989; Planning (General Development) Order (Northern Ireland), 1993	AREA PLAN Covers the whole or substantial part of area of a district council; strategic policies providing a framework for local plans and for development control; 10-15 year horizon
	PLANNING POLICY STATEMENTS (proposed) (5) To provide guidance for plan preparation and development control	LOCAL PLAN Covers part of the area of one or more district councils; detailed policies to guide development control
	DEVELOPMENT CONTROL ADVICE NOTES (14) Detailed guidance for development control	SUBJECT PLAN Covers any area for a particular planning topic
	A PLANNING STRATEGY FOR RURAL NORTHERN IRELAND 1993 Strategic objectives and detailed development control policies	
	PROPOSED SUB-REGIONAL DEVELOPMENT STRATEGY FOR BELFAST	

WALES

UK Parliament	Secretary of State for Wales	Unitary Councils/National Parks
PRIMARY LEGISLATION The Town & Country Planning Act, 1990	SECONDARY LEGISLATION eg: Town & Country Planning (General Permitted Development) Order, 1996 Town & Country Planning (General Development Procedure) Order, 1996 Town & Country Planning (Use Classes) Order, 1992	UNITARY DEVELOPMENT PLANS Same instrument as applies in metropolitan districts in England with Part I: Strategy and Part II: Detailed policies
	PLANNING GUIDANCE WALES (2) • Unitary Development Plans • Planning Policy	This system was introduced in 1996 to replace the previous two-tier system of structure plans and local plans, which will remain in force until superseded
	TECHNICAL ADVICE NOTES	
	STRATEGIC PLANNING GUIDANCE IN WALES	

Figure 4.3 The Planning Policy Framework in Scotland, Northern Ireland and Wales

(C) DEVELOPMENT PLANS

FROM ZONING TO PLANNING

The main instrument of land use control in Britain during the first half of this century was the planning scheme. This was, in effect, development control by zoning. The zones shown in planning schemes indicated the development which would be permitted. The scheme document and maps showed where land was zoned for industry, for open space, for residential development at, for example, no more than eight houses to the acre, and so on. Under the 1932 Act, planning schemes had to be approved by the minister and by resolution of both Houses of Parliament. This was a very time-consuming process, but it gave schemes the force of law and, with it, the advantage of absolute certainty about what development would be permitted. But therein lay one of its gravest shortcomings: certainty for the developer meant inflexibility for the local authority.

One way of circumventing this was for planning authorities to take advantage of the lengthy and cumbersome procedure for preparing and obtaining approval to their schemes by remaining at the draft stage for as long as possible. Yet this had the opposite danger: flexibility obtained at the sacrifice of accountability. Whilst zoning flourished as the dominant form of planning in most of the rest of the world, Britain introduced in 1947 a system which is markedly different, and which attempts to strike a distinctive balance between flexibility and commitment.

1947-STYLE DEVELOPMENT PLANS

The approach adopted in Britain, which is in many important ways the same in the 1990s as it was in the 1950s, is fundamentally a discretionary one in which decisions on particular development proposals are made as they arise against the policy background of a generalised plan. The 1947 Act defined a development plan as 'a plan indicating the manner in which a local planning authority propose that land in their area should be used'.

Unlike the prewar operative scheme, the development plan did not of itself imply that permission would be granted for particular developments even if they appeared to be in harmony with the plan. Though a developer was able to find out from the plan where particular uses were likely to be permitted, his specific proposals had to be considered by the local planning authority. When considering applications, the authority was expressly directed to 'have regard to the provisions of the development plan', but the plan was not binding and, indeed, authorities were instructed to have regard not only to the development plan but also to 'any other material considerations'. Furthermore, in granting permission to develop, local authorities could impose 'such conditions as they think fit'.

However, though the local planning authorities had considerable latitude in deciding whether to approve applications, it was intended that the planning objectives for their areas should be clearly set out in development plans. The development plan consisted of a report of survey, providing background to the plan but having no statutory effect; a written statement, providing a short summary of the main proposals but no explanation or argument to support them; and detailed maps at various scales. The maps indicated development proposals for a twenty-year period and the intended pattern of land use, together with a programme of the stages by which the proposed development would be realised. The plans were approved by the minister (with or without modifications) following a local public inquiry. Initially, a three-year target was set for submission of the plans, but only twenty-two authorities met this, and it was not until the early 1960s that they were all approved.

By this time, the requirement to review plans on a five-yearly cycle had brought forward amendments, many taking the form of more detailed plans for particular areas. These had to follow the same process of inquiry and ministerial approval as the original plans, and many authorities were still engaged on the first review in the mid-1960s. Furthermore, although the system of development control guided by development plans operated fairly well for two decades without significant change, 1947-style plans did not prove flexible in the face of the very different conditions of the 1960s. The statutory requirement for determining and mapping land use led inexorably towards greater detail and precision. Increasingly, the system became out of tune with contemporary needs and was bogged down in details and cumbersome procedures. The quality of planning suffered, and delays were beginning to bring the system into disrepute. As a result, public acceptability, which is the basic foundation of any planning system, was jeopardised.

It was within this context that the Planning Advisory Group (PAG) was set up in May 1964 to review the broad structure of the planning system and, in particular, development plans. In its report, published in 1965, PAG concluded that plans had 'acquired the appearance of certainty and stability which is misleading since the primary use zonings may themselves permit a wide variety of use within a particular allocation, and it is impossible to forecast every land requirement over many years ahead'.

The report proposed a further fundamental change to the planning system, one which would distinguish between strategic issues and detailed tactical issues. Only plans dealing with the former would be submitted for ministerial approval: the latter would be for local decisions within the framework of the approved policy. Legislative effect to the PAG proposals was given by the Town and Country Planning Acts of 1968 (for England and Wales) and 1969 (for Scotland), creating a two-tier system of structure plans and local plans.

DEVELOPMENT PLANS IN ENGLAND SINCE 1968

The essential features of the 1968 system are still in place today, though there have been numerous incremental changes. Structure plans provide a strategic tier of development plan and, until 1985, were prepared for the whole of England by county councils (and the two national park boards). They were originally subject to the Secretary of State's approval, but since 1992 have been adopted by the planning authority itself. They consist of a written statement and key diagram setting out the broad land use policies (but not detailed land allocations) for the area, measures for the improvement of the physical environment and policies for the management of traffic. Accompanying these is an explanatory memorandum in which the authority summarises the reasons which justify the policies and general proposals in the plan.

The DoE's view of the functions of the structure plan has been fairly consistent over the years, but the scope of the structure plan has been narrowed considerably. The functions, as now set out in PPG 12, are to:

- provide the strategic policy framework for planning and development control locally;
- ensure that the provision for development is realistic and consistent with national and regional policy; and
- secure consistency between local plans for neighbouring areas.

The policies relating to the form, function, content and procedure of structure plans have evolved since they were first introduced: they now consistently emphasise their use as a *strategic* land use planning instrument, as indicated by the use of a key diagram rather than a map, avoiding the identification of particular parcels of land. This limits debate to the general questions of strategic location rather than the use of specific sites. General land use policies can thus be determined before detailed land use allocations are made, albeit not always to the liking of those affected by later more detailed plans.

Local plans provide detailed guidance on land use. They consist of a written statement, a proposals map and other appropriate illustrations. The written statement sets out the policies for the control of development, including the allocation of land for specific purposes. The proposals map must be on an Ordnance Survey base, thus showing the effects of the plan to precise and identifiable boundaries.

Under the 1968 system, there were three types of local plan: general plans (referred to as 'district plans' before 1982), action area plans and subject plans. *General local plans* were prepared 'where the strategic policies in the structure plan need to be developed in more detail'. Unlike the structure plan, which was prepared by all relevant authorities for the whole of their area, local authorities were advised that local plans would not be needed in all areas, for example where there was little pressure for development and no need to stimulate growth. This discretion was used: a small number of authorities prepared a single plan for the whole area, others prepared one or more plans for parts of their area, while others prepared none at all.

An *action area local plan* dealt with an area intended for comprehensive development, redevelopment or improvement by public authorities or private enterprise, and where implementation was to be given priority over a comparatively short period of time. *Subject plans*, as their name suggests, dealt with specific planning issues over an extensive area. Minerals and green belt local plans were numerous, but there were less common ones, such as Humberside County Council's *coastal camping and caravans local plan* and its *intensive livestock units local plan*.

Local plans were never subject to Secretary of State approval, but were simply adopted by the planning authority (although the Secretary of State has rarely used powers to call-in plans and to require modifications). The original rationale for this was that a local plan would be prepared within the framework of a structure plan; and since structure plans would be approved by the Secretary of State, local authorities could safely be left to the detailed elaboration of local plans. This went to the very kernel of the philosophy underlying the 1968 legislation, namely that central government should be concerned only with strategic issues, and that local matters should be the clear responsibility of local authorities.

This division of plan-making functions was predicated on the reorganisation of local government into unitary authorities, with a single authority being responsible for preparing both a broad strategic plan and any necessary detailed plans. The Secretary of State, having approved the structure plan, could safely leave the detailed elaboration of its policies at the local level to the (same) local authority *without the necessity of further approval*. But the 1972 Local Government Act established two main types of local authority in England and Wales, and divided planning functions between them. Thus, while counties were made responsible for broad planning strategy (structure planning), districts were independently responsible for local planning (and most matters of planning control). Clearly, once the responsibility for local plans was allocated to a different authority, the institutional (and political) link between the two levels of planning was in jeopardy. Counties and districts are independent political entities which may have very different ideas on the way in which the general policies in a structure plan are to be put into practice. Furthermore, there is an inevitable temptation for counties to formulate their 'policy and general proposals' in greater detail than would be the case if there were no division of functions. In this way, they may assume greater control over the implementation of policy.

The division of responsibilities has indeed given rise to considerable conflict which was compounded by the delays in the preparation of structure plans. Two devices were 'designed to promote effective co-operation in the planning field and to minimise delay, dispute and duplication': the development plan scheme and the certificate of conformity. The *development plan scheme* (later replaced by the *local plan scheme*) set out the agreed programme for the preparation and amendment of local plans. With the introduction of a mandatory requirement to produce district-wide local plans such schemes have been made redundant. However, the 1986 Act required local authorities to keep a register of development plan policies and the new regulations require a similar index of information in respect of the development plan.

The other device designed to ensure compatibility between the structure plan and local plans was the *certificate of conformity*. After initial consultation and before deposit of the proposed plan, it was a requirement for the county council to issue a certificate indicating that the plan conformed generally with the provisions of the approved structure plan. In some cases, disputes over conformity have given rise to delays in the adoption of local plans, demonstrating the difficulty of making a distinction between county strategy and district tactics.

EVALUATION OF THE 1968 DEVELOPMENT PLANS

There has probably never been a time when development plans, of whatever vintage, did not have their critics – and some of the criticisms have never changed. A decade after the start of the new system, Bruton (1980: 135) summarised the problems as 'delay and lack of flexibility; an over-concentration on detail; [and] ambiguity in regard to wider policy issues'.

Plans were very slow in coming forward to statutory approval and adoption. By 1977, only seventeen of the necessary eighty-nine structure plans for England and Wales had been submitted, and seven approved. By 1980, of seventy-nine English structure plans which were expected, sixty-four had been submitted and thirty-eight approved. The first structure plan cycle took fourteen years to complete: over the years from 1981 to 1985 the time taken from the submission of structure plans to their final approval averaged twenty-eight months. One of the main reasons for this long delay was that many of the written statements and explanatory memoranda were very lengthy: in the first round, several contained more than 100,000 words. They also contained too many policies – typically more than a hundred, many of which the DoE considered to be irrelevant to structure plans: 'building design standards, storage of cycles, the costs of waste collection,

the development of cooperatives, racial or sexual disadvantage, standards of highway maintenance, parking charges, the location of picnic sites and so-called nuclear-free zones'.

The disputes and delays over structure plan approval also held back the adoption of local plans where a very slow start was made in the first ten years of the new system; indeed, the first local plan was not adopted until 1975. However, the rate of deposit and adoption increased sharply after most of the initial round of structure plans was completed and, by March 1987, 495 local plans had been adopted in England and Wales (Coon 1988). Unfortunately, many of the plans were out of date by the time their processing was complete. This is not surprising when it is noted that the average time taken to prepare and adopt a local plan was about five years.

It is against this background that the popularity of non-statutory planning (or 'informal policy') has to be viewed. The terms relate to policy documents which are prepared and used by a local planning authority for development control but which have not been processed through the full statutory system. The studies of Bruton and Nicholson (1985) estimated that non-statutory policy documents outnumbered statutory plans by about ten to one. This non-statutory policy took many forms, from single issue or area policy notes to comprehensive but informal plans. Some informal policy may have been subject to public consultation, and may indeed have progressed through some stages of the statutory procedure; but some was prepared for internal use, and not intended to be made public. Central government has increasingly pressed for the elimination of non-statutory policy, except where it might be legitimately described as supplementary planning guidance (SPG), intended to assist applicants. The value of SPG is that it avoids excessive detail in the statutory plan and, for this reason, it is supported by the DoE – particularly 'when it has been the subject of public consultation and a council resolution' (PPG 12: 3.19). But a distinction is drawn between SPG and 'bottom drawer' plans and policies which have not been subject to proper procedures.

The attitude of central government to this issue has been vacillating and confused. The status of statutory plans reached a low point in 1985 when the White Paper, *Lifting the Burden*, denigrated both structure and local plans, and criticised the procedures for preparing plans as 'too slow and cumbersome'. More flexibility was also called for – somewhat at odds with previous advice, which had sought to reduce administrative discretion in the system by a planning framework which offered more certainty, clarity and consistency to private sector investors (Healey 1986).

The experience of structure planning was disappointing, and there was at this time growing confusion about its role. Certainly, it had not lived up to the expectations of the PAG report. Though it undoubtedly provided a forum for debate about strategy, it did not provide the firm lead that was promised. The uncertainties and complications of structure planning in practice carried over to local planning, and contributed, in some areas, to a professional culture that was at best indifferent to statutory plans (Shelton 1991). There were more positive attitudes in other areas. Where the stakes involved in development applications were high, as in London and counties such as Hertfordshire (where full statutory plan cover was completed during the 1980s), statutory plan-making was vigorously pursued. Also, despite turbulent economic conditions, the plans proved to be reasonably robust and effective in implementing policy and defending council decisions at appeal.

Research carried out during the 1980s produced useful findings about the operation of development plans. One important study examined a series of plans in the South-East, the West Midlands and Greater Manchester (Healey *et al.* 1988). This concluded that plans had proved to be effective in guiding and supporting decisions and in providing a framework for the protection of land. They were particularly useful in shaping private sector decisions, especially in the urban fringe. Conversely, the difficulty encountered in controlling public sector investment in housing, economic development, inner city policy and infrastructure provision was

shown to be an impediment to effective implementation of strategy. The research team argued that this criticism was not one that plans alone could address. In similar vein, Carter *et al.* (1991) highlighted the 'considerable confusion' about the relationship between development plans and expenditure-based plans, such as those for housing (HIPs) and transport (TPPs).

Another significant project on the role of the planning framework in the 1980s was undertaken by Davies and his colleagues (1986a, 1986b, 1986c) at the University of Reading. This focused on the relationship between development plans, development control and appeals. It is discussed in detail in the next chapter, but here it is interesting to note that the general conclusions proved to be prophetic of the way the system has been modified over recent years. They concluded that plans might play only a small part in guiding development control decisions overall, but were much more important when a case went to appeal – what they termed the 'pinch points' of the system. They suggested that this reflected the system's chief virtue: its ability to enable a sensitive response to local conditions. It was recommended that the DoE should encourage local authorities to provide better written policy cover; to reduce its complexity by incorporating as much as possible in statutory plans; and to facilitate more speedy adoption.

Further support has been lent to these arguments by subsequent research (Rydin *et al.* 1990; Collins and McConnell 1988). Together, these studies argued for recognition of the value of plans and for a stronger development plan framework, but also for flexibility to allow local variation in form and content.

PLANNING IN LONDON AND THE METROPOLITAN COUNTIES SINCE 1985

The Thatcher government's precipitate decision to abolish the GLC and the MCCs forced hasty action about the planning system in these areas. This was simple in the extreme: London boroughs and the metropolitan districts became 'unitary' planning authorities. Thus, in precisely those parts of the country where there is a particular need for a two-tier planning system, it was lost.

Initially, the government had proposed that the borough and district authorities should have responsibility for both structure and local plans, but later it was decided that this would be too cumbersome. Instead, a new *unitary development plan* was proposed, together with a joint planning committee for Greater London (a role which was undertaken by the London Planning Advisory Committee). Legislative effect was given to this in the Local Government Act 1985. The intention was that, after consultations, the Secretary of State would provide strategic guidance to assist in the preparation of the unitary development plans. Little advice was given to the districts about their input to the development of strategic guidance except that it was to be produced on a cooperative and voluntary basis by the districts themselves.

Unitary development plans (UDPs) are in two parts, as explained in PPG 12:

> Part I is analogous to the structure plan in non-metropolitan areas. It consists of a written statement of the authority's general policies for the development and use of land in their area. The broad development and land-use strategy of Part I provides a framework for the authority's detailed proposals in Part II, which is analogous to the local plan in non-metropolitan areas. Part II contains a written statement of the authority's proposals for the development and use of land; a map showing these proposals on an Ordnance Survey base; and a reasoned justification of the general policies in Part II of the plan. The proposals in Part II of a plan must be in general conformity with the policies in Part I.

Advice has made it clear that the plans will be prepared simultaneously, that they should be presented as one document, and that they should have a ten-year horizon. Furthermore, a UDP is adopted by the district council and is not subject to the approval of the Secretary of State (although reserve powers of central intervention have been maintained). Whilst a UDP is being prepared, any exist-ing plans which were in force on 1 April 1986 are to be treated as 'the development plan'.

There was a good deal of initial scepticism about these new arrangements, though they are more closely allied to the 1965 thinking of PAG than the system that was then put into place. There were particular concerns about the future of strategic thinking in the metropolitan areas, difficulties of cooperation between districts, and problems of participation and coping with the statutory right to objection in plans which embrace such large areas (Nadin and Wood 1988). For the districts themselves, many of these worries have proved unfounded. It has been possible to accommodate policy and political differences among districts, but this has been very much on a lowest common denominator level (Hill 1991; G. Williams *et al.* 1992). It has also proved possible, perhaps even desirable, to produce the strategy and detail concurrently. However, there remain serious concerns about the extent to which the public, interest groups and even some professionals can engage effectively in the process.

THE FUTURE OF DEVELOPMENT PLANS

As early as 1977, proposals were being made for a review of the 1968 system; but, at this time, review was considered premature since only twenty-four structure plans had been submitted, and only seven of these had been approved. By 1985, prompted by the concern for 'freeing' enterprise from unnecessary restraints, the White Paper *Lifting the Burden* included an announcement that the government was 'giving further consideration to whether there should be changes in the content and procedures of development plans and in the relationship between development plans and development control'. The following year a Consultation Paper was published on possible changes to simplify and improve the development plan system. The analysis this provided of the weaknesses of the current system was decidedly thin, though it did make reference to the research mentioned above. The paper proposed, *inter alia*, the abolition of structure plans in England and Wales (but not in Scotland, where they had 'not in general given rise to the same problems as have been experienced south of the Border'); a wider coverage of regional and sub-regional planning guidance to be issued by the Secretary of State after consultations and public comment; the introduction of 'statements of county planning policies' on a limited range of issues (to be specified by the Secretary of State) which would not form part of the statutory development plan; and new-style single-tier district development plans covering the whole of each district.

The context for preparation and discussion of these proposals centred on the growing dissatisfaction of many different interests about the operation of the planning system and, in particular, the making of many *ad hoc* and apparently inconsistent decisions by both the Secretary of State and local authorities. The lobby for change created some unusual bedfellows (both the development and the conservation lobbies). The common concern was for more certainty in the system and a reduction in the growing number of speculative applications; the former because of the increased level of speculation and risk, the latter because of the erosion of important environments. There was also some dissatisfaction amongst government supporters about decisions taken centrally which went against local (often Conservative) opinion. Local authorities were concerned that more of their decisions were being overruled, and complained at the lack of clarity in central policy. By comparison, matters looked better in Scotland and in the emerging, although as yet untried, system in the metropolitan counties.

Many of the 500 responses to the Consultation Paper argued very strongly against the proposed abolition of structure plans. In November 1988, PPG 12 was published; it urged local authorities to extend statutory plan coverage, normally by district-wide plans, and to replace non-statutory policy which it described as 'insufficient and weak'. Strategic green belt boundaries were singled out as requiring further specification in detailed local plans. In return, the government offered an enhanced status for plans.

Early in 1989, the White Paper *The Future of Development Plans* (Cm 569) contained a set of proposals very similar to those which had been circulated earlier. The only difference in the plan framework was the addition of a mandatory provision for all counties to prepare minerals development plans. However, PPG 15, published in 1990 (now incorporated in the revised PPG 12), urged county councils to press ahead with the revision and updating of structure plans and to cooperate on the elaboration of regional guidance. It seemed at the time that the considerable lobbying by major pressure groups had been successful in bringing about a reprieve for structure plans.

The main objective was to encourage some simplification of structure plans, but it also marked a change in the government's attitude to regional guidance, which was now fully accepted. The counties were encouraged to review their structure plans and to take them forward to 2006, with the promise that the delays after submission would be reduced. The counties, for their part, were to ensure that plans were less bulky and concentrated on strategic issues.

Concerted lobbying from virtually all sides of the planning debate, perhaps assisted by further changes in ministerial responsibilities at the DoE, were to have some further success later that year. Chris Patten's short tour of duty at the DoE will now be remembered for his announcement of the intention to retain a statutory strategic tier of development plan, and to end the requirement that the Secretary of State must himself approve all structure plans and alterations to them. Shortly afterwards, the Planning and Compensation Bill was published incorporating the necessary legislative changes. The provisions relating to the planning framework were widely welcomed. Indeed, in response to pressure from the Opposition and its own backbenchers, provisions were added to further increase the status of the statutory plans in development control.

DEVELOPMENT PLANS IN ENGLAND SINCE 1992

The Planning and Compensation Act 1991 came hard on the heels of the Town and Country Planning Act 1990 which consolidated the law in England and Wales. It made four major changes to the planning framework.

The first, brought about by an Opposition amendment and accepted by the government in the last stages of the Bill, was to make the plan the primary consideration in development control. In commending the amendment, Sir George Young coined a phrase in saying that 'the approach shall leave no doubt about the importance of the plan-led system'. This shift in the relationship between plans and control, enacted by the insertion of section 54A into the 1990 Act could have potentially far-reaching implications for the nature of the planning system, but it is too soon to be sure about this.

Two other changes involved making mandatory the adoption of *district-wide local plans*, and abolishing the requirement for central approval of structure plans. Central government has retained its powers of intervention, and the examination in public is still to be used to debate the plan after deposit. Finally, the Act introduced a mandatory requirement for counties to produce minerals plans and waste plans for the whole of their areas. There will be no further small area local plans, action area plans or subject plans other than for minerals and waste.

In the first part of the 1990s, it seemed that the framework of local planning policy in England and Wales was to become more coherent. The 1968 system had left it to local authorities to decide whether a local plan (or a number of overlapping plans) should be prepared. Given that most local authorities had little coverage of statutory plans (but many other informal policy documents) and others had produced a mix of interlinked subject and small area-based policy documents, those investigating policy did so with not a little uncertainty. Now the prospect was for a much clearer system. Those needing to know about planning policy would make reference to the structure plan, the

district-wide local plan and the minerals and waste plans, with some certainty that they existed.

The impact of local government reorganisation is to introduce much greater variety in the arrangements for development planning, especially in England as illustrated in Figure 4. 2. Where the two-tier system remains, the planning framework is not affected: counties prepare the structure plan and waste and minerals plans (or one plan for both topics), and districts prepare the district-wide local plan. Where new unitary authorities are created, they are effectively county councils, but will in almost all cases prepare joint structure plans with the neighbouring county councils. The joint arrangements are summarised in Table 4. 1. The exceptions are Herefordshire and the Isle of Wight, which are to prepare unitary development plans (as in the metropolitan districts). All the other unitary authorities will prepare their own district-wide local plans. The county will continue to prepare waste and minerals plans (or one plan for both topics) for its area. The metropolitan districts are, of course, unaffected by local government reorganisation and will continue with their unitary development plans.

DEVELOPMENT PLANS IN NORTHERN IRELAND

The formal change to a discretionary system of development plans and control did not come to Northern Ireland until 1972. Prior to this, the system was much the same as for the rest of Britain before 1947, with local authorities able to prepare planning schemes. Practice was similar also in that very little progress was made on the preparation and approval of such schemes, and a system of interim development control operated. The 1972 Order introduced the development plan, with similar status to those in the rest of the UK. There are three types of development plan (area, local and subject plans) which are produced and adopted by the DoENI. Area plans which can cover the whole or a substantial part of one or more district council

Table 4.1 Structure Plan Areas in England

Previous structure plan authority	New joint arrangements
Avon County Council	Bristol North Somerset (formerly Woodspring) Bath and NE Somerset (formerly Wansdyke and Bath) South Gloucestershire (formerly Northavon and Kingswood)
Bedfordshire County Council	Luton Bedfordshire County Council
Berkshire County Council	Bracknell Forest Newbury Reading Slough Windsor and Maidenhead Wokingham
Buckinghamshire County Council	Milton Keynes Buckinghamshire County Council
Cambridgeshire County Council	Peterborough Cambridgeshire County Council
Cheshire County Council	Halton Warrington Cheshire County Council
Cleveland County Council	Middlesbrough Hartlepool Redcar and Cleveland (formerly Langbaurgh-on-Tees) Stockton on Tees
Derbyshire County Council	Derby City Derbyshire County Council
Devon County Council	Plymouth Torbay Devon County Council
Dorset County Council	Bournemouth Poole Dorset County Council
Durham County Council	Darlington Durham County Council
East Sussex County Council	Brighton and Hove East Sussex County Council

Previous structure plan authority	New joint arrangements
Essex County Council	Southend Thurrock Essex County Council
Hampshire County Council	Portsmouth Southampton Hampshire County Council
Hereford and Worcester	Worcestershire County Council Herefordshire will prepare a UDP
Humberside County Council	Kingston upon Hull North-East Lincolnshire (formerly Cleethorpes and Great Grimsby) North Lincolnshire (formerly Glandford, Scunthorpe and part of Boothferry) East Riding (formerly East Yorks., Beverley, Holderness and part of Boothferry)
Isle of Wight	Isle of Wight will prepare a UDP
Kent County Council	Medway Towns (formerly Rochester and Gillingham) Kent County Council
Lancashire County Council	Blackburn Blackpool Lancashire County Council
Leicestershire	Leicester City Rutland Leicestershire County Council
North Yorkshire County Council	York North Yorkshire County Council
Nottinghamshire County Council	Nottingham City Nottinghamshire County Council
Shropshire County Council	The Wrekin Shropshire County Council
Staffordshire County Council	Stoke on Trent City Staffordshire County Council
Wiltshire County Council	Thamesdown Wiltshire County Council

Notes
1 The remaining counties are the sole structure plan authorities. The Peak District National Park and Lake District National Park are also structure plan authorities. From April 1997 all national parks become the sole planning authority for their area.
2 Herefordshire and the Isle of Wight Unitary Districts are the only two districts in England which will prepare a UDP and therefore will not be engaged in structure planning.

areas are the main reference for development control, and include both strategic and detailed policies.

The provisions of the 1991 Act, including those on the primacy of the development plan, do not apply in Northern Ireland. However, the general tenor of government policy on the plan-led system has filtered through and there has been a tendency for plans to become more detailed.

Northern Ireland is not affected by changes in local government. Area and local plans will continue to be prepared by the six divisional offices of the DoENI. However, the planning framework may be affected by the recommendations of the House of Commons Northern Ireland Affairs Committee in its 1996 report on *The Planning System in Northern Ireland*. The Committee expressed serious concerns about the lack of a clear strategy for the Province as a whole, and the inadequacy of the development plans system. At the time of writing, there was no indication what action (if any) would be taken on its recommendation for 'a more coherently thought out strategy and a defined hierarchy of complementary policies and plans' in place of 'a collection of piecemeal policies'.

DEVELOPMENT PLANS IN SCOTLAND

The Scottish system differs in several significant ways from that in England and Wales, but the two-tier system of development plans and the procedures for the adoption and approval were broadly similar until 1996. Some differences can be attributed to the particular geographical characteristics of Scotland; others may legitimately be attributed to a canny move to avoid some of the difficulties of the English system. Because of the different administrative structure and larger planning areas in Scotland, there is a slightly different emphasis in the functions of structure plans (PAN 37: 7) which are to indicate policies and proposals concerning the scale and general location of new development, and to provide a regional policy framework for accommodating development. Progress on the approval of structure plans has been a significant problem, with an average of seventeen months needed for approval from the Secretary of State.

Changes to the development plan system itself have followed closely those introduced south of the border. For example, the procedure for making alterations to structure plans and local plans has been made simpler. Also, certain adjustments have been made to the division of planning responsibilities between regions and districts.

The 1991 Act has brought to Scotland some of the same changes made in England and Wales, notably the enhanced status of development plans in development control and insertion of section 18A into the 0972 Act, with the same effect as section 54A in England and Wales; calls for more succinct statements of policy; and the emphasis on 'physical land use development' (PAN 37). However, in a number of ways the Scottish development plan system remains distinctive. The structure plan still has to be approved by the Secretary of State. The survey still plays a part in the approval, and must be put on deposit and accompany the deposited plan in the submission.

Evaluation of the system also reveals similarity with the situation in England and Wales. Local gnvernment reorganisation delayed the production of plans, and it was not until 1989 that full structure plan cover was achieved. Progress on local plans has been better overall, mainly because of the mandatory requirement.

Local government reorganisation created unitary authorities in Scotland in 1996 and although the two-tier system of structure and local plans was retained, joint working is now necessary for the production of some structure plans; the arrangements are summarised in Table 4.2. The Scottish Office has designated seventeen structure plan areas, six of which cover more than one unitary authority. Local planning continues unchanged in the new unitary districts.

DEVELOPMENT PLANS IN WALES

In Wales, the system of development plans was virtually the same as that for England until 1996. One important variation was that the responsibility for waste rested with the districts (rather than counties) and thus waste policies were included in local plans rather than in separate county-wide subject plans.

Local government reorganisation created unitary councils in 1996, and the plan framework was amended to require each authority (including the national parks) to prepare a unitary development plan. The Welsh UDP will have a similar form to the UDPs in the English metropolitan districts with a Part I and Part II. The councils may prepare UDPs jointly, and the Part II element may be prepared by area committees. Arrangements for the transition to the new framework have been put in place. Welsh local authorities were able to seek approval from the Secretary of State to continue through to the adoption of plans already in preparation. With changes to local authority boundaries, local plans (including

Table 4.2 Structure Plan Areas in Scotland
In Scotland new structure plan areas have been created, taking into account the complete revision of local authority boundaries.

Single authority structure plan areas	Joint structure plan areas
Argyle and Bute	Aberdeen City
Borders	Aberdeenshire
Dumfries and Galloway	
Falkirk	East Ayrshire
Fife	North Ayrshire
Highland	South Ayrshire
Moray	
Orkney Islands	Angus
Perthshire and Kinross	Dundee
Shetland Islands	
Western Isles	Stirling
	Clackmannanshire
	East Lothian
	Edinburgh
	Midlothian
	West Lothian
	Dumbarton and Clydebank
	East Dunbartonshire
	East Renfrewshire
	Glasgow City
	Inverclyde
	North Lanarkshire
	Renfrewshire

some yet to be adopted) may cover only part of an authority's area or be split between two.

THE IMPACT OF THE PLAN-LED SYSTEM

There were high hopes in official circles for the new planning regime. It was anticipated that the planning system would become simpler and more responsive, reducing costs for both the private sector and local authorities and making it easier for people to be involved in the planning process. Central government set a target of the end of 1996 for the adoption of the new round of plans, and considerable efforts were made by local authorities to produce the plans expeditiously. Nevertheless, there has been significant slippage. One reason for this is the increasing complexity of plans. In 1988, an Inspector would have spent an average of seven weeks holding and reporting on a local plan inquiry (Planning Inspectorate 1992). By 1995 that average was forecast to increase to fifty weeks (Planning Inspectorate, 1995). This is an indication of both the larger area covered by plans and the increasing participation of interest groups. Objections to plans are now counted in their hundreds (and sometimes even thousands), whilst few plans would have been subject to this level of objection ten years earlier. Interest groups and individuals affected by plans now recognise the significance of plans for later decision-making. The DoE too has become increasingly active in scrutinising plans prior to the inquiry and has recruited additional staff to check them for consistency with central government policy and regional guidance. The result has been more departmental objections to plans and frequent and sometimes lengthy requests for changes. This represents a sharp increase in central government involvement in local planning.

THE CONTENT OF PLANS

The 1947 legislation was largely concerned with land use: 'a development plan means a plan indicat-ing the manner in which a local planning authority propose that land in their area should be used'. The 1968 Act signalled a major shift in focus: emphasis was laid on major economic and social forces and on broad policies or strategies for large areas. In formulating structure plans, local planning authorities were to pay particular attention 'to current policies with respect to the economic planning and development of the region as a whole' and to the likely availability of resources. It was held that land use planning could not be undertaken satisfactorily in isolation from the social and economic objectives which it served. Thus the plans were to encompass such matters as the distribution of population and employment, housing, education and leisure.

This broader concept of planning did not survive, however, and by 1980 central government had moved back to a predominantly land use approach. This radical departure from the ideas of 1968 and the contraction of the scope of structure plans has been widely documented (Cross and Bristow 1983; Healey 1986). Central government also intervened to restrict plan content significantly. Thornley (1991: 124) provides a useful summary of what he describes as the 'attack on structure plans'. In the 1980 Manchester Structure Plan, for example, the Secretary of State 'deleted more than 40 per cent of the policies, and a further 20 per cent were substantially modified'. Thornley argues that such actions reflected the government's intention to allow market forces to operate at the cost of social and other wider objectives.

Central control over plan content has continued into the 1990s but, as structure plans no longer require central approval, LPAs now have more freedom, though they have been warned not to adopt policies which conflict with national guidance. Departmental advice about plan content has thus become increasingly specific and restrictive.

Local plans are not subject to the formal constraints of central government but in practice they reflect them. From her extensive study of development plans, Healey (1983: 189) points out that, whilst local plans have embraced wide-ranging social and economic objectives, their proposals nevertheless are 'primarily about land allocation'. Moreover, whilst local plans vary substantially in form, and 'appear local in orientation and specific to particular areas and issues', there is considerable consistency in scope and content. Consistency arises from the need for central government support for policy and because of the limited planning powers provided by legislation. Perhaps a deeper cause lies in the professional training and culture of planners, which is rooted in land use and physical concerns.

A consequence of the increasing attention that many organisations now give to plans in the light of the 'plan-led system' has been the production of 'model policies'. The idea of model policies for local plans was proposed in the early 1970s (Fudge *et al.* 1982) but received little support. The idea is that policy wording can be taken 'off the shelf' rather than written anew for each local plan. In practice, planning officers do draw on examples from other authorities and colleagues, but model policies have been introduced on a more systematic basis over recent years by a number of organisations. The amount of advice given to planning authorities is somewhat overwhelming, and the detail suggested often goes beyond what is appropriate in local plans. General guidance on policy and plan preparation has also been given by professional bodies such as the District Planning Officers' Society and the RTPI.

THE STATUTORY PROCEDURES

A particularly helpful feature of the 1991 Act is that it brought the procedures for the various types of plan (in England and Wales) much closer into line with each other. The general procedure is illustrated in Figure 4.4. Essentially, the procedures comprise 'safeguards' to ensure the accountability of government in the exercise of plan-making. This is particularly important in the UK, where there is no constitutional safeguard of private property or other rights (other than that provided by the European Convention on Human Rights) and where there is wide administrative discretion in decision-making. The procedures also provide for increased involve-

ment of other organisations and the public in policy formulation. The process of open discussion and formal adoption lends authority and standing to plans, and provides an element of legitimacy even though the plans are not subject to direct ministerial approval (PAG 1965: 6.19).

In the following discussion, the focus is on the key safeguards, the main criticisms of the procedure and recent amendments. The knotty questions about the extent to which the public and other objectors are effectively able to make use of the safeguards and how this influences plan content are dealt with in the final chapter. The main safeguards in plan adoption are: the opportunity to be consulted in the formative stages of plan preparation; the need for authorities to consider conformity between plans and regional and national guidance; the right to make objections to both strategic and detailed plans, and to have objections to the latter heard before an independent inspector; a further right to object to any proposed modifications or where the authority proposes to reject the recommendation of an inspector or a panel; the overarching right of the Secretary of State to intervene and to direct modifications; and the right to challenge the plan in the courts.

The central focus of the formal adoption procedure is the hearing. In the case of a local plan or UDP, this is a public local inquiry; in the case of a structure plan, it is an examination in public. At the inquiry, an independent inspector hears 'objections', whereas the examination in public (EIP) is a 'probing discussion' of selected matters which the authority needs to consider before taking the structure plan forward.

Anyone can object to a development plan, and the planning authority has a duty to consider all objections. For local plans and UDPs, objectors also have a right to present their case to the inquiry. The EIP deals with only those matters which the authority considers need examination in public, and the planning authority determines who shall participate in the examination (whether or not they have made objections or representations).

Not surprisingly, the extensive procedures for

plan adoption have been criticised as cumbersome and time-consuming. The plan-making process takes about five and half years on average (Steel *et al*. 1995), and the time taken in the formal adoption procedure has increased as the district-wide format has become the norm and as the number of objections has increased. It is important to note, however, that the greatest proportion of time taken in the process is still in the preparation of a draft plan prior to deposit. Various changes have been made to the procedure since its introduction in 1968 in an attempt to simplify and to speed it up. Cumulatively, these changes are substantial. The first change was a reduction in the requirements for public participation at the outset of the plan process. The Local Government and Planning Act 1980 introduced an expedited procedure which, in certain circumstances, allowed the local plan to be adopted in advance of a structure plan review. The Housing and Planning Act 1986 gave powers to the Secretary of State to request modifications to plans (in addition to the seldom-used powers to call-in plans). Most recently, the 1991 Act abandoned the need for the six-week consultation period prior to the deposit for objections (although the Secretary of State has urged that there should be wide publicity). At the same time, an extra opportunity has been provided for objections after the inquiry where the planning authority does not accept the inspector's recommendations.

Many other recommendations for reform have been proposed, including abolition of the inquiry altogether, or making the inspector's recommendations binding on the local authority. Research on the efficiency and effectiveness of local plan inquiries (Steel *et al*. 1995) suggests that this would be difficult to implement and probably counterproductive. The report pointed up the weakness in the capacity (and sometimes willingness) of many planning authorities to manage the procedure according to the guidance laid down. It also noted other factors which can be critical to the effectiveness of the procedure, including the form and content of the plan, especially the level of detail and the types of policies proposed.

In 1994, the DoE issued a Consultation Paper setting out proposals for further change. This was followed in 1996 by proposed revisions to the Code of Conduct for inquiries and EIPs, and regulations governing the preparation of plans. The proposals include removal of excessive detail from plans, more effective consultation early in the process, more emphasis on dealing with objections in writing, and shorter reports from inspectors.

At the same time, the RTPI began research on inquiry procedures, the TCPA mounted an inquiry into development planning, and the local government associations published their own consultation paper with recommendations for improving the effectiveness of the planning system. The Inspectorate had already made various changes in its practice and produced guidance for local authorities in preparing for the inquiry (which is discussed further in Chapter 12). The outcome of this flurry of activity is unknown at the time of writing. However, the various papers all place emphasis on procedural changes, and relatively little attention is given to the form and content of plans: this is surprising given the massive changes in the context for development planning since this was last reviewed.

Nevertheless, the basic procedure has proved resilient; perhaps because it nicely balances the concerns of local authorities (who typically call for more freedom of action in order to speed the process), and the concerns of objectors of all kinds (who naturally desire more influence in the local planning process). But the criticisms have been consistent and certainly a factor in the lack of enthusiasm of planning authorities for statutory plan-making. Piecemeal decision-making and non-statutory policy were dominant until the change in government policy over plan-making in the late 1980s.

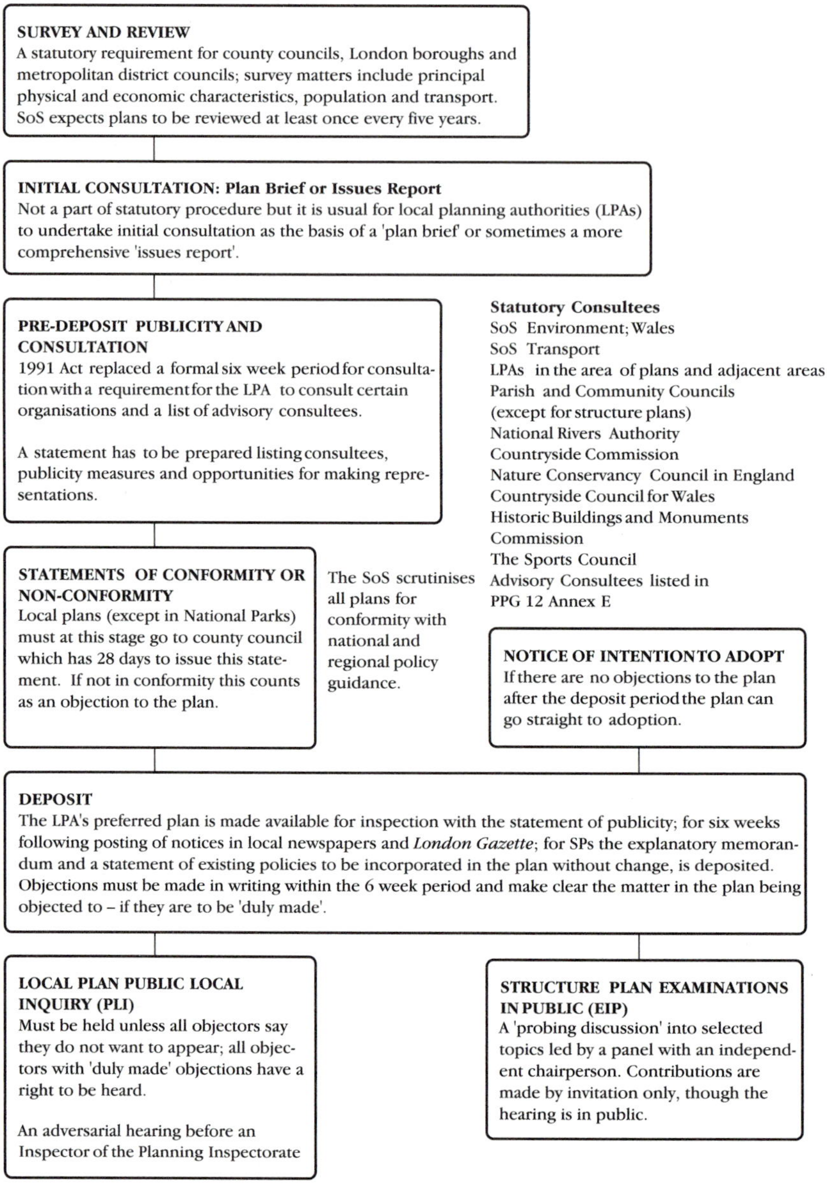

SURVEY AND REVIEW
A statutory requirement for county councils, London boroughs and metropolitan district councils; survey matters include principal physical and economic characteristics, population and transport. SoS expects plans to be reviewed at least once every five years.

INITIAL CONSULTATION: Plan Brief or Issues Report
Not a part of statutory procedure but it is usual for local planning authorities (LPAs) to undertake initial consultation as the basis of a 'plan brief' or sometimes a more comprehensive 'issues report'.

PRE-DEPOSIT PUBLICITY AND CONSULTATION
1991 Act replaced a formal six week period for consultation with a requirement for the LPA to consult certain organisations and a list of advisory consultees.

A statement has to be prepared listing consultees, publicity measures and opportunities for making representations.

Statutory Consultees
SoS Environment; Wales
SoS Transport
LPAs in the area of plans and adjacent areas
Parish and Community Councils
(except for structure plans)
National Rivers Authority
Countryside Commission
Nature Conservancy Council in England
Countryside Council for Wales
Historic Buildings and Monuments Commission
The Sports Council
Advisory Consultees listed in PPG 12 Annex E

STATEMENTS OF CONFORMITY OR NON-CONFORMITY
Local plans (except in National Parks) must at this stage go to county council which has 28 days to issue this statement. If not in conformity this counts as an objection to the plan.

The SoS scrutinises all plans for conformity with national and regional policy guidance.

NOTICE OF INTENTION TO ADOPT
If there are no objections to the plan after the deposit period the plan can go straight to adoption.

DEPOSIT
The LPA's preferred plan is made available for inspection with the statement of publicity; for six weeks following posting of notices in local newspapers and *London Gazette*; for SPs the explanatory memorandum and a statement of existing policies to be incorporated in the plan without change, is deposited. Objections must be made in writing within the 6 week period and make clear the matter in the plan being objected to – if they are to be 'duly made'.

LOCAL PLAN PUBLIC LOCAL INQUIRY (PLI)
Must be held unless all objectors say they do not want to appear; all objectors with 'duly made' objections have a right to be heard.

An adversarial hearing before an Inspector of the Planning Inspectorate

STRUCTURE PLAN EXAMINATIONS IN PUBLIC (EIP)
A 'probing discussion' into selected topics led by a panel with an independent chairperson. Contributions are made by invitation only, though the hearing is in public.

Figure 4.4 The Procedure for the Adoption of Development Plans in England

INSPECTOR'S REPORT
Makes recommendations to local planning authority on how plan could be modified to meet objections, including written objections not heard at the inquiry

PANEL'S REPORT
Makes recommendations to LPA on how plan could be modified in respect of matters selected for discussion at EIP only

STATEMENT OF DECISIONS AND REASONS
LPA are not obliged to accept all recommendations (although 9 out of 10 usually are) but they must give reasons for their decisions in each case paying special attention to recommendations rejected

If modifications recommended and accepted

If some or all recommended modifications not accepted

If no modifications recommended

LIST OF MODIFICATIONS AND REASONS

LIST OF MODIFICATIONS LIST OF RECOMMENDATIONS NOT ACCEPTED

LPA can also make 'additional modifications' that do no materially affect plan content, for example to correct and update the content

Any modification which makes a material change to the plan must be listed

Anyone may object to the absence of modifications recommended in the reports

NOTICE OF INTENTION TO ADOPT
If there are no objections the plan may be adopted after the six week period of deposit

DEPOSIT
The Inspector's/Panel's report and the statement of decisions must be placed on deposit for six weeks with any list of modifications and/or recommendations not accepted; notices are served on objectors

Statement of decisions and reasons and Inspector's/Panel's report is made available for inspection

If objectors raise new issues

If objectors do not raise new issues

If no objections

SECOND INQUIRY OR RE-OPENED EIP
This will only take place where entirely new issues, e.g. a new proposal, or if LPA propose to withdraw a modification

STATEMENT OF DECISIONS

ADOPTION OF PLAN
The plan is adopted by resolution of the Council, notices are published in the *London Gazette* and local newspapers and sent to those who asked to be notified

CHALLENGE IN THE COURTS
There is a right to challenge the plan but only on the grounds that the proposals are not within the powers of the 1990 Act or that regulations have not been complied with

Figure 4.4 continued

(D) ZONING INSTRUMENTS

There are two examples of attempts to reintroduce the zoning approach in the UK planning system, in an effort to simplify planning: enterprise zones and simplified planning zones. Both reflect economic rather than land use planning objectives.

ENTERPRISE ZONES

A major plank in the Conservative government's response to economic recession in the early 1980s was the proposed reduction in the 'burden' of regulation on business and enterprise. In enterprise zones (EZs), amendments to the planning regime were part of a much wider range of advantages offered, although the statement announcing their introduction was made with a strong side-swipe at mainstream traditional town and country planning, claiming that private initiative had been stifled by rules and regulations.

The same Conservative spokesperson, Sir Geoffrey Howe, credited the notion to Peter Hall, who in turn, identified the origins of the concept in a 1969 article (Banham *et al.* 1969). The germ of the idea thus arises from the notion of virtually complete freedom from state control and intervention. Peter Hall (1991) has reviewed the ways in which this notion was transposed and 'sanitised' into the enterprise zone initiative in Britain. He explains how the theoretical justification for such a proposal has itself been questioned and how, in its implementation, 'there has been a huge gap between the grand sweep of the original concept and the reality of what was actually achieved'.

The 'freedoms' of enterprise zones, as introduced by the Local Government, Planning and Land Act 1980, were exemption from rates on industrial and commercial property, 100 per cent tax allowances for capital expenditure, reduced demands for information from government, and simpler planning procedures or 'zoning'. An enterprise zone scheme has the effect of granting planning permissions in advance for such developments as the scheme specifies.

EZ schemes are prepared by authorities invited to do so by the Secretary of State (district councils, London boroughs and development corporations). Subject to any directions contained in the invitation, the authority is free to determine what planning concessions are to be offered. A draft scheme has to be given adequate publicity; 'persons who may be expected to want to make representations' are to be made aware of their entitlement to do so; and any representations have to be considered by the authority (but no public inquiry is required). The authority then proceeds to adopt the scheme, which may be approved by the Secretary of State after the expiry of a six-week period, during which its validity may be challenged. There is no requirement for conformity with existing plans. On the contrary, if any existing plans conflict with the scheme, they have to be amended as necessary. Nor is there a requirement for an inquiry into objections to the scheme.

By the end of 1996, thirty-two enterprise zones had been designated, most of which had completed their ten-year lifespan. The full list is given in Table 4.3. This is a much larger number than originally envisaged, but since 1984 designation has been much more selective, and exceptional reasons cited in each case, notably the very sharp damaging effects of shipyard and coal mine closures.

The enterprise zone initiative has been closely monitored, and findings show in some cases a dramatic increase in development activity. The Corby EZ, for example, was virtually fully committed after seven years with 5,600 jobs and 3.15 million square feet of new floorspace. Overall, however, it seems that most of the new development and employment in the zones would have occurred in any case, though not necessarily within the designated areas (PA Cambridge Economic Consultants 1987). Other studies have shown that the liberalisation of land use planning controls made only a minor contribution to any success. The extent of the zone regimes amounts only to a total land area of less than 40 square kilometres, and the simplified planning controls have not proved to be a major incentive. Case studies showed that a very considerable amount of negotiation (whether termed 'planning' or not) still

Table 4.3 Enterprise Zones in the UK

	Area (acres)	Year
England		
Allerdale, Workington, Cumbria	87	1983
Corby, Northants	113	1981
Dudley, West Midlands	263	1981 and 1984
Glandford, Humberside	50	1984
Hartlepool, Cleveland	109	1981
Isle of Dogs, London	147	1982
Middlesbrough	79	1983
NE Lancashire	114	1983
NW Kent	125	1983 and 1986
Rotherham	105	1983
Salford/Trafford	352	1981
Scunthorpe, Humberside	105	1983
Speke, Liverpool	138	1981
Telford, Shropshire	113	1984
Tyneside	454	1981
Wakefield	90	1981 and 1983
Wellingborough, Northants	54	1983
Sunderland	150	1990
East Midlands	n.a.	1995
Dearne Valley	n.a.	1995
North East Derbyshire	n.a.	1995
Tyne Riverside	240	1996
Wales		
Delyn, Clwyd	118	1983
Milford Haven, Dyfed	146	1984
Swansea, West Glamorgan	314	1981 and 1985
Scotland		
Clydebank	230	1981
Invergordon, Highland	60	1983
Tayside	120	1984
Inverclyde	274	1989
Lanarkshire	507	1993
Northern Ireland		
Belfast	207	1981
Londonderry	109	1983

had to take place, both between the developers and local authorities and also between developers and other agencies. Moreover, it has become apparent that the zones are not as distinctive as had been envisaged: planning controls are often retained along zone boundaries; special industrial uses, including noxious and dangerous processes, are still subject to control; and, on occasion, environmental improvements are often written into declaration reports.

Nevertheless, planning is *seen* as an important obstacle by developers, partly because the outcome is never quite certain. What the scheme offers is the advantage of certainty. The conditions are set out and known from the beginning. Much more important than any change in 'planning' has been the substantial financial benefits: tax allowances and the direct public investment in infrastructure and land reclamation.

SIMPLIFIED PLANNING ZONES

Whatever the research on enterprise zones may have concluded, the government was so enamoured of the idea that it introduced a new type of *simplified planning zone* (SPZ) based upon it. The general notion of zoning as an alternative to the development plan had been rejected, but the DoE did see a limited role for zoning in particular locations where greater certainty, and some flexibility in the detail of development proposals, would contribute to economic development objectives. An SPZ is the local equivalent to a development order made by the Secretary of State. It replaces the normal discretionary planning system with advance acceptance of specified types of development.

Two broad types of scheme are possible: the *specific scheme* which lists certain uses to be permitted, and the *general scheme* which gives a wide permission but excludes certain uses. Conditions can be made in advance, and certain matters can be reserved for detailed consideration through the normal planning process. SPZs cannot be adopted in national parks, the Broads, AONBs, SSSIs, approved green belts, conservation areas, or other protected areas.

During the process of consultation there was a marked change in the focus of the SPZ proposal: in place of the concern for reducing the negative impacts of planning control, there was a more positive concern for the potential advantages of SPZ designation for urban regeneration. Further emphasis was given in a Scottish circular to the potential employment-generating benefits of SPZs, their relationship to other grant-funded regeneration initiatives, and the need for schemes to be closely linked to the development plan. SPZs are now considered to be particularly appropriate for older industrial sites (especially those in single ownership) where there is a need to promote regeneration and use the SPZ designation to promote sites (Lloyd 1992).

The introduction of the SPZ provisions has excited very limited interest, and progress has been slow. Adopted schemes are listed in Table 4.4. According to a 1991 report by Arup Economic

Table 4.4 Adopted Simplified Planning Zones in the UK

Authority	Zone	Date adopted	Size (hectares)	Previous use
England				
Derby	Ley	June 1988	8.4	Foundries
Wellingborough	Park Farm South	November 1993	35.0	Agriculture
Gedling	Victoria Park	November 1991	32.0	Railway sidings
Corby	Willowbrook	July 1988	178.0	Steelworks
Knowsley	Birds Eye Site	April 1995	10.5	Food factory
Tyne & Wear	Willington Quay/ Wallsend	March 1994	46.5	Oil industry
Slough	Slough Trading Estate	April 1993	164.0	Business and industry
Scotland				
Highland	Dingwall	1989	4.75	Agriculture
Monklands	Coatbridge	September 1991	180	Steelworks
Falkirk	Grangemouth Docks	March 1992	50.0[a]	Docks
Highland	Alness and Invergordon	December 1994	137.0	Docks, industrial and green field
Wales				
Gwynedd	Penrhyndeudraeth	December 1994	11.0	Agriculture
Flintshire		November 1995	191.0	Industrial plant

Note: [a] 39 hectares land, 11 hectares water

Consultants, such interest as there was tended to come from authorities with experience of (or failure to obtain) enterprise zones: these authorities had fewer fears about the loss of normal development control powers over the quality of development. In all important respects the procedures for adoption of SPZs were identical to those of local plan preparation and adoption. The prospect of taking a scheme through these lengthy procedures was daunting, and it rapidly became clear that they were (in the words of the research report) 'undoubtedly cumbersome'.

Following consultation, the 1991 Planning and Compensation Act introduced changes to the procedure in line with those for development plans, including making optional public consultation before deposit, but also allowing the authority to consider objections in writing with or without the assistance of an inspector and to dispense with the public local inquiry. (Details of the revised procedures are set out in a 1992 revision of PPG 5.) The procedural changes have had little effect (Blackhall 1993, 1994). Some of the reasons are perhaps obvious. There is little difference between the allocation of land in a development plan and an SPZ: both indicate the type of development that is acceptable. Moreover, the extra 'certainty' provided by an SPZ designation is to some extent illusory since formal relationships are replaced by informal discussions. Additionally, decisions on the fulfilment of conditions and negotiations on reserved matters may still be needed. Furthermore, where a local authority is promoting urban regeneration it is likely that there will be a sympathetic approach to development proposals and a fast-track procedure for dealing with planning applications. Indeed, in situations where a developer has been identified, even the normal timescale for grant of planning permission is likely to take less time than setting up an SPZ.

Overall, as Allmendinger (1996a and 1996b) argues, the SPZ concept largely failed because it lacked clear and consistent objectives. It sought to offer deregulation and more certainty for developers but, in fact, led to greater uncertainty when put into practice. Above all, the recession at the end of the 1980s undermined property-led development on which the idea rested.

It might be argued, therefore, that SPZs offer little in the way of simplified planning, and the limited response from the private sector supports this view. There is only a very small number of zones, and these operate in a narrow range of circumstances. The conclusion must be that the reintroduction of zoning into the British planning system through SPZs has been unsuccessful.

British planners view zoning as a rather strange and crude tool: they are accustomed to dealing with proposals 'on their merits'. The provision of 'planning permission in advance' has little role to play within the current planning system. The original intention to move the balance between flexibility and commitment in the direction of the latter is now being served by the enhanced status of the development plan. It would seem, therefore, at least for the time being, that zoning is not likely to find fertile ground in the UK.

FURTHER READING

General

A good starting point for investigation of the policy framework in England is DoE PPG 1: *General Policy and Principles* (1992) and PPG 12: *Development Plans and Regional Planning Guidance* (and their equivalents in the other countries of the UK). There are no textbooks which cover all the material in this chapter, but selected sources on particular aspects are noted below.

Supranational Planning

Williams (1996) *European Union Spatial Policy and Planning* provides an excellent introduction to European planning instruments. A good summary of the emergence of planning at the European level is by Fit and Kragt (1994) 'The long road to European spatial planning'. See also Davies (1994) 'Towards a European planning system?', and Healey and Williams (1993) 'European urban planning systems'.

Directorate-General XVI: Regional Policy and Cohesion has produced a series of reports on planning in Europe which are listed at the end of this book. Of these, *Europe 2000+* and its predecessor, *Europe 2000*, are essential reading.

There is increasing interest in the various planning systems in Europe. The European Commission's *Compendium of Spatial Planning Systems and Policies* is the most comprehensive source. Other recent sources include Newman and Thornley (1996) *Urban Planning in Europe*; Healey *et al.* (1994) *Trends in Development Plan Making in European Planning*; Schmidt-Eichstaedt (1996) *Land Use Planning and Building Permission in the European Union*; and the updated ISOCARP *International Manual of Planning Practice* (Lyddon and dal Cin 1996).

National and Regional Planning

In addition to the guidance notes themselves, there are several reviews of national guidance including Land Use Consultants (1995a) *The Effectiveness of Planning Policy Guidance Notes*, and Raemaekers *et al.* (1994) *Planning Guidance for Scotland*. A comprehensive review of regional guidance is given in the symposium edited by Minay (1992), with an update by Roberts (1996) 'Regional planning guidance in England and Wales'. On Scottish guidance, see Rowan-Robinson and Lloyd (1991) 'National planning guidelines: a strategic opportunity wasting away'; and on guidance in England and Wales, see Alden and Offord (1996) 'Regional planning guidance', and Quinn (1996) 'Central government planning policy'.

There are many references on regional planning, including the standard textbook by Glasson (1992a) *An Introduction to Regional Planning*. Wannop (1995) *The Regional Imperative* provides a review with international comparisons. The resurgence of interest in strategic planning over recent years is analysed in Glasson (1992b)'The fall and rise of regional planning in the economically advanced nations', and Breheny (1991) 'The renaissance of strategic planning'. For a review of experience in the south-east see SERPLAN (1992) *SERPLAN: Thirty Years of Regional Planning 1962–1992*. Current practice in

Northern Ireland is reflected in the DoENI (1993) *A Planning Strategy for Rural Northern Ireland*.

Development Plans

PPG 12 (and its equivalents elsewhere in the UK) is the basic source. The classic text is Cross and Bristow (1983) *English Structure Planning*, though practice has changed considerably in recent years. More recent references are Phelps (1995) 'Structure plans – the conduct and conventions of examinations in public', and Jarvis (1996) 'Structure planning policy and strategic planning guidance in Wales'.

There is more material on local planning, although much of it concentrates on procedures. Adams (1994) *Urban Planning and the Development Process* is a welcome addition to the standard texts, but earlier ones remain interesting and useful: Healey (1983) *Local Plans in British Land Use Planning*; Bruton and Nicholson (1987) *Local Planning in Practice*; and Fudge *et al.* (1983) *Speed, Economy and Effectiveness in Local Plan Preparation and Adoption*. Other sources on local planning include Healey (1986) 'The role of development plans in the British planning system'; Healey (1990) 'Places, people and politics: plan-making in the 1990s'; and several chapters of case studies in Greed (1996a) *Implementing Town Planning*.

An important text summarising a major evaluation of the impact of development plans is Healey *et al.* (1988) *Land Use Planning and the Mediation of Urban Change*. See also Davies *et al.* (1986) *The Relationship between Development Plans and Development Control*, and MacGregor and Ross (1995) 'Master or servant? The changing role of the development plan in the British planning system'.

Up-to-date summaries of procedures are given in the latest editions of Telling and Duxbury (1996) *Planning Law and Procedure*, and Moore (1995) *A Practical Approach to Planning Law*. Steel *et al.* (1995) *The Efficiency and Effectiveness of Local Plan Inquiries* examine the procedures in practice. For Northern Ireland, see Dowling (1995) *Northern Ireland Planning Law*; and for Scotland: Collar (1994) *Green's Concise Scots Law: Planning*, and McAllister and McMaster (1994) *Scottish Planning Law*.

Enterprise Zones

A retrospective by Peter Hall, 'The British enterprise zones', is included in a compendium edited by Green (1991): *Enterprise Zones*. An overview of the monitoring of EZs is given by PA Cambridge Economic Consultants (1995) *Final Evaluation of Enterprise Zones*. See also Gunther and Leathers (1987) 'British enterprise zones: a critical assessment', and Thornley (1993) *Urban Planning under Thatcherism*.

Simplified Planning Zones

The number of publications on SPZs outnumbers the zones. See Arup Economic Consultants (1991) *Simplified Planning Zones: Progress and Procedures*; Blackhall (1993) *The Performance of Simplified Planning Zones*; Lloyd (1992) 'Simplified planning zones, land development, and planning policy in Scotland'; and Allmendinger (1996) 'Twilight zones'. Official guidance is given in PPG 6, PAN 31 and TAN (W) 3, all entitled *Simplified Planning Zones*.

5

THE CONTROL OF DEVELOPMENT

Development control is a process by which society, represented by locally elected councils, regulates changes in the use and appearance of the environment. As such it is of critical importance. Decisions taken in the planning process have long-term consequences and are usually irreversible. Well-considered decisions can enhance and enrich the environment. Poor decisions will be endured long after the decision-takers have died.

Audit Commission, *Building in Quality* 1992

THE SCOPE OF CONTROL

Most forms of development (as statutorily defined) are subject to the prior approval of the local planning authority, though certain categories are excluded from control. Local planning authorities have considerable discretion in giving approval. The legislation requires them to 'have regard to the provisions of the development plan, so far as material to the application', but also to 'any other material considerations'. Thus the authority can approve a proposal that does not accord with the provisions of the plan. The 1991 Act also increased the significance of plans, such that the plan becomes the first and primary point of reference in decision making, effectively introducing a presumption in favour of development proposals which are in accordance with the development plan.

Planning decisions of a local planning authority can be one of three kinds: unconditional permission, permission subject to conditions, or refusal. The practical scope of these powers is discussed in a later section (p. 111); here it is necessary merely to stress that an applicant has the right of appeal to the Secretary of State against conditional permissions

and refusals. If the action of the LPA is thought to be *ultra vires* (beyond their legal powers), there is also a right of recourse to the courts. Furthermore, planning applications which raise issues which are of more than local importance, or are of a particular technical nature, can be 'called-in' for decision by the Secretary of State. The general development control process in England is illustrated in Figure 5.1. Readers should beware that this is only a general guide and reference should be made to the further reading listed at the end of this chapter on matters of detail.

Development control necessarily involves measures for enforcement. This is provided by procedures which require anyone who carries out development without permission or in breach of conditions to consult with the LPA and, in certain circumstances, to 'undo' the development, even if this involves the demolition of a new building. A *stop notice* can also be used to put a rapid end to the carrying out or continuation of development which is in breach of planning control, when serious environmental problems are being caused by the unauthorised activity.

These are very strong powers, and it is clearly important to establish the meaning of *development*, particularly since the term has a legal connotation far wider than in ordinary language.

THE DEFINITION OF DEVELOPMENT

→ Broad

In brief, development is 'the carrying out of building, engineering, mining or other operations in, on, over or under land, or the making of any material change in the use of any buildings or other land' (and, since the 1991 Act, now covers some categories of demolition). There are many legal niceties attendant upon this definition with which it is fortunately not necessary to deal in the present outline. Some account of the breadth of the definition is, nevertheless, needed. 'Building operations', for instance, include rebuilding, structural alterations of or additions to buildings and, somewhat curiously, 'other operations normally undertaken by a person carrying on business as a builder'; but maintenance, improvement and alteration works which affect only the interior of the building or which do not materially affect the external appearance of the building are specifically excluded.

The second half of the definition introduces a quite different concept: development here means not a physical operation, but a change in the use of a piece of land or a building. To constitute 'development', the change has to be *material*, that is, substantial: a concept which it is clearly difficult to define, and which, indeed, is not defined in the legislation. A change in *kind* (for example from a house to a shop) is material, but a change in *degree* is material only if the change is substantial. For instance, the fact that lodgers are taken privately in a family dwelling house does not of itself constitute a material change so long as the main use of the house remains that of a private residence. On the other hand, the change from a private residence with lodgers to a declared guest house, boarding house or private hotel would be material. Difficulties arise with changes of use involving part of a building, with ancillary uses, and with the distinction between a material change of use and a mere interruption.

This is by no means the end of the matter, but enough has been stated to show the breadth of the definition of development and the technical complexities to which it can give rise. Reference must, nevertheless, be made to one further matter. Experi-

ence has shown that complicated definitions are necessary if adequate development control is to be achieved, but the same tortuous technique can be used to exclude matters over which control is not necessary. First, there are certain matters which are specifically declared *not* to constitute development (for example, internal alterations to buildings, works of road maintenance, or improvement carried out by a local highway authority within the boundaries of a road). Second, there are others which, though possibly constituting development, are declared not to require planning permission. Third, there is provision for the Secretary of State to make a *General Development Order* (GDO) specifying classes of 'permitted' development, a *Use Classes Order* (UCO) specifying groups of uses within which interchange is permissible, and *Special Development Orders* for specific locations or categories of development.

The distinction between the GDO and the UCO is that the former lists activities which, though constituting development, do not require permission from the LPA, while the UCO lists categories of use within which any changes do not constitute development (Home 1992).

Until 1995, the GDO contained both permitted development rights and procedural matters (relating to planning applications). In 1995 these were separated (following the Scottish model introduced in 1992). There is therefore now a *General Permitted Development Order* (GPDO) and a *General Development Procedure Order* (GDPO). Though these new orders are predominantly consolidations, they contain a number of changes.

THE USE CLASSES ORDER AND THE GENERAL DEVELOPMENT ORDERS

The Use Classes Order prescribes classes of use within which change can take place without constituting development (Table 5.1). Thus, class A1 covers shops used for all or any of a list of ten purposes, including the retail sale of goods (other than hot food); the sale of sandwiches or other cold food for consumption off the premises; for hairdressing; for the direction of funerals; and for the display

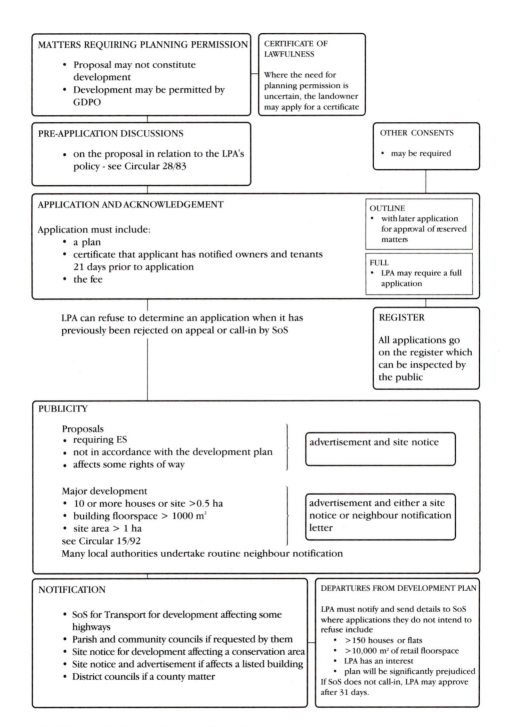

MATTERS REQUIRING PLANNING PERMISSION

- Proposal may not constitute development
- Development may be permitted by GDPO

CERTIFICATE OF LAWFULNESS

Where the need for planning permission is uncertain, the landowner may apply for a certificate

PRE-APPLICATION DISCUSSIONS

- on the proposal in relation to the LPA's policy - see Circular 28/83

OTHER CONSENTS

- may be required

APPLICATION AND ACKNOWLEDGEMENT

Application must include:
- a plan
- certificate that applicant has notified owners and tenants 21 days prior to application
- the fee

OUTLINE
- with later application for approval of reserved matters

FULL
- LPA may require a full application

LPA can refuse to determine an application when it has previously been rejected on appeal or call-in by SoS

REGISTER

All applications go on the register which can be inspected by the public

PUBLICITY

Proposals
- requiring ES
- not in accordance with the development plan
- affects some rights of way

advertisement and site notice

Major development
- 10 or more houses or site >0.5 ha
- building floorspace > 1000 m²
- site area > 1 ha
see Circular 15/92
Many local authorities undertake routine neighbour notification

advertisement and either a site notice or neighbour notification letter

NOTIFICATION

- SoS for Transport for development affecting some highways
- Parish and community councils if requested by them
- Site notice for development affecting a conservation area
- Site notice and advertisement if affects a listed building
- District councils if a county matter

DEPARTURES FROM DEVELOPMENT PLAN

LPA must notify and send details to SoS where applications they do not intend to refuse include
- >150 houses or flats
- >10,000 m² of retail floorspace
- LPA has an interest
- plan will be significantly prejudiced
If SoS does not call-in, LPA may approve after 31 days.

Figure 5.1 The Planning Application Process in England

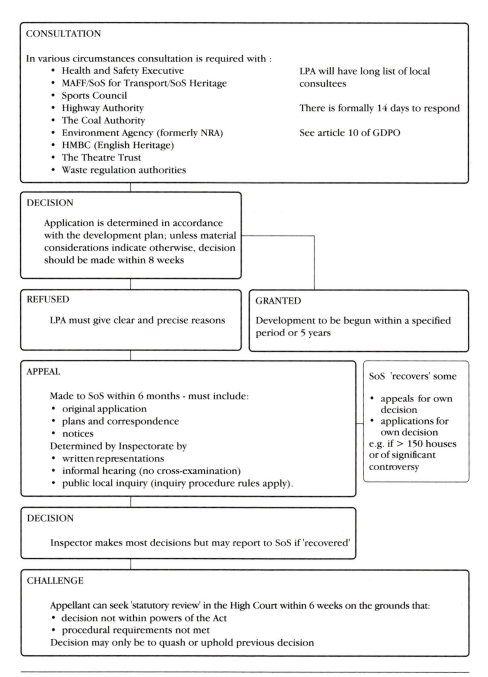

CONSULTATION

In various circumstances consultation is required with :
- Health and Safety Executive
- MAFF/SoS for Transport/SoS Heritage
- Sports Council
- Highway Authority
- The Coal Authority
- Environment Agency (formerly NRA)
- HMBC (English Heritage)
- The Theatre Trust
- Waste regulation authorities

LPA will have long list of local consultees

There is formally 14 days to respond

See article 10 of GDPO

DECISION

Application is determined in accordance with the development plan; unless material considerations indicate otherwise, decision should be made within 8 weeks

REFUSED

LPA must give clear and precise reasons

GRANTED

Development to be begun within a specified period or 5 years

APPEAL

Made to SoS within 6 months - must include:
- original application
- plans and correspondence
- notices

Determined by Inspectorate by
- written representations
- informal hearing (no cross-examination)
- public local inquiry (inquiry procedure rules apply).

SoS 'recovers' some
- appeals for own decision
- applications for own decision
e.g. if > 150 houses or of significant controversy

DECISION

Inspector makes most decisions but may report to SoS if 'recovered'

CHALLENGE

Appellant can seek 'statutory review' in the High Court within 6 weeks on the grounds that:
- decision not within powers of the Act
- procedural requirements not met

Decision may only be to quash or uphold previous decision

For a more comprehensive explanation see Moore (1995) and Duxbury (1996).
For variations in Scotland see McAllister and McMaster (1994) and Collar (1994).
For variations in Northern Ireland see Dowling (1995).

Figure 5.1 continued

Table 5.1 Summary and Comparison of the Use Classes Orders

England and Wales (Town and Country Planning (Use Classes) Order 1987 (SI No. 764) (as amended)			Scotland (The Town and Country Planning (Use Classes) (Scotland) Order 1989 (SI No. 147) (as amended)		Northern Ireland (Planning) (Use Classes) Order (NI) 1989 (as amended)	
Class	Use	Development permitted	Class	Development permitted	Class	Development permitted
A1	Shops	from A3, A2 if premises have a display window at ground floor level, or for the display or sale of motor vehicles	1 Shops	from sale and display of motor vehicles	1 Shops	from a betting office or from food or drink
A2	Financial and professional	to A1 from A3	2 Financial, professional and other services	to 1	2 Financial, professional and other services	from a betting office or from food or drink; to 1 if it has a display window at ground floor level
A3	Food and drink	to A1 and A2	3 Food and drink	to 1 and 2	3 Business	
B1	Business	to B8 (max. 235m^2) from B2 and B8 (max. 235m^2)	4 Business	to 11 (max. 235m^2)	4 Light industrial	to 11
B2	General industrial	to B1 and B8 (max. 235m^2)	5 General industrial	to 4 and 11 (max. 235m^2)	5 General industrial	to 4 or 11 (max. 235m^2)
			7–10 Special industry groups		7–10 Special industrial groups	
B8	Storage or distribution	to B1 (max. 235m^2) from B1 or B2 (max. 235m^2)	11 Storage or distribution	to 4	11 Storage and distribution	to 4 (max. 235m^2)
C1	Hotel, boarding house, and guest house		12 Hotels and hostels (not including public houses)		12 Guest houses and hostels	to 14
C2	Residential institutions		13 Residential institutions		13 Residential institutions	to 14
C3	Dwelling houses		14 Houses		14 Dwelling houses	
D1	Non-residential institutions		15 Non-residential institutions		15 Non-residential institutions	
D2	Assembly and leisure		16 Assembly and leisure		16 Assembly and leisure	

Note: The sub-division of residential dwellings into two or more separate dwellings is a change of use of goods for sale. Class A3 covers 'use for the sale of food or drink for consumption on the premises or of hot food for consumption off the premises'. As a result of these classes, a shop can be changed from a hairdresser to a funeral parlour or a sweet shop (or vice versa), but it cannot be changed (unless planning permission is obtained) to a restaurant or a hot food take-away, which are in a different class. The classes, it should be stressed, refer only to changes of use, not to any building work, and the Order gives no freedom to change from one class to another. Whether such a change constitutes development depends on whether the change is 'material'.

The General Permitted Development Order gives the developer a little more freedom by listing classes of 'permitted development' – or, to be more precise, it gives advance general permission for certain classes of development, typically of a minor character. If a proposed development falls within these classes, no application for planning permission is necessary: the GPDO itself constitutes the permission. The Order includes minor alterations to residential buildings and the erection of certain agricultural buildings (other than dwelling houses). It also permits certain changes of use within the UCO, such as a change from an A3 use (the food and drink class) to an A1 use (shop), but not – because of the possible environmental implications – the other way round. While the use changes allowed by the UCO are all 'bilateral' (any change of use within a class is reversible without constituting development), the GPDO builds upon this structure by specifying a number of 'unilateral' changes *between* classes for which permission is not required. The rationale here is that the permitted changes generally constitute an environmental improvement.

The cynic may perhaps be forgiven for commenting that the freedom given by the UCO and the GPDO is so hedged by restrictions, and frequently so difficult to comprehend, that it would be safer to assume that any operation constitutes development and requires planning permission (though it may be noted with relief that painting is not normally subject to control, unless it is 'for purpose of advertisement, announcement or direction'). The legislators

have been helpful here. Application can be made to the LPA for a *certificate of lawfulness* of a proposed use or development. This enables a developer to ascertain whether or not planning permission is required.

The Orders were modified by the Conservative government as part of its policy of 'lifting the burden' on business. Some of the changes have proved to be very controversial. For example, it is now allowable to change a restaurant to a fast hot-food take-away, or to change a public house to more profitable uses such as professional offices and other uses appropriate to a shopping area. The impact of these changes on such matters as local amenity and traffic generation is detailed in a 1992 report by Sandra Bell for the London Boroughs Association. The government response to the representations of local government are set out in the report: it is maintained that 'the advantages of the present arrangements in terms of the certainty and flexibility they provide for the commercial sector, and the reduction in intervention and bureaucracy, far outweigh the disadvantages'. A clearer illustration of the political nature of planning would be difficult to find (though more persuasive are the photographs in the LBA report depicting the detrimental effects of 'change of use').

VARIATIONS IN NORTHERN IRELAND AND SCOTLAND

Development control operates in a similar way across the whole of the UK, although it is established by separate law and policy in Northern Ireland and Scotland, and there is a separate 'national' policy context in Wales. A comparison of the Use Classes Orders shows minor variations which also exist in other parts of the system. The most important difference is that in Northern Ireland development control is operated by the Planning Service, an executive agency of the Department of Environment for Northern Ireland, which operates through six divisional planning offices. Local authorities in Northern Ireland have only a consultative role. The Planning Service makes recommendations to the local district councils who can request the Service to reconsider. The Service may reconsider the application but, if there is no agreement, the matter is referred to the Chief Executive's Office and a decision is made by the Management Board (senior civil servants). Appeals in Northern Ireland are heard by the Planning Appeals Commission; this is an independent body whose members are appointed by the Secretary of State for Northern Ireland. The Commission also hears inquiries into major planning applications and development plans.

The provisions of the 1991 Planning and Compensation Act (which have made important changes to the procedure for enforcement, the control of demolition, and the relationship between plans and decisions) have not been implemented in Northern Ireland. Another variation is in neighbour notification, where Northern Ireland has a more thorough system. This is guided by a non-statutory notification scheme requiring, for example, advertisement of all applications. Scotland's legislation also differs in this respect, and requires the applicant (rather than the local authority) to serve notices on neighbours. Neighbour notification is discussed further in Chapter 12.

SPECIAL DEVELOPMENT ORDERS

While the GPDO is applicable generally, Special Development Orders (SDOs) relate to particular areas or particular types of development. Thus, an urban development corporation SDO grants permission for the carrying out of development (approved by the Secretary of State) by an urban development corporation within the designated area. Other SDOs deal with development control in national parks, areas of outstanding natural beauty and conservation areas, and with such special uses as the oxide fuel processing plant at Sellafield (Windscale), and sites for nuclear waste disposal.

SDOs (like other Orders) are subject to parliamentary debate and annulment by resolution of either House. This can provide an opportunity for testing opinion on controversial proposals such as the reprocessing of nuclear fuels at Windscale or the designation of Stansted as the site of the third London airport.

An unusual use of an SDO was the Vauxhall Cross proposal in London which, though receiving architectural acclaim, was fiercely opposed by the local authority. This granted planning permission for a very large development including more than one million square feet of offices and 260,000 square feet of dwellings on the eastern end of Vauxhall Bridge. Though, in fact, the development did not proceed, the use of the SDO procedure for such a purpose involved a high degree of central involvement in local planning decisions. The case submitted by the DoE was that 'the purpose of making fuller use of SDOs would not be to make any general relaxation in development control, but to stimulate planned development in acceptable locations, and speed up the planning process' (Thornley 1993: 163). This is tantamount to saying that the central government knows best.

WITHDRAWAL OF PERMITTED DEVELOPMENT RIGHTS

The development rights which are permitted by the GPDO can be withdrawn by a Direction made under Article 4 of the Order (and hence are known as *Article 4 Directions*). The effect of such a direction is not to prohibit development, but to require that a planning application is made for development proposals in a particular location. The direction can apply either to a particular area (such as a conservation area) or, unusually, to a particular type of development (such as caravan sites) throughout a local authority area. Since the direction involves taking away a legal right, compensation may be payable.

The most common use of an Article 4 Direction is in areas where special protection is considered desirable, as with a dwelling house in a rural area of exceptional beauty, a national park, or a conservation area. Without the direction, an extension of the house would be permitted up to the limits specified in the GPDO. The majority of Article 4 Directions in fact relate to 'householder' rights in conservation areas. They are also used in national parks and other designated areas to control temporary uses of land (such as camping and caravanning) which would otherwise be permitted (Tym 1995a).

MATERIAL CONSIDERATIONS

Crucial to the development control process is the concept of *material considerations*. These are exactly what the term suggests: considerations that are material to the taking of a development control decision. However, the concept is one that gives rise to considerable confusion – which itself stems from the discretionary nature of the planning system. The legislation provides no guide, and the courts have had to decide where the limits of discretion lie. Since planning is concerned with the use of land, purely personal considerations are not generally material (though they might become so in a finely balanced case). Similarly, in the words of PPG 1 (para 6), 'it is not the function of the planning system to interfere with or inhibit competition between users or investors in land, or to regulate the over-all provision and character of space for particular uses for other than land use planning reasons'. It follows that planning permission cannot be refused on the grounds that there is already sufficient provision (of, for example, hotels, petrol stations or shops). However, a special consideration arises with out-of-town shopping centres: it is explicitly advised (in PPG 6, para. 33) that account should be taken 'of the possible impact (including the cumulative impact with other recent or proposed retail developments) on the vitality and viability of any nearby town centre' or 'on the rural economy (including the role of village shops)'.

The courts have held that a very wide range of matters can be material. Indeed, in the oft-quoted *Stringer* case, it was stated that 'any consideration which relates to the use and development of land is capable of being a planning consideration' (*Stringer* v. *Minister of Housing and Local Government* 1971). Whether a particular consideration falling within that broad class in any given case is material will depend on the circumstances. In another important case, the House of Lords formulated a threefold 'planning test': to be valid a planning decision had to (a) have a planning purpose; (b) relate to the permitted development; and (c) be reasonable (*Newbury District Council* v. *Secretary of State for the Environment* 1981).

These and similar judicial statements provide only the flimsiest of guidelines, and a succession of cases has shown that the issues are legion: the suitability of the site and its accessibility; environmental factors; the historical, aesthetic and landscape nature of the site; economic and social benefits of the development; considerations of energy and 'sustainable development'; implications for transport; impact on small business; and so on. Given this wide range of possibilities, one looks to 'planning guidance' for a lead. Here the problem is partly that there is so much of it that one can find a policy justification for virtually anything. Planning law is not neatly codified: it lies all over the place. As long as a minister has stated a policy somewhere, it seems that it is acceptable as a material consideration:

> The courts have accepted that policies should take into account white papers, circulars, policy guidance, previous decisions, written parliamentary answers, and even after dinner speeches. There is an increasing tendency to announce policy at professional conferences . . .
>
> (Read and Wood 1994: 13)

In short, there are few limits to the matters which can be regarded as material to a planning decision, and few clear (but many unclear) guidelines.

CONDITIONAL PERMISSIONS

A local planning authority can grant planning permission subject to conditions. This can be a very useful way of permitting development which would otherwise be undesirable. Thus, a garage may be approved in a residential area on condition that the hours of business are limited. Residential development may be permitted in an area designated as a green belt subject to the condition that the houses are occupied only by agricultural workers.

The power to impose conditions is a very wide one. The legislation allows LPAs to grant permission subject to 'such conditions as they think fit'. However, this does not mean 'as they please'. The conditions must be appropriate from a planning point of view: 'the planning authority are not at liberty to use their power for an ulterior object, however desirable that object may seem to them to be in the public interest. If they mistake or misuse their powers, however *bona fide*, the court can interfere by declaration and injunction' (*Pyx Granite Co. Ltd* v. *Minister of Housing and Local Government* 1981).

DoE Circular 1/85, *The Use of Conditions in Planning Permissions*, stresses that, in addition to satisfying the legal criteria for validity, 'conditions should not be imposed unless they are both necessary and effective, and do not place unjustifiable burdens on applicants'. Conditions should be 'necessary, relevant to planning, relevant to the development to be permitted, enforceable, precise, and reasonable in all other respects'. As might be expected, there is considerable debate on the meaning of these terms. A striking example of a condition which was quite unreasonable was dealt with in the *Newbury* case. There the district council gave permission for the use of two former aircraft hangars for storage, subject to the condition that they be demolished after a period of ten years. The House of Lords held that since there was no connection between the proposed use and the condition, it was *ultra vires*. The pronouncement of Viscount Dilhorne on the matter is often quoted: 'The conditions imposed must be for a planning purpose and not for any ulterior one . . . and they must fairly and reasonably relate to the development permitted. Also they must not be so unreasonable that no reasonable planning authority could have imposed them.' (This is commonly referred to as a three-part test: planning purpose; relation to permitted development; and reasonableness.)

Up to 1968, there was no general time limit within which development had to take place: unless a specific condition was imposed, development for which planning permission had been given could take place at any time. The 1968 Act, however, made all planning permissions subject to a condition that development is commenced within five years. If the work is not begun within this time limit, the permission lapses. However, the Secretary of State or the local planning authority can vary the period, and there is no bar to the renewal of permission after that period has elapsed (whether it be more or less than five years).

The purpose of this provision is to prevent the accumulation of unused permissions and to discourage the speculative landhoarder. Accumulated unused permissions could constitute a difficult problem for some LPAs: they create uncertainty and could make an authority reluctant to grant further permissions, which might result in, for example, too great a strain on public services. The provision is directed towards the bringing forward of development for which permission has been granted, and thus to enable new allocations of land for development to be made against a reasonably certain background of pending development.

The provision relates, however, only to the beginning of development, and this has in the past been deemed to include digging a trench or putting a peg in the ground. But (if the permission is not a pre-1968 Act one) the trench-digger may be brought up against a further provision: the serving of a *completion notice*. Such a notice states that the planning permission lapses after the expiration of a specified period (of not less than one year). Any work carried out after then becomes liable to enforcement procedures.

Planning 'conditions' of a different nature, involving 'planning gain' and *planning agreements*, are discussed in the context of land values in Chapter 6.

It is sometimes convenient for an applicant or the LPA (or both) to deal with an application in outline. *Outline planning permission* gives the applicant permission in principle to carry out development subject to *reserved matters* which will be decided at a later stage. This is a useful device to enable a developer to proceed with the preparation of detailed plans with the security that they will be not be opposed in principle.

FEES FOR PLANNING APPLICATIONS

Fees for planning applications were introduced in 1980. This represented a marked break with planning traditions, which had held (at least implicitly) that development control is of general communal benefit and directly analogous to other forms of public control for which no charges are made to individuals. The Thatcher administration had a very different view; to quote from the parliamentary debate:

> We do not believe that the community as a whole should continue to pay for all sorts of things that it has paid for in the past . . . In the general review that has taken place to see where we can reduce spending from the public purse . . . we came to the conclusion that the cost of development control was an area where some part of the cost should be recovered.
>
> (Standing Committee Debates on the Local Government Planning and Land (No. 2) Bill, col. 245, 22 April 1980)

The 1980 Bill provided additionally for fees for appeals, but this was dropped in the face of widespread objections from both sides of the House.

The current regulations were made in 1989, and amended in 1991 and 1992. The fee structure is subject to change over time, and a detailed schedule is therefore not appropriate. However, to illustrate, under the current regulations the fee for residential development is £120 per dwelling (up to a maximum of £6,000 for fifty or more); applications relating to commercial and industrial buildings are charged according to the gross floor space to be created, with £60 for up to 40 square metres, and £120 for each additional 75 square metres, up to a maximum of £6,000. The government's ultimate aim is to recover the full administrative costs of dealing with planning applications.

The power to charge fees for planning applications does not extend to the pre-application discussion stage. A decision of the Court of Appeal allowing such charges was reversed by the House of Lords in 1991.

PLANNING APPEALS

An unsuccessful planning applicant can appeal to the Secretary of State, and a large number in fact do so. Appeals decided during 1994–5 (England and Wales) numbered 16,177, of which about one-third were allowed. Here, wide powers are available to the Secretary of State. These include the reversal of a local authority's decision or the addition, deletion or modification of conditions. The conditions can be made more onerous or, in an extreme case, the Secretary of State may even go to the extent of refusing planning permission altogether if it is decided that the local authority should not have granted it with the conditions imposed.

Though each planning appeal is considered and determined on its merits, the cumulative effect is an emergence of the Department's views on a wide range of planning matters. The effect of these on the policy of individual authorities may be difficult to assess, but clearly they are likely to have a very real influence. Local planning authorities are unlikely to refuse planning consents for a particular type of development if they are convinced that the Department would uphold an appeal.

It is not, of course, every planning appeal that raises an issue of policy. Yet until 1969 all had to be dealt with by the Department's inspectorate, though the majority were settled by correspondence after an informal visit to the site and without a local inquiry. (This is the *written representations procedure*.) The Franks Committee on Administrative Tribunals and Inquiries argued that it was not satisfactory 'that a government department should be occupied with appeal work of this volume, particularly as many of the appeals relate to minor and purely local matters, in which little or no departmental policy entered' (Franks Report 1957: 85).

In view of the delay which was inevitable in this appeals system and the huge administrative burden it placed on the Department, considerable thought was given to possible alternatives. The solution adopted by the 1968 Planning Act was for the determination of certain classes of appeals by inspectors. The classes are defined by regulation and can thus be amended in the light of experience. The trend has been to increase the range, and now the great majority of appeals are decided in this way. The Department has, in fact, divested itself of responsibility for adjudicating on all but a small number of planning appeals. Only matters of major importance are 'recovered' for decision by the Secretary of State.

Since the beginning of the 1980s, the government has been particularly concerned at 'freeing' the economy from unnecessary regulation. White Papers were issued with rousing titles such as *Lifting the Burden* (1985) and *Building Businesses Not Barriers* (Cmnd 9794, 1986). Perversely, the number of appeals received rose during this decade, reaching a peak of 32,281 in 1989–90, since when there has been a significant fall. The increase in appeals had the effect of slowing down the appeals procedures. As the government was committed to speeding up planning processes, this position was embarrassing (to put the matter no more strongly). In seeking ways to deal with this, reviews of the appeals system were undertaken within the Department, the first of which was completed in 1985 and published together with an 'action plan' in 1986. This was an internal 'efficiency scrutiny' of the written representation system, which showed that some delays were the fault of the Department, while others were the result of factors outside the Department's control, including the way in which the applicant and his agent dealt with the appeal. In short, there was no simple answer to the problem, and the report made some forty detailed recommendations.

In recent years, increased emphasis has been placed on ensuring that development plans provide a clear guide to applicants and a firm basis for planning decisions. This, in turn, requires that there is adequate input into the plan preparation process. It is envisaged that this will reduce the number of appeals. In the words of the Minister of State for Housing and Planning, Sir George Young:

> The ultimate success of the plan-led system will be measured by smaller proportions of refusals, because developers come to recognise the need to participate in plan preparation and then work within the framework it provides. Another key indicator will be the appeal success rate; we will be less likely to allow appeals if we have satisfied ourselves on the content of the plan, and the planning decision accords with the plan.

CALL-IN OF PLANNING APPLICATIONS

The power to 'call-in' a planning application for decision by the Secretary of State is quite separate from that of determining an appeal against an adverse decision of a LPA. This power is not circumscribed: the Secretary of State may call-in *any* application. There are no statutory criteria or restrictions, and no prescribed procedures for handling representations from the public. Though there is no general statement of policy as to which applications will normally be called-in, there are several categories which are particularly liable.

In the first place, all applications for development involving a substantial departure from the provisions of a development plan which the LPA intends to grant must be sent to the Secretary of State together with a statement of the reasons why it wishes to grant the permission. This procedure enables the Secretary of State to decide whether the development is sufficiently important to warrant its being called-in. Second, mineral workings often raise problems of more than local importance, and the national need for particular minerals has to be balanced against planning issues. Such matters cannot be adequately considered by local planning authorities and, in any case, involve technical considerations requiring expert opinion of a character more easily available to the Department. For these reasons, a large proportion of applications for permission to work minerals have been called-in. Furthermore, there is a general direction calling-in all applications for the winning and working of ironstone in certain counties where there are large-scale ironstone workings. Third, the power of call-in is generally used when the matter at stake is (as in the case of minerals) of more than local importance or interest. (The Royal Fine Art Commission has, in its terms of reference, the power 'to call the attention of any of our departments of state . . . to any project or development which [it considers] may appear to affect amenities of a national or public character'. The Commission has requested the use of call-in on a number of occasions, not always successfully.)

In answer to a 1995 parliamentary question on whether he intended to take design considerations into account when deciding whether or not to call-in planning applications, the Secretary of State outlined current policy:

My general approach is not to interfere with the jurisdiction of local planning authorities unless it is necessary to do so. I will therefore be very selective about calling in applications to determine myself and will, in general, only take this step if planning issues of more than local importance are involved. Each case must be considered on its individual merits. An application for development which raises significant architectural and urban design issues is one example of the type of case which may be of more than local importance. Other examples include cases which, in my opinion, could have wide effects beyond their immediate locality, which give rise to substantial regional or national controversy, which conflict with national policy on important matters, and those where the interests of foreign governments may be involved.

When an application is called in, the Secretary of State must, if either the applicant or the local planning authority so desire, hold a hearing or public inquiry. The public inquiry is more usual, particularly in important cases. Two 1994 examples concerned Paignton Zoo and Portsmouth Football Club.

The Paignton Zoo case raised a number of interesting issues, and it is worth summarising some of the major points. The proposed development included the refurbishment of the zoo, the development of a 65,000 sq ft retail food store, parking spaces for 600 cars, and a petrol filling station. The proposals clearly raised major issues of policy, including those set out in PPG 6 *Town Centres and Retail Development* and PPG 21 *Tourism*. There were several conflicting considerations, including the likely effect of the retail development on the town centre and the precarious economic position of the zoo which, so it was argued, was 'likely to close unless it receives a capital injection of the size that only this proposal is likely to provide, thereby causing a loss to the local economy of approximately £6 million per annum and a significant loss of jobs'. The Secretary of State decided that these and other

benefits to tourism and the local economy (together with highway improvements) more than outweighed any harm which might be done to the vitality of the town centre, and he therefore granted planning permission. In the words of the decision, 'the harm likely to arise from the proposals is less clear cut than the effects that would result from the decline and possible closure of the zoo . . . ; the balance of advantage lies in favour of allowing the proposal . . . ; the zoo's leading role in the local economy places it in a virtually unique position . . . ; [but the decision] should not be regarded as a precedent for other businesses seeking to achieve financial stability' (*JPL* 1995: 657).

The Portsmouth Football Club case revolved round the Club's 'incontestable need' to improve its ground and its inability to do so without other development to subsidise it. The proposal included an all-seater stadium, retail and leisure development, and a new railway station. However, in view of the powerful objections, it was rejected by the Secretary of State. In addition to the inadequacy of the case presented by the Club (in terms of financial viability, site development, traffic implications, etc.), a very important consideration was the proximity of a Ramsar site and European Special Protection Area (which raised questions 'about compliance with international obligations') (see p. 221). Also of significance was the conflict of the retail proposals with PPG guidelines. Reading between the lines, it would seem that the combination of an inadequately presented case and substantial objections was fatal (*JPL* 1995: 764–9).

Since 1968, the Secretary of State has had power to refer development proposals of a far-reaching or novel character to an *ad hoc* Planning Inquiry Commission. This power has never been used: the Roskill Commission on the third London airport was set up under non-statutory powers, while the Greater London Development Plan Inquiry was established under the *general* powers to hold local inquiries provided by the Town and Country Planning Act. For major inquiries such as Windscale, Belvoir, Stansted and Sizewell, this special form of inquiry might have been considered particularly apt: that it was not so considered gave rise to much debate. (The issue is discussed further in the final chapter.)

ENFORCEMENT OF PLANNING CONTROL

If the machinery of planning control is to be effective, some means of enforcement is essential. Under the prewar system of interim development control there were no such effective means. A developer could go ahead without applying for planning permission, or could even ignore a refusal of permission. He took the risk of being compelled to 'undo' his development (for example, demolish a newly built house) when, and if, the planning scheme was approved, but this was a risk which was usually worth taking. And if the development was inexpensive and lucrative (for example, a petrol station or a greyhound racing track) the risk was virtually no deterrent at all. This flaw in the prewar system was remedied by the strengthening of enforcement provisions.

These are required not only for the obvious purpose of implementing planning policy, but also to ensure that there is continuing public support for, and confidence in, the planning system. To quote PPG 18:

> The integrity of the development control process depends on the LPA's readiness to take effective action when it is essential. Public acceptance of the development control process is quickly undermined if unauthorised development, which is unacceptable on planning merits, is allowed to proceed without any apparent attempt by the LPA to intervene before serious harm results from it.

Development undertaken without permission is not an offence in itself, but ignoring an *enforcement notice* is, and there is a maximum fine following conviction of £20,000. (In determining the amount of the fine, the court is required to 'have regard to any financial benefit which has accrued'.)

There is a right of appeal against an enforcement notice. Appeals can be made on several grounds; for example, that permission ought to be granted, that permission has been granted (e.g. by the GPDO), or that no permission is required. There is also a limited right of appeal on a point of law to the High Court.

Enforcement provisions were radically changed by the 1991 Planning and Compensation Act, following a comprehensive review by Robert Carnwath QC, published in 1989. In addition to the long-standing provision for enforcement notices, a LPA now has power to issue a *planning contravention notice.* This enables it to obtain information about a suspected breach of planning control and to seek the cooperation of the person thought to be in breach of planning control. This optional procedure is intended to enable discussions to take place on whether planning permission (with or without conditions) is required. If agreement is not forthcoming (whether or not a contravention notice is served) an enforcement notice may be issued, but only 'if it is expedient' to do so 'having regard to the provisions of the development plan and to any other material considerations'. In short, the local authority must be satisfied that enforcement is necessary in the interests of good planning. In view of the government's commitment to fostering business enterprise (discussed further below, pp. 124–5), LPAs are advised in PPG 18 to consider the financial impact on small businesses of conforming with planning requirements: 'Nevertheless, effective action is likely to be the only appropriate remedy if the business activity is causing irreparable harm.' Development 'in breach of planning control' (development carried out without planning permission or without compliance with a planning condition) might be undertaken in good faith or in ignorance. In such a case, application can be made for retrospective permission. It is unlikely that a local authority would grant unconditional permission for a development against which it had served a planning contravention notice, but it might be willing to give conditional approval.

The 1991 Act also introduced a *breach of condition notice* as a remedy for contravention of a planning condition. This is a simple procedure against which there is no appeal, though there may be some legal complexities which will prevent its widespread use

(Cocks 1991). Further, there is a new provision enabling a LPA to seek an injunction in the High Court or County Court to restrain 'any actual or apprehended breach of planning control'. In Scotland, the provision is for an interdict by the Court of Session or the Sheriff.

Where there is an urgent need to stop activities that are being carried on in breach of planning control, a LPA can serve a *stop notice*. This is an attempt to prevent delays in the other enforcement procedures (and advantage being taken of these delays) which could result in the local authority being faced with a *fait accompli*. Development carried out in contravention of a stop notice constitutes an offence.

The provisions for enforcement are complex, and the reader is referred to the discussion in Chapter 5 of the 1989 Carnwath Report. The position can be exacerbated by the lowly esteem in which the enforcement system (and those who staff it) are often held. Several commentators have termed enforcement 'the weakest link in the planning chain', both south and north of the border. Fortunately, the majority of alleged contraventions of planning control are dealt with satisfactorily and without any recourse to legal action, but the minority have a disproportionate effect on the credibility of the planning process as a whole.

REVOCATION, MODIFICATION AND DISCONTINUANCE

The powers of development control possessed by local authorities go considerably further than the granting or withholding of planning permission. They can interfere with existing uses and revoke a permission already given, even if the development has actually been carried out.

A *revocation order* or *modification order* is made when the development has not been undertaken (or before a change of use has taken place). The local authority must 'have regard to the development plan and to any other material considerations', and an opposed order has to be confirmed by the Secretary of State.

Compensation is payable for abortive expenditure and any loss or damage due to the order. Such orders are rarely made.

Quite distinct from these powers is the much wider power to make a *discontinuance order*. This power is expressed in extremely wide language: an order can be made 'if it appears to a local planning authority that it is expedient in the interests of the proper planning of their area (including the interests of amenity)'. Again, confirmation by the Secretary of State is required, and compensation is payable for depreciation, disturbance and expenses incurred in carrying out the works in compliance with the order.

An order will be confirmed only if the case is a strong one. In rejecting a discontinuance order on a scrap metal business in an attractive residential area, for instance, it was explained that 'the fact that such a business is out of place in an attractive residential area must be weighed in the light of an important distinction between the withdrawal of existing use rights, as sought in the discontinuance order, and the refusal of new rights'. In this particular case, it was decided not to override 'the proper interests of the objector as long as his business is maintained on an inoffensive scale' (*JPL* 1962: 753). Other cases have established the principle that a stronger case is needed to justify action to bring about the discontinuance of a use than would be needed to warrant a refusal of permission in the first instance.

British planning legislation does not assume that existing non-conforming uses must disappear if planning policy is to be made effective. This may be an avowed policy, but the planning Acts explicitly permit the continuance of existing uses.

DEVELOPMENT BY GOVERNMENT DEPARTMENTS AND STATUTORY UNDERTAKERS

Development by government departments does not require planning permission, but there have been special arrangements for consultations since 1950. Increased public and professional concern about their inadequacy led to revised, but still non-statutory, arrangements culminating in DoE Circular 18/84. This asserts clearly that, before proceeding with development, government departments will consult LPAs when the proposed development is one for which specific planning permission would, in normal circumstances, be required.

Development undertaken by statutory undertakers is also subject to special planning procedures. Where a development requires the authorisation of a government department (as do developments involving compulsory purchase orders, work requiring loan sanction, and developments on which government grants are paid) the authorisation is usually accompanied by *deemed planning permission*. Much of the regular development of statutory undertakers and local authorities (for example, road works, the laying of underground mains and cables) is *permitted development* under the GDO. Statutory undertakers wishing to carry out development which is neither permitted development nor authorised by a government department have to apply for planning permission to the local planning authority in the normal way, but special provisions apply to *operational land*. The original justification for this special position of statutory undertakers was that they are under an obligation to provide services to the public and could not, like a private firm in planning difficulties, go elsewhere.

DEVELOPMENT BY LOCAL AUTHORITIES

Until 1992, LPAs were also deemed to have permission for any development which they themselves undertook in their area, as long as it accorded with the provisions of the development plan; otherwise they had to advertise their proposals and invite objections. These 'self-donated' planning permissions are problematic: though local authorities are guardians of the local public interest, they can face a conflict of interest in dealing with their own proposals for development. Pragmatic consideration of the merits of a case involving their own role as developers can easily distort a planning judgment. Examples include attempts by authorities to dispose of surplus school playing fields with the benefit of permission for development; and competing applications for superstore development when one of the sites is owned by the authority itself. Because of these difficulties, new regulations were issued in 1992 which require LPAs to make planning applications in the same way as other applicants, and generally to follow the same procedures (Moore 1995: 249–56). The Scottish Local Government Ombudsman has for long complained about 'the ease with which planning authorities breach their own plans, particularly considering the time, effort, and consultation which goes into them'. The wider issue of adequate procedures for dealing with planning proposals which involve a departure from the development plan has been dealt with in Scotland by a Direction and the publication of a Planning Advice Note (PAN 41: *Development Plan Departures*, 1994).

CONTROL OF ADVERTISEMENTS

The need to control advertisements has long been accepted. Indeed, the first Advertisements Regulation Act of 1907 antedated by two years the first Planning Act. But even when amended and extended (in 1925 and 1932), the control was quite inadequate. Not only were the powers permissive: they were also limited. For instance, under the 1932 Act, the right of appeal (on the ground that an advertisement did not injure the amenities of the area) was to the Magistrates' Court – hardly an appropriate body for such a purpose. The 1947 Act set out to remedy the deficiencies. There are, however, particular difficulties in establishing a legal code for the control of advertisements. Advertisements may range in size from a small window notice to a massive hoarding; they vary in purpose from a bus stop sign to a demand to buy a certain make of detergent; they could be situated alongside a cathedral, in a busy shopping street, or in a particularly beautiful rural setting; they might be pleasant or obnoxious to look at; they might be temporary or permanent; and so on. The task of devising a code which takes all the relevant factors into account and, at the same time, achieves a balance between the conflicting interests of legitimate advertising and 'amenity' presents real problems. Advertisers themselves frequently complain that decisions in apparently similar cases have not been consistent with each other. The official departmental view is that no case is exactly like another, and hard and fast rules cannot be applied: each case has to be considered on its individual merits in the light of the tests of amenity and – the other factor to be taken into account – public safety.

The control of advertisements is exercised by regulations. The Secretary of State has very wide powers of making regulations 'in the interests of amenity or public safety'. The question of public safety is rather simpler than that of amenity, though there is ample scope for disagreement: the relevant issue is whether an advertisement is likely to cause danger to road users or to 'any person who may use any road, railway, waterway (including coastal waters), docks, harbour or airfield'. In particular, account has to be taken of the likelihood of whether an advertisement 'is likely to obscure, or hinder the ready interpretation of, any road traffic sign, railway signal, or aid to navigation by water or air'. Amenity includes 'the general characteristics of the locality, including the presence of any feature of historic, architectural, cultural or similar interest'. The definition of an advertisement is not quite as complicated as that of development, but it is very wide:

> Advertisement means any word, letter, model, sign, placard, board, notice, awning, blind, device or representation, whether illuminated or not, in the nature of, and employed wholly or partly for the purposes of, advertisement, announcement or direction and . . . includes any hoarding or similar structure used, or designed or adapted for use, and anything else principally used, or designed or adapted principally for use, for the display of advertisements.

It is helpfully added that the definition excludes anything 'employed as a memorial or as a railway signal'.

Various classes of advertisement are excepted from all control: those displayed on a balloon; on enclosed land; within a building; and on or in a vehicle. Also excepted are traffic signs, election signs and national flags. As one might expect, there are some interesting refinements of these categories, which can be ignored for present purposes (though we might note, in passing, that a vehicle must be kept moving or, to use the more exact legal language, must be normally employed as a moving vehicle). With these exceptions, no advertisements may be displayed without *consent*. However, certain categories of advertisement can be displayed without *express consent*; so long as the local authority take no action, they are *deemed* to have received consent. These include bus-stop signs and timetables, hotel and inn signs, professional or business plates, 'To Let' and 'For Sale' signs, election notices, statutory advertisements and traffic signs.

It needs to be stressed that amenity and public safety are the only two criteria for control. The content or subject of an advertisement is not relevant, and a local authority cannot refuse express consent on grounds of morality, offensiveness or taste. Thus an advertisement which contained the words 'Chish and Fips' was considered by the Secretary of State, on appeal, to be questionable on grounds of taste, but not detrimental to amenity: the appeal was allowed (*JPL* 1959: 736).

If an advertisement displayed with deemed consent becomes unsafe, unsightly or in any way 'a substantial injury to the amenity of the locality or a danger to members of the public', the LPA can serve a *discontinuance order*. There is the normal right of appeal to the Secretary of State. Advertisements displayed with express consent can be subject to revocation or modification, again with the normal rights of appeal.

Complex though this may seem, it is not all that there is to advertisement control. In some areas, for example conservation areas, national parks or areas of outstanding natural beauty, it may be desirable to prohibit virtually all advertisements of the poster type and seriously restrict other advertisements including those normally displayed by the ordinary trader. Accordingly, local planning authorities have power to define *areas of special advertisement control* where very strict controls are operated. Within such areas, the general rule is that no advertisement may be displayed; such advertisements as are given express consent are considered exceptions to this general rule.

These special controls originated primarily from the need to deal with the legacy of advertising hoardings which were such a familiar sight in the 1930s. It is now argued that they are obsolete, and can be replaced by simpler controls. Added justification is given to this argument by the fact that, in 1995, nearly a half of the area of England and Wales was defined by local planning authorities as being within areas of special control. A 1996 DoE consultation paper (*Outdoor Advertisement Control – Areas of Special Control of Advertisements*) maintained that many orders were out of date since they no longer corresponded to the current limits of the built environment, while the system was either obscure or widely misunderstood by the public.

CONTROL OF MINERAL WORKING

The reconciliation of economic and amenity interests in mineral working is an obvious matter for planning authorities (MPAs). It would, however, be misleading to give the impression that the function of planning authorities is simply to fight a continual battle for the preservation of amenity. Planning is concerned with competing pressures on land and with the resolution of conflicting demands. Amenity is only one of the factors to be taken into account. The general policy framework is set out in MPG 1. It is interesting to compare the current policy, as set out in the 1994 MPG with that of the earlier (1988) version (both of which are illustrated in the box). The 1994 version places a significantly greater emphasis on conservation and environmental considerations.

Planning powers provide for making the essential survey of resources and potentialities, the allocation of land in development plans, and the control (by means of planning permission) of mineral workings.

Objectives of Sustainable Development for Minerals Planning (1994)

1 To conserve minerals as far as possible, whilst ensuring an adequate supply to meet the needs of society for minerals;

2 to minimise production of waste and to encourage efficient use of materials, including appropriate use of high quality materials, and recycling of wastes;

3 to encourage sensitive working practices during minerals extraction and to preserve or enhance the overall quality of the environment once extraction has ceased;

4 to protect areas of designated landscape or nature conservation from development, other than in exceptional circumstances where it has been demonstrated that development is in the public interest.

Previous Statement of Objectives (1988)

(a) To ensure that the needs of society for minerals are satisfied with due regard to the protection of the environment;

(b) to ensure that any environmental damage or loss of amenity caused by mineral operations and ancillary operations is kept at an acceptable level;

(c) to ensure that land taken for mineral operations is reclaimed at the earliest opportunity and is capable of an acceptable use after working has come to an end;

(d) to prevent the unnecessary sterilisation of mineral resources.

(MPG 1, 1994 and 1988)

The MPA has to assess the amount of land required for mineral working, and this requires an assessment of the future demand likely to be made on production in their area. Obviously, this involves extensive and continuing consultation with mineral operators. All MPAs are now required to prepare *minerals plans* (which may be produced jointly with their waste plan).

Powers to control mineral workings stem from the definition of development, which includes 'the carrying out of . . . mining . . . operations in, on, over or under land'. However, a special form of control is necessary to deal with the unique nature of mineral operations. Unlike other types of development, mining operations are not the means by which a new use comes into being; they are a continuing end in themselves, often for a very long time. They do not adapt land for a desired end-use: on the contrary, they are essentially harmful and may make land unfit for any later use. They also have unusual location characteristics: they have to be mined where they exist. For these reasons, the normal planning controls are replaced by a unique set of regulations.

Two major features of the minerals control system are that it takes into account the fact that mineral operations can continue for a long period of time, and that measures are needed to restore that land when operations cease. It is, therefore, necessary for MPAs to have the power to review and modify permissions and to require restoration. Under current legislation, MPAs have a duty to review all mineral sites in their areas. This includes those which were 'grandfathered' in by the 1947 Act. These old sites, of which there may be around a thousand in England and Wales, often lack adequate records. They present the particular problem that they can include large unworked extensions which are covered by the permission; if worked, these could have serious adverse effects on the environment. The provisions relating to these sites are even more complicated than those pertaining to the generality of mineral operations, and they have been significantly altered by the 1995 Environment Act. Details are set out in MPG 14.

Policies for restoration (and what the Act quaintly calls 'aftercare') have become progressively more stringent, mainly in response to what the Stevens Report (1976) referred to as a great change in standards and attitudes to mineral exploitation. A lengthy guidance note (MPG 7) fully explains restoration policies and options. In view of the ongoing nature of mineral operations, particular importance is attached to schemes of progressive restoration which are phased in with the gradual working out of the site. (A very effective policy is to make new working dependent upon satisfactory restoration of used sites.) A good idea of the current policy is gained from the following quotation from MPG 7:

> Standards of reclamation have generally improved over recent years. Continuation of this trend will enable a wider range of sites to be restored to appropriate standards, leading to the release of land which has not so far been made available for mineral working. If there is serious doubt whether satisfactory reclamation can be achieved at a particular site, then there must also be a doubt whether permission for mineral working should be given.

Mineral deposits are frequently located on high-grade agricultural land or on sites of particular amenity or attraction. It is not surprising, therefore, that much restoration is for agricultural, amenity and recreational uses.

CARAVANS

During the 1950s, the housing shortage led to a boom in unauthorised caravan sites. The controversy and litigation to which this gave rise led to the introduction of special controls over caravan sites (by Part I of the Caravan Sites and Control of Development Act 1960). This legislation has remained as a separate code and is not consolidated in the Town and Country Planning Act of 1990. (The Caravan Sites Act 1968, which deals mainly with the protection of caravan dwellers and gypsies from eviction, is similarly separate.)

The 1960 Act gave local authorities new powers to control caravan sites, including a requirement that all caravan sites had to be licensed before they could start operating (thus partly closing loopholes in the planning and public health legislation). These controls over caravan sites operate in addition to the normal planning system: thus both planning permission and a licence have to be obtained. Most of the Act dealt with control, but local authorities were given wide powers to provide caravan sites.

Local authorities face strong pressure from their ratepayers 'to preserve local amenities and property values', to which caravans are seen as a threat. The DoE may be clear as to what 'planning policy recognises', but the reality differs considerably from the official statement.

Holiday caravans are subject to the same planning and licensing controls as residential caravans. To ensure that a site is used only for holidays (and not for 'residential purposes'), planning permission can include a condition limiting the use of a site to the holiday season. Conditions may also be imposed to require the caravans to be removed at the end of each season or to require a number of pitches on a site to be reserved for touring caravans.

One group of caravanners is particularly unpopular: gypsies, or, to give them their less romantic description, 'persons of nomadic life, whatever their race or origin' (but excluding 'members of an organised group of travelling showmen, or persons engaged in travelling circuses, travelling together as such'). The basic problem is that no one wants gypsies around: 'all too often the settled community is concerned chiefly to persuade, or even force, the gypsy families to move on'. In an attempt to deal with the problem, a Private Member's Bill was presented by Eric Lubbock and was passed as the Caravan Sites Act 1968. This gave local authorities in England and Wales (but not in Scotland) the duty to provide adequate sites for gypsies 'residing in or resorting to' their areas.

A 1977 report on the operation of the Act prepared by Sir John Cripps (stimulated by sporadic violence on gypsy encampments) underlined the lack of progress. Cripps's message was clear: the living conditions of many gypsies was scandalous, and no improvement in the slow rate of progress could be expected without a high level of commit-

ment by central government. A major element in this was a proposed 100 per cent grant to local authorities of the capital cost of providing gypsy caravan sites. This was accepted and such grants were introduced in 1979. Circular 28/77 clearly conveyed the government policy of the time to give gypsies special protection in the planning system: it even accepted the necessity of establishing gypsy sites in protected areas such as green belts and AONBs. It was anticipated that caravan sites would be located in such protected areas, especially when close to the urban fringe.

The problems, however, persisted; indeed, despite significant increases in the provision of sites and in government expenditure on gypsy site grants (between 1978 and 1994, some £87 million was spent on grant-aid for site provision), the problem persists: in fact it has increased. Whereas in 1965 there were an estimated 3,400 gypsy caravans in England and Wales, by 1992 the number had increased to 13,500. Of these, 4,500 were on unauthorised sites – roughly the same number as in 1981. (The figures are not very reliable, but are sufficient to indicate the scale of the problem – and to justify public concern!) There has been a succession of inquiries and reports, but public opinion has precluded an effective resolution of the problems. Indeed, attitudes have hardened in recent years, and a 1992 consultation paper *Reform of the Caravan Sites Act 1968* heralded a marked shift in policy. The paper lamented the fact that 'the problem has grown faster than its remedy', and that it is now compounded by 'new age travellers' (more popularly known as 'hippies').

Under the new regime, the obligation of local authorities to provide gypsy sites has been abolished ('privatised'?), and central government grants for gypsy caravan sites are being phased out. However, local authorities should 'continue to indicate the regard they have had to meeting gypsies' accommodation needs', with 'broad strategic policies' in structure plans and detailed policies in local plans (DoE Circular 1/94). However, gypsy sites will not be appropriate in green belts or other protected areas.

Significantly, the legislation implementing the

new policy is not of a planning character: it is the Criminal Justice and Public Order Act 1994. In addition to repealing the obligations imposed on local authorities by the Caravan Sites Act, it provides stronger powers to remove 'unauthorised persons', though the DoE Circular espouses a policy of toleration towards gypsies on unauthorised sites.

'The public visibility of gypsies has grown, while the tolerance of the settled community to them has declined' (Home 1993). As a result, it can be expected that appeals relating to gypsy sites will increase. The first appeal decisions suggest that inspectors may pay special attention to personal circumstances (particularly of children and the elderly) in justifying the grant of planning permission for gypsy sites. There may also be complaints to the European Commission of Human Rights: the first case to reach Strasbourg has been held to be admissible. Clearly, the abolition of 'the privileged position of gypsies' will not end this sorry saga: it merely opens a new chapter.

PURCHASE AND BLIGHT NOTICES

A planning refusal does not of itself confer any right to compensation. On the other hand, revocations of planning permission or interference with existing uses do rank for compensation, since they involve the taking away of a legal right. In cases where, as a result of a planning decision, land becomes 'incapable of reasonably beneficial use', the owner can serve a *purchase notice* upon the local authority requiring it to buy the property. In all cases, ministerial confirmation is required. The circumstances in which a purchase notice can be served include:

- refusal or conditional grant of planning permission;
- revocation or modification of planning permission;
- discontinuance of use.

In considering whether the land has any *beneficial use*, 'relevant factors are the physical state of the land, its

size, shape and surroundings, and the general patterns of land-uses in the area; a use of relatively low value may be regarded as reasonably beneficial if such a use is common for similar land in the vicinity' (DoE Circular 13/83).

A purchase notice is not intended to apply in a case where an owner is simply prevented from realising the full potential value of his or her land. This would imply the acceptance in principle of paying compensation for virtually all refusals and conditional permissions. It is only if the existing and permitted uses of the land are so seriously affected as to render the land incapable of reasonably beneficial use that the owner can take advantage of the purchase notice procedure.

There are circumstances, other than the threat of public acquisition, in which planning controls so affect the value of the land to the owner that some means of reducing the hardship is clearly desirable. For example, the allocation of land in a development plan for a school or for a road will probably reduce the value of houses on the land or even make them completely unsaleable. In such cases, the affected owner can serve a *blight notice* on the local authority requiring the purchase of the property at an 'unblighted' price. These provisions are restricted to owner occupiers of houses and small businesses who can show that they have made reasonable attempts to sell their property but have found it impossible to do so except at a substantially depreciated price because of certain defined planning actions. These include land designated for compulsory purchase, or allocated or defined by a development plan for any functions of a government department, local authority or statutory undertaker, and land on which the Secretary of State has given written notice of his intention to provide a trunk road or a *special road* (i.e. a motorway).

The subject of planning blight takes us into the much broader area of the law relating to compensation. This is an extremely complex field, and only an indication of three major provisions can be attempted here.

First, there is a statutory right to compensation for a fall in the value of property arising from the use of highways, aerodromes and other public works which have immunity from actions for *nuisance*. The depreciation has to be caused by physical factors such as noise, fumes, dust and vibration, and the compensation is payable by the authority responsible for the works.

Second, there is a range of powers under the heading 'mitigation of injurious effect of public works'. Examples include: sound insulation; the purchase of owner occupied property which is severely affected by construction work or by the use of a new or improved highway; the erection of physical barriers (such as walls, screens or mounds of earth) on or alongside roads to reduce the effects of traffic noise on people living nearby; the planting of trees and the grassing of areas; and the development or redevelopment of land for the specific purpose of improving the surroundings of a highway 'in a manner desirable by reason of its construction, improvement, existence or use'.

Third, provision is made for *home loss payments* as a mark of recognition of the special hardship created by compulsory dispossession of one's home. Since the payments are for this purpose, they are quite separate from, and are not dependent upon, any right to compensation or the *disturbance payment* which is described below. Logically, they apply to tenants as well as to owner occupiers, and are given for all displacements whether by compulsory purchase or any action under the Housing Acts. These provisions were slightly extended in the 1991 Planning and Compensation Act.

Additionally, there is a general entitlement to a *disturbance payment* for persons who are not entitled to compensation. Local authorities have a duty 'to secure the provision of suitable alternative accommodation where this is not otherwise available on reasonable terms, for any person displaced from residential accommodation' by acquisition, redevelopment, demolition, closing orders and so on.

DEVELOPMENT BY SMALL BUSINESSES

The changing nature of 'town and country planning' is nowhere more apparent than in the importance

attached in recent years, first, to business activity and, later, to environmental issues. DoE Circular 22/80 cancelled relevant previous circulars, and local authorities were 'asked to pick out for priority handling those applications which in their judgement will contribute most to national and local economic activity'. Particular emphasis was laid on small businesses which the government were 'particularly keen to encourage'. Indeed, in striking contrast to earlier ideas about the separation of industry and housing, 'the characteristics of industry and commerce . . . have changed . . . There are many businesses that can be carried on in rural and residential areas without causing unacceptable disturbance . . . The rigid application of 'zoning' policies (where indeed it continues) can have a very damaging effect.'

Moreover, far from 'planning out' non-conforming industry (which was a worthy planning aim in earlier years), such industry 'substantially eases the problems of starting and maintaining small scale businesses if permission can be given for such uses to be established in redundant buildings such as disused agricultural buildings, industrial, warehouse, or commercial premises, on derelict sites, or in unsuitable housing'.

There are many generations of qualified town planners who would have failed their examinations had they suggested such a thing. Yet the circular as published was a considerably milder version of one which was circulated among local authority associations in July 1980. This came under heavy fire for implying that small firms that set up without planning permission should have enforcement orders issued against them only if alternative premises were available. A shadow of this remained in a section on enforcement and discontinuance. While stressing that nothing in the circular should be taken as condoning a wilful breach of planning law, there is a highly significant qualification:

the power to issue an enforcement notice alleging that there has been a breach of planning control is discretionary and is only to be used if the authority 'consider it expedient to do so having regard to the provision of the development plan and to any other material considerations'. This permissive power should be used, in regard to either operational development or material changes of use, only where planning reasons clearly warrant such action, and there is no alternative to enforcement proceedings. Where the activity involved is one which would not give rise to insuperable planning objections if it were carried out somewhere else, then the planning authority should do all it can to help in finding suitable alternative premises before initiating enforcement action.

Similarly, 'but with even more force', discontinuance orders were appropriate 'only if there appears to be an overriding justification on planning grounds'.

The theme was developed in later circulars and in the original 1988 version of PPG 4; but by the beginning of the 1990s, environmental issues had risen to prominence, and rather less emphasis was given to business. The revised PPG 4, issued in 1992, opens as follows:

> One of the government's key aims is to encourage continued economic development in a way which is compatible with its stated environmental objectives. Economic growth and a high quality environment have to be pursued together.

Development control, so the argument continued, should not place 'unjustifiable obstacles' in the way of development, but 'nevertheless, planning decisions must reconcile necessary development with environmental protection and other development plan policies'.

The point does not need labouring, though it is an important one: planning policies reflect the political concerns of the day. Economic development is still a priority, but it is now modified by an increasing regard for the quality of the environment.

EFFICIENCY IN DEVELOPMENT CONTROL

There has been a succession of attempts on the part of central government to 'streamline the planning process' and to make it more 'efficient'. The reasons for these have differed. In 1981, government concern was with the economic costs of control, with cutting public expenditure and with 'freeing' private initiative from unnecessary bureaucratic controls. In the early 1970s, the concern was with the enormous increase in planning applications and planning appeals which, of course, stemmed from the property boom of the period. This is very clearly illustrated in Figure 5.2 which shows the sharp decline in both applications and appeals in recent years. The resultant delay created a political situation which was dealt with in traditional style by setting up an inquiry. This was undertaken by George Dobry QC. His report is now part of planning history, but it is live history: the issues are still very much with us in the 1990s, and there is no guarantee that the current resolution of them will prove sufficiently resilient to withstand the unpredictable changes in the context within which they operate. It is therefore useful to look briefly at Dobry's analysis.

The starting point for Dobry's inquiry was the lengthening delay in the processing of planning applications, but he was quick to point out that 'not all delay is unacceptable: it is the price we must pay for the democratic planning of the environment'. Moreover, his review took account of factors which were very different from those relevant to 'streamlining the planning machine': the increasing pressure for public consultation and participation in the planning process; and the 'dissatisfaction on the part of applicants because they often do not understand why particular decisions have been made, or why it is necessary for what may seem small matters to be the concern of the planning machinery at all'. Additionally, he noted that 'many people feel that the system has not done enough to protect what is good in an environment or to ensure that new development is of a sufficiently high quality'.

Dobry therefore had a difficult task of reconciling apparently irreconcilable objectives: to expedite planning procedures while at the same time facilitating greater public participation and devising a system which would produce better environmental results. His solutions attempted to provide more speed for developers, more participation for the public *and* better-quality development and conservation. This was to be effected by the division of applica-

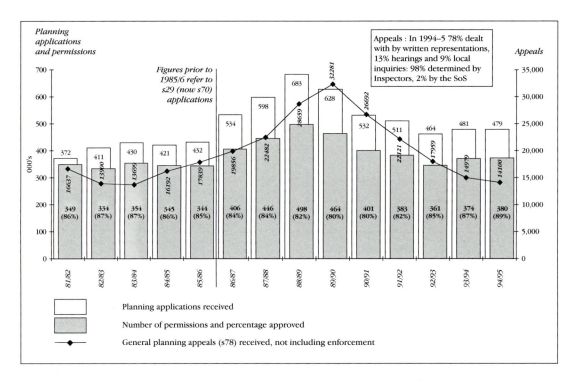

Figure 5.2 Planning Applications, Appeals and Decisions, England, 1981–2 to 1994–5

tions into minor and major. Despite the inherent difficulties of determining this in advance (at least to the satisfaction of the public and the local amenity societies), it is nevertheless a fact that some 80 per cent of all applications are granted, and that many of these *are* simple and straightforward. Dobry's proposal, in essence, was that simple applications should be distinguished and treated expeditiously by officials, though with the opportunity for some participation and with a safety channel to allow them to be transferred to the major category if this should prove appropriate.

This, so Dobry believed, would relieve the overloaded planning machine to deal more thoroughly with the major and/or controversial applications (which he suggested would constitute less than a half and, hopefully, only a third of the total).

Dobry's scheme was an heroic attempt to improve the planning control system to everyone's satisfaction (Jowell 1975). Inevitably, therefore, it disappointed everybody. For example, though he made a number of proposals to increase public participation, his overriding concern for expediting procedures forced him to compress 'simple' applications into an impracticable time scale.

The Dobry inquiry was instigated by a Conservative government at a time when the property market was booming. On its completion the boom had collapsed and a Labour government had published their outline proposals for the community land legislation. Thus, the planning scene had changed fundamentally. The government rejected all Dobry's major recommendations for changes in the system, though it was stressed that their objectives could typically be achieved if local authorities adopted efficient working methods. Dobry's view that 'it is not so much the system which is wrong but the way in which it is used' was endorsed, and his *Final Report* (1975) was commended 'to students of our planning system as an invaluable compendium of information about the working of the existing development control process, and to local authorities and developers as a source of advice on the best way to operate within it'.

The Conservative government which was elected in 1979 lost no time in preparing a revised development control policy. A draft circular was sent out for comment in mid-1980. This created alarm among the planning profession, partly because of its substantive proposals but also partly because of its abrasive style. 'The Most Savage Attack Yet' expostulated *Municipal Engineering*, while *Planner News* remonstrated that the results of the circular 'could be disastrous'.

An example of the matters to which objection was taken was the call for relaxation of controls over private sector housing: 'Local authorities should not lay down requirements on the mix of house types, provision of garages, internal standards, sizes of private gardens, location of houses on plots and in relation to each other, provision of private open space.'

The circular also stated that planning authorities should not attempt to compel developers to adopt designs which were unpopular with customers or clients 'and they shouldn't attempt to control such details as shapes of windows or doors or the colour of bricks'.

The revised circular, as published (22/80), was written with a gentler touch, but much of the message was very similar. The emphasis was on securing a 'speeding up of the system' and on ensuring that 'development is only prevented or restricted when this serves a clear planning purpose and the economic effects have been taken into account'.

Regular publication of performance figures (the percentage of planning applications decided within eight weeks) became the standard by which the efficiency of the development control system is measured. Quarterly figures have been published since 1979, and are used by both the government and the development industry to bolster criticisms of the system.

The policy 'to simplify the system and improve its efficiency' (to use the words of the 1985 White

Paper, *Lifting the Burden*) continued with revised circulars, new White Papers, and the introduction of planning mechanisms which reduced or bypassed local government control, such as simplified planning zones and urban development corporations. However, towards the end of the 1980s, a greater emphasis on 'quality' emerged as environmental awareness and concern increased. A particularly striking example is provided by the 1992 Audit Commission report on development control, significantly entitled *Building in Quality*. Though the major emphasis is still on the process of planning control rather than its outcome, there is a very clear recognition of the importance of the latter. It is noted that there had been a preoccupation with the speed of processing planning applications 'ignoring the mix of applications, the variety of development control functions, and the quality of outcomes'. But there had been no 'shared and explicit' concept of quality, yet 'The quality of outcomes is more important than the quality of the process because buildings will be seen long after memories of the decision process have lapsed, but it is far harder to assess.' Quality of development control is seen as involving an 'adding of value' by the local authority. What that 'added value' may be is dependent upon the authority's overall objectives: 'in areas under heavy development pressure or in rural areas, environmental, traffic, or ecological considerations may be paramount'; in Wales, 'the impact of the development on the Welsh language can be a consideration'.

This is a far cry from the character of the pronouncements made earlier by government spokespersons. At the least, it is refreshing to see a departure from the crude 'eight weeks' yardstick of quality.

GOOD DESIGN

The national heritage is created in part by good design. Much of the built heritage is worth preserving because it is well designed. It is therefore of more than contemporary concern that new buildings should be well designed. Nevertheless, the extent to which 'good' design can be fostered by the planning system (or any other system) is problematic.

Good design is an elusive quality which cannot easily be defined. In the words of Sir William Holford (1953), 'design cannot be taught by correspondence; words are inadequate, and being inadequate may then become misleading, or even dangerous. For the competent designer a handbook on design is unnecessary, and for the incompetent it is almost useless as a medium of instruction.' Yet local authorities have to pass judgement on the design merits of thousands of planning proposals each year, and there is continuous pressure from professional bodies for higher design standards to be imposed.

There is a long and inconclusive history to design control (well set out by John Punter, in various publications between 1985 and 1994). A 1959 statement by the MHLG stressed that it was impossible to lay down rules to define good design. Developers were recommended to seek the advice of an architect (presumably a good one!). The policy should be to avoid stifling initiative or experiment in design, but 'shoddy or badly proportioned or out of place designs' should be rejected – with clear reasons being given.

The reader is referred to Punter's work for the fascinating details of the continuing story, recounting the personal achievements of Duncan Sandys, particularly in founding the Civic Trust in 1957, and later in promoting the Civic Amenities Act; the high buildings controversy ('sunlight equals health'); the problem of protecting views of St Paul's Cathedral; the arguments over the Shell Tower (which prompted the quip that the best view of the Shell Tower was to be obtained from its roof); the publication of Worskett's *The Character of Towns* (1969); the unpublished Matthew-Skillington Report on *Promotion of High Standards of Architectural Design* which led to the appointment of a Chief Architect in the Property Services Agency; the property boom and a spate of books bearing titles such as *The Rape of Britain* (Amery and Cruikshank 1975) and *The Sack of Bath* (Fergusson 1973); the Essex County Council *Design Guide for Residential Areas*

(1973) – 'the most influential local planning authority publication ever'; the attempt (in 1978) to prevent the building of the National Westminster Tower; and so on.

New contexts emerged with the return of the Conservative government in 1979, and there was the unprincely attack in 1984 by Prince Charles on the 'monstrous carbuncle' of the proposed extension to the National Gallery, and the 'giant glass stump' of the Mies van der Rohe office block adjacent to the Mansion House (described by one architect – who will not be named – as a building which would be 'unsurpassed in elegance and economy of form'). The Prince followed up his criticisms with *A Vision of Britain* (1989): 'a personal view of architecture', spelling out, with telling illustrations, how 'we can do better'.

In his case study of office development control in Reading, Punter (1985) demonstrates the interesting point that it is only since the late 1970s that the local authority 'have begun to influence the full aesthetic impact of office buildings, though they have controlled height, floorspace and functional considerations since 1947'. Moreover:

> Aesthetic considerations do not operate in a vacuum: they are merely one set of considerations amongst many in deciding whether a development gets planning permission. In the case of office development, despite its visual impact, the control of floorspace and the provision of associated facilities and land uses have been higher order goals in Reading . . . Aesthetic considerations are inevitably the first to be sacrificed in the cause of 'speed and efficiency' in decision-making, by clients, developers, architects and planners . . . There is a lack of design and architectural skills within the control section of the planning authority, but while its presence would strengthen the planning effort, its scope would be severely constrained by wider policy constraints, by general manpower shortages and, most of all perhaps, by the relevant circulars and the appeal process.

This dismal conclusion is corroborated by the studies of Booth and Beer (1983). They found that, although nearly two-thirds of all permissions granted 'carried conditions that were intended to modify the landscape design, layout and architec-

Superquarries to Supply South-East England Aggregate Requirements, and (1994) *Coastal Superquarries: Options for Wharf Facilities on the Lower Thames*. On oil and gas, see DoE Circular 2/85, *Planning Control over Oil and Gas Operations*. On coal, see MPG 3 *Coal Mining and Colliery Spoil Disposal*, 1994; and CPRE (1994e) *Coal Mining and Colliery Spoil Disposal: Revision of Minerals Planning Guidance Note 3*. An excellent up-to-date review of mineral resource planning and sustainability is given by Owens and Colwell (1996) *Rocks and Hard Places*.

Caravans and Gypsies

The appalling living conditions of gypsies, combined with public attitudes towards them, have presented difficult political problems for governments: see Adams *et al.* (1976) *Gypsies and Government Policy in Britain*, and Gentleman (1993) *Counting Travellers in Scotland*. The main official reference is DoE Circular 1/94 *Gypsy Sites and Planning*. Cases coming before the European Commission on Human Rights are reported in the *Journal of Planning and Environment Law*: see *JPL* 1994: 536, and 1995: 633.

Development by Small Businesses

A research report by Roger Tym & Partners (1989a) deals with *The Effect on Small Firms of Refusals of Planning Permission*.

Efficiency in Development Control

The 1992 Audit Commission report, *Building in Quality: A Study of Development Control*, criticises the emphasis which is placed on the speed of the decision-making process in development control.

Design

An annex to PPG 1 deals with 'design considerations'. Punter's work is notable in this field. See his 1990 book *Design Control in Bristol 1940–1990*; two earlier articles in the *TPR* (1986–7) are perhaps

more readily accessible: 'A history of aesthetic control: the control of the external appearance of development in England and Wales, 1986 and 1997. See also Booth and Beer (1983) 'Development control and design quality'. There is a huge library on particular design issues. See, for example, Countryside Commission (1993) *Design in the Countryside*; DoE (1994) *Quality in Town and Country* (Discussion Paper); and (1995) *Quality in Town and Country: Urban Design Campaign*; National Audit Office (1994) *Environmental Factors in Road Planning and Design*; NIDoE (1994) *A Design Guide for Rural Northern Ireland*; Hillman (1990) *Planning for Beauty: The Case for Design Guidelines*; Scottish Office (1994) *Fitting New Housing Development into the Landscape* (PAN 44); and Bishop (1994) 'Planning for better rural design'. See also Owen (1991) *Planning Settlements Naturally* and Barton *et al.* (1995) *Sustainable Settlements*.

Amenity

For a discussion of statutory provisions in relation to 'amenity', see Sheail (1992) 'The *amenity* clause', and also an unusual historical study of the development of the notion of amenity in Millichap (1995a) 'Law, myth and community: a reinterpretation of planning's justification and rationale'.

Plan-led Development

The introduction of section 54A (s.18A of the Scottish Act) has given rise to much speculation which only time will settle. (Note that there is no equivalent in the Northern Ireland legislation.) A review of the issues is given by MacGregor and Ross (1995) 'Master or servant? The changing role of the development plan in the British planning system'. See also Purdue (1991) 'Green belts and the presumption in favour of development'; Gatenby and Williams (1992) 'Section 54A: the legal and practical implications'; M. Harrison (1992) 'A presumption in favour of planning permission?'; Tewdwr-Jones (1994) 'Policy implications of the plan-led system; and Herbert-Young (1995) 'Reflections on section 54A and *plan-led* decision-making'.

6

LAND POLICIES

It is clear that under a system of well-conceived planning the resolution of competing claims and the allocation of land for the various requirements must proceed on the basis of selecting the most suitable land for the particular purpose, irrespective of the existing values which may attach to the individual parcels of land.

Uthwatt Report 1942

THE UTHWATT REPORT

It was the task of the Uthwatt Committee, from whose report this quotation is taken, to devise a scheme which would make this possible. Effective planning necessarily controls, limits, or even completely destroys, the market value of particular pieces of land. Is the owner to be compensated for this loss in value? If so, how is the compensation to be calculated? And is any 'balancing' payment to be extracted from owners whose land appreciates in value as a result of planning measures?

This problem of compensation and betterment arises fundamentally 'from the existing legal position with regard to the use of land, which attempts largely to preserve, in a highly developed economy, the purely individualistic approach to land ownership'. This 'individualistic approach', however, has been increasingly modified during the past hundred years. The rights of ownership were restricted in the interests of public health: owners had (by law) to ensure, for example, that their properties were in good sanitary condition, that new buildings conformed to certain building standards, that streets were of a minimum width, and so on. It was accepted that these restrictions were necessary in

the interests of the community: *salus populi est suprema lex*, and that private owners should be compelled to comply with them even at cost to themselves:

> All these restrictions, whether carrying a right to compensation or not, are imposed in the public interest, and the essence of the compensation problem as regards the imposition of restrictions appears to be this – at what point does the public interest become such that a private individual ought to be compelled to comply, at his own cost, with a restriction or requirement designed to secure that public interest? The history of the imposition of obligations without compensation has been to push that point progressively further on and to add to the list of requirements considered to be essential to the well-being of the community.
>
> (Uthwatt Report, para. 33)

But clearly there is a point beyond which restrictions cannot reasonably be imposed on the grounds of good neighbourliness without payment of compensation – and 'general consideration of regional or national policy require so great a restriction on the landowner's use of his land as to amount to a taking away from him of a proprietary interest in the land'.

This, however, is not the end of the matter. Planning sets out to achieve a selection of the most suitable pieces of land for particular uses. Some land will therefore be zoned for a use which is profitable for the owner, whereas other land will be zoned for a use with a low, or even nil, private value. It is this difficulty of *development value* which raises the compensation problem in its most acute form. The expectations (or hopes) of owners extend over a far larger area than is likely to be developed. This *potential* development value is therefore speculative, but until the individual owners are proved to be wrong in their assessments (and how can this be done?) all owners of land with a potential value can make a case for compensation on the assumption that their particular pieces of land would in fact be chosen for development if planning restrictions were not imposed. Yet this *floating value* might never have settled on their land, and obviously the aggregate of the values claimed by the individual owners is likely to be greatly in excess of a total valuation of all pieces of land. As Haar (1951: 99) has nicely put it, the situation is akin to that of a sweepstake: a single ticket fetches much more than its mathematically calculated value, for the simple reason that the grand prize may fall to any one holder.

Furthermore, the public control of land use necessarily involves the shifting of land values from certain pieces of land to other pieces: the value of some land is decreased, while that of other land is increased. Planning controls, so it was argued, do not destroy land values: in the words of the Uthwatt Committee, 'neither the total demand for development nor its average annual rate is materially affected, if at all, by planning ordinances'. Nevertheless, the owner of the land on which development is prohibited will claim compensation for the full potential development of his land, irrespective of the fact that the value may shift to another site.

In theory, it is logical to balance the compensation paid to aggrieved owners by collecting a betterment charge on owners who benefit from planning controls (Hagman and Misczynski 1978). But previous experience with the collection of betterment had not been encouraging. The principle was first established in an Act of 1662 which authorised the levying of a capital sum or an annual rent in respect of the 'melioration' of properties following street widenings in London. There were similar provisions in Acts providing for the rebuilding of London after the Great Fire. The principle was revived and extended in the Planning Acts of 1909 and 1932. These allowed a local authority to claim, first, 50 per cent, and then (in the later Act), 75 per cent of the amount by which any property increased in value as the result of the operation of a planning scheme. In fact, these provisions were largely ineffective since it proved extremely difficult to determine with any certainty which properties had increased in value as a result of a scheme or, where there was a reasonable degree of certainty, how much of the increase in value was directly attributable to the scheme and how much to other factors. The Uthwatt Committee noted that there were only three cases in which betterment had actually been paid under the planning acts.

The Uthwatt Committee concluded that the solution to these problems lay in changing the system of land ownership under which land had a development value dependent upon the prospects of its profitable use. They maintained that no new code for the assessment of compensation or the collection of betterment would be adequate if this 'individualistic' system remained. The system itself had inherent 'contradictions provoking a conflict between private and public interest and hindering the proper operation of the planning machinery'. A new system was needed which would avoid these contradictions and which so unified existing rights in land as to 'enable shifts of value to operate within the same ownership'.

The logic of this line of reasoning led to a consideration of land nationalisation. This the Committee rejected on the grounds that it would arouse keen political controversy, would involve insuperable financial problems, and would necessitate the establishment of a complicated national administrative machinery. In their view the solution to the problem lay in the nationalisation, not of the land itself, but of all development rights in undeveloped land.

THE 1947 ACT

Essentially, this is precisely what the 1947 Town and Country Planning Act did. Effectively, development rights and their associated values were nationalised. No development was to take place without permission from the local planning authority. If permission were refused, no compensation would be paid (except in a limited range of special cases). If permission were granted, any resulting increase in land value was to be subject to a development charge. The view was taken that 'owners who lose development value as a result of the passing of the Bill are not on that account entitled to compensation'. This cut through the insoluble problem posed in previous attempts to collect betterment values created by public action. Betterment had been conceived as 'any increase in the value of land (including the buildings thereon) arising from central or local government action, whether positive, for example by the execution of public works or improvements, or negative, for example by the imposition of restrictions on the other land'. The 1947 Act went further: all betterment was created by the community, and it was unreal and undesirable (as well as virtually impossible) to distinguish between values created, for example, by particular planning schemes and those due to other factors such as the general activities of the community or the general level of prosperity.

If rigorous logic had been followed, no payment at all would have been made for the transfer of development value to the state but this, as the Uthwatt Committee had pointed out, would result in considerable hardship in individual cases. A £300 million fund was therefore established for making 'payments' (as distinct from 'compensation') to owners who could successfully claim that their land had some development value on the appointed day – the day on which the provisions of the Bill which prevented landowners from realising development values came into force. Considerable discussion took place during the passage of the Bill through Parliament on the sum fixed for the payments, and it was strongly opposed on the ground that it was too

small. The truth of the matter was that, in the absence of relevant reliable information, any global sum had to be determined in a somewhat arbitrary way; but in any case it was not intended that everybody should be paid the full value of their claims. Landowners would submit claims to a centralised agency, the Central Land Board, for *loss of development value*; that is, the difference between the *unrestricted value* (the market value without the restrictions introduced by the Act) and the *existing use value* (the value subject to these restrictions). When all the claims had been received and examined, the £300 million would be divided between claimants at whatever proportion of their 1948 value that total would allow. (In the event the estimate of £300 million was not as far out as critics feared: the total of all claims finally amounted to £380 million.)

The original intention was to have a flexible rate of development charge. In some cases 100 per cent would be levied, but in others a lower rate would be more appropriate in order to encourage development. However, when the regulations came to be made, the government maintained that the policy which had been set out during the passage of the Bill through the House was unworkable. The only explanation given for this was that:

> The whole conception is that the value of land is divided into two parts – the value restricted to its existing use and the development value. The market value is the sum of the two. If, by the action of the State, the development value is no longer in the possession of the owner of the land, then all he has left is the existing use value. Moreover, the fund of £300 million is being provided for the purposes of compensating the owner of land for this reduced value . . . therefore the owner of land can have no possible claim to any part of the development value and it is logical and right that the State should, where development takes place, make a charge which represents the amount of the development value.

The very idea of variable development charges was rejected, and a flat-rate 100 per cent levy introduced.

These provisions, of which only the barest summary has been given here, were very complex and, together with the inevitable uncertainty as to when compensation would be paid and how much it would be, resulted in a general feeling of uncertainty and discontent which did not augur well for the scheme. The principles, however, were clear. To recapitulate, all development rights and values were vested in the state: no development could take place without permission from the local planning authority and then only on payment of a betterment charge to the Central Land Board. The nationalisation of development rights was effected by the 'promised' payments in lieu of compensation. As a result, land-owners only owned the existing use rights of their land and it thus followed, first, that if permission to develop was refused no compensation was payable, and, second, that the price paid to public authorities for the compulsory acquisition of land would be equal to the existing use value, that is, its value excluding any allowance for future development.

THE 1947 SCHEME IN OPERATION

The scheme did not work as smoothly as was expected. In their first annual report, the Central Land Board 'noted with concern some weeks after the Act came into operation that, despite the liability for development charge, land was being widely offered and, still worse, taken at prices including the full development value'. This remained a problem throughout the lifetime of the scheme, though the magnitude of the problem still remains a matter of some controversy. Nevertheless, it is clear that the conditions were such that developers were prepared to pay more than existing use prices for land. This was largely due to the severe restrictions which were imposed on building. Building licences were very scarce, and developers who were able to obtain them were willing to pay a high price for land upon which to build.

It was to prevent such problems that the Central Land Board had been given powers of compulsory purchase at the 'correct' price. These powers were used, not as a general means of facilitating the supply of land at existing use prices, but selectively, 'as a warning to owners of land in general'. (An example given in the 1949–50 report of the Central Land Board was a plot of land offered for private purchase at £300 and compulsorily acquired by the Board for £10; it was resold, inclusive of development charge for the erection of a house, at £180.) These compulsory purchase powers were used only where an owner had actually offered his land for sale at a price above existing use value. Thus, purchase by the Board would have done nothing to facilitate an increase in the total supply of land for development even if cases had been much more numerous. In fact, however, their very rarity served only to make the procedure arbitrary in the extreme and, indeed, may have added to the reluctance of owners to offer land for development at all.

The Conservative government which took office in 1951 was intent on raising the level of construction activity and particularly the rate of private housebuilding. Though, within the limits of building activity set by the Labour government, it is unlikely that the development charge procedure seriously affected the supply of land, it is probable that the Conservative government's plans for private building would have been jeopardised by it. This was one factor which led the new government to consider repealing development charges. There is no doubt that these charges were unpopular, particularly since they were payable in cash and in full, whereas payment on the claims on the £300 million fund were deferred and uncertain in amount. The position was slightly eased after the announcement that the Central Land Board would accept claims as security for the charge of up to 80 per cent of their agreed value. However, the basic difficulty remained: purchasers of land were compelled to pay a premium above the existing-use value in order to persuade an owner to sell. A development charge of 100 per cent therefore constituted a permanent addition to the cost of development.

Further problems began to loom ahead as the final date for payments from the £300 million fund (1 July 1953) drew near. First, the payment of this sum of money over a short period would have a considerable inflationary effect. Second, all claimants on the fund would receive payment whether or not they had actually suffered any loss as a result of the 1947 Act. (Some would have already recovered the development value of the land by selling at a high price; others may never have wished to develop their land, and, indeed, might even have bought it for the express purpose of preventing its development.) But the main difficulty was that if compensation were paid out on this 'once for all' basis, 'it would be exceedingly difficult for any future government ever to make radical changes in the financial provisions, however badly they were working. For all the holders of claims on the fund would have to be compensated for loss of development value – those who will be allowed to develop their land as well as those who will not.' (This and the following quotations are from the 1952 White Paper *Town and Country Planning Act 1947: Amendment of the Financial Provisions*, Cmd 8699.)

Some amendment of the 1947 Act was clearly desirable, but though there might have been agreement on this, there was no agreement on what the amendments should be. There was a real fear that an amendment which satisfied developers would seriously weaken or even wreck the planning machine: the scheme was part of a complex of planning controls which might easily be upset and result in a return to the very problems which the 1947 Act was designed to solve.

Various proposals were canvassed, but the most popular was a reduction in the rate of development charge. The intention was to provide an incentive to owners to sell their land at a price which took account of the developer's liability to pay the (reduced) charge. The government took the view that this was not possible: 'vendors of land, like vendors of any other commodity, will always get the best price that they can, and the development charge, however small, would in effect be passed on, in whole or in part, to the ultimate user of the land'. Furthermore, the government's objective was not merely one of easing the market in land: it was particularly concerned to encourage more private development, and even a low rate of development charge would act as a brake. On the (implicit) assumption that market prices for land would rise,

the time would inevitably come when the charge would begin to exceed greatly the corresponding claim on the £300 million fund. Finally, it was felt that once the rate of development charge was reduced there would be no clear principle as to the level at which it should continue to be levied – 'the process of reduction, once begun, would be difficult to stop'. In short, the government held that the financial provisions of the 1947 Act were inherently unsatisfactory and could not be sufficiently improved by a mere modification: what was needed was a complete abolition of development charges.

THE 1954 SCHEME

The abolition of development charges was made on the ground that they had proved 'too unreliable an instrument to act as the lynch-pin of a permanent settlement'. But, at the same time, if the main part of the planning system was to remain, some limit to the liability to compensation for planning restrictions was essential. Otherwise effective planning controls would be prohibitively expensive: the cost of compensation for restrictions, if paid at the market value, would be crippling. The solution arrived at was to compensate only 'for loss of development value which accrued in the past up to the point where the 1947 axe fell – but not for loss of development value accruing in the future'.

There were some clear advantages in this scheme: not only was the state's liability for compensation limited, but it was to be paid only if and when the owner of land suffered from planning restrictions. The compensation would be the *admitted claim* on the £300 million fund (plus one-seventh for accrued interest on the amount of the claim). But not all admitted claims were to be met, even where loss of development value was caused by refusal of planning permission or by conditions attached to a permission. The 1932 Act had clearly established the principle that compensation should not be paid for restrictions imposed in the interest of 'good neighbourliness' and this principle was extended. No compensation was payable for refusal to allow a change in the use of a building; or for restrictions regarding density, layout, construction, design and so on; or for refusal to permit development which would place an undue burden on the community (for example, in the provision of services). Some of these matters clearly fall within the 'good neighbour' concept, while others are based on the principle that compensation is not to be paid merely because maximum exploitation has been prevented, so long as development of a reasonably remunerative character is allowed.

The 1954 scheme did not put anything in place of the development charge: the collection of betterment was now left to the blunt instruments of general taxation. Hence the attempt to 'hold the scales evenly between those who were allowed to develop their land and those who were not' was abandoned, but the use of 1947 development values as a 'permanent basis for compensation' safeguarded the public purse. But this created a dual market in land. Compensation both for planning restrictions (in cases where a claim had been admitted) and for compulsory purchase by public authorities was to be paid on the basis of existing use plus any admitted 1947 development value, but private sales would be at current market prices. The difference between these two values might be very substantial, particularly where development took place of a far more valuable character than had been anticipated in 1947. Furthermore, with the passage of time land values generally would increase, especially if inflation continued. Whatever theoretical justification there might be for a dual market it would appear increasingly unjust.

Moreover, there is a real distinction between the hardships inflicted by a refusal of planning permission (that is, the loss of the development value of land) and that caused by the loss of the land itself (that is, compulsory purchase). In the first case, the owner retains the existing use value of his land and is worse off only in comparison with owners who have been fortunate in owning land on which development is permitted and who can therefore realise a capital gain. But in the second case, compulsory acquisition at less than market price involves an actual loss since the owner is not only deprived of his property, he is also compensated at a price which might be less than he paid for it and which would almost certainly be insufficient to purchase a similar parcel of land in the open market.

To recapitulate, the effect of the complicated network of legislation which was now (1954) in force was basically to create two values for land according to whether it was sold in the open market or acquired by a public authority. In the former case, there were no restrictions and thus land changed hands at the full market price; but in the latter case, the public authority would pay only the existing (1947) use value plus any agreed claim for loss of 1947 development value. This was a most unsatisfactory outcome. As land prices increased, due partly to planning controls, the gap between existing use and market values widened, particularly in suburban areas near green belt land. The greater the amount of planning control, the greater did the gap become. Thus, owners who were forced to sell their land to public authorities considered themselves to be very badly treated in comparison with those who were able to sell at the enhanced prices resulting in part from planning restrictions on other sites. The inherent uncertainties of future public acquisitions – no plan can be so definite and inflexible as to determine which sites will (or might) be needed in the future for public purposes – made this distinction appear arbitrary and unjust. The abolition of the development charge served to increase the inequity.

The contradictions and anomalies in the 1954 scheme were obvious. It was only a matter of time before public opinion demanded further amending legislation.

THE 1959 ACT: THE RETURN TO MARKET VALUE

Opposition to this state of affairs increased with the growth of private pressures for development following the abolition of building licences. Eventually the government was forced to take action. The resulting legislation (the Town and Country Planning Act

1959) restored *fair market price* as the basis of compensation for compulsory acquisition. This, in the government's view, was the only practicable way of rectifying the injustices of the dual market for land. An owner now obtained (in theory at least) the same price for his land irrespective of whether he sold it to a private individual or to a public authority.

These provisions thus removed a source of grievance, but they did nothing towards solving the fundamental problems of compensation and betterment, and the result proved extremely costly to public authorities. If this had been a reflection of basic principles of justice there could have been little cause for complaint but, in fact, an examination of the position shows clearly that this was not the case.

In the first place, the 1959 Act (like previous legislation) accepted the principle that development rights should be vested in the state. This followed from the fact that no compensation was payable for the loss of development value in cases where planning permission was refused. But if development rights belong to the state, surely so should the associated development values. Consider, for example, the case of two owners of agricultural land on the periphery of a town, both of whom applied for planning permission to develop for housing purposes – the first being given permission and the second refused on the ground that the site in question was to form part of a green belt. The former benefited from the full market value of his site in residential use, whereas the latter could benefit only from its existing use value. No question of compensation arises since the development rights already belonged to the state, but the first owner had these given back to him without payment. There was an obvious injustice here which could have led eventually to a demand that the 'penalised' owner should be compensated.

Second, as has already been stressed, the comprehensive nature of the planning system has a marked effect on values. The use for which planning permission has been, or will be, given is a very important factor in the determination of value. Furthermore, the value of a given site is increased not only by the

development permitted on that site, but also by the development not permitted on other sites. In the example given above, for instance, the value of the site for which planning permission for housing development was given might be increased by virtue of the fact that it was refused on the second site.

THE LAND COMMISSION 1967–71

Mounting criticism of the inadequacy of the 1959 Act led to a number of proposals for a tax on betterment, by way either of a capital gains tax or of a betterment levy. The Labour government which was returned to power in 1964 introduced both. The 1967 Finance Act introduced a capital gains tax and the 1967 Land Commission Act introduced a new betterment levy. Broadly, the distinguishing principle was that capital gains tax was charged on increases in the current use value of land only, while betterment levy was charged on increases in development value. Though the Land Commission was abolished by the Conservative government in 1971, a summary account of its powers and operations is appropriate. The rationale underlying the Land Commission Act was set out in a 1965 White Paper (Cmnd 2771):

> In the government's view it is wrong that planning decisions about land use should so often result in the realising of unearned increments by the owners of the land to which they apply, and that desirable development should be frustrated by owners withholding their land in the hope of higher prices. The two main objectives of the government's land policy are, therefore:
>
> (i) to secure that the right land is available at the right time for the implementation of national, regional and local plans;
> (ii) to secure that a substantial part of the development value created by the community returns to the community and that the burden of the cost of land for essential purposes is reduced.

To enable these two objectives to be achieved, a Land Commission was established (with headquarters located at Newcastle upon Tyne, in line with the dispersal-of-offices policy). The Commission could buy land either by agreement or compulsorily, and it was given very wide powers for this purpose. The second objective was met by the introduction of a betterment levy on development value. This was necessary not only to secure that a substantial part of the development 'returned to the community', but also to prevent a two-price system as existed under the 1954 Act. The levy was deducted from the price paid by the Commission on its own purchases and was paid by owners when they sold land privately. A landowner thus theoretically received the same amount for his land whether he sold it privately, to the Land Commission, or to another public authority.

Though the Commission could buy by agreement, it had to have effective powers of compulsory purchase if it was 'to ensure that the right land is made available at the right time'. There were two reasons for this. First, though the levy was at a rate (initially 40 per cent) thought to be adequate to leave enough of the development value to provide 'a reasonable incentive', some owners of land might still be unwilling to sell. Second, though the net price obtained by the owner of land should have been the same irrespective of whether the body to whom it was sold was private or public, some owners might have been unwilling to sell to the Commission.

The Act provided two sets of compulsory powers. One was the normal powers available to local authorities, with the usual machinery for appeals and a public inquiry. The second set of compulsory powers were not to become operative until the 'second appointed day' and were to be brought into effect only if it appeared 'that it is necessary in the public interest to enable the Commission to obtain authority for the compulsory acquisition of land by a simplified procedure'. They were intended to provide a rapid procedure under which objectors would have no right to state their case at a public inquiry, and the Commission was not required to disclose the purpose for which the land was needed. The purpose here was to deal quickly and effectively with landowners who were holding up development. In fact,

they did not become operative during the lifetime of the Commission.

The levy differed from the development charge of the 1947 Act in two important ways. First, it did not take all the development value. The Act did not specify what the rate was to be, but it was made clear that the initial rate of 40 per cent would be increased to 45 per cent and then to 50 per cent 'at reasonably short intervals'. (It never was.)

The second difference from the development charge was that, though the levy would normally be paid to the seller, if 'when the land comes to be developed, it still has some development value on which levy has not been taken in previous sales, that residual value will be subject to levy at the time of development'. Thus (ignoring a few complications and qualifications), if a piece of land was worth £500 in its existing use but was sold for £3,500 with planning permission, the levy was applied to the difference, that is, £3,000; the levy, at the initial rate, was £1,200. If, however, the land were sold (at existing-use value plus a 'hope' value that planning permission might be obtained) at £1,000, while the full development value was £3,500, the levy would be paid by both seller and purchaser: £200 by the former and £1,000 by the latter.

The Land Commission's first task was to assess the availability of, and demand for, land for house building, particularly in the areas of greatest pressure. In its first annual report, it pointed to the difficulties in some areas, particularly the South-East and the West Midlands, where the available land was limited to only a few years' supply. Most of this land could not, in fact, be made available for early development. Much of it was in small parcels; some was not suitable for development at all because of physical difficulties; and, of the remainder, a great deal was already in the hands of builders. Thus there was little that could be acquired and developed immediately by those other builders who had an urgent need for land. All this highlighted the need for more land to be allocated by planning authorities for development.

The Land Commission had to work within the framework of the planning system and was subject to the same planning control as private developers. The intention was that the Commission would work harmoniously with local planning authorities and form an important addition to the planning machinery. As the Commission pointed out, despite the sophistication of the British planning system, it was designed to control land use rather than to promote the development of land. The Commission's role was to ensure that land allocated for development was in fact developed, by channelling it to those who would develop it. It could use its powers of compulsory acquisition to amalgamate land which was in separate ownerships and acquire land whose owners could not be traced. It could purchase land from owners who refused to sell for development or from builders who wished to retain it for future development.

In its first report, the Land Commission gently referred to the importance of their role in acting 'as a spur to those local planning authorities whose plans have not kept pace with the demand for various kinds of development'. Though it hoped that planning authorities would allocate sufficient land, it warned that in some cases it might have to take the initiative and, if local authorities refused planning permission, go to appeal. In its second interim report, a much stronger line was taken. It pointed out that, in the pressure areas, it had only modest success in achieving a steady flow of land on to the market. This was largely because these were areas in which planning authorities were aiming to contain urban growth and preserve open country.

In 1969–70, the Land Commission purchased 1,000 acres by agreement and a further 240 acres compulsorily. But the use of these compulsory powers was on the increase, and a further 2,500 acres were subject to compulsory purchase at March 1970.

It is not easy to appraise what success the Land Commission achieved. It was only beginning to get into its stride in 1970 when a new government was returned which was pledged to its abolition on the grounds that it 'had no place in a free society'. This pledge was fulfilled in 1971 and thus the Land Commission went the same way as its predecessor, the Central Land Board.

THE CONSERVATIVE YEARS 1970–4

Land prices were rising during the late 1960s (with an increase of 55 per cent between 1967 and 1970), but the early 1970s witnessed a veritable price explosion. Using 1967 as a base (100), prices rose to 287 in 1972 and 458 in 1973. Average plot prices rose from £908 in 1970 to £2,676 in 1973 (DoE, *Housing and Construction Statistics 1969–1979*, Table 3).

Not surprisingly, considerable pressure was put on the Conservative government to take some action to cope with the problem, though it was neither clear nor agreed what the basic problem was (Hallett 1977: 135). The favourite explanation, however, was 'speculative hoarding', and it was this which became the target for government action (in addition to a series of measures designed to speed up the release and development of land). A White Paper, *Widening the Choice: The Next Steps in Housing* (Cmnd 5280), set out proposals for a *land hoarding charge*. This was to be levied 'for failure to complete development within a specified period from the grant of planning permission'. After this 'completion period' (of four years from the granting of outline planning permission or three years in the case of full planning permission), the charge was to be imposed at an annual rate of 30 per cent of the capital value of the land.

The scheme was clearly a long-term one and, to deal with the urgent problem ('urgent' in political if not in any other terms), a *development gains tax* and a *first letting tax* were introduced.

The development gains tax provided for gains from land sales by individuals to be treated, not as capital gains, but as income (and thus subject to high marginal rates). The first letting tax, as its name implies, was a tax levied on the first letting of shops, offices or industrial premises. In concept, it was an equivalent to the capital gains tax which would have been levied had the building been sold.

Both taxes came into operation at the time when the land and property boom turned into a slump. Indeed, it has been suggested that they contributed to it (Hallett 1977: 137).

THE COMMUNITY LAND SCHEME

The Labour government which was returned to power in March 1974 lost little time in producing its anticipated White Paper *Land* (Cmnd 5730). The objectives of this were 'to enable the community to control the development of land in accordance with its needs and priorities' and 'to restore to the community the increase in value of land arising from its efforts'. The keynote was 'positive planning', which was to be achieved by public ownership of development land. In England and Scotland, the agency for purchasing development land was to be local government (thus avoiding the inter-agency conflict which arose between local authorities and the Land Commission). In Wales, however, with its smaller local authorities, an *ad hoc* agency was to be created (this became the Land Authority for Wales).

In order 'to restore to the community the increase in value of land arising from its efforts', it was proposed that 'the ultimate basis on which the community will buy all land will be current use value'. Sale of the land to developers, on the other hand, would be at market value. Thus all development value would accrue to the community. Provisionally, however, development values were to be recouped by a development land tax.

The ensuing legislation came in two parts: the 1975 Community Land Act provided wide powers for compulsory land acquisition, while the Development Land Tax Act 1976 provided for the taxation of development values. Thus the twin purposes of 'positive planning' and of 'returning development values to the community' were to be served.

The Community Land Scheme was complex, and became increasingly so as regulations, directions and circulars followed the passing of the two Acts. The intention was for it to be phased in gradually, thus enabling programmes to be developed in line with available resources of finance, manpower and expertise.

In the first stage, which started on the 'first appointed day' (6 April 1976), local authorities had a general duty 'to have regard to the desirability of bringing development land into public owner-ship'. In doing so, they had 'to pay particular regard to the location and nature of development necessary to meet the planning needs of their areas'. To assist them in carrying out this role, they had new and wider powers to buy land to make it available for development. Following the passing of the Development Land Tax Act of 1976, all land acquisitions by authorities were made at a price *net* of any tax payable by the sellers of development land.

The second stage was to be introduced as authorities built up resources and expertise. The Secretary of State would make orders providing that land for development of the kind designated in the order, and in the area specified by the order, would pass through public ownership before development took place. These *duty orders* were to be brought in to match the varying rates at which authorities became ready to take on such responsibilities.

When duty orders had been made covering the whole of Great Britain, the 'second appointed day' (or SAD Day as critics dubbed it) could be brought in. This would have had the effect of changing the basis of compensation for land publicly acquired from a market value (net of tax) basis to a current use value basis, that is, its value in its existing use, taking no account of any increase in value actually or potentially conferred by the grant of a planning permission for new development.

The scheme, like its two predecessors, had little chance to prove itself (Grant 1979; Emms 1980). The economic climate of the first two years of its operation could hardly have been worse, and the consequent public expenditure crisis resulted in a central control which limited it severely.

PLANNING AGREEMENTS AND OBLIGATIONS

The abandonment of attempts to collect betterment was one of a number of factors which, in the early 1980s, stimulated an already established trend for increasing the levying of charges on developers. Other influences included a general move from a regulatory to a negotiatory style of development control, increased delays in the planning system and the financial difficulties of local authorities (Jowell 1977a).

Planning authorities have had power to make 'agreements' since 1932, but it was not until the property boom of the early 1970s that they became widely used – or, as some argue, abused. The term *planning gain* is popularly used, but with two different meanings. The term can denote facilities which are an integral part of a development; but it can also mean 'benefits' which have little or no relationship to the development and which the local authority require as the price of planning permission. There has been very extensive debate on this issue, and the list of relevant publications is very long. Unfortunately, neither these publications nor statutory changes and ministerial exhortations have done much to settle the arguments. The extremes range from the Property Advisory Group's (1981) categorical statement that planning gain has no place in the planning control system, to Mather's (1988) proposal that planning gain should be formalised by allowing local authorities to sell or auction planning consents. Essentially, the issue is the extent to which local authorities can legitimately require developers to shoulder the wider costs of development: the needed infrastructure, schools and other local services.

The extremes are easy to identify: the cost of local roads in a development are clearly legitimate, while financial contributions to the cost of running a central library are not. But, of course, most items fall well within these extremes. Byrne (1989) provides a useful selection of agreements, and concludes that the majority are legitimate – a view corroborated by studies commissioned by DoE (Eve 1992) and by the Scottish Office (Rowan-Robinson and Durman 1992a). These studies effectively demolish the argument that there is widespread extortion by way of planning gain. In England, less than 1 per cent of planning decisions involve planning agreements; the largest proportion are concerned with regulatory matters (contracts, plans and drawings, building materials, etc.); and over a half deal with occupancy conditions (for example, restrictions

required for sheltered housing, agricultural dwellings, social housing). Agreements serve an important function in securing the provision of infrastructure necessitated by a development (particularly local roads), and for environmental improvement (such as landscaping). Only a very small number of agreements are concerned with wider planning objectives. In Scotland, 'most agreements are a useful adjunct to the development control process; abuse of power does not present a problem; and for the most part, the benefits secured by agreements have been related to the development proposed: where they have not, the benefits have been of a relatively minor order' (Rowan-Robinson and Durman 1992a: 73).

The statutory provisions relating to agreements were amended by the Planning and Compensation Act 1991. Agreements were replaced by 'obligations' and can now be unilateral – not involving any 'agreement' at all between a local authority and a developer. Though the wider debate has been on the ethics of planning gain, this provision in fact deals only with a narrow legal difficulty. A DoE Consultation Paper issued in August 1989 explained that a logjam could arise where the Secretary of State decided that a planning appeal should be allowed if a certain condition were met, but there was no legal basis for imposing the condition (typically because it involved off-site infrastructure). The new provision allows a developer to make an agreement to provide the necessary off-site works even if the local authority is not prepared to be a party to the agreement. This seems a small point on which to base the change from 'agreements' to 'obligations': it is possible that a more important function of the new provision is to give the appearance of a change in policy which will curb the alleged excesses of planning gain.

In fact, nothing could be further from the reality. DoE Circular 7/91 had already made it clear that local authorities could negotiate with developers for the provision of social housing. This represented a major extension of the arena of planning agreements. But, in the debates on the 1991 Bill, the Minister (Sir George Young) went further:

I think we are all agreed that planning gain is a useful part of the planning system and should be preserved and even encouraged . . . A planning gain would do more than merely provide facilities that would normally have been provided at public expense. It would provide facilities that the public would never have afforded.

Similarly, the RICS in their response (1991) to the White Paper *This Common Inheritance* (Cm 1200) expressed the hope that agreements would be extended: 'It is hoped that consideration can be given to an increased use of agreements where major developments are proposed so that the community can gain some off-setting benefit, particularly when there is a loss of amenity.'

As the report on the Scottish study (Rowan-Robinson and Durman 1992a) notes, these views (from such eminent sources) amounted to a major change in opinion since the Property Advisory Group (1981) declared the pursuit of planning gain to be unacceptable.

At the root of this is a significant change in the expected roles of the private and public sectors in land development. Whereas it used to be the case that the responsibilities of developers were clearly limited, it has become generally (even if not unanimously) accepted that the public sector is financially unable to meet the associated costs in the traditional way. The move from a regulatory to a negotiatory style of control is another aspect of this, as has been the willingness of developers to shoulder these costs.

Economic and social factors now loom large in planning decisions; and the courts have clearly stated that financial issues can be 'material considerations' in planning, as long as they are secondary to planning matters. Thus, in the case of office development granted (contrary to the local plan) to enable the redevelopment of the Covent Garden Opera House to be financially viable, it was argued:

> Financial constraints on the economic viability of a desirable planning development are unavoidable facts of life in an imperfect world. It would be unreal and contrary to common sense to insist that they must be excluded from the range of considerations which may properly be regarded as material in determining planning applications . . . Provided that the ultimate determination is based on planning grounds and not on some ulterior motive, and that it is not irrational, there would be no basis for holding it to be invalid in law solely on the ground that it has taken account of, and adjusted itself to, the financial realities of the overall situation.
> (R. v *Westminster City Council ex parte Monahan*, JPL 1989: 107)

Social factors may present greater difficulties, as when Lord Widgery held that the London Borough of Hillingdon could not impose a condition that the occupants of a private housing development should be people on the council's waiting list (*Rex* v *London Borough of Hillingdon ex parte Royco Homes Ltd* [1974] 2 All ER 643). Nevertheless, the matter is not settled – as is instanced by the debate on the role of planning policies (as distinct from housing policies) in the provision of affordable housing.

PLANNING AND AFFORDABLE HOUSING

The stance of the DoE on the role of planning in relation to affordable housing has been far from clear for some time. On the one hand, 'planning conditions and agreements cannot normally be used to impose restrictions on tenure, price or ownership', but 'they can properly be used to restrict the occupation of property to people falling within particular categories of need'. Both statements are from Circular 7/91 on *Planning and Affordable Housing* which was an early attempt to wrestle with this politically difficult issue. Even more curious (during the regime of a Conservative government) is the policy of 'exceptional release' of land, outside the provisions of the development plan, for 'local needs' housing. This is an explicit 'use of the planning system to subsidise the provision of low cost housing through containment of land value' (RTPI, *Planning Policy and Social Housing*, 1992: 5).

The extent to which authorities can achieve planning benefits depends, of course, on their bargaining power, which in turn may be related to current (and local) economic conditions. The situation varies over time and by region. In some circumstances, 'getting a developer to build anything is, in our eyes, a planning gain' (quoted in Jowell 1977a: 428); in others, the local pressures for development are so strong that local authorities can secure considerable benefits.

The growth of planning agreements gives rise to a number of concerns. The ethics of bargaining are debatable; there is scope for unjustifiable coercion; and equal treatment as between applicants can be abandoned in favour of charging what the market will bear at any particular time. Additionally, bargaining is a closed, private activity which sits uneasily astride the current emphasis on open government and public participation.

Much of the difficulty in this area may arise from the discretionary nature of the British planning control system in which negotiation is an important feature. However, studies of US land use regulation (which supposedly emphasises property rights and reduces development uncertainties) show that negotiation is equally prevalent there (Cullingworth 1993). An essential issue is that, while development rights in land are nationalised, their associated values are privately owned. Much of the case for 'planning gain' is that it is a means of capturing some of this value for the public benefit.

LAND POLICIES IN THE 1980s

Though the Community Land Act was repealed by the Local Government, Planning and Land Act 1980, local authorities still retained considerable powers of compulsory acquisition of land. They could acquire, with the consent of the Secretary of State:

any land in their area which –

(a) is suitable for and required in order to secure the carrying out of development, redevelopment, or improvement; or

(b) is required for a purpose which it is necessary to achieve in the interests of the proper planning of an area in which the land is situated.

These powers (which are still possessed by local authorities under section 226 of the 1990 Act) specifically provided for compulsory acquisition of land for disposal to a private developer. Indeed, the government made it clear that these 'planning purposes' powers (which could be of particular importance in bringing land on to the market) were generally to be used to assist the private sector.

Additionally, the Secretary of State has some formidable powers. First is the reserve power to direct a local authority to make an assessment of land available and suitable for residential development. Second are the powers to acquire any land 'necessary for the public service' include the authorisation of acquisitions 'to meet the interests of proper planning of the area, or to secure the best or most economic development or use of land'. (Ironically these provisions are a modified re-enactment of a section of the repealed Community Land Act.) However, little use has been made of these powers; instead, reliance has been placed on ensuring that LPAs make sufficient land 'available'.

Considerable debate has taken place on the adequacy of 'land availability' policies. The problems are partly financial (providing the necessary infrastructure), but mainly political. This is particularly the case for a Conservative government aiming at privatisation and the reduction in controls, an ample supply of land for private development, and the retention of land-use planning at the local level.

Patsy Healey (1983: 269) has argued that the crux of the political dilemma lies in the traditional Conservative support in the shire counties:

This support combines a concern to preserve the attractive environments in which they live and a commitment to local democracy at the smallest scale. The 1972 Local Government Act ensured continuing Conservative control of suburban and rural areas. Conservative governments thus face a dilemma. At national level they may be concerned to shift land policies more towards production than consumption purposes. Yet they must not lose the support of the environmental lobby or local Conservative councillors. In other words, the ideology of limited intervention which the current Conservative administration espouses sits uneasily with its need to respond to the demands of industrial and property production and to those for environmental conservation and local control over land policy.

Of course, the debate is not carried on in these terms. Instead, there are numerous surveys and a barrage of figures. The calculation of land availability, in particular, has moved to centre stage.

LAND AVAILABILITY STUDIES

It was a major objective of the architects of the postwar planning system to ensure that land required for development would become available – if necessary by the use of compulsory purchase powers. As previous discussion has shown, things did not work out like this despite three attempts (in 1947, 1967 and 1975). Except in special cases, such as new towns and comprehensive development areas, there has been little use of compulsory purchase powers. Thus the land 'allocations' in plans remained just that – allocations on paper. There is no necessary relationship between the allocation of land and its *availability*. It is therefore not surprising that there has been considerable controversy over the extent to which allocated land is in fact available for development. In Hooper's (1980) words: 'The planning system and the house building industry operate not only with a different definition, but with a different conception, of land availability – the former based on public control over land use, the latter on market orientation to the ownership of land.' The early land availability studies foundered on this difference, but the government continued to press their importance. The 1980 Local Government, Planning and Land Act even gave powers to the Secretary of State to *direct* a local authority to 'make an assessment of land which is in its area and which is in its opinion available and suitable for development for residential purposes'. Circular 9/80 set out a detailed methodology for studies and urged cooperation between local authorities and housebuilders in establishing the local situation.

Dissatisfied with the quality of many land availability studies, the DoE commissioned Coopers & Lybrand to carry out a study 'to assess and report on the varying assessments and assumptions about new housing made by the planning authorities and house builders, and to assess the extent to which both the provision in plans and land which is made available for housing takes account of the requirements of the market for new private sector housing' (Coopers & Lybrand 1985). Their conclusion is clear: 'there is no doubt that most of the structure plans of the 1970s paid little attention to the market demand for housing. The structure plans tended to be based on a *survey–analysis–plan* approach which required a rather determinist view of the issues which they examined'. But there is a deeper issue: 'market demand for housing was not *and probably cannot* [emphasis added] be estimated in an area and over time; this precludes its incorporation into structure plans and prevents any meaningful quantitative comparisons to be made with plan figures'.

This constitutes a fundamental challenge to the basis of the British planning system, and raises a host of thorny questions about the nature and efficacy of the system. Surprisingly, the nine members of the steering group for the Coopers & Lybrand study were not members of the Conservative Party Central Office or the Adam Smith Institute but, with one exception (the representative of the House Builders' Federation), were government officials, mostly from the DoE. Moreover, they had no hesitation in stressing the point that builders do not operate or think in terms familiar to planners: instead they look to 'market signals'. It follows, so they argued, that the planning system should concern itself with ways of improving the process of responding to demand. Three types of change were proposed. First, there should be clearer signals from the market. (A working paper includes a list of possible indicators, together with the suggestion that those operating in the market should work with planners to develop useful indicators.) Second, 'the plans themselves should consciously take note of such demand factors and, of critical importance, the plans should be sufficiently flexible to be able to respond to demand in the course of implementation – including the identification of criteria which, when met, should signal a need for a review. Plans should ensure that sites of varying size and location are available.'

Finally, the report maintained that clearer guidance from central government was required, and that appeal decisions should be consistent with exhortations. Review of structure plans should be simplified and accelerated, and other major aspects of the planning process should be strengthened – 'including some possible limit on the extent of public participation'. The 'rationale for some of the more rigid land constraints which flow from national policies' also needs review, particularly the planning presumption against development of agricultural land.

A study by Duncan Maclennan (1986), commissioned by the Scottish Development Department (SDD) was equally critical of Scottish practice. In estimating demand, Scottish authorities had used inadequate techniques: in particular, they had omitted economic factors. Builders' estimates, whilst stressing economic considerations, lacked a sound quantitative basis. He continued:

> The importance of current omissions is clearly illustrated by recourse to some basic economics. In the short run, say a single year, quantity demanded is largely determined by market price. Over the longer period changes in population, income, etc., shift this relationship in measurable ways. Price and income effects on housing demand are demonstrably important but they are ignored in structure plans. In consequence 1981 Census-based demand estimates could understate 1990 demand by as much as one quarter. Supply side estimates ignore long term changes in land and construction costs and in consequence probably overestimate the long run level of effective demand. Thus the omission of economic factors leads to a divergence of demand estimates between builders and regions thus increasing the potential for unnecessary conflict.

Maclennan was more optimistic about improvements in forecasting than the Coopers & Lybrand team, partly because of Scotland's unique source of house price information (known as the Register of Sasines). Nevertheless, speedy results were not to be expected, since much research remained to be done.

In reviewing the controversy over land availability studies, one is struck by their curious remoteness from the real world of land assembly and development. Though a greater understanding was probably achieved between planners and builders, the studies themselves proved to be of less value than the DoE had envisaged. Hooper (1985: 126) concluded that, 'whilst preserving a façade of cooperation between the main agents involved in the land conversion process', the approach in fact only served 'to obscure the fundamental issues underlying land policy in relation to residential development'. Cuddy and Hollingsworth (1985), however, while noting that land policy tended to be of a 'stable door' character (with major changes taking place before a policy was fully implemented) concluded that the land availability exercises were worthwhile since they promoted a dialogue on uncertain futures where 'bargains' were struck between local authorities and housebuilders. The 'bargains' cannot be firm since both the parties face uncertainties, but the outcome is of less importance than the process of negotiation itself. In Barrett's words:

> [They] see what is going on in such studies as a process of negotiation in which neither party to the negotiation has control over its own delivery system. It is not a simple win/lose game, but a strategic negotiation for flexibility of manoeuvre in an uncertain environment. Both parties stand to gain from not committing themselves to a firm outcome 'contract' yet maintaining the negotiation process . . . The main value of the studies lies in the process itself as a basis for reviewing performance and as a means of learning more about the short/medium term flow of land through the development process.
>
> (Barrett and Healey 1985)

This may be the latent function of land availability studies, but the DoE remains focused on its manifest function, as is evident from PPG 3 *Housing*. Since a commissioned report from Roger Tym and Partners (1991) repeats the long-standing criticisms that the studies take insufficient account of ownership and marketability constraints, it seems that history is repeating itself. The studies now seem to be a part of the planning system, whether they have any tangible use or not! Whether they have any effect on restraining land price increases is yet another question.

THE LAND AUTHORITY FOR WALES

A surprising feature of the Welsh planning scene is the survival of the Land Authority for Wales (LAW), an *ad hoc* body originally established under the Community Land Act. This provides precisely what is missing from the English scene: a body with a long-term and wide-ranging view of the land situation, with powers and resources to act positively in order to solve land availability problems.

It makes land available for development in the Principality, particularly where the private sector experiences difficulties in acquiring land. It can also acquire land (compulsorily if necessary). The Authority obtains planning permission for development and, in many cases, provides the necessary infrastructure. By virtue of its powers and financial position, it is able to purchase large sites and phase their development. Though its main function is to make land available for private housing, it also promotes the regeneration of town centres. It plays a major role in Welsh land availability studies. Its land assembly activities are illustrated by the figures for 1994–5: in that year its turnover exceeded £23 million, including land sales of 663 acres. Purchases totalled 314 acres; and gross profits amounted to nearly £6 million.

Grant (1982: 521) has commented that this represents 'The clearest and least restrictive legislative authorisation for positive planning that now exists'. He also explains why LAW exists:

> Two factors in particular contributed to the Authority's success . . . and prevented their abolition. First, they are a single purpose authority, and, unlike the local planning authorities of England and Scotland, were able, and indeed required, to pursue positive planning as their first priority. Second, the authority are not themselves a planning authority. They operate within the confines of the planning policies administered by the Welsh local planning authorities, and, where their planning applications are refused they have the usual right of appeal to the Secretary of State for Wales. They have managed therefore to avoid the suspicion of conflict of interest which has often attached to the positive planning efforts of local planning authorities. In practice they have also assisted

Welsh local authorities in land availability studies and with advice on land disposal for development.

Praise has come from other quarters. For example, Chubb (1988: para. 6.6) points to LAW's advantage in being able to 'pursue its objectives single mindedly, without conflicts of priority . . . which cause difficulties for local authorities'. The House Builders Federation (1987) commends LAW's success in obtaining land for development and securing planning permissions for private builders, often on appeal against local authorities. It argues that there is a strong case for a similar land development agency for London and perhaps other areas.

NEW SETTLEMENTS

The conclusion of the new towns programmes, coupled with increasing concern with the 'land for housing' problem naturally prompted debate on additional new towns. The TCPA had traditionally maintained that these should be a major plank in regional policy, but, during the 1980s, against the background of a buoyant housing market, proposals came from the private sector for private enterprise new towns that would fill the gap left by the completion of the existing new towns. The best known of these came from the now disbanded Consortium Developments (Hebbert 1992). This was a group of house builders who were advised by Conran Roche, a privatised group of officials from the Milton Keynes Development Corporation. They proposed a ring of new villages around the South-East which would form 'balanced communities' developed to high standards of design:

> Consortium Developments Ltd, by working on a relatively large scale, can negotiate a keen price that allows investment in a quality product. High quality infrastructure in the paving and road surfaces, high quality landscaping, sensitive design of public spaces, variety in both form and tenure of housing provision, and a wide range of supporting facilities.
>
> (Roche 1986: 312)

These were words in the direct tradition of the new towns movement (Hardy 1991a, 1991b), but their spokespersons now had to contend with a sophisti-cated planning machine. Proposals for Foxley Wood in Hampshire, Stone Bassett in Oxfordshire, West-mere in Cambridgeshire, and Tillingham Hall in Essex were all rejected on appeal. As Hebbert (1992: 178) comments, their experience 'demon-strated that even the presence of the most radical free enterprise British government of recent times is no guarantor of profitable large scale private devel-opments in green field sites'. However, they have not been completely ruled out, and PPG 3 provides a list of the requirements for proposals (which it is noted 'have almost invariably been deeply contro-versial'). Politically, the importance attached to 'local choice' effectively means that any proposal for a new settlement is likely to be killed.

In the meantime, the DoE commissioned a study of *Alternative Development Patterns: New Settlements* (Breheny *et al.* 1993). Much of this is of a technical nature, comparing the costs and benefits of different forms of development, curiously omitting all refer-ence to the major earlier study by the National Institute of Economic and Social Research (Stone 1973). This is a difficult and complex matter, since so much depends on site-specific issues. The authors neatly point up the difficulties by stressing that their analysis is 'intended to focus discussion rather than present a definitive assessment'. But central government is urged to come off the fence and to give a clear statement on the management of urban growth. It is unequivocally stated that 'unless much tougher containment policies are introduced – at the very time when concerns are being expressed over urban intensification – it is inevitable that signifi-cant greenfield/village development will take place in the UK'.

GREEN BELTS

The policy of maintaining an adequate supply of land for housing can be difficult to reconcile with policies relating to green belts and the safeguarding of agricultural land. Though 1987 saw a major policy shift on the latter (which is discussed later,

p. 153), green belts have, for a variety of reasons, remained a strong policy issue for both central and local government, as well, of course, for the envir-onmental lobby. A wealth of material is available in the House of Commons Environment Committee's 1984 report *Green Belt and Land for Housing* and in Martin Elson's detailed studies (1986 and in Elson *et al.* 1993).

Green belt policy emerged in 1955 after the expression of considerable concern at the implica-tions for urban growth of the expanded house build-ing programme. Unusually, the policy can be identified with a particular minister: Duncan Sandys (who later made another contribution to planning with the promotion of the Civic Trust and the Civic Amenities Act). Sandys' personal commitment involved disagreement with his senior civil servants, who advised that it would arouse opposition from the urban local authorities and private developers who would be forced to seek sites beyond the green belt. Experience with the Town Development Act (which provided for negotiated schemes of 'overspill' from congested urban areas to towns wishing to expand) did not suggest that it would be easy to find sufficient sites. Sandys, however, was adamant, and a circular was issued asking local planning authorities to consider the formal designation of clearly defined green belts wherever this was desir-able in order to check the physical growth of a large built-up area; to prevent neighbouring towns from merging into one another; or to preserve the special character of a town.

The policy had widespread appeal, not only to county councils which now had another weapon in their armoury to fight expansionist urban authori-ties, but also more widely. One planning officer commented that 'probably no planning circular and all that it implies has ever been so popular with the public. The idea has caught on and is supported by people of all shades of interest.' Another noted that 'the very expression *green belt* sounds like something an ordinary man may find it worthwhile to be interested in who may find no appeal whatever in 'the distribution of industrial population' or 'decentralisation' . . . Green belt has a natural faculty for engendering support' (Elson 1986: 269).

The green belt also formed a tangible focal point for what is now called the environmental lobby. However, initially, its biggest support came from the planning profession which in those days still saw planning in terms of tidy spatial ordering of land uses. Desmond Heap, in his 1955 presidential address to the (then) Town Planning Institute went so far as to declare that the preservation of green belts was 'the very *raison d'être* of town and country planning'. Their popularity, however, has not made it any easier to reconcile conservation and development. The land availability studies are an attempt to do precisely this.

The green belt policy commands even wider support today than it did in the 1950s. Elson (1986: 264) concludes his study with a discussion of why this is so:

> It acts to foster rather than hinder the material and non-material interests of most groups involved in the planning process, although it may be to the short term tactical advantage of some not to recognise the fact. To *central government* it assists in the essential tasks on interest mediation and compromise which planning

policy-making represents . . . To *local government* it delivers a desirable mix of policy control with discretion. To *local residents* of the outer city it remains their best form of protection against rapid change. To the *inner city local authority* it offers at least the promise of retaining some economic activities that would otherwise leave the area; and to the *inner city resident* it offers the prospect, as well as often the reality, of countryside recreation and relaxation. To the *agriculturist* it offers a basic form of protection against urban influences, and for the *minerals industry* it retains accessible, cheap, and exploitable natural resources. *Industrial developers* and *housebuilders* complain bitterly about the rate at which land is fed into the development pipeline, yet at the same time are dependent on planning to provide a degree of certainty and support for profitable investment.

Planning may be an attempt to reconcile the irreconcilable, but green belt is one of the most successful all-purpose tools invented with which to try.

The latest policy statement on green belts (the revised PPG 2 of 1995) confirms the validity and permanence of the green belts, which now cover over one and a half million hectares (12 per cent) of England. The general location of the green belts is shown in Figure 6.1. Table 6.1 illustrates the growth in the area of green belt land in England;

Table 6.1 Green Belts in England, by Region, 1979, 1989 and 1993 (sq km)

Region	Green belt	1979	1989	1993
Northern	Tyne & Wear	400	504	465
Yorks and Humberside	S. and W. Yorkshire	1,263	2,232	2,259
	York	0	248	237
East Midlands	Burton/Swadlincote	0	7	7
	Nottingham/Derby	0	608	608
Eastern	Cambridge	17	108	261
South-East	London	3,068	4,847	4,856
	S. W. Hants./S. E. Dorset	0	856	854
	Oxford	251	349	348
South-West	Avon	628	743	706
	Glos/Cheltenham	57	75	81
West Midlands	West Midlands	1,425	2,092	2,093
	Stoke-on-Trent	0	366	365
North-West	Gr. Manchester/Merseyside/ Cheshire and Lancs	15	2,451	2,417
Total		7,125	15,485	15,557

Source: DoE, *Digest of Environmental Statistics*, no. 18, 1996, Table 8.4

Table 6.2 Green Belts, Scotland, 1993 (sq km)

Green belt	Area
Aberdeen	237
Ayr/Prestwick	29
Falkirk/Grangemouth	35
Glasgow	1201
Lothian/Edinburgh	146
Total	1648

Table 6.2 gives the area of green belt in Scotland, which is discussed separately below. There is no formal green belt policy in Wales.

The major policies remain as they were when Sandys insisted on promoting the green belt policy:

> The government attaches great importance to green belts, which have been an essential element of planning policy for some four decades. The purposes of green belt policy and the related development control policies set out in 1955 remain valid today with remarkably little alteration.
>
> (PPG 2, 1988: para. 1.1)

Elson's 1993 study, which was undertaken at a time when the earlier (1988) PPG 2 was operative, concluded that the green belts had been successful in checking unrestricted sprawl and in preventing the merging of towns. Green belt boundary alterations in development plans had affected less than 0.3 per cent of green belts in the areas studied over an eight-year period. Most planning approvals in green belts had been for small-scale changes which had no significant effect on the open rural appearance of green belts. The appeal system had strongly upheld green belt policy.

The relationship between green belt restraint and the preservation of the special character of historic towns was much more difficult to evaluate. Though the idea had 'a well-established pedigree' and though the green belt boundaries were particularly tight, there was little evidence to connect policy and outcomes. Also, it was difficult to assess how far green belts had assisted in urban regeneration. Though the green belts did 'focus development interest on sites in urban areas', local authorities tended to regard the creation of jobs 'as more important than any land development objective *per se*'. Indeed, urban regeneration was often seen as requiring 'the selective release of employment sites in the green belt'. The supply of adequate sites within urban areas was not sufficient for development needs (though it might be increased by an expanded programme of land reclamation). Moreover, refusal to

---

Green Belt Policy

Purposes

- to check the unrestricted sprawl of large built-up areas;
- to prevent neighbouring towns from merging into one another;
- to assist in safeguarding the countryside from encroachment;
- to preserve the setting and special character of historic towns; and
- to assist in urban regeneration, by encouraging the recycling of derelict and other urban land.

Use of land in green belts

- to provide opportunities for access to the open countryside for the urban population;
- to provide opportunities for outdoor sport and outdoor recreation near urban areas;
- to retain attractive landscapes, and enhance landscapes near to where people live;
- to improve damaged and derelict land around towns;
- to secure nature conservation interest;
- to retain land in agricultural, forestry and related uses.

(DoE, PPG 2, 1995)

allow development on the periphery of an urban area could lead to leap-frogging beyond the green belt, or to development by the intensification of uses in towns located within the green belt. A note is made of the 'suggestion' that 'the inner city will rarely be a substitute location for uses seeking planning permission on the urban fringe':

> The housing market potential in the two locations is quite different (in terms of the size and price range of houses which may be marketed for example), and many of those developing other uses require the better accessibility (normally by private car) which a peripheral or outer location affords.
>
> (Elson *et al.* 1993: para. 2.37)

There was seen to be a clear need for further research here.

In Scotland, green belts have been established around Aberdeen, Ayr/Prestwick, Edinburgh, Falkirk/Grangemouth, and Glasgow. Interestingly, the Dundee green belt has been replaced by a general countryside policy (Regional Studies Association 1990: 22). Scottish green belts have somewhat wider purposes than those in England: these include maintaining the identity of towns by establishing a clear definition of their physical boundaries and preventing coalescence; providing for countryside recreation and institutional uses of various kinds; and maintaining the landscape setting of towns. There is a greater emphasis on the environmental functions of the green belts, and recreation is included as a primary objective. The title of the Scottish circular is significant: *Development in the Countryside and Green Belts* underlines the links between general countryside policies and green belts: 'As a result, a much more integrated approach to the planning of green belt and non-green belt areas is achieved in Scotland.' The Regional Studies Association (1990) study commends the Scottish approach, arguing that 'green belts have become an outmoded and largely irrelevant mechanism for handling the complexity of future change in the city's countryside'.

Green belts are the first article of the British planning creed. They are hallowed by use, popular support and fears of what would happen if they were 'weakened'. Fierce arguments are raged by a wide range of groups from national bodies such as the CPRE to local green-belt residents. There are, however, other issues which do not attract the same concern, such as the costs imposed by green belts, and the inadequacy of a planning policy which lays such great emphasis on protection and such minor emphasis on instruments for meeting development needs. On this line of argument, green belts should be part of a more comprehensive land use/transportation policy.

DEVELOPMENT INVOLVING AGRICULTURAL LAND

The 'loss' of agricultural land has been an issue of debate throughout the postwar period. The valiant work of the late Robin Best (1981) has failed to dispel popular images, yet his figures showed that the annual average transfer of agricultural land to urban land use in England and Wales was only 9,300 hectares between 1975 and 1980 (compared with 17,500 between 1945 and 1950, and 25,100 between 1931 and 1939). The total urban area is about 1,640,000 hectares: about 11 per cent of the total. More recent estimates from DoE show (for England) an annual rate of transfer of agricultural land to urban, industrial and recreational use of 14,000 hectares between 1950 and 1969, falling to 12,000 in the decade 1970–9, and still further to 5,000 hectares in the 1980s (though alternative estimates by the CPRE put the figure much higher).

Nevertheless, concern about the 'loss of agricultural land' has had a major impact on development plans, on planning decisions and on appeal decisions. This stems partly from local resistance to change and partly from the postwar approach to agricultural policy. The decision was taken to develop a strong and healthy agricultural sector. The 1947 Agriculture Act demanded 'a stable and efficient industry, capable of providing such part of the nation's food as in the national interest it is desirable to produce'. It was against this background that farming was largely exempted from planning control.

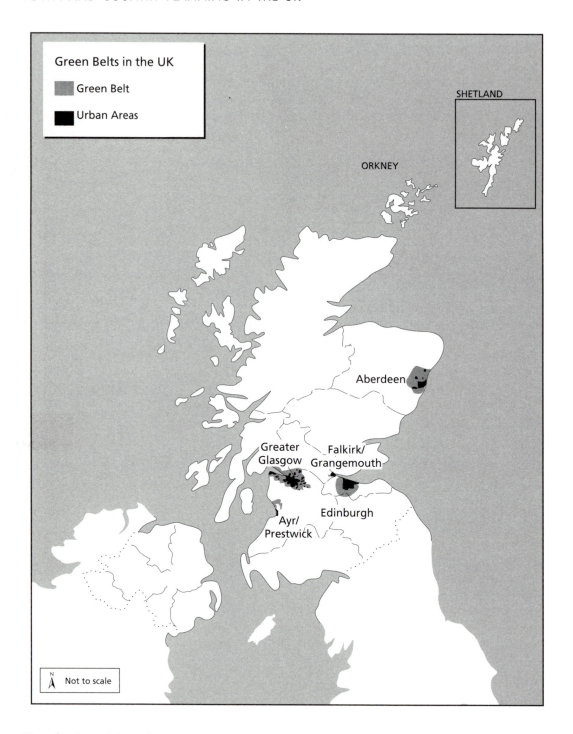

Figure 6.1 Green Belts in the UK

Figure 6.1 continued

The 1976 DoE circular (75/76) on the safeguarding of agricultural land defined the general policy as being 'to ensure that, as far as possible, land of a higher agricultural quality is not taken for development where land of a lower quality is available, and that the amount of land taken should be no greater than is reasonably required for carrying out the development in accordance with proper standards'. This was based on the 1975 White Paper, *Food from our Own Resources*, which proposed a significant increase in home food production to reduce dependence on imports which had risen sharply in cost. It also drew a graphic picture of the continuing loss of agricultural land, estimated at 144,000 acres a year which, if continued, 'would imply a substantial reduction in the available agricultural land'.

Times change, and Circular 16/87 noted that 'at present, by contrast, there are substantial surpluses of the main agricultural products in western countries. The need now is to foster the diversification of the rural economy so as to open up wider and more varied employment opportunities'. The Circular continued:

> The agricultural quality of the land and the need to control the rate at which land is taken for development are among the factors to be considered [in assessing planning applications affecting agricultural land], together with the need to facilitate development and economic activity that provides jobs, and the continuing need to protect the countryside for its own sake rather than primarily for the productive value of the land.

The circular was a mere twelve paragraphs long, but it represented a dramatic change in policy. This change was due, of course, to the increasing agricultural overproduction – in Europe generally as well as in Britain.

This is further underlined by several initiatives of the Ministry of Agriculture, Fisheries and Food (MAFF). First, section 17 (1) of the Agriculture Act 1986 placed a duty on the Minister of Agriculture to

> have regard to and endeavour to achieve a reasonable balance between the following considerations –

> (a) the promotion and maintenance of a stable and efficient agricultural industry;
> (b) the economic and social interests of rural areas;
> (c) the conservation and enhancement of the natural beauty and amenity of the countryside (including its flora and fauna and geological and physiographical features) and of any features of archeological interest there; and
> (d) the promotion of the enjoyment of the countryside by the public.

Second, the same Act made provision for *environmentally sensitive areas* (ESAs), where annual grants are payable by this ministry to farmers to enable them to follow farming practices which will achieve conservation objectives. Additionally, several 'diversification' programmes have been mounted, for example to encourage farmers to plant new woodland on productive agricultural land, and to promote ancillary businesses on or adjacent to farms. These and similar programmes are discussed in Chapter 9.

VACANT AND DERELICT LAND

Much land that was once useful and productive has become waste land, particularly in the inner cities and in mining areas. It is unsightly, unwanted and, at worst, derelict and dangerous. The planning system is not designed to deal with such land easily: its essential characteristic is to allocate land between competing uses. Where there are no pressures for development, there is a severe limit to what can be done, especially when the amount of waste land is large, as it is in older industrial areas. Major efforts have been made to deal with the problems. Between 1988 and 1993, some 9,500 hectares of derelict land were reclaimed. Unfortunately, a large amount of 'new dereliction' is continually being created, and the result is that the total amount remains high. Though the amount of derelict land in England decreased by 2 per cent between 1988 and 1993, the 1993 Derelict Land Survey showed a total of 39,600 hectares of derelict land at the latter date (Table 6.3).

Reclamation policies have changed over the years.

Table 6.3 Derelict Land, England, 1988–93 (sq km)

Type of dereliction	1988	1993	Reclaimed 1988–93
Spoil heaps	119	92	21
Excavations and pits	60	58	9
Military dereliction	26	33	5
Derelict railway land	64	56	17
Mining subsidence	10	7	2
General industrial dereliction	85	97	28
Other forms of dereliction	41	53	12
Total	405	396	95

Source: DoE, Digest of Environmental Statistics 1996, Table 8.5

Originally, the objective was to remove eyesores and potential dangers caused by spoil heaps and other waste. Much of this was located in rural areas, and the policy was to return the land to agriculture or forestry, or to make it available for public open space (known in the jargon as a 'soft end use'). Since the 1980s, emphasis has shifted to 'hard end uses' such as industrial, commercial or residential development, particularly in older urban areas.

A range of policy instruments to deal with derelict land have been developed. Some of these have been part of broader policies in relation to urban regeneration (through urban development corporations, enterprise zones and the urban programme). At one time considerable emphasis was placed on promoting the reclamation of publicly owned land. Since this has some interesting policy aspects it is worth examining.

Registers of Unused Publicly Owned Land

Policies can be founded on myths as well as on an adequate understanding of problems. So it was with the land registers established by the 1980 Local Government, Planning and Land Act. The myth was that one of the major causes of urban dereliction was the hoarding of land by public authorities. By requiring local authorities and other public bodies to 'register' their land, it was expected that it would

find its way into the development process. In the words of the Secretary of State:

> The publication of land registers for the greater part of the country represents a major opportunity to secure the better use of massive acreages of underused land. Given the resources also available for dealing with dereliction, it is important that action should be taken to dispose of the land to the best advantage as soon as possible . . . It is up to builders and developers to examine registers and seek out the owners to make an offer for any sale in which they are interested.
> (HC Debates (Hansard) vol. 21, col. 40, 29 March 1982)

In fact, with the reality being much more complicated than the perception, the registers were of little effect. Indeed, an evaluation undertaken for the DoE (Whitbread et al. 1991: 57) concluded that the registers were 'an ineffective instrument which was not used in a single case [in the sample studied] to help bring land forward for development'.

Research shows that land vacancy is typically a transient feature of the environment. Though much of it has been vacant for a long time (two-thirds of a sample had been vacant for more than twelve years) some of this idle land – perhaps a third – can be used when subsidies are paid to overcome physical constraints. However, some vacant land – perhaps two-thirds – is so because of institutional factors, owners' intentions or poor demand. As the evaluation study explained:

Many sites remain vacant for non-physical reasons. Some are delayed by the legitimate workings of the planning system, and by legal and other institutional difficulties. Existing policy instruments can do little to overcome these difficulties. Others are delayed by owners', particularly private sector owners', intentions that they should remain vacant for various (largely obscure) reasons.

(Whitbread *et al.* 1991: para. 3.147)

An earlier report suggested a long list of reasons why vacant land is not put to temporary uses: expenditure by the owner in meeting fire, safety and insurance requirements, in proving access, and in site clearance; temporary tenants tend to be unreliable and to cause environmental problems; demand from temporary users is deficient and uncertain, and often provides landowners with a very low financial return; there are often problems in securing vacant possession; landowners may be unaware of the potential of temporary uses; or they may think that keeping sites vacant preserves existing use rights, or puts pressure on local authorities to grant planning consent for development (Cameron *et al.* 1988). Much of this land is in private ownership and, short of compulsory acquisition (which is unpopular with a Conservative government), there is little that 'policy' can do to speed up the reuse of the land.

But these are the failures: more striking are the undoubted successes of other policy instruments, of which the evaluation study counted thirteen (ranging from grant aid to planning and promotional action by local authorities). Foremost among these was the Derelict Land Grant. However, as with City Grant, this has now been replaced by the English Partnerships Land Reclamation Programme. Before discussing this, it is relevant to note another policy initiative which failed and which has been overtaken by this same programme. This concerns contaminated land.

Contaminated Land

There is no clear line between vacant, derelict and contaminated land (or neglected, underused, waste and despoiled land). The terms are used in different ways, sometimes for different purposes, sometimes with the same or similar meanings. Contaminated land is particularly difficult to define, but the term is commonly used to imply the existence of a hazard to public health. The 1995 Environment Act introduced a new definition which incorporates this long-standing idea. Though there is an overlap with 'derelict' land, there are important differences. A chemical waste tip may be both derelict and contaminated; a disused chalk quarry may be derelict but not contaminated; an active chemical factory may be contaminated but not derelict. It is the additional health danger which is the characteristic feature of contaminated land; and this also implies a severe degree of pollution and, typically, an increased difficulty in abating it. However, the health risk arises only in relation to the use to which the land is to be put. A piece of land may pose no risk if used for one purpose, but a severe risk if it is used for another. The site of an oil refinery may be contaminated, but that is of no consequence if no other use is intended (and assuming that there are no effects beyond the site): 'A scrapyard contaminated by metal traces would constitute a hazard for subsequent agricultural use, but the contamination would be of no account in the construction of an office block' (this, and the following quotations, are from the HC Environment Committee report on *Contaminated Land*, 1990).

Partly because of a characteristically pragmatic approach, there has never been an attempt to quantify the amount of contaminated land in Britain. Instead of identifying contaminated land and then determining appropriate policies for dealing with it, the British approach has been to regard contamination as a general concept which is given substance only in relation to particular sites and particular end-uses. The nature of policy flows from this: 'Policy is to ensure that the quality of land is fit for the purpose to which it is being or will be used. There is no requirement for land to be brought up to a minimum quality standard regardless of use, unless that land poses a threat to the public health or the environment.' The House of Commons Environment Committee considered this approach to be inadequate since (in its judgement) there is land which is so contaminated that it is 'a threat to health and the environment both on site and in the surrounding area'. The Committee also recommended that local authorities should be given a duty 'to seek out and compile registers of contaminated land'.

There was a remarkably swift response to this: the Environment Protection Bill was amended to place such a duty on district councils and London boroughs. The implementation of this duty, however, rapidly ran into severe difficulties, and the initial proposals had to be drastically changed. The crux of the problem lay in the concept of 'contamination'. Instead of referring to land that is contaminated, the Act relates to 'land which is being or has been put to any use which may cause that land to become contaminated with noxious substances'. This very inclusive definition was made particularly onerous in the initial draft regulations because of the very large number of contaminative uses which were specified. There was strong criticism that the registers would create widespread blight and, in an attempt to pacify objectors, the number of specified uses was greatly reduced.

The problem underlying all this is that it is relatively simple to register land that is *possibly* contaminated, but extremely laborious and costly to identify land that is in fact contaminated. Even at the low rate of £15,000 per hectare, it would cost around £600 million merely to investigate the 40,000 hectares of land identified in the 1988 Derelict Land Survey. To cover all relevant land would cost many times this amount, and would take many years to complete (Thompson 1992: 22).

Another objection to the initial regulations was that they prohibited the deregistration of sites. This was defended on two grounds:

> One is that factual information on the site's history (which cannot by definition change) will be necessary when any future change of use is proposed. The other is that contamination from the site may have migrated to adjacent sites; owners, regulatory authorities and developers are expected to use registers to identify such sources of contamination.

It was because of difficulties such as these that the government was eventually forced to abandon the scheme as originally envisaged. The difficulties of changing from the traditional British reactive approach to a genuinely pro-active approach are manifest (Harrison 1992: 809). However, a renewed attempt is made in the Housing and Urban Development Act of 1993 and the Environment Act of 1995 (Tromans and Turrall-Clarke 1994 and 1996).

ENGLISH PARTNERSHIPS

The 1993 Act provides the powers and machinery for implementing 'a new approach to vacant land' – which includes unused, under-used or ineffectively used urban land, land which is contaminated, derelict, neglected or unsightly, or land which is likely to be affected by subsidence. A new agency has been established for this purpose: this is statutorily termed the Urban Regeneration Agency but, to underline the role which it sees itself as playing, it has taken the non-statutory name of English Partnerships. (Its remit applies only to England: similar functions are undertaken in Scotland by Scottish Enterprise and in Wales by the Welsh Development Agency.)

The agency aims to facilitate a 'targeted' approach, to 'speed up the provision of good quality land', and to 'act entrepreneurially to encourage redevelopment by bringing together all parties with a role to play'. Since about a half of vacant urban land is owned by public bodies (and 'is not always subject to the normal incentives and disciplines of the market that apply to private sector land'), the Secretary of State is given power to vest such land in the agency 'where this would help to promote its development'. The agency has extensive powers of land assembly and preparation, and is intended to draw together the existing activities and funding of English Estates (a DTI-sponsored body charged with developing and managing industrial estates) with the derelict-land and city-grant regimes of the DoE.

In line with current philosophy, the new agency concentrates on enabling activities (hence its quickly assumed pseudonym). It acts primarily as

---

DoE Guidance to English Partnerships

The Agency's preferred mode of operation should be to act wherever possible as an enabling body, achieving its objective by helping others, particularly the private sector, to regenerate land and buildings. To that end, it may assemble, plan and service sites and provide the vision, energy and, where necessary, financial support which will encourage others to proceed with projects which meet its objectives. Where such facilitation mechanisms are insufficient in themselves to produce the desired result, the Agency may as a last resort carry out development, having first satisfied itself that to do so will not displace or disadvantage private sector development of investment in the area . . .

In deciding which land is suitable for regeneration the Agency should pay particular attention to the following types of need:

(a) areas of urban deprivation or areas of localised high unemployment where bringing vacant, unused and derelict land into use can contribute significantly to regeneration;

(b) derelict land in both urban and rural areas which should be reclaimed for health and safety reasons or because it is blighting the area.

(Guidance from the Secretary of State for the Environment to English Partnerships, 1994)

---

a one-stop shop for grant aid and facilitates development by land assembly. Though it has compulsory purchase powers, it will generally acquire land on the open market. It is not intended that the agency will maintain a long-term interest in any development, though it will have the power to manage the estates it develops.

While the 1993 Act is focused on the reuse of land which is contaminated, the 1995 Environment Act introduces a new regime for the identification and 'remediation' of contaminated land. Responsibilities are shared between local government and the newly established Environment Agencies (Lewis 1995). These are discussed in the following chapter.

FURTHER READING

Land Values

There has been surprisingly little study of the operation of the various experiments in capturing land values for the public benefit. A long and detailed account of the legislative history is given in Cullingworth (1981) *Land Values, Compensation and Better-* *ment (Environmental Planning 1939–1969*, vol. 4). More digestible accounts are provided by McKay and Cox (1979) *The Politics of Urban Change* (ch. 3); and Cox (1984) *Adversary Politics and Land*.

Planning Gain

By contrast, there has been a large number of studies of planning agreements, planning obligations and planning gain. Selected titles (listed by date of publication) are Rowan-Robinson and Young (1989) *Planning by Agreement in Scotland*; Callies and Grant (1991) 'Paying for growth and planning gain: an Anglo-American comparison'; Eve (1992) *Use of Planning Agreements*; Healey *et al.* (1992b) 'Rationales for planning gain'; and Bunnell (1995) 'Planning gain in theory and practice – negotiation of agreements in Cambridgeshire'. The most comprehensive study to date is Healey *et al.* (1995b) *Negotiating Development: Rationales and Practice for Development Obligations and Planning Gain*.

Planning and Affordable Housing

The use of planning powers to require the provision of affordable housing has attracted much debate.

See, for example, Kirkwood and Edwards (1993) 'Affordable housing policy – desirable but unlawful?'; Barlow *et al.* (1994a) *Planning for Affordable Housing*; Elson *et al.* (1996) *Green Belts and Affordable Housing: Can We Have Both?*

On the even stranger 'exceptions' policy (the exceptional release of land for 'local needs' housing), see Annex A to PPG 3; Bishop and Hooper (1991) *Planning for Social Housing* (which contains a good bibliography of publications up to this date); and Williams *et al.* (1991) *Evaluating the Low Cost Rural Housing Initiative*.

More generally on housing need issues, see HC Environment Committee (1996) *Housing Need*. The most comprehensive, up-to-date analysis of English housing conditions is given in Green *et al.* (1996) *Housing in England 1994/95*.

Land Availability Studies

Land availability studies have attracted much comment and argument. Current policy is set out in PPG 3 and the regional planning guidance notes. For Wales, TAN 1 deals with *Joint Housing Land Available Studies*. The *Journal of Planning and Environment Law* is a good source of relevant articles; see, for example, Hooper (1979) 'Land availability'; Humber (1980) 'Land availability – another view'; Hooper *et al.* (1988) 'Housing land availability'. See also Rydin (1986) *Housing Land Policy*; and Bramley (1989) *Land Supply, Planning, and Private Housebuilding*. More recent publications include: Jackson *et al.* (1994) *The Supply of Land for Housing: Changing Local Authority Mechanisms*; Bramley *et al.* (1995) *Planning, the Market and Private House-building*; and Bramley and Watkins (1996) *Steering the Housing Market: New Building and the Changing Planning System*. More generally, see Williams and Wood (1994) *Urban Land and Property Markets in the United Kingdom*.

The effect of planning on the land market has been the subject of a long-standing debate; see, for example, Evans (1988) *No Room! No Room! The Costs of the British Town and Country Planning System*, and Evans (1991) 'Rabbit hutches on postage stamps'. Less tendentious (and therefore less interesting) is the study commissioned by the DoE on *The Relationship between House Prices and Land Supply* (Eve 1992). At the time of writing, the most recent discussion was by Monk *et al.* (1996) 'Land-use planning, land supply, and house prices'. There is a huge American literature on the subject which is referred to in Cullingworth 1993: ch. 6.

Green Belts

Two major publications on green belts are Elson (1986) *Green Belts: Conflict Mediation in the Urban Fringe*, and the report of a study for DoE led by Elson (1993) *The Effectiveness of Green Belts*. Broader in scope is the classic study by Peter Hall *et al.* (1973) *The Containment of Urban England*. On green belts in Scotland, see Regional Studies Association (1990) *Beyond Green Belts*; and Pacione (1991) 'Development pressure and the production of the built environment in the urban fringe'.

New Settlements

An account of the long-standing British campaign for new settlements is included in Hardy (1991a) *From Garden Cities to New Towns*, and (1991b) *From New Towns to Green Politics* (which is a two-volume history of the TCPA). A volume in the 'Official History' of environmental planning 1939–69 by Cullingworth (1979) provides a detailed dead-pan record of government policy over this thirty-year period. Breheny *et al.* (1993) *Alternative Development Patterns: New Settlements* (Department of the Environment) provide a more up-to-date picture.

Vacant, Derelict, and Contaminated Land

The research on vacant land includes: Cameron *et al.* (1988) *Vacant Urban Land: A Literature Review*; and Whitbread *et al.* (1991) *Tackling Vacant Land: An Evaluation of Policy Instruments*. The latter provides a review of previous research. For a broader overview of urban land policies, see Chubb (1988) *Urban Land Markets in the United Kingdom*.

Derelict land is dealt with in a number of DoE reports, including *The Strategic Approach to Derelict Land Reclamation* (1992); *Assessment of the Effectiveness of Derelict Land Grant in Reclaiming Land for Development* (1994); *Derelict Land Survey 1993* (1995); *Derelict Land Prevention and the Planning System* (1995).

Policy on contaminated land is succinctly set out in PPG 23 *Planning and Pollution Control*. A commentary on this is given by Graham (1996) 'Contaminated land investigations – how will they work under PPG 23?' An exhaustive legal guide is Tromans and Turrall-Clarke (1994 with 1996 supplement) *Contaminated Land*.

7

PLANNING AND THE ENVIRONMENT

The United Kingdom is determined to make sustainable development the touchstone of its policies. We recognise that this means a change of attitudes throughout the nation. That change cannot be achieved overnight, but that gives no grounds for defeatism – it should act as a spur to action.

Sustainable Development: The UK Strategy (Cm 2426), 1994

THE ENVIRONMENT

In one sense, all town and country planning is concerned with the environment. However, the reverse is not true, and it is difficult to decide where to draw the boundaries. The difficulty is increased by the shifting of responsibilities from local government to *ad hoc* bodies, and by the flood of new legislation, prompted in part by the EU. Further complications arise because of the rate of organisational change which has taken place in recent years – including the establishment of Her Majesty's Inspectorate of Pollution and the National Rivers Authority, followed by their merger into new environmental agencies for England and Wales and for Scotland. But the most important factor of all has been the increased concern for the environment which has brought about changes such as these.

The implications for 'town and country planning' are still working themselves out, not always with clarity. Thus, it has been a long-standing feature of planning control that permission is given unless there are good reasons for refusal. It is for the local planning authority to demonstrate (to the Secretary of State if necessary) that an application should be refused. With 'environmental' procedures, however, the onus shifts somewhat: the developer's proposals

have to be demonstrably acceptable, and permission can be refused if they are not. Though official pronouncements and advice are coy in acknowledging this, it is clear that environmental factors can be decisive in a planning decision and that applicants may even be required to discuss the merits of alternative sites. In the words of PPG 23 (para. 3.16), environmental statements, which must accompany particular applications (discussed later, pp. 181–3) 'may – and as a matter of practice should – include an outline discussion of the main alternatives studied by the developer and an indication of the reasons for choosing the development proposed, taking account of environmental effects'.

The scope of the requirements for environmental assessment go beyond the developments covered by the planning legislation. Local authorities have specific powers in relation to some environmental issues such as certain aspects of pollution, waste and noise, but they are not environmental planning authorities. Other specific 'pollution control regimes' exist for this purpose, but there is no clear dividing line. A related issue here is that of *sustainability* – a concept around which much environmental policy revolves.

Planning Controls and Pollution Controls

The dividing line between planning and pollution controls is not always clear cut. Both seek to protect the environment. Matters which will be relevant to a pollution control authorisation or licence may also be material considerations to be taken into account in a planning decision. The weight to be attached to such matters will depend on the scope of the pollution control system in each particular case . . .

In deciding whether to grant planning permission, planning authorities must be satisfied that planning permission can be granted on land-use grounds, and that concerns about potential releases can be left for the pollution control authority to take into account in considering the application for the authorisation or licence. Alternatively, they may conclude that the wider impact of potential releases on the development and use of land is unacceptable in all the circumstances on planning grounds, despite the grant, or potential grant, of a pollution control authorisation or licence.

(PPG 23: para. 1.34–36)

SUSTAINABILITY

Words cast a spell which can, at one and the same time, command respect and create great confusion. No word illustrates this better than the ubiquitous 'sustainability'. So overworked has this word become that it has ceased to have any communicable meaning. It has descended into that political bog where words seduce or obscure. Nevertheless, the term flourishes, and many earnest arguments are spiced by its use. Perhaps it will eventually go out of fashion as its uselessness in communication becomes all too apparent. In the meantime, however, because it reflects so well the general feeling that environmental betterment should be a goal of political action, it will be with us for some time. And so, many academics, environmental groups and government officials are devoting earnest effort to establishing what the term means – or what it should mean. This is not an idle activity: there is a broad political commitment to the idea of sustainability, even if there is no agreement on what it means.

There are innumerable definitions of sustainability. One famous poetic rendering is by Chief Seattle: 'We do not inherit the world from our ancestors: we borrow it from our children.' This encapsulates the essential idea, which is more prosaically expressed in the well-known formulation of the 1987 Brundtland Report: 'Sustainable development is development that meets the needs of the present without compromising the ability of future generations to meet their own needs.'

Shiva (1992: 192) has pointed to two very different uses of the concept. One ('the real meaning') relates to the primacy of nature: 'sustaining nature implies the integrity of nature's processes, cycles and rhythms'. This is to be contrasted with 'market sustainability', which is concerned with conserving resources for development purposes, and, if they become depleted, finding substitutes. On this latter approach, sustainability is convertible into substitutability and hence a cash nexus. The distinction is given eloquent expression in the words of a native American elder who, in epitomising the non-convertibility of money into life said: 'Only when you have felled the last tree, caught the last fish, and polluted the last river, will you realize that you can't eat money' (Shiva 1992: 193).

On the other hand, there are some formidable (if not popular) economic arguments which more prosaically point to the differences between notions of sustainability, optimality and ethical superiority. The fact that a particular path of development is unsustainable does not necessarily mean that it is undesirable or sub-optimal. In the words of Becker-man (1995: 126), 'most definitions of sustainable development tend to incorporate some ethical injunction without apparently any recognition of the need to demonstrate why that particular ethical injunction is better than many others that one could think up'.

It has to be accepted that the term eludes precise definition, as is clearly apparent from the monumental 1995 report of the Select Committee of the House of Lords on Sustainable Development. Interestingly, the British Government Panel on Sustainable Development, in its first report (1995) commented that the term was 'not so much an idea as a convoy of ideas'. It is a rallying cry, a demand that environmental issues need to be taken into account; but it provides little guide to action.

LOCAL SUSTAINABILITY AND AGENDA 21

The problems to which the term gives rise are neatly illustrated by two quotations: 'in Strathclyde, the steering committee could see the relevance of transport and health to sustainability, but crime was a bridge too far'; 'In Cardiff, the forum grappled with the difficulty of deciding whether the Cardiff Bay Barrage would move the area towards or away from sustainability'.

What is immediately striking about the discussions from which these quotations are taken is that the debate involves intensely political issues: it is not a matter restricted to environmental experts. Interestingly, this was well recognised at the 1992 Rio Earth Summit which gave a major impetus to the elaboration of 'sustainable' policies. Agreement was given to *Agenda 21*: a comprehensive worldwide programme for sustainable development in the twenty-first century. In formulating this programme, major emphasis was placed on a very wide degree of participation. In the UK this is organised at central and local government levels. The former resulted in the *Sustainable Development Strategy* of 1994. At the local level, *Local Agenda 21* calls for each local authority to prepare and adopt a local sustainable development strategy. These local efforts have been aided by the work and publications of the Local Government Management Board. It is one of these publications that is the source of the quotations from Strathclyde and Cardiff given above (LGMB 1995b).

A major feature of the consultation programme at the local level is that it involves much more than the term 'consultation' often means. Groups have been established in local areas to debate the meaning of sustainability and to determine how progress towards it can be achieved and assessed ('you can only manage what you can measure'). These local endeavours are designed to produce policies and indicators which are locally appropriate. The research has underlined the importance of this local 'ownership'. There is a positive and a negative aspect to this. Positively, 'Agenda 21 is as much concerned with the *process* of sustainable development – participative, empowering, consensus-seeking, and democratic – as it is with *content*'; and 'social processes of securing agreement on and commitment to sustainability aims are indispensable' (LGMB 1995b) even where the requirements for sustainability are determined externally. In short, the changes in attitudes and behaviour which will be required by policies of sustainability will come about only if they are acceptable. The negative side to this is the widespread distrust of both local and central government which research has uncovered (Macnaghten *et al.* 1995). Agenda 21 emphasises equality and economic, social and political rights. The research team similarly stresses the importance of 'social justice and economic issues'. Among the top concerns are poverty, unemployment and deterioration in the quality of life and the health of local communities.

As this discussion indicates, the popularity of the concept of 'sustainability' is related to broader concerns about planning, the environment (social, political and economic as well as physical) and government. Planning is thus subject to some strong forces of change. Its framework is now broad and complex. The boundaries between land use

Definitions of Sustainability

Sustainable development: development that meets the needs of the present without compromising the ability of future generations to meet their own needs. It contains within it two key concepts:

- the concept of 'needs', in particular the essential needs of the world's poor, to which overriding priority should be given;
- the idea of limitations imposed by the state of technology and social organisation on the environment's ability to meet present and future needs.

World Commission on Environment and Development: *Our Common Future* (Brundtland Report); 1987

The [Brundtland definition] is totally useless since 'needs' is a subjective concept. People at different points in time, or at different income levels, or with different cultural or national backgrounds, will differ about what 'needs' they regard as important. Hence, the injunction to enable future generations to meet their needs does not provide any clear guidance as to what has to be preserved in order that future generations may do so.

Beckerman, *Small is Stupid*, 1995

Sustainable development is about integrating the needs of environmental protection, social equity and economic opportunity into all decision making.

Environ (Leicester): *Indicators of Sustainable Development in Leicester*, 1995

Our standard definition of sustainable development will be non-declining per capita utility – because of its self-evident appeal as a criterion for intergenerational equity.

Pezzy, *Economic Analysis of Sustainable Growth and Sustainable Development*, 1989

Sustainability means making sure that substitute resources are made available as non-renewable resources become physically scarce, and it means ensuring that the environmental impacts of using those resources are kept within the Earth's carrying capacity to assimilate those impacts.

Pearce, *Blueprint 3: Measuring Sustainable Development*, 1993

---

planning and other spheres of government are blurred. The bewildering changes in governmental organisation which have characterised recent years are one indication of this; and it seems likely that such changes will continue: to borrow a phrase, there is now little likelihood of a 'stable state' (Schon 1971). Two of the factors which will propel this require further discussion before detailing current environmental policies and programmes: the growth of environmental politics in Britain and the impact of European policies.

ENVIRONMENTAL POLITICS

Environmental politics has only recently become an energetic force on the British scene. Its rise has been prompted by a miscellany of matters, including the oil crises of the 1970s which prompted a new look at resource depletion; fears of environmental disasters (global warming, the ozone layer, etc.) which seemed more credible after catastrophes such as Seveso, Bhopal, Chernobyl and, at home, Windscale and Flixborough. Environmental tragedies became eminently newsworthy and received constant press attention. There was (and remains) a constant apprehension that catastrophe is likely to strike at any time: as in California, where 'the big one' (the forecast earthquake of enormous power) is confidently expected. The environment has become part of the political coinage, and the parties vie with each other in producing convincing statements not only of their concern but also of their workable programmes of action.

This Common Inheritance (Cm 3188) includes, on its inside cover, an erudite statement which makes a declaration of the government's environmental sanctity:

> This White Paper is printed on recycled paper comprising about 50 per cent de-inked fibre and about 50 per cent unbleached best white unprinted waste, depending on availability. Up to 15 per cent virgin pulp from managed forests may have been used to strengthen the stock, depending on the quality of the recycled material. The latex bonding in the coating of the paper is fully recyclable.

Such statements, though usually briefer and clearer, are now commonplace: 'recycled', 'recyclable', 'environmentally friendly' are now part of the advertising stock-in-trade, though not always justifiably.

Curiously, part of the growth of environmental consciousness was due initially to the lack of government concern. The environment was rarely the subject of political battles. Yet England has been a world pioneer on a number of environmental issues. The Alkali Inspectorate, which was established in 1863, was the world's first environmental agency. Some of the earliest voluntary organisations had their origin in England: for example, the Commons, Open Spaces and Footpaths Preservation Society in 1865 and the National Trust in 1895 – the latter an organisation that (with over two million members) has grown to be the largest conservation organisation in Europe. The 1947 Town and Country Planning Act introduced a remarkably comprehensive land use planning system (even though, in the circumstances of the time, much of rural land use was purposely omitted). Legislation on clean air has a long history, its major landmark being the 1956

Act, which was passed following the killer smog of 1952. England also had the first cabinet-level environment department (the Department of the Environment was established in 1970), although its name was, for many years, more impressive than its achievements. Yet these historical events stand as lonely peaks in an otherwise flat plain: until recently, the environment has not been a salient political issue (McCormick 1991; Robinson 1992).

Part of the reason for this has been the idiosyncratic nature of British pollution control: instead of the formal, legalistic and adversarial styles common elsewhere, Britain has traditionally operated a system of comfortable negotiation between government technicians and industry. This curiously informal and secretive system avoids confrontation and legalistic procedures (McAuslan 1991). The objective has been to achieve the *best practicable means* (affectionately known as BPM) of dealing with pollution problems – 'means' that will go as far as seems reasonable towards meeting desirable standards but which do not involve too great a strain on the polluter's resources. This system of voluntary compliance is a striking feature of other British regulatory systems. (It also appears in a different guise in countryside policy where, to quote from *This Common Inheritance* (Cm 3188, para. 7.3), 'the government works in partnership with [countryside] owners and managers to protect it through voluntary effort'. This approach has a long history: indeed, it has been the cornerstone of industrial air pollution control since the Alkali Act of 1874. Its modern version has been expanded to BPEO: *best practicable environmental option* which retains the element of negotiation but involves a wider consideration of environmental factors and an openness which was foreign to its predecessor (Royal Commision on Environmental Pollution 1988).

Environmental groups do not have the clout of the major economic interests, and therefore are not involved in the initial stages of policy-making, but they do have some powers of reaction to proposals that governments may have worked out with the more powerful interests. The Control of Pollution Act 1974, for example, 'was essentially shaped by industry and local government'; and the Country-side and Wildlife Act 1981 was similarly the result of the efforts of 'powerful farming and landowning lobbies' (McCormick 1991: 12). In both these cases, environmental groups fared badly, but they did have some effect – and they were accepted as legitimate actors on the political stage. Another success was achieved in 1986 when the environmental lobby prevailed upon the Minister of Agriculture to exempt conservation advice from the charges being introduced generally for agricultural advice: the consequence was that 'this previously peripheral role of the Ministry's advisory staff has become much more central' (Lowe and Flynn 1989: 278).

Mention also needs to be made of the importance of the system of parliamentary select committees. Though these committees are often regarded as inef-fectual, they have been of great value to environ-mental groups. They have provided a new public platform and a route for exerting pressure on Parlia-ment. In particular, the Environment Committee has become a respected source of alternative wisdom and relatively accessible information.

Another feature of British environmental politics is the active character of some of the important interest groups. Some of these are not merely inter-est groups: they own and manage extensive areas of land, and they fulfil a range of executive responsi-bilities. The National Trust and the Royal Society for the Protection of Birds, for instance, own and manage large areas of protected land. Such bodies are also characteristically charities and therefore debarred from overt political activity. Lobbying is thus not only well mannered: it is also discrete. The emphasis may be more on education than propa-ganda, though the distinction can be a fine one.

Governments may try to outflank environmental groups, but increasingly they cannot ignore them, particularly with their new access to power via the EU. Some thirty British groups, together with eighty from other countries, are members of the European Environmental Bureau, which gives them access to the European Commission and the Council of Ministers (Deimann 1994). The British groups have been able to make good use of their experience in lobbying. According to Lowe and Flynn (1989: 272), they 'have adapted more easily than many of their counterparts to the successive rounds of consultation and detailed redrafting of directives and regulations that characterise Commu-nity decision-making'. One illustration of their activity within Britain was the role they played in thwarting the Conservative government's original proposals for the privatisation of the regional water authorities. Initially, it was envisaged that the pri-vatised water companies would take over responsi-bilities for pollution control. The CPRE took legal advice on this, and it was submitted that it would be illegal under EU law for water pollution control to be the responsibility of private organisations. The government was obliged to change its plans radi-cally: the solution adopted was to set up a new body, the National Rivers Authority, to protect water quality. The CPRE have since published a *Cam-paigners' Guide* (1992a), which is one of the clearest, most succinct and easily accessible guides to EU environmental law.

Mrs Thatcher was initially averse to environmen-tal concerns, which she viewed as a brake on enter-prise. On this, as with much of what she sought to do, she neither obtained nor sought consensus. Nevertheless, her administrations followed tradi-tional British practice in responding 'pragmatically and flexibly, even opportunistically, when environ-mental issues have threatened to become too con-tentious' (Lowe and Flynn 1989: 273). And, of course, there was her remarkable conversion to the environmental cause in 1988 when she surprised everybody by testifying her personal 'commitment to science and the environment'. With resounding words, she rallied her followers to environmental-ism, declaring that Conservatives were 'not merely friends of the earth' but also 'its guardians and trustees for generations to come'. She continued: 'The core of Tory philosophy and the case for pro-tecting the environment are the same. No genera-tion has a freehold on the Earth. All we have is a life tenancy – with a full repairing lease. And this government intends to meet the terms of that lease in full.'

Whatever might have been the reasons for Mrs Thatcher's conversion, it moved environmental policy to centre stage (Robinson 1992: 178). It did not, however, resolve the dilemma faced by the Conservative government: how to reconcile its enterprise philosophy with a concern for good environmental management. The former is characterised by a market orientation, with profits as the reward; the latter revolves around much broader ideas. Market forces do not necessarily work well with environmental protection. Quite the contrary: individuals may be rewarded for actions which harm the environment (and, indeed, they may be subsidised to do so – as with some agricultural policies). Attempts can be made to adjust or influence the market (for example, with devices such as environmentally sensitive areas, woodland schemes or other financial mechanisms), but there is a limit to the extent to which a government wedded to market ideals can provide incentives for actions which protect the environment – and certainly incentives at the level necessary to provide adequate rewards (Milton 1991). An increasingly accepted solution is a firmer use of the *polluter pays principle* – officially embraced by the EU. Other alternatives involve increased regulation; and here again there is the very handy excuse of the EU.

THE IMPACT OF THE EUROPEAN UNION

There can be no doubt that the EU has had a major impact on British environmental policy. Indeed, it is not much of an exaggeration to say that much of the government's policy has been dictated by its directives (Milton 1991: 11). This is so despite the fact that the Treaty of Rome imposed no environmental obligations on member states, and the Community initially had no environmental policies. Indeed, Article 2 of the Treaty provided that sustained rather than sustainable growth was the aim: 'a continuous and balanced expansion'. The international scene changed in the late 1960s and early 1970s, and was significantly influenced by the UN Conference on the Human Environment which was held in Stockholm in 1972. In the same year, the EC determined that economic expansion should not be 'an end in itself', and that 'special attention will be paid to protection of the environment' (Robins 1991: 7). In 1973, the first EC *Action Programme on the Environment* was agreed, covering the period 1973–6. Further programmes followed: at the time of writing, the most recent (the fifth) covers the period 1993–2000. The Single European Act of 1987 placed this activity on a firm constitutional basis and, significantly, added the important provision that 'environmental protection requirements shall be a component of the Community's other policies' (Haigh 1990: 11).

The environmental action programmes have had increasing impact on policy and practice in member states. They are 'forward planning' documents for emerging policies to be implemented by the EU and followed by national, regional and local governments. Whilst they have no binding status, many of the proposals result in directives and other action. The Fifth Action Programme has brought a more comprehensive and long-term approach. The overriding aim of the programme is to ensure that all EU policies have an explicit environmental dimension. It stresses the potential of spatial planning instruments. EU documents are not noted for their brevity and the programme documents are far too wordy to reproduce, but the following gives some flavour of their character. It also illustrates the importance attached to spatial planning instruments:

> The community will further encourage activities at local and regional level on issues vital to attain sustainable development, in particular to territorial approaches addressing the urban environment, the rural environment, coastal and island zones, cultural heritage and nature conservation areas. To this purpose, particular attention will be given to: further promoting the potential of spatial planning as an instrument to facilitate sustainable development . . . ; developing a comprehensive approach to urban issues . . . [and] developing a demonstration programme on integrated management of coastal zones . . . to facilitate sustainable development, and promote the potential of the European Spatial Development Perspective in addressing the territorial impact of sectoral policy,

and the need for greater integration of land use and transport planning.

(EC 1992: *Towards Sustainability*)

However, these are general policy statements. The workhorses of the EU are its regulations and directives. Regulations become law throughout the EU as soon as they become operative. They create legal rights and obligations without any further action by national governments, though they may be implemented by domestic legislation. The great majority of EU environmental laws are in the form of directives. These are 'binding as to the result to be achieved, upon each member state to which they are addressed'. This leaves the individual states free to choose the manner in which directives are translated into national law. Since different countries have different legislative and administrative styles (and, of course, their particular political situations) it is not surprising that this gives rise to difficulties. Indeed, it is unusual for directives to be transposed into national legislation by the due date – which is typically two months after adoption by the Council of Ministers. Nevertheless, they must be implemented 'in a way which fully meets the requirements of clarity and certainty in legal situations'. States cannot rely on administrative practices carried out under existing legislation (Wägenbaur 1991). Moreover, if a directive is not implemented by national law, it is possible for legal action to be taken by private parties to seek enforcement. A case in point is the action taken by Friends of the Earth alleging that the water supplied by Thames Water Utilities did not comply with the requirements of the directive on drinking water. Furthermore, EU law must be followed even if national law contradicts it.

In spite of the seemingly very strong powers possessed by the EU, it seems likely that much more is to come: 'the environment is still regarded as a peripheral issue for the Community'. Only 0.1 per cent of the EU budget is spent on environmental projects. Whereas the agriculture directorate has a staff of over 1,000, the environment directorate has less than 150 (Robins 1991: 8). The advent of the European Environment Agency will presumably expand the Commission's activities; but current

agreements limit its responsibilities to monitoring: it will have no enforcement powers. It also needs to be noted that, as with much on the national scene, EU environmental policy has often been 'disaster driven' – for example, the introduction of measures on sea pollution following the wreck of the *Amoco Cadiz* (Freestone 1991: 138).

The constraints placed on the UK government must be stressed. It is particularly striking when a Conservative government is in power:

> A number of factors have combined to imprint on European environmental policy a temper of progressive reform somewhat at odds with the outlook of the Thatcher government. These factors include the regulatory and integrating inclinations of the European Commission, the corporatist interests of transnational business, the strength of the environmental and consumer movements across Western Europe, the pro-environmental stance of some of the leading member states, and the imperative need for community safeguards over sensitive aspects of public welfare to foster popular confidence in the process of economic integration.
>
> (Lowe and Flynn 1989: 277)

Several of the major issues in environmental regulation are discussed in this chapter, and the influence of the EU noted. First, however, it is worth noting how the UK approach differs from other countries in Europe, particularly Germany. An important difference in principle (differences in practice may be less marked) is that of 'anticipation' as distinct from reaction. Whereas the UK has taken the view that environmental problems should be defined in terms of their measurable impacts, other countries have gone beyond this and have anticipated problems before the degree of environmental damage can be ascertained.

THE PRECAUTIONARY PRINCIPLE

A particularly striking example of the precautionary principle is the German *vorsorgeprinzip*. The term is not easily translatable: it is commonly taken to mean the principle of 'prevention' or 'anticipation', but this fails to capture its full meaning, as do the terms 'precaution' or 'foresight'. The German word connotes a 'notion of good husbandry which represents what one might also call best practice'. *Vorsorgeprinzip* is also different from the principle of prevention which forms the basis of the EU policy. To quote from the *Second Action Programme on the Environment*:

> The best environment policy consists in preventing the creation of pollution or nuisances at source rather than subsequently trying to counteract their effects. To this end, technical progress must be conceived and directed so as to take into account the concern for the protection of the environment and for the improvement of the quality of life, at the lowest cost to the Community.

Möltke (1988) comments that this principle, though 'practicable and generally also economically the most reasonable approach . . . gives no guidance as to the degree of prevention, whereas this is an essential aspect of the *vorsorgeprinzip*': '*Vorsorgeprinzip* is more than just prevention as an efficient means to an end but rather prevention as an end of itself.' The aim is, therefore, to establish pollution control policy, not merely as a means of reducing economic or social cost but also as a means of preserving wider ecosystems. The principle has no economic qualifications attached to it, although these tend to appear in practice. Typically, the European approach involves the avoidance of 'excessive cost'. This, of course, is no easier to define than concepts such as 'reasonable cost', but it is clearly intended to be more demanding. Shed of its more philosophical overtones, the issue is fundamentally 'whether to protect environmental systems before science can determine whether damage will result, or whether to apply controls only with respect to a known likelihood of environmental disturbance' (O'Riordan and Weale 1989: 290).

BATNEEC, BPEO AND BPM

The currently favoured concept in the EU is BATNEEC: *best available technology not entailing excessive cost*. The term has been adopted in the English Environmental Protection Act 1990, but there are differences of interpretation among European countries. Also in currency, in this directory of acronyms, is the BPEO (*best practicable environmental option*) concept proposed by the Royal Commission on Environmental Pollution. This was seen as an extension of the concept of *best practicable means* (BPM), which dates back to the Alkali Act 1874 'and which has been the cornerstone of industrial air pollution control in England and Wales since that time' (Royal Commission on Environmental Pollution 1988: para. 1.3). Central to this principle is the recognition of the need for a coordinated approach to pollution control — an approach that takes into account the danger of the transfer of pollutants from one medium to another, as well as the need for prevention. It is defined by the Royal Commission in these terms:

> A BPEO is the outcome of a systematic consultative and decision-making procedure which emphasises the protection and conservation of the environment across land, air and water. The BPEO procedure establishes, for a given set of objectives, the option that provides the most benefit or least damage to the environment as a whole, at acceptable cost, in the long term as well as in the short term.

Costs in environmental protection are elusive, complex and controversial. The Conservative government laid great stress on deregulation and, in harmony with the EC, on the *polluter pays principle*. It is an article of faith (which has considerable economic support) that 'economic instruments are an inherently more flexible and cost effective way of achieving environmental goals'.

ECONOMIC INSTRUMENTS OF ENVIRONMENTAL POLICY

Public opinion is in favour of regulatory standards because of their apparent fairness: all are required to meet the same target. Polluters may also like them because of the certainty which they give to the market. In fact, the fairness is illusory. Fixed standards impose quite different costs on different firms. Some can meet standards easily, particularly if their machinery is modern. Older firms, by contrast, may need to invest heavily in new plant to meet a standard; and they will understandably seek to negotiate a less onerous standard. More important in terms of effective environmental improvement, firms will tend not to seek anything beyond the regulatory standard even if they can achieve a higher standard at relatively low cost. They will have no incentive to do so if they do not thereby obtain other benefits.

There are considerable advantages to be derived from designing pollution controls in a way that gives firms economic incentives to reduce pollution to the maximum extent. An incentive can make a firm take a totally different approach to its waste. If, for example, a tax is levied for every ton of waste produced, a firm will not be satisfied with calculating the economics of compliance: it will be motivated to review its processes to reduce its waste to the minimum. It has an inducement to calculate the real cost which its waste involves – a cost which otherwise is borne by the environment.

Of course, this is not to suggest that the interests of individual firms and of environmental policy are now uniformly harmonious: most will obey the law and follow their self interest: 'It is not the job of companies to decide what values ought to be attached to natural resources and what the priorities of environmental policy ought to be' (Cairncross 1993: 299).

The use of regulatory instruments and an absence of an economic incentive to reduce pollution does not mean that firms simply abide by the dictates of the regulation. Far from it, the incentive is to avoid the costs of compliance. Thus the regulatory agency has to demand detailed records, inspect the record-keeping system, carry out site inspections, and undertake other such control functions. If, however, the resources of the regulatory body are not sufficient to enable this to be done, some firms at least will be tempted to circumvent or even ignore some part of a regulation. Since administrative resources are typically inadequate, this is a significant issue. The laxer the day-to-day controls and the higher the costs of compliance, the greater will this temptation be. Overstretched agencies may well know that

some firms are in default, but they may have some difficulty proving it, or they may have to accept a firm's assurance that it is doing the best it can or, given the pressure of work, they may simply leave the relevant file in the pending tray. Particularly bad cases may be prosecuted, but this takes even more time and resources, and the courts can be unpredictable. There is no need to labour the point further. The incentives to adhere to a regulation are weak for the individual firm; the costs of enforcement are high for the agency.

The simplest economic instrument is a tax on pollution, levied at a rate determined, for example, in relation to the damage caused by the pollutants and the costs of cleanup. Such a tax could be levied on, for instance, lead or carbon content. (Several European countries have such a carbon tax.) The tax provides an immediate incentive to firms to reduce their use of the pollutant – and it is a continuing incentive. The difficulty arises in setting an equitable rate – a problem which also arises with marketable pollution rights.

Economic incentives can be applied to some types of waste with a deposit-refund system. This is essentially the same as the charges on returnable bottles, though rather more complicated. The producer of something which would become a waste after it has been used in a manufacturing process (a solvent, for instance) would be required to pay a charge for each unit produced. This would increase its price (thereby introducing an incentive for reduction in its use). A refund of the charge would be payable to anyone who returned the solvent after its use. This system has the advantage of providing a disincentive to illegal tipping. The same system can be applied to motor vehicles.

The Advisory Committee on Business and the Environment reports annually (to the President of the Board of Trade and the Secretary of State for the Environment) on economic instruments. Its recent reports have dealt with tradable permits for water pollution, the landfill tax, and the promotion of alternative fuels. Proposals for a landfill tax were published in 1995 and implemented in 1996. Proposals for alternative fuels, as well as various economic instruments, were discussed by the Royal Commission on Environmental Pollution in its report on *Transport and the Environment* (Cm 2674, 1994). Some of these, such as road pricing, have been debated for many years, but the technical and political difficulties constitute a major obstacle.

Nevertheless, there is no doubt that there is considerable scope for the greater use of economic instruments in environmental policy, though there are areas where there is no alternative to regulation, such as requiring catalytic converters to reduce car emissions. Sometimes, a combination of regulation and pricing is used, as with the regulation of the lead content of petrol and the lower tax rate on unleaded petrol.

ENVIRONMENTAL AGENCIES

The 1995 Environment Act provided for the establishment of an Environment Agency for England and Wales and an equivalent Scottish Environment Protection Agency. The idea of such an agency had been resisted by the government for a number of years, and the announcement by the Prime Minister in 1991 that it had been decided to set up 'a new agency for environmental protection and enhancement' represented a major change in policy. The reasons for this change were explored by the House of Commons Environment Committee which, having had its proposals for an agency previously rejected, wondered why there had been such a change of heart: 'in just two years, the government's policy had shifted from outright rejection of the notion of such an agency, to enthusiastic acceptance'. In reply, the Secretary of State, Michael Heseltine, referred to 'evolutionary change' and the fresh views of a new minister (i.e. himself), but he laid stress on the importance of relationships with industry:

> The relationships that we are developing with our industrial base, particularly through the Advisory Committee [on Business and the Environment] chaired by Mr John Collins of Shell with industry and commerce, has brought home to us very forcibly that

industry is deeply affected by these increasingly comprehensive regulatory processes. The concept of a one-stop shop, where one group of experts is available for discussion and negotiation with the industrial and commercial world is particularly evident as we see the consequences of integrated pollution control flowing through.
> (HC Environment Committee 1992: *The Government's Proposals for an Environment Agency*)

Heseltine's point was echoed in the evidence submitted by business and commerce: there was considerable support for the 'one-stop shop' for businesses whose activities involve pollution. The Institute of Directors, for example, saw the Agency as 'a step towards overcoming the administrative nightmare that would otherwise have resulted from businesses having to seek a multiplicity of permits to transport and dispose of waste'. In its view, the local authority powers over air pollution control should be similarly integrated. The Environment Committee itself argued that the Agency should have more functions than proposed by the government. The TCPA went further and argued the need for integration between environmental planning and land use planning: the new Agency should be introduced in conjunction with the latest reorganisation of local government.

Another factor in the debate was the importance of having an agency that was able to negotiate with the EU from a position of strength. There were dangers that pressures would develop for the new European Environmental Agency to have enforcement (as well as monitoring) powers. The HC Committee considered that it was 'imperative that the United Kingdom's Environmental Agencies be seen to be effective at an early stage in order to maintain a high degree of subsidiarity and to avoid any further attempts to erode United Kingdom sovereignty in this regard'.

Against this background, the Environment Agency has taken over the responsibilities of bodies which were established by a reorganisation only a few years earlier. In England and Wales, these were the National Rivers Authority, Her Majesty's Inspectorate of Pollution, and the local waste regulation authorities. In Scotland, they were the river purification authorities, HM Industrial Pollution Inspectorate, and the waste regulation and local air pollution responsibilities of the district and islands councils. In Northern Ireland, the DoENI has all the responsibilities for environmental protection except waste disposal, which lies with the local authorities.

The Agency is a non-departmental public body, operating on *Next Steps* lines: the management has a large degree of freedom within the framework of ministerial guidance and its management framework. This framework is based on the government's overall environmental strategy (set out in *This Common Inheritance* (Cm 1200), its annual reports, and the 1994 *UK Strategy for Sustainable Development* (Cm 2426). In essence, this is a commitment to the goal of sustainable development: reconciling the twin objectives of economic development and environmental protection. However much debate there may be on what such terms really mean, an important implication is that the Agency has to take an integrated approach to its responsibilities: this, indeed, is its essential *raison d'être*.

The functions of the Agency are very wide and include responsibilities in relation to industrial pollution, waste and water. The objectives, as set out in its management strategy, are reproduced in the box.

CLEAN AIR

Concern about air pollution is not new: as early as 1273 action was taken in Britain to protect the environment from polluted air: a royal proclamation of that year prohibited the use of coal in London. (It was not effective, despite the dire penalties: it is recorded that a man was sent to the scaffold in 1306 for burning coal instead of charcoal.)

Those who pollute the air are no longer sent to the gallows, but, though gentler methods are now preferred, it was not until the disastrous London smog of 1952 (resulting in 4,000 deaths) that really effective action was taken. The Clean Air Acts of 1956 and 1968 prohibited the emission of dark smoke, provided for the control of the emission of grit and dust from furnaces, and established a system for local authority approval of chimney heights.

Environment Agency Objectives

Within the overall aim of contributing to sustainable development:

1 to adopt across all its functions, an integrated approach to environmental protection and enhancement which considers impacts of substances and activities on all environmental media and on natural resources;

2 to work with all relevant sectors of society, including regulated organisations, to develop approaches which deliver environmental requirements and goals without imposing excessive costs (in relation to benefits gained) on regulated organisations or society as a whole;

3 to adopt clear and effective procedures for serving its customers, including by developing single points of contact through which regulated organisations can deal with the agency;

4 to operate to high professional standards, based on sound science, information and analysis of the environment and of processes which affect it;

5 to organise its activities in ways which reflect good environmental and management practice and provide value for money for those who pay its charges and taxpayers as a whole;

6 to provide clear and readily available advice and information on its work;

7 to develop a close and responsive relationship with the public, local authorities and other representatives of local communities, and regulated organisations.

The Environment Agency

However, the principal source of air pollution at the time was domestic smoke, and it was in connection with this that the most extensive powers were introduced. Local authorities were empowered to establish *smoke control areas* in which the emission of smoke from chimneys constitutes an offence. All areas of the country which have had high levels of sulphur dioxide (SO_2) are now covered by this system.

Air quality has improved considerably over the last three to four decades: smoke emissions have fallen by 85 per cent since 1960; the notorious big-city smogs are a thing of the past; and hours of winter sunshine in central London have increased by 70 per cent. This is partly a success story of clean air policy, but also a result of the demise of the household coal fire and its replacement largely by central heating (now installed in over fourth-fifths of houses). In matters of the environment, however, problems are never 'solved': they are merely replaced by new ones. This is certainly the case with pollution. Pollutants from new industrial processes, from vastly increased use of chemicals and from vehicle emissions have presented problems of a serious and often baffling nature.

The governmental response has included the reorganisation of the machinery of controls, the gradual introduction of economic instruments for the control of pollution, and the expansion of research in the surprisingly large areas of scientific uncertainty.

The *Sustainable Development Strategy* (Cm 2426) warned that it would not be easy to achieve a progressive improvement in urban air quality, mainly because of the increase in road traffic. Though improvements are currently being made, the increase in traffic will 'at some point after 2010 begin to erode and outweigh the effect of reduction in emission levels'. Moreover, severe problems in the shorter run can be expected in 'hotspots', particularly in congested urban centres. Additionally, continuing research suggests that some pollutants may have health effects at lower levels of concentration than had previously been thought. The approach adopted to deal with these problems involves the development of a new framework of national air quality standards and targets; the introduction of new systems for local air quality management based on designated *air quality management areas*; and continued control of emissions, particularly from vehicles.

Air quality standards are not easy to determine: the scientific base is too inadequate, and a great deal of judgement amounting at times to guess work is necessary. The governmental response to this has been to work towards two measures: a long-term goal, and an operational gauge which would indicate air quality conditions so low as to require an immediate response. (Confusingly, these are both termed 'standards'.) Local authorities have an important role to play in working towards the target. In the first place, they 'will be expected to have regard to targets in preparing land use development plans and in carrying our other duties such as transport planning'. Second, in areas of high pollution they are responsible for *local air quality management plans* 'to overcome local problems in the most cost-effective manner' (see DoE, *Air Quality: Meeting the Challenge*, 1995). Legislative provisions were made in the 1995 Environment Act. At the time of writing, a national strategy (incorporating standards, targets and a timetable for their achievement) was in preparation.

Given the importance of traffic emissions in reducing air quality, it is not surprising that they are a focus of much policy debate. Unfortunately, the debate is more apparent than the policy: any effective policy relating to road traffic is fraught with political difficulties. Here, of course, environmental issues overlap with transport matters; these are discussed within the latter context in Chapter 11. However, to conclude this section on air quality, a summary is given of the specifically environmental initiatives.

An initial spur to policy was the EU directives on emission standards: these have had a noticeable effect, as has the introduction of unleaded petrol. A tax differential in favour of this type of petrol was introduced in 1986. Another tax policy ostensibly directed towards reducing emissions (by reducing the use of cars) is the annual increase in road fuel duties. Such policies, together with the higher efficiency standards of new cars, have so far kept pace with the effects of increased traffic. This is not expected to continue, however, without major changes in policy on transport (Royal Commission on Environmental Pollution (Cm 2674) 1994).

CLEAN WATER

In a country surrounded by water and with an annual rainfall of around 1,100 mm, one might expect that there would be little of a quantitative shortage of water in Britain. However, rain falls unevenly over both area and time. In the mountainous areas of the Lake District, Scotland and Wales, average annual rainfall exceeds 2,400 mm, while in the Thames estuary it is less than 500 mm. Annual variations can be very marked (though there is no overall trend), and droughts can cause severe shortages particularly in areas of low rainfall and high population (though the picture is complicated by the amount of reuse – which is high in the Thames region).

A number of factors have led to an increased concern about the adequacy of future water supplies: the drought of 1988–92 and the acute shortages during the summer of 1995; public reactions to water privatisation; and, more substantively, a better understanding of the issues stimulated at least in part by the establishment of the National Rivers Authority (under the Water Act 1989). There has been a stream of official reports and consultation papers, and also a development of academic studies in this area.

The NRA, now merged in the Environment Agency (EA), took over many of the regulatory powers of the former regional water authorities though, unlike those authorities, it has no operational responsibilities (these are carried out by ten water service companies). The EA has statutory functions in relation to water resources and the control of pollution in inland, underground and coastal waters. It can take preventive action to stop water pollution, take remedial steps where pol-

lution has already occurred, and recover the reasonable costs of doing so from a polluter. It also has certain powers to prevent flooding, as well as responsibilities for the licensing of salmon and freshwater fisheries, for navigation, and for conservancy and harbour authority functions. These powers have been broadened and strengthened by the Environment Act 1995.

The NRA prided itself as being 'the strongest environmental protection agency in Europe', and a very effective 'Guardian of the Water Environment'. It developed a sophisticated and relatively public regulatory system which involved the setting of water quality objectives and a requirement that consent be obtained for discharges of trade and sewage effluent to controlled waters. Extensive monitoring programmes include surveys of the quality of rivers, estuaries and coastal waters. The highly detailed figures produced from these surveys are not easy to summarise or to interpret. There appears to be a modest improvement in the overall quality of rivers, but the number of pollution incidents has increased. It is thought that this is 'largely due to heightened public concern about pollution and specific encouragement by the NRA to get the public to report incidents' (DoE, *Digest of Environmental Statistics* no. 17, 1995, para. 3.44). Bathing water quality is improving but, in 1994, 18 per cent of UK bathing waters failed to meet the mandatory coliform bacteria standards of the EU Bathing Waters Directive.

The relationships between 'land use planning and the water sector' (the title of an important paper by Slater *et al.* 1994) have received remarkably little attention until recently, no doubt because of the general adequacy of water supplies, their cheapness and the continued benefit of the huge investment made in Victorian times. A major programme of new investment is now clearly needed, not only to replace outworn facilities, but also to meet new demands for water, for environmental protection and for sustainability (CPRE 1993c). At the same time, increased concerns about water supply have come from a variety of sources: developers' complaints about the level of infrastructure charges;

NRA objections to the DoE household projections for the South-East on the grounds of insufficient water and sewerage capacity; increasing public resistance (particularly in the South-East) to new reservoirs and sewage plants; public concerns about variations in the level and cost of services; and an increased awareness of the importance of the relationship between water and land use planning, particularly on the part of the NRA (Slater *et al*. 1994: 376).

The new perception of water as a valuable resource (as distinct from a plentiful and costless natural bounty) has raised a number of major policy issues including leakage control, metering and other methods of demand management. In the words of David Pearce, 'placing water use and planning on a more economic footing might well be expected to produce the classic "double dividend": it will not only greatly facilitate efficient and sustainable water use in the future, but it will also significantly reduce costs now. The economic value of the United Kingdom's fresh water needs to be recognised' (Pearce 1993: 77).

WASTE: THE DUTY OF CARE

The UK produces over 400 million tonnes of waste each year (Table 7.1). Most of this is agricultural and mining and quarrying waste. Much of the remainder is 'controlled waste', i.e. waste that is controlled by the provisions of the Environmental Protection Act. The definition of waste gives rise to problems of a byzantine character: the lengthy DoE Circular 11/94 explains all.

Before the 1990 Act, the statutory responsibility for both waste management and waste disposal was vested in the English county councils, the Welsh district councils and the Scottish district and island councils. Generally, the quality of service was judged to be related to the size of authority: the larger authorities had a greater range of competence and also were less subject to parochial pressures. There was also a wider concern that waste regulation and disposal was the responsibility of the same authority. The situation was generally accepted as being unsatisfactory, and the Environmental Protection Act of 1990 introduced a new regime of waste control which included the separation of the regulatory and operational functions. Administratively, 'the poacher will no longer be the gamekeeper' (Lewis 1992). Following the changes made by the Environment Act 1995, waste regulation functions are the responsibility of the new Environment Agencies. Waste collection remains with local government.

The 1990 Act imposes a *duty of care* on all who are concerned with controlled waste. This duty, similar to that imposed on employers by the Health and Safety at Work Act 1974, is designed to ensure that

Table 7.1 Estimated Annual Waste Arising in the UK, by sector, 1990

Sector	Million tonnes	%
Construction and demolition waste	70	16
Other industrial waste	70	16
Sewage sludge	35	8
Dredged spoils	35	8
Agricultural waste	80	18
Mining and quarrying waste	110	25
Households	20	5
Commercial	15	3
Total	435	100

Source: Making Waste Work (Cm 3040), 1995

waste is properly managed. In particular, it aims at making fly-tipping more difficult. The duty requires anyone who has control of waste at any stage, from its original production to its final disposal, to take responsibility for its safe and legal handling.

In the same way that energy conservation is an important feature of energy policy, so waste minimisation is of waste management. Increased attention is being paid to this as well as to the potential for recycling. A scheme of recycling credits was introduced by the government in 1992. These payments, made by local authorities to recyclers, reflect the savings made on disposal costs because of the recycling. The government's targets (for the year 2000) are for 25 per cent of household waste to be recycled (compared to less than 10 per cent in 1996); recovery of 58 per cent of packaging waste; and (by 2006) an increase in the use of waste and recycled materials for aggregates (from 30 million tonnes to 55 million tonnes a year).

It seems, however, that existing measures are not sufficient to enable such targets to be reached, and new economic instruments are being considered, such as charging householders for the amount of waste they throw away.

Before the Planning and Compensation Act 1991, policies for waste formed part of county structure plans (and, in some cases, separate 'subject' waste plans). The 1991 Act introduced 'waste local plans' in England: these contain 'detailed policies in respect of development which involves the depositing of refuse or waste materials other than mineral waste'. Minerals come under a different provision, but, since a significant proportion of waste arises from mineral workings, waste and mineral plans can be combined. Waste policies deal with all types of waste, including scrapyards, clinical and other types of waste incinerator, landfill sites, waste storage facilities, recycling and waste reception centres, concrete crushing and blacktop reprocessing facilities, and bottle banks. In drawing up their waste local plans, LPAs are required to have regard to national and regional policies. The 1995 strategy for sustainable waste management (*Making Waste Work*, Cm 3040) is a material consideration for this purpose and for determining planning applications. (This is a non-statutory document which will eventually be superseded by a statutory waste strategy; this is expected in 1997 or 1998.)

Local authorities have the responsibility of ensur-

Waste Management Policy

The government's policy is that:

(a) subject to the best practicable environmental option (BPEO) in each case, waste management should be based on a hierarchy in which the order of preference is:

   (i) *reduction* – by using technology which requires less material in products and produces less waste in manufacture, and by producing longer-lasting products with lower pollution potential;

   (ii) *re-use* – for example, returnable bottles and reusable transit package;

   (iii) *recovery* – finding beneficial uses for waste including: *recycling* it to produce a usable product; *composting* it to create products such as soil conditioners and growing media for plants; *recovering energy* from it either by burning it or by using landfill gas; and

   (iv) *disposal* – by incineration or landfill without energy recovery; and

(b) each of these options should be managed, and where necessary regulated, to prevent pollution of the environment or harm to human health.

(DoE Circular 11/94)

ing that waste facilities are not developed in locations where they would be harmful or otherwise unacceptable for land use reasons. But they also have an important positive planning role in 'ensuring that there is adequate scope for the provision of the right facilities in the right places'. In this, they are required to have regard to 'the proximity principle' and the 'regional self-sufficiency principle'. These stem from the desirability of waste recovery or disposal being close to the place where it is produced. This 'encourages communities to take more responsibility for the waste which they – either themselves as householders, or their local industry – produce. It is their problem, not someone else's.' It also limits environmental damage due to the transportation of waste.

INTEGRATED POLLUTION CONTROL

Among the innovations made in environmental policy, a particularly interesting one is that of *integrated pollution control* (IPC). This is the administrative apparatus for implementing the *best practical environmental option* (BPEO). It contrasts with the customary British method of operating different controls in isolation, with separate approaches to individual forms of pollution. The crucial problem with this is that pollution does not abide by the boundaries of air, land and water: pollution is mobile. In the jargon, it is a 'cross-media' problem. Thus:

> Substances discharged into one medium may have damaging effects in another. For example, chemical fertilizers and sewage sludge spread on agricultural land, or toxic waste buried in land-fill disposal sites, may leach into watercourses. Sulphur dioxide and oxides of nitrogen emitted into the atmosphere from conventional power stations, large industrial combustion plants, and motor vehicles can produce acid rain that erodes buildings and kills trees and fish. Likewise, atmospheric emissions of carbon dioxide from the burning of fossil fuels contribute to the greenhouse effect, which may lead to global warming and a consequent rise in sea level . . . Moreover, unless pollutants can be eliminated altogether, they are simply transformed or transferred elsewhere, and measures taken to reduce pollution in one environmental med-

ium may create problems for another. For instance, when flue gas desulphurization equipment is fitted to power stations in order to reduce air pollution from emissions of sulphur dioxide, the 'scrubbing' process produces by-products of contaminated water and lime that must be discharged into the aqueous environment or disposed of as waste on land.

> (Gibson 1991: 19)

Using BPEO involves choosing the best way of dealing with pollution, and this can be done only if there is an administrative organisation with sufficiently broad powers to take an 'integrated' approach. The 1990 Act made provision for HMIP to fulfil this purpose (now merged into the even more integrated Environment Agency). Under the Act, certain prescribed polluting processes require authorisation by the Agency. To obtain this, the operator must show that the *best available techniques not entailing excessive cost* (BATNEEC) are being used:

1. for preventing the release of prescribed substances into an environmental medium, or, where that is not practicable, for reducing the release to a minimum; and
2. for rendering harmless any other substance which could cause harm if released into any environmental medium.

Where a process involves the release of harmful substances to more than one medium, BPEO must be adopted. Additionally, certain statutory environmental standards ('quality objectives') have to be met.

NOISE

'Quiet costs money . . . a machine manufacturer will try to make a quieter product only if he is forced to, either by legislation or because customers want quiet machines and will choose a rival product for a lower noise level.' So stated the Wilson Committee in 1963. This, in one sense, is the crux of the problem of noise. More, and more-powerful, cars, aircraft, portable radios and the like must receive strong public opprobrium before manufacturers –

and users – will be concerned with their noise level. Similarly, legislative measures and their implementation require public support before effective action can be taken.

As with other aspects of environmental quality, attitudes to noise and its control have changed in recent years, partly as a result of the advent of new sources of noise such as portable music centres, personal stereos and electrical DIY and garden equipment, as well as greatly increased traffic. (Developments in electronics have also provided easier methods of obtaining data on noise.) The increased concern about noise is reflected in a succession of inquiries and new planning guidance (PPG 24). More substantively, two Acts have been passed to provide stronger measures for dealing with the problems. The Noise and Statutory Nuisance Act, which was passed in 1993, strengthened local authority powers to deal with burglar alarms, noisy vehicles and equipment, and various other noise nuisances. Second, the Noise Act of 1996 provided a summary procedure for dealing with noise at night (11 p.m. to 7 a.m.). This includes powers for local authorities to serve a warning notice and to seize equipment which is the source of offending noise. The 1996 Act does not require local authorities to use its provisions, but the situation is to be reviewed in the light of experience. Some of the procedures are rather tortuous, but this is common with nuisance law. A significant amount of research on various aspects of noise continues: reports have been published on aircraft noise and sleep disturbance (DoT 1994) and neighbourhood noise (BRE 1994).

There are three ways in which noise is regulated: by setting limits to noise at source (as with aircraft, motor cycles and lawnmowers); separating noise from people (as with subsidised double glazing in houses affected by serious noise from aircraft or from new roads); and exercising controls over noise nuisance. Where intolerable noise cannot be reduced and reduces property values, an action can be pursued at common law or, in the case of certain public works, compensation can be obtained under the Land Compensation Act 1973.

As an example of the latter, some 1,700 claims in the area of Stansted airport have been made, and compensation paid in amounts ranging from £1,000 to £57,000 for depreciation of value. By mid-1995, about 150 claims had been settled and a total of £500,000 paid in compensation (*Saffron Walden Weekly News*, 24 August, 1995). The compensation is payable for 'physical factors arising from commencement of use of the aprons and taxiways associated with development at Stansted Airport'.

Noise from neighbours is the most common source of noise nuisance and complaints. The number of such complaints increased by six times over the ten years from 1976 (Batho Report 1990: para. 3.3). This is a difficult problem to deal with, and official encouragement is being given to various types of neighbourhood action, such as 'quiet neighbourhood', 'neighbourhood noise watch', noise mediation and similar schemes (Oliver and Waite 1989). There is provision under the Control of Pollution Act 1974 for the designation by local authorities of *noise abatement zones*, though the statutory procedures for these are cumbersome and, in any case, they are not well suited to dealing with neighbourhood noise in residential areas (though they are useful for regulating industrial and commercial areas). Further measures to deal with neighbourhood noise were canvassed in a 1995 consultation paper. These included the controversial proposal that a new criminal offence should be created for night-time noise disturbance. This and other statutory provisions were under consideration at the time of writing.

Transportation noise takes many forms and is being tackled in various ways (conveniently summarised in Chapter 4 of the Royal Commission on Environmental Pollution's 1994 report on *Transport and the Environment*, Cm 2674). Road traffic noise is the most serious, in the sense that it affects the most people. Here emphasis is being put on the development of quieter road surfaces and vehicles. The Batho report also proposed extending the compensation scheme for people affected by high noise levels, including those resulting from traffic management schemes. Aircraft noise has long been subject to controls both nationally and (with the UK in the lead) internationally. The principal London airports are required by statute to provide sound insulation in homes seriously affected by aircraft noise, and similar non-statutory schemes apply to major airports in the provinces. Older jets are being phased out, and controls are also exercised over flight paths and times.

ENVIRONMENTAL ASSESSMENT

As environmental issues have become more complex, ways have been sought to measure the impacts of development. Cost-benefit analysis was at one time seen as a good guide to action. By taking into account non-priced benefits, such as the saving of time and the reduction in accidents, it can 'prove' that developments such as the Victoria Underground line are justified. Useful though this technique is for incorporating certain non-market issues into the decision-making process, it has serious limitations. In particular (quite apart from the problems of valuing 'time'), some things are beyond price, while others have quite different 'values' for different groups of the population. Reducing everything to a monetary price ignores factors such as these. Alternatives such as Lichfield's (1956) *planning balance sheet* and Hill's (1968) *goals achievement matrix* attempt to take a much wider range of factors into account.

Environmental assessment (EA) is a procedure introduced into the British planning system as a result of an EC Directive. Though it might appear that environmental assessment is nothing new on the British planning scene (hasn't this always been done with important projects?), it is in fact conceptually different in that it involves a highly systematic quantitative and qualitative review of proposed projects – though early indications suggest that the practice is somewhat different (Wood and Jones 1991). Nevertheless, unlike some European countries, Britain has had, since the 1947 Act, a relatively sophisticated system which involves a case-by-case review of development proposals. Indeed, there was some controversy between the UK government and the EC on a number of major projects about the

need for formal EA in addition to the extensive reviews under the standard system. A summary of the procedure is given in Figure 7.1. A good statement of the particular character of EA is given in a DoE guide to procedures:

> What is new about EA is the emphasis on systematic analysis, using the best practicable techniques and best available sources of information, and on the presentation of information in a form which provides a focus for public scrutiny of the project and enables the importance of the predicted effects, and the scope for modifying or mitigating them, to be properly evaluated by the planning authority before a decision is given.
>
> (DoE, 1989, *Environmental Assessment: A Guide to Procedures*: 3)

It is important to appreciate that EA is a *process*. The production of an *environmental statement* (ES) is one part of this. (The common term *environmental impact statement* (EIS) is, in fact, an American import, though the meaning is the same.) The process involves the gathering of information on the environmental effects of a development. This information comes from a variety of sources: the developer, the local planning authority, statutory consultees (such as the Countryside Commission), and third parties (including environmental groups).

For some types of development an EA is mandatory. These are listed in schedule 1 of the regulations (and are therefore inevitably known as 'schedule 1 projects'). These include large developments such as power stations, airports, installations for the storage of radioactive waste, motorways, ports and such like. Projects for which EA *may* be required ('schedule 2 projects') are those which have *significant* environmental impacts. There are three main types of development where it is considered that an EA is needed:

1 for major projects which are of more than local importance, principally in terms of physical size;
2 'occasionally' for projects proposed for particularly sensitive or vulnerable locations, for example, a national park or a SSSI; and
3 'in a small number of cases' for projects with unusually complex or potentially adverse effects, where expert analysis is desirable, for example, with the discharge of pollutants.

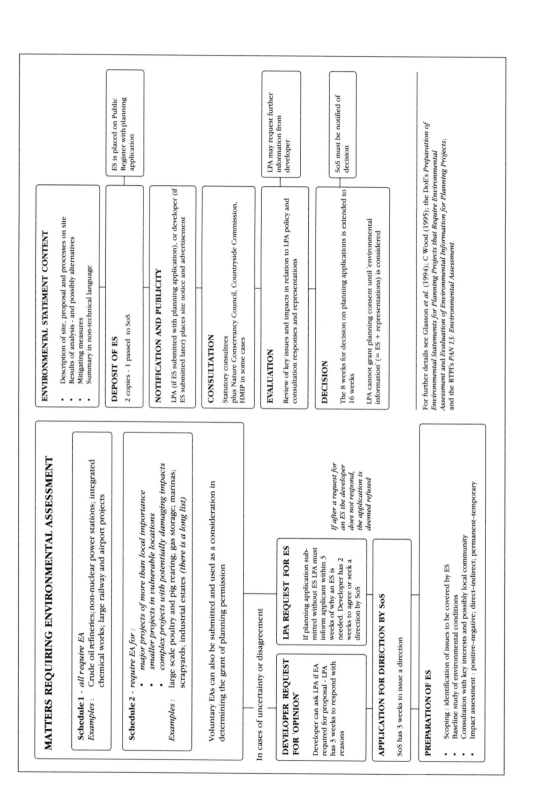

MATTERS REQUIRING ENVIRONMENTAL ASSESSMENT

Schedule 1 - *all require EA*
Examples : Crude oil refineries; non-nuclear power stations; integrated chemical works; large railway and airport projects

Schedule 2 - *require EA for* :
- *major projects of more than local importance*
- *smaller projects in vulnerable locations*
- *complex projects with potentially damaging impacts*
Examples : large scale poultry and pig rearing; gas storage; marinas; scrapyards; industrial estates (there is a long list)

Voluntary EAs can also be submitted and used as a consideration in determining the grant of planning permission

In cases of uncertainty or disagreement

| **DEVELOPER REQUEST FOR 'OPINION'** | **LPA REQUEST FOR ES** |

Developer can ask LPA if EA required for proposal - LPA has 3 weeks to respond with reasons

If planning application submitted without ES LPA must inform applicant within 3 weeks of why an ES is needed. Developer has 2 weeks to agree or seek a direction by SoS

If after a request for an ES the developer does not respond, the application is deemed refused

APPLICATION FOR DIRECTION BY SoS

SoS has 3 weeks to issue a direction

PREPARATION OF ES
- Scoping : identification of issues to be covered by ES
- Baseline study of environmental conditions
- Consultation with key interests and possibly local community
- Impact assessment : positive–negative; direct–indirect; permanent–temporary

ENVIRONMENTAL STATEMENT CONTENT
- Description of site, proposal and processes on site
- Results of analysis - and possibly alternatives
- Mitigating measures
- Summary in non-technical language

DEPOSIT OF ES

2 copies - 1 passed to SoS

ES is placed on Public Register with planning application

NOTIFICATION AND PUBLICITY

LPA (if ES submitted with planning application), or developer (if ES submitted later) places site notice and advertisement

CONSULTATION

Statutory consultees
plus Nature Conservancy Council, Countryside Commission, HMIP in some cases

EVALUATION

Review of key issues and impacts in relation to LPA policy and consultation responses and representations

LPA may request further information from developer

DECISION

The 8 weeks for decision on planning applications is extended to 16 weeks

LPA cannot grant planning consent until 'environmental information' (= ES + representations) is considered

SoS must be notified of decision

For further details see Glasson *et al.* (1994); C Wood (1995); the DoE's *Preparation of Environmental Statements for Planning Projects that Require Environmental Assessment and Evaluation of Environmental Information for Planning Projects*; and the RTPI's *PAN 13: Environmental Assessment*.

Figure 7.1 The Environmental Impact Assessment Process

Major Events in Environmental Planning

1972 UN Conference on the Human Environment, Stockholm

1973 First EC Action Programme on the Environment

1985 First EC Directive on Environmental Assessment

1987 World Commission on Environment and Development: Brundtland Report – *Our Common Future*

1990 *This Common Inheritance: Britain's Environmental Strategy* (Cm 1200)

1992 UN Conference on Environment and Development (the Earth Summit), Rio:

Agenda 21: a comprehensive world-wide programme for sustainable development in the twenty-first century

Climate Change Convention: international agreement to establish a framework for reducing risks of global warming by limiting 'greenhouse gases'

Biodiversity Convention: international agreement to protect diversity of species and habitats

Statement of Forest Principles for management, conservation and sustainable development of the world's forests

UN CSD: establishment of UN Commission on Sustainable Development

1994 *Sustainable Development: The UK Strategy* (Cm 2426)

1996 UN Habitat II Conference, Istanbul

There is a marked resemblance between this and the circumstances in which the Secretary of State may exercise the powers of 'call-in' – they both relate to developments of particular importance which require more than a normal scrutiny for planning and environmental purposes.

In 1995, *permitted development rights* were withdrawn from projects listed in schedule 1, and also for projects having likely significant environmental effects (DoE Circular 3/95).

PENALTIES FOR POLLUTION

A striking feature of the recent environmental legislation is the severity of the penalties for polluting (Harris 1992a). One feature is particularly noteworthy: the use of 'strict liability'. Generally, under English law, the prosecution has the burden of proving that a defendant is guilty beyond reasonable doubt. The 1990 Act, however, provides that where

it is alleged that BATNEEC has not been used in a prescribed operation, 'it shall be for the accused to prove that there was no better available technique not entailing excessive cost than was in fact used'. This makes an offence one of 'strict liability', in contrast to the traditional one of 'fault-based'. Though its use is likely to be rare, it is indicative of the change in official attitudes to pollution (documented in Rowan-Robinson and Ross 1994). It will also involve highly technical matters which may present severe difficulties for the existing courts. Indeed, some have argued that there is a need for a specialised court (Carnwath 1992).

FURTHER READING

For a history of pollution control and much else on the origins of 'environmental policy', see Ashby and Anderson (1981) *The Politics of Clean Air*. Also strongly recommended is Ashby's (1978) reflective

Reconciling Man with the Environment. A detailed legal sourcebook is Tromans *et al.* (1997) *Encyclopaedia of Environmental Law* (looseleaf; updated regularly). Less daunting is Hughes (third edition, 1996) *Environmental Law.*

Sustainability

The Brundtland Report, (Report of the World Commission on Environment and Development, 1987) *Our Common Future*, is perhaps the most quoted source on sustainability, though its interest is increasingly historical. Among the very many books and articles dealing with the subject are: Blowers (1993) *Planning for a Sustainable Environment*; Breheny (1992) *Sustainable Development and Urban Form*; Cairncross (1993) *Costing the Earth* (ch. 1); Cooper and Palmer (1992) *The Environment in Question*; Khan (1995) 'Sustainable development: the key concepts, issues and implications'; and Owens (1994a) 'Land, limits and sustainability'.

On the urban environment and sustainability, see Elkin *et al.* (1991) *Reviving the City: Towards Sustainable Development*; Breheny (1992) *Sustainable Development and Urban Form*; Gibbs (1994) 'Towards the sustainable city'; Haughton and Hunter (1994) *Sustainable Cities*. Beckerman (1995) gives an iconoclastic appraisal of 'environmental alarmism' in *Small is Stupid: Blowing the Whistle on the Greens*.

Major official references on the environment include *This Common Inheritance: Britain's Environmental Strategy* (Cm 1200) (1990) and the *UK Annual Report* on this. In response to the UN Rio Conference on Environment and Development, the UK government published in 1994 four major environmental strategies: *Sustainable Development: The UK Strategy* (Cm 2426), and the annual progress report; *Biodiversity: The UK Action Plan* (Cm 2428); *Climate Change: the UK Programme* (Cm 2427); and *Sustainable Forestry: The UK Programme* (Cm 2429).

On Agenda 21, see the publications of the Local Government Management Board, particularly (1993b) *Local Agenda 21: Principles and Process – A Step by Step Guide*; see also Wilks and Peter (1995) 'Think globally, act locally: Implementing Agenda 21 in Britain'.

Environmental Politics

The subject of environmental politics is discussed by Lowe and Goyder (1983) *Environmental Groups in Politics*; Lowe and Flynn (1989) 'Environmental politics in the 1980s'; McCormick (1991) *British Politics and the Environment*; and Robinson (1992) *The Greening of British Politics*. Newby (1990) 'Ecology, amenity and society' shows that environmental politics are not simply a modern fad.

The Precautionary Principle

There is a useful collection of papers edited by O'Riordan and Cameron (1994) *Interpreting the Precautionary Principle*. Beckerman's book, noted above, gives short shrift to the concept. The best description of the principle is by Möltke (1988): 'The *vorsorgeprinzip* in West German environmental policy', printed as an appendix to the twelfth report of the RCEP, *Best Practicable Environmental Option* (Cm 310).

Economic Instruments of Environmental Policy

Cairncross (1993) *Costing the Earth* (ch. 4) is a good non-technical discussion. More technical is Tietenberg (1990) 'Economic instruments for environmental regulation'. A DoE guide is *Making Markets Work for the Environment*. There is also a discussion of economic instruments in Cullingworth (1997) *Planning in the USA* (ch. 16) on which parts of the text are based.

Clean Air

Current policies relating to clean air are set out in DoE (1995) *Air Quality: Meeting the Challenge: The Government's Strategic Policies for Air Quality Management* (on which much of the account in the text is based). See also Elsom (1996) *Smog Alert: Managing Urban Air Quality*.

Clean Water

Official publications on water include: DoE (1992) *Using Water Wisely*; Scottish Office (1992) *Water and Sewerage in Scotland: Investing for our Future: A Consultation Paper*; NRA (1994) *Water: Nature's Precious Resources: An Environmentally Sustainable Water Resources Development Strategy for England and Wales*. See also Slater *et al.* (1994) 'Land use planning and the water sector'.

Waste

The 'strategy of sustainable waste management in England and Wales' is set out in *Making Waste Work* (Cm 3040, 1995). Other official publications include DOE (1994) *Waste Management Licensing: The Framework Directive on Waste*, DoE Circular 11/94; DoE (1996) *Waste Management: The Duty of Care: A Code of Practice*, revised edition; and RCEP (1985) *Eleventh Report: Managing Waste: The Duty of Care*. On Scottish policies, see NPPG 10 (1996) *Planning and Waste Management*. For a critical review of the policy of encouraging the recycling of paper products, see Collins (1996) 'Recycling and the environmental debate'.

Integrated Pollution Control

In addition to the RCEP reports, there is a technical guide published by DoE (revised edition, 1996) *Integrated Pollution Control: A Practical Guide*. See also Owens (1989) 'Integrated pollution control in the United Kingdom'; Gibson (1991) 'The integra-

tion of pollution control'; Harris (1992b) 'Integrated pollution control in practice'; and Layfield (1992) 'The Environmental Protection Act 1990: the system of integrated pollution control'.

Noise

PPG 24 (1994) deals with *Planning and Noise*. The Batho Report (Noise Review Working Party Report, 1990) examined a wide range of issues concerned with noise. Later reports have dealt with particular aspects such as the Mitchell Report (1991) on *Railway Noise and the Insulation of Dwellings*, and the Building Research Establishment report (1994) on *The Noise Climate Around Our Homes*. New policy guidance on *The Control of Noise at Surface Mineral Workings* was published as MPG 11 (1993). On neighbourhood noise, in addition to BRE reports, see Oliver and Waite (1989) 'Controlling neighbourhood noise: a new approach'.

Environmental Assessment

The principal texts on environmental impact assessment are Glasson *et al.* (1994) *Introduction to Environmental Impact Assessment*; and C. Wood (1995) *Environmental Impact Assessment: A Comparative Review*. Two DoE publications are (revised edition 1994) *Environmental Assessment: A Guide to Procedures*, and (1994) *Evaluation of Environmental Information for Planning Projects*. See also DoE (1993) *Environmental Appraisal of Development Plans: A Good Practice Guide*. Standard texts on cost-benefit analysis are Mishan (1976) *Elements of Cost-Benefit Analysis*, and Schofield (1987) *Cost-Benefit Analysis in Urban and Regional Planning*.

8

HERITAGE PLANNING

Buildings, towns, monuments and other historic sites give us a sense of place. They remind us of our past, of how our forebears lived, and how our culture and society developed. They tell us what earlier generations aspired to and achieved.

This Common Inheritance (Cm 1200) 1990

HERITAGE RESPONSIBILITIES

'Heritage' is the fashionable word for the national inheritance of historic buildings and features of the landscape. Much has happened in heritage planning over the last two decades in addition to the acclamation of the term. Particularly striking are the institutional changes: the establishment of the Historic Buildings and Monuments Commission (English Heritage) by the National Heritage Act of 1983; the different arrangements for Scotland and Wales, with executive agencies being located in the Scottish Office (Historic Scotland) and the Welsh Office (Cadw: Welsh Historic Monuments); and the transfer of heritage responsibilities from the DoE to the Department of National Heritage (DNH).

Major institutional change is often a politically adept technique of seeming to be doing something substantive while only giving the appearance of so doing. In this case, however, the institutional change has been part of a new commitment to preserving and enhancing the historic legacy. This has been greatly facilitated by the advent of funds from the National Lottery. Awards are administered by the National Heritage Memorial Fund. They amounted to £285 million in 1995.

Many governmental and voluntary organisations play a role in heritage planning. In England, there are the National Heritage Fund, the Royal Armouries, the Historic Royal Palaces Agency and the Royal Commission on the Historical Monuments of England. Scotland and Wales both have an Ancient Monuments Board and a Historic Buildings Council which act as advisory bodies to the departments, and also Royal Commissions on Historical Monuments, which have responsibility for compiling a record of the built, archaeological and landscape heritage (not to be confused with the Historic Buildings and Monuments Commission). In Northern Ireland, historic preservation policy dates from 1972 when the Planning (Northern Ireland) Order established the Historic Buildings Council and legislated for the listing of buildings and the designation of conservation areas (Hendry 1993).

CONSERVATION

Britain has a remarkable wealth of historic buildings, but changing economic and social conditions often turn this legacy into a liability. The cost of maintenance, the financial attractions of redevelopment, the need for urban renewal, the roads programme and similar factors often threaten buildings which are of architectural or historic interest.

This is a field in which voluntary organisations have been particularly active. The first of these dates back to 1877 when William Morris (horrified at the proposed 'restoration' of Tewkesbury Abbey) inspired the founding of the Society for the Protection of Ancient Buildings (Ross 1996). In 1895, another prominent figure in Victorian history, Octavia Hill, founded the National Trust (its full name is the National Trust for Places of Historic Interest or Natural Beauty). Other organisations were established as threats to the heritage increased. The Georgian Group was founded in 1937 after the Commissioners for Crown Lands demolished Nash's Regent Street and threatened to do the same with Carlton House Terrace. The widespread destruction of Victorian and Edwardian buildings led to the creation of the Victorian Society in 1958. The Twentieth Century Society, which was set up in 1979 (originally as the Thirties Society) to safeguard inter-war architecture, nearly saved the Firestone factory on the Great West Road, but was thwarted by the developers (Trafalgar House) who moved the bulldozers in over the August 1980 holiday weekend before the procedure for 'spot listing' had been completed. There are now numerous such heritage organisations, several of which have statutory consultee status on proposals to demolish listed buildings.

The first state action came in 1882 with the Ancient Monuments Act, but this was important chiefly because it acknowledged the interest of the state in the preservation of ancient monuments. Such preservation as was achieved under this Act (and similar Acts passed in the following thirty years) resulted from the goodwill and cooperation of private owners.

A major landmark in the evolution of policy in this area was the establishment, in 1908, of the three Royal Commissions on the Historical Monuments (of England, Scotland and Wales). They had (and still have) the same purpose, exemplified by the original terms of reference of the English Commission:

To make an inventory of the Ancient and Historical Monuments and constructions connected with or illustrative of the contemporary culture, civilisation and conditions of life of the people of England, from the earliest times to the year 1700 and to specify those that seem most worthy of preservation.

The quotation is instructive: the emphasis is on preservation and on 'ancient'. There was no concern for anything built after 1700: a prejudice which, as Ross (1991: 14) notes, was typical of the time. Slowly changing attitudes were reflected in 1921 when the year 1714 was substituted for 1700! The date was advanced to 1850 after the end of the Second World War, and in 1963 an end-date was abolished.

The Commissions were established to record monuments, not to safeguard them. It was not until 1913 that general powers were provided to enable local authorities or the Commissioners of Works to purchase an ancient monument or (a surprising innovation in an era of sacrosanct property rights) to assume 'guardianship' of a monument, thereby preventing destruction or damage while leaving 'ownership' in private hands.

Major legislative changes were made in the 1940s though, in practice, the most important innovation was the establishment of a national survey of historic buildings. This was a huge job (quite beyond the capabilities of the slow-moving Royal Commissions). It was undertaken, county by county, by so-called 'investigators'. (Ross gives an interesting account of how this mammoth job was done, often on a voluntary or near-voluntary basis.) The survey took twenty-two years and, even then, it was incomplete. This first survey, which ended in 1969, gave statutory protection to almost 120,000 buildings and non-statutory recognition (but not protection) to a further 137,000 buildings. Given the attitudes of the time, Victorian architecture was almost totally neglected. (The Victorian Society did not come into existence until 1957, after the demolition of the Euston Arch.)

Statutory protection, however, is not sufficient by itself: the owners of historic buildings often need financial assistance if the cost of maintaining old structures is to be met. The issue was highlighted by the 1950 Gowers Report on *Houses of Outstanding Historic and Architectural Interest*. The Historic Buildings and Ancient Monuments Act of 1953 followed. This established Historic Building Councils for England, Scotland and Wales, and introduced grants for preserving houses which were inhabited or 'capable of occupation'.

Further big changes were made in 1983, and later most of the provisions relating to heritage properties were consolidated in the Planning (Listed Buildings and Conservation Areas) Act 1990, though curiously those relating to ancient monuments are still separate in the Ancient Monuments and Archaeological Areas Act 1979.

In considering the role of this regulatory system, it is important to appreciate what is meant by the term 'conservation'. Though often used synonymously with 'preservation', there is an important difference. Preservation implies maintaining the original in an unchanged state, but conservation embraces elements of change and even enhancement. To provide an economic base for the conservation of an old building, new uses often have to be sought. It is quite impossible to conserve all buildings in their original state irrespective of cost, and there frequently has to be a compromise between 'the value of the old and the needs of the new' (Ross 1991: 92). Thus 'new uses for old buildings' is a major factor in conservation, and it necessarily implies a degree of change, even if this is restricted to the interior. (It was no doubt with this necessity for change in mind that the CPRE altered the second word in its name from 'preservation' to 'protection'.) Again, for conservation purposes it may be necessary to enhance a site to cater for public enjoyment. The difference is more than one of name.

ANCIENT MONUMENTS

The term *ancient monument* is defined very widely: it is 'any scheduled monument' and 'any other monument which in the opinion of the Secretary of State is of public interest by reason of the historic, archi-

tectural, traditional, artistic or archaeological interest attaching to it'. This is so broad a definition that it could include almost any building, structure or site of archaeological interest made or occupied at any time. It includes, for instance, a preserved Second World War airfield complex at East Fortune (near Haddington, in Lothian): this houses the Museum of Flight.

The legislation requires the Secretary of State to prepare a schedule of monuments 'of national importance'. This 'scheduling' is a selective and continuing process. It has been in process for over a century and, for many years, proceeded at a very slow rate. (The term 'schedule' originates from the Ancient Monument Protection Act 1882, which provided for the protection of twenty-nine monuments which were set out in a *schedule* to the Act; the term has persisted.)

English Heritage has a *Monuments Protection Programme* which is evaluating all known archaeological remains. This is expected to increase significantly the number of scheduled monuments. There are thought to be some 600,000 archaeological sites in England which 'are of outstanding national importance and should be afforded statutory protection through scheduling' (DNH Annual Report 1996: 14). The official aim is to review 70–80 per cent of the total by the year 2003. In 1996 it was reported that progress was on target (ibid.). In Scotland, though it is considered that the most outstanding monuments have been scheduled, there are over 10,000 monuments which might be scheduled that have yet to be assessed. At the current rate of progress (according to the 1995 NAO report on *Protecting and Presenting Scotland's Heritage Properties*), the schedule will not be complete for at least a further twenty-five years.

There are 25,400 scheduled sites in the UK: 15,300 in England, 6,200 in Scotland, 2,800 in Wales and 1,100 in Northern Ireland (COI, *Britain 1996*: 358). Estimates differ on the number of additional sites which are worthy of scheduling. Though the figure of 600,000 is frequently quoted, this is only a rough guess. Some have suggested that the figure should be much higher, and since there is such a huge number of known archaeological sites and monuments, it is not surprising that estimates differ. The problem of estimating is further complicated by the fact that some figures refer to register entries (which cover more than one site), and others relate to individual sites.

It is recognised that the present schedule is not only very incomplete but is also an inadequate and unrepresentative sample of the archaeological heritage. Both the DoE and the Welsh Office (in their almost identical versions of PPG 16) advise that 'where nationally important archaeological remains, whether scheduled or not, and their settings, are affected by proposed development there should be a presumption in favour of their physical preservation'. The fact that a monument is scheduled does not mean that it will automatically be preserved under all circumstances. It simply ensures that full consideration is given to the case for preservation if any proposal is made which will affect it. Any works have to be approved by the Secretary of State (who receives advice from English Heritage). Such approval is known as *scheduled monument consent*. Where consent is refused, compensation is payable (under certain limited circumstances) if the owner thereby suffers loss. In practice, the great majority of applications for consent are approved, often with conditions attached. The issue here is seen as one of balancing the need to protect the heritage with the rights and responsibilities of farmers, developers, statutory undertakers and other landowners. The legislation also empowers the Secretary of State to acquire (if necessary by compulsion) an ancient monument 'for the purpose of securing its preservation' – a power which applies to any ancient monument, not solely those which have been scheduled.

Though most heritage properties remain in private ownership, a small number are managed by the Heritage Departments – officially known as being 'in care'. These are generally of important historical, archaeological and architectural significance. A very high proportion are of great antiquity, including prehistoric field monuments such as Maiden Castle; prehistoric structures such as Stonehenge; Roman monuments such as Wroxeter and parts of Hadrian's Wall; and a large number of medieval buildings. Properties in care of Historic Scotland include Edinburgh Castle, Stirling Castle, Fort George, and Urquhart Castle (near Loch Ness). Welsh Historic Monuments manage Chepstow Castle, the Blaenavon Ironworks, the Welsh Slate Museum (Llanberis), Neath Abbey and Tintern Abbey. In Northern Ireland, 166 monuments are in the care of the DoENI, including Londonderry's City Walls, Newtownards Priory, Enniskillen Castle, Tully Castle, Carrickfergus Castle and, perhaps surprisingly, the Carrickfergus Gas Works.

ARCHAEOLOGY

The *areas of archaeological importance* in the 1979 Act represented a new concept. These are areas which the Secretary of State considers 'to merit treatment as such': no further definition is provided. In these areas, developers are required to give six weeks' notice (an *operations notice*) of any works affecting the area, and the 'investigating authority' (e.g. the local authority or a university) can delay operations for a total period of up to six months. The powers have been used very sparingly, and only five areas have been designated, comprising the historic centres of Canterbury, Chester, Exeter, Hereford and York. The 1996 consultation paper *Protecting Our Heritage* argues that the powers are now redundant and should be repealed.

On a wider scale, there are similar (though less stringent) provisions which provide an opportunity for the investigation and recording of archaeological remains prior to proposed development. This is known as *rescue archaeology*. It is very different from the scheduling of ancient monuments where the essential aim is preservation. Rescue work in central London has been well publicised. Interest in archaeology has grown greatly in recent years, as has understanding, competence and professionalism. PPG 16 makes it clear that there is a presumption in favour of the preservation of important remains, whether or not they are scheduled. There is thus a measure of protection over the large number of unscheduled sites which are on the lists maintained by county archaeological officers. (These are known as SMRs: *county sites and monuments records*.) Such sites are a 'material consideration' in dealing with planning applications.

Many local authorities make provision in their development plans for the protection of archaeological interests (Redman 1990: 89), often with good cooperation from large developers. The image of archaeological excavations in areas undergoing redevelopment is one of outright war. Given the conflicting interests involved this is not surprising. What is perhaps surprising is the extent to which some developers are prepared to go to assist rescue archaeology, and even to fund it. Sometimes, this goes far beyond any statutory requirement (Hobley 1987: 25). A useful mechanism for liaison is provided by the British Archaeologists and Developers Liaison Group. This is an organisation promoted by the British Property Federation and the Standing Conference of Archaeological Unit Managers (itself a representative body of some seventy-five professional archaeological units). This has issued a code of practice.

However, rescue archaeology is, at best, of limited benefit: it is certainly far inferior to preservation *in situ*. The cost of this, however, can be enormous and, given the incredible range of archaeological remains in Britain, some selection is inevitable.

LISTED BUILDINGS

Under planning legislation, and quite separate from the provisions relating to monuments, the central departments maintain lists of buildings of *special architectural or historic interest*. The preparation of these lists has been a mammoth task which has progressed slowly because of inadequate funding and (the underlying factor) a lack of public interest. Added impetus, however, was given by events such as the declaration of European Architectural Heritage Year in 1975 and, less joyfully but perhaps more effectively, by the already mentioned demolition of the Firestone factory. The latter precipitated action by the Secretary of State, Michael Heseltine, to accelerate the survey. 'It is no exaggeration to say that Firestone was a sacrificial lamb' (Ross 1991: 44).

Although the national listing survey is now substantially complete, listing is a continuing process, not only for additional buildings but also for updating information on existing listed buildings, particularly in terms of their condition. In addition, individual buildings can be spot-listed. This arises because of individual requests, often precipitated by the threat of alteration or demolition. The majority of these requests are made by local authorities (ideally well in advance of development proposals being submitted for planning permission) so that 'any application for listed building consent can be considered in tandem with the planning application for the new development' (Ross 1996: 81).

There are two objectives in listing. First, it is intended to provide guidance to local planning authorities in carrying out their planning functions. For example, in planning for redevelopment, local authorities will take into account listed buildings in the area. Second, and more directly effective, when a building is listed, no demolition or alteration which

Table 8.1 Scheduled Monuments and Listed Buildings, 1995

	Listed buildings	Scheduled monuments	In care
England	447,400	15,300	400
Scotland	41,500	6,200	330
Wales	17,200	2,800	129
N. Ireland	8,600	1,100	181
Total UK	514,700	25,400	1,040

would materially alter it can be undertaken by the owner without the approval of the local authority. This is technically termed *listed building consent*.

The procedure for obtaining listed building consent is summarised in Figure 8.1. Applications have to be advertised, and any representation must be taken into account by the local authority before it reaches its decision. Where demolition is involved,

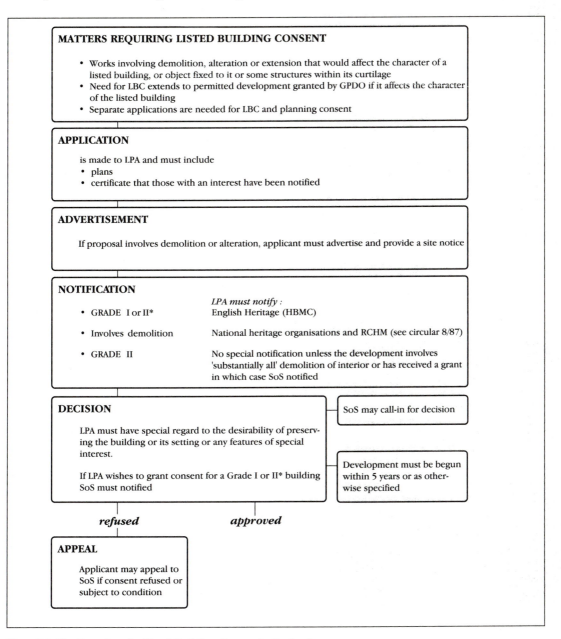

MATTERS REQUIRING LISTED BUILDING CONSENT

- Works involving demolition, alteration or extension that would affect the character of a listed building, or object fixed to it or some structures within its curtilage
- Need for LBC extends to permitted development granted by GPDO if it affects the character of the listed building
- Separate applications are needed for LBC and planning consent

APPLICATION

is made to LPA and must include
- plans
- certificate that those with an interest have been notified

ADVERTISEMENT

If proposal involves demolition or alteration, applicant must advertise and provide a site notice

NOTIFICATION

	LPA must notify :
• GRADE I or II*	English Heritage (HBMC)
• Involves demolition	National heritage organisations and RCHM (see circular 8/87)
• GRADE II	No special notification unless the development involves 'substantially all' demolition of interior or has received a grant in which case SoS notified

DECISION

LPA must have special regard to the desirability of preserving the building or its setting or any features of special interest.

If LPA wishes to grant consent for a Grade I or II* building SoS must notified

SoS may call-in for decision

Development must be begun within 5 years or as otherwise specified

refused   *approved*

APPEAL

Applicant may appeal to SoS if consent refused or subject to condition

Figure 8.1 The Procedure for Listed Building Consent in England

English local authorities have to notify English Heritage, the appropriate local amenity society and a number of other bodies, namely, the Ancient Monuments Society, the Council for British Archaeology, the Georgian Group, the Society for the Protection of Ancient Buildings, the Victorian Society and the Royal Commission on Historical Monuments. Scottish and Welsh authorities are required to notify their respective Royal Commissions on Ancient and Historical Monuments. Again any representations have to be taken into account when the application is being considered.

If, after all this, the local authority intends to grant consent for the demolition (or, in certain cases, the alteration) of a listed building, it has to refer the application to the DoE so that it can be considered for 'call-in' and decision by the Secretary of State. English Heritage advise on this, and in most cases the DoE accept its advice. It is the Department's policy that consent applications should generally be decided by local authorities, and only a very small number are referred.

Conditions can be imposed on a listed building consent in the same way as is done with planning permissions. The type of conditions that can be imposed are set out in DoE Circular 8/87; they include the preservation of particular features, the making good of damage caused by works of alteration, and the granting of access (before work commences) to a named body to enable a photographic record or measured drawings to be made.

All these provisions apply to listed buildings, but local authorities can serve a *building preservation notice* on an unlisted building. This has the effect of protecting the building for six months, thus giving time for the DoE to consider (on the advice of English Heritage) whether or not it should be listed. For owners and developers who wish to be assured that they will not be unexpectedly made subject to listing, application can be made to the LPA for a *certificate of immunity from listing*.

With a listed building, the presumption is in favour of preservation. It is an offence to demolish or to alter a listed building unless listed building consent has been obtained. This is different from the

general position in relation to planning permission where an offence arises only after the enforcement procedure has been invoked. Fines for illegal works to listed buildings are related to the financial benefit expected by the offender.

The legislation also provides a deterrent against deliberate neglect of historic buildings. This was one way in which astute owners could circumvent the earlier statutory provisions: a building could be neglected to such an extent that demolition was unavoidable, thus giving the owner the possibility of reaping the development value of the site. In such cases, the local authority can now compulsorily acquire the building at a restricted price, technically known as *minimum compensation*. If the Secretary of State approves, the compensation is assessed on the assumption that neither planning permission nor listed building consent would be given for any works to the building except those for restoring it to, and maintaining it in, a proper state of repair; in short, all development value is excluded.

The strength of these powers (and others not detailed here) reflects the concern which is felt at the loss of historic buildings. However, they are not all of this penal nature. Indeed, ministerial guidance has emphasised the need for a positive and comprehensive approach. Grants are available towards the cost of repair and maintenance. Furthermore, an owner of a building who is refused listed building consent can, in certain circumstances, serve a notice on the local authority requiring it to purchase the property. This is known as a *listed building purchase notice*. The issue to be decided here is whether the land has become 'incapable of reasonably beneficial use'. It is not sufficient to show that it is of less use to the owner in its present state than if developed. Local authorities can also purchase properties by agreement, possibly with Exchequer aid. Exceptionally, a neglected building can be compulsorily acquired. There is only one case of this: the St Ann's Hotel building in Buxton which is part of a late eighteenth-century crescent. It had had a long history of neglect which continued through various ownerships. After all alternatives had been exhausted, the Secretary of State served a compulsory purchase order in 1993.

LISTING AND MARKET VALUES

A major issue with listed buildings is whether a contemporary use can be found for them, and whether alterations for such a use would be compatible with the character which it is desired to preserve. Research (which is continuing with support from English Heritage) generally shows that listed office buildings have a 'market performance' which is generally as good as other buildings, and sometimes better. On the other hand, listing can reduce market value, particularly of small buildings in areas of high development outside conservation areas. However, the reduction is a one-time cost which is borne by the owner at the time of listing: future 'market performance' is not affected. But listing can also increase values because of the 'prestige' thereby accorded; and this can also raise neighbouring values. Like all such issues, much depends on local factors.

In deciding whether or not to list a building, the Secretary of State is required to have regard only to the special architectural or historic interest. No account can be taken of economic issues (such as the condition of the building and the cost of conserving it, or the possibilities of finding a viable use for the building). Nor can the personal circumstances of the owner be considered. (Such issues become relevant only when an application is made for listed building consent to demolish or alter a listed building.)

The issue is raised in the 1996 consultation paper *Protecting Our Heritage* and, though views are sought, the government view is clearly that no change is appropriate:

In the government's view, the logic of the current regime is clear and defensible – the purpose of listing a building is to register its quality. The fact of listing does not rule out any future change to the building. Rather, it ensures that any proposals affecting it can be fully assessed, having regard to all the circumstances prevailing at the time, including economic and financial considerations. The government will consider alternative views on this issue but will not easily be persuaded that a change of approach would be desirable.

CONSERVATION AREAS

Of particular importance in heritage planning is the emphasis on areas, as distinct from individual buildings, of architectural or historic interest. Statutory recognition of the area concept was introduced by the Civic Amenities Act 1967 (promoted as a private member's bill by Duncan Sandys, President of the Civic Trust, and passed with government backing). Local planning authorities have a duty 'to determine which parts of their area are areas of special architectural or historic interest, the character of which it is desirable to preserve or enhance', and to designate such areas as *conservation areas*. When a conservation area has been designated, special attention has to be paid in all planning decisions to the preservation or enhancement of its character and appearance. Demolition of all buildings (unlisted as well as listed) is controlled, and there are special provisions for preserving trees; but owners of unlisted buildings have 'permitted development rights': they are not subject to the restrictions applied to owners of listed buildings. However, LPAs can withdraw these permitted development rights by use of an *Article 4 direction* (discussed in Chapter 5). Indeed, this is the common use of such Directions (Tym & Partners 1995a). They are typically intended to prevent piecemeal erosion of the character of an area through the cumulative effects of numerous small changes. Local planning authorities also have a duty to seek 'the preservation and enhancement' of conservation areas. Though some authorities take this duty seriously, it is generally poorly implemented, often on the grounds of inadequate resources.

The statutory provisions relating to the establishment of conservation areas are remarkably loose: there is no formal designation procedure, there is no requirement for a formal public inquiry (though proposals have to be put before a public meeting), and there is no specification of what qualifies for conservation area status. Circular 8/87 notes that 'these areas will naturally be of many different kinds':

> They may be large or small, from whole town areas to squares, terraces and smaller groups of buildings. They will often be centred on listed buildings, but not always. Pleasant groups of other buildings, open spaces, trees, and historic street patterns, a village green or features of historic or archaeological interest may also contribute to the special character of an area. Areas appropriate for designation as conservation areas will be found in almost every town and many villages. It is the character of areas, rather than individual buildings, that [section 60 of the Planning (Listed Buildings and Conservation Areas) Act 1990] seeks to preserve or enhance.

(The 1996 consultation paper proposes that the designation of new conservation areas should 'include a statement identifying the specific features of the area that it is considered desirable to preserve or enhance'. It also seeks views on whether there should be a statutory requirement for consultation with local residents and interested parties prior to a decision on designation.)

The number of conservation areas has grown dramatically: in 1995, there were over 8,000 in England, 400 in Wales, 674 in Scotland and 40 in Northern Ireland. Over a million buildings are in these areas. Indeed, it has been suggested that perhaps a 'saturation point' has been reached in that the resources are simply not available for 'enhancing' such a large number of areas (Morton 1991; Suddards and Morton 1991). There seems to be a widespread view that more attention should be given to managing existing conservation areas, and less to designating additional ones (Larkham and Jones 1993).

A special type of conservation area is the *town scheme*. (There were 248 of these in England in 1995.) This is an historic area in which conservation and improvement is jointly funded by a local authority and English Heritage. While other grant schemes focus on buildings, this form of grant-aid serves wider environmental purposes:

> Town schemes cover not only the centres of the classic historic towns but also hundreds of smaller towns, and range from small domestic terraces to large warehouses. Increasingly, additional demands are coming from the run-down inner city areas, where grants at a comparatively low level can give a significant stimulus to urban regeneration and improve the quality of life for those living or working there.

English Heritage grants are also administered through *conservation area partnerships*. These were introduced in 1993 to provide a new and flexible framework for targeting English Heritage resources on areas of greatest need.

In spite of all these (and other) provisions, many listed buildings are at risk. In Scotland (according to the 1995 National Audit Office report on *Protecting and Presenting Scotland's Heritage Properties*), the on-going *Buildings at Risk Register* in 1994 contained 860 listed properties which were unoccupied or derelict and which had a dubious future (over 2 per cent of the total). The position is relatively worse in England: an English Heritage report showed that 36,700 listed buildings (7 per cent of the total) are at risk from neglect; twice as many are in a vulnerable condition and need repair if they are not to fall into the 'at risk' category. Of course, most listed buildings are in private ownership, and the owners may well not feel the respect for their buildings which preservationists do; or they may simply be unable to afford to maintain them adequately. Advice, grants and default measures cannot achieve all that might be hoped and, though a precious building can be taken into public ownership, this is essentially a matter of last resort.

CRITERIA FOR LISTING HISTORIC BUILDINGS

Criteria for listing historic buildings are divided into four groups according to the date of building:

Date of Building	Principles of selection
Before 1700	All buildings which survive in anything like their original condition are listed.
1700–1840	Most buildings are listed, though selection is necessary.
1840–1914	Only buildings of definite

quality and character are listed and the selection is designed to include the best works of the principal architects.

Post-1914   Selected buildings of high quality are listed.

Table 8.2 Listed Building Categories

England and Wales		Northern Ireland		Scotland	
Grade	Criteria	Category	Criteria	Category	Criteria
I	Buildings of outstanding or exceptional interest	A	Of national importance	A	Buildings of national or international importance, either architectural or historic, or fine and little-altered examples of some particular period, style or building type
II*	Particularly important buildings of more than special interest but not in the outstanding class	B+	Of national importance but with minor detracting features or of national importance with some exceptional features	B	Buildings of regional or more than local importance, or major examples of some period, style or building type which may have been somewhat altered
		B1	Of national or local importance, or good examples of some period or style		
		B2			
II	Buildings of special interest which warrant every effort being made to preserve them	C	Of positive architectural interest or historic interest but are not 'special' and including those that contribute to the value of groups of buildings	C(S)	Buildings of local importance; lesser examples of any period, style or building type, whether as originally constructed or as the result of subsequent alteration; simple well-proportioned traditional buildings often forming part of a group

In Scotland, the grouping is: pre-1840; 1840–1914; 1914–1945; and post-1945.

In choosing buildings, particular attention is paid: to 'special value within certain types, either for architectural or planning reasons or as illustrating social and economic history'; to technological innovation or virtuosity (for instance, cast-iron pre-fabrication or the early use of concrete); to any association with well-known characters or events; and 'group value', especially as examples of town planning such as squares, terraces or model villages (DoE Circular 8/87). Buildings are graded according to their relative importance. The grading systems are set out in Table 8.2.

Scotland has for long had a rolling thirty-year rule under which any building of that age could be considered for listing. This was initially thought to be too problematic in relation to the much larger number of buildings that would be covered by such a rule in England. How was the quality of buildings to be assessed over such a short time period? Would apparent 'successes' soon be seen as 'failures' – and vice versa? There have been marked changes in attitudes to different architectural styles. Perhaps the most famous comes from Paris, where the Eiffel Tower was once described in terms of 'the grotesque mercantile imaginings of a constructor of machines', but is now 'the beloved signature of the Parisian skyline and an officially designated monument to boot' (Costonis 1989: 64). Many buildings have been demolished which would today attract vociferous defence. On the other hand, some more recent architecture would have difficulty in finding a place in the hearts of those who support the protection of good interwar buildings. Clearly, this is an area where attitudes differ and firm guidelines are far from easy to determine – as is also the case with contemporary design and amenity guidelines.

Matters were suddenly accelerated when Sir Albert Richardson's Bracken House in the City of London was threatened with demolition. The Secretary of State decided to list this grade II*, thus copying the Scottish principle that buildings over thirty years old could be listed. At the same time, going one better than the Scots, it was decided that outstanding buildings that were only ten years old could be listed if there was an immediate threat to them. By the end of 1995 listing of postwar buildings amounted to '189 separate items on 111 sites across England' (as reported in a 1996 English Heritage leaflet *Something Worth Keeping?* published in connection with a programme of public consultation on the listing of further postwar buildings).

PUBLIC PARTICIPATION IN LISTING

Theoretically, anyone can propose that a building be listed but, in practice, buildings have usually been proposed by English Heritage and decided by the DNH. There was no advance publicity about this system. It was therefore a surprise when the Secretary of State announced that the system was to be opened up, not only for proposals from the public, but also for comments on proposals for listing. In 1995, following an earlier initiative the public were invited to comment on proposals from English Heritage for the listing of forty modern buildings and thirty-seven textile mills in the Manchester area. Among the former were the Centre Point office block in central London, Millbank Tower, the John Lewis warehouse at Stevenage and the signal box at Birmingham New Street station.

There was concern that such a highly publicised process of listing might incite owners to demolish their earmarked buildings at speed (as the Firestone building had been). Spot listing is one answer to this (if it is done quickly enough), or the imposition of a building preservation order (though this renders the local authority liable to compensation if the building is not in fact eventually listed). A better solution would be a new power for an instant listing which carried no compensation penalties for the local authority. The 1996 consultation paper discusses a proposal for a new power for the Secretary of State to provisionally list an endangered building to allow consultation to take place. Such a procedure (as with the current listing regime) would not involve any compensation.

There is an enormous range in the character of listed buildings, of which one of the most novel is the beloved old red telephone kiosk. British Telecom is replacing these with models of contemporary design. Opposition and lobbying resulted in a number of kiosks of unusual design being listed. Listings were later extended, but with strict criteria of special architectural or historic interest. There are now well over a thousand listed kiosks.

As these examples show, public opinion (when aroused) can play an important part in this planning field. The same is true with listed buildings under threat, as is well exemplified by the successful campaign to save St Pancras Station. There is a wide degree of public support for conservation, which has been heightened by the increased concern for environmental issues (Lane and Vaughan 1992).

Some questions remain, however, and they are likely to come to the fore as local planning authorities continue to develop their competencies in the area. One question is the justification for the existence of two regimes: one for the listing of historic buildings and the other for the scheduling of ancient monuments and archaeological remains. With historic buildings, some categories are automatically protected, while ancient monuments are protected only if they accord with some concept of national importance. (Some buildings may be both scheduled and listed!) Other questions relate to the division of responsibilities between planning authorities and central government and, in particular, the degree of the integration between heritage planning and the other functions of local planning authorities (Redman 1990; Scrase 1991).

WORLD HERITAGE SITES

The UNESCO World Heritage Convention established a World Heritage List of sites which UN member states are pledged to protect. There are thirteen sites in the UK. The inclusion of a site on the World Heritage List carries no additional statutory controls though, of course, it underlines its outstanding importance. This is a relevant material consideration in planning control. Local planning policies should, in the words of PPG 15, 'reflect the fact that all these sites have been designated for their outstanding universal value, and they should place great weight on the need to protect them for the benefit of future generations as well as our own'. Significant development proposals affecting a World Heritage Site generally require an environmental assessment.

CHURCHES

The situation regarding 'ecclesiastical buildings' is exceptional and also complicated. In essence, there is what is technically termed 'the ecclesiastical exemption' from listed building and conservation area controls.

By way of background, it is worth noting the statistics which were given to the HC Environment Committee during its 1986–7 inquiry: the Church of England then owned 16,700 churches, of which

World Heritage Sites in the UK

- Hadrian's Wall Military Zone
- Durham Castle and Cathedral
- St Kilda
- Fountains Abbey and St Mary's, Studley Royal
- The Castles and Town Walls of Edward I in Gwynedd
- Ironbridge Gorge
- Stonehenge, Avebury and associated sites
- Giant's Causeway and Causeway Coast
- Blenheim Palace
- Palace of Westminster and Westminster Abbey
- City of Bath
- Tower of London
- Canterbury Cathedral, St Augustine's Abbey and St Martin's Church

no less than 8,500 were pre-Reformation and 12,000 were statutorily listed (2,675 in the highest grade). The Church introduced measures to control demolition more than 700 years ago, and has been regularly inspecting churches for 300 years. It spends a large amount each year on the upkeep and maintenance of its buildings (mainly funded by its congregations). The result is that 'a listed Church of England church has a chance of avoiding demolition nearly three times better than a listed secular building'.

There are two parallel statutory systems of control over Church of England churches: the Church's system and the secular system. The Church's system is much stricter and more comprehensive. It involves regular inspection of every church, and embraces not only the fabric of the buildings but also their contents and churchyards. There are two separate statutory procedures applying to parish churches (whether listed or unlisted), according to whether they are in use or redundant.

Churches in use are subject to an elaborate system of inspecting and reporting at the local level, and to monitoring at higher levels: by Diocesan Advisory Committees at diocesan level, and by the Council for the Care of Churches at the national level.

Redundant churches are safeguarded by the Pastoral Measure 1983 which provides procedures for deciding whether a church is still required for worship, and, if not, what the future of the building should be. The *Churches Conservation Trust* (formerly the Redundant Churches Fund) finances the management, maintenance and repair of churches judged of sufficient architectural or historic importance. The fund receives 70 per cent of its funds from the DNH and the remainder from the Church Commissioners.

Until recently, cathedrals were outside any planning procedure and, despite their huge popularity with visitors (and contribution to tourism), were not eligible for grant aid. A separate system of controls over building works was introduced by the Care of Churches Measure 1990. This is administered by the individual cathedrals jointly with a Cathedrals Fabric Commission in consultation with English Heritage which also provides grant aid.

All church buildings are subject to normal planning control over, for example, changes of use and significant alterations. They are also listed in exactly the same way as other buildings of special historic or architectural interest. However, because of the Church's separate statutory procedure, listed building consent is not required for churches in use. Such consent is required, however, for alterations to redundant churches, though not if demolition is carried out pursuant to a scheme under the Pastoral Measure 1983.

A government review of the ecclesiastical exemption completed at the beginning of 1993 led to a decision to extend it to churches of all denominations where an acceptable system of control operates on principles set out in a code of practice. The code includes the requirement that all works to a listed church building which would affect its character are submitted for the approval of an independent body, and that there is consultation with the LPA, English Heritage and national amenity societies.

THE NATIONAL HERITAGE MEMORIAL FUND

In 1946, a National Land Fund was established from the sale of war stores as a memorial to those who lost their lives during the Second World War. Some £50 million of Exchequer moneys were allocated to the fund for use to assist organisations whose purpose was to promote appreciation and enjoyment of the countryside. The money was used for the purchase by the Secretary of State of buildings of outstanding architectural or historic interest, together with their contents. The fund was raided by the Exchequer in 1957 and became moribund. However, following the outcry over the controversial sale of the assets of the Mentmore estate in 1977, a National Heritage Memorial Fund was established by the National Heritage Act of 1980. This is dedicated as 'a memorial to those who have died for the United Kingdom'. Grants have included £1 million to the British Museum for the purchase of the Hoxne hoard of Roman treasure, £300,000 to Cambridge University Library for the acquisition of the Royal Commonwealth Society Library, and £4.9 million for the purchase of Croome Park, Worcester.

In addition to normal Exchequer payments into the fund, further payments can be made in relation to property accepted in satisfaction of tax debts. The fund can make loans or grants for any property (in the widest sense of the term) which is 'of outstanding scenic, historic, architectural or scientific interest'. The Fund distributes the heritage tranche of the National Lottery funds.

PRESERVATION OF TREES AND WOODLANDS

Trees are a delight in themselves; they also have the remarkable quality of hiding developments which are best out of sight. Trees are clearly, so far as town and country planning is concerned, a matter of amenity. Indeed, the powers which local authorities have with regard to trees can be exercised only if it is 'expedient in the interests of amenity'. Where a local authority is satisfied that it is expedient, it can make a *tree preservation order*, applicable to trees, groups of trees or woodlands. Such an order can prohibit the cutting down, topping or lopping of trees except with the consent of the local planning authority.

Mere preservation, however, can lead eventually to decay and thus defeat its object. To prevent this, a local authority can make replanting obligatory when it gives permission for trees to be felled. The aim is to avoid any clash between good forestry and the claims of amenity. But the timber of woodlands always has a claim to be treated as a commercial crop, and though the making of a tree preservation order does not necessarily involve the owner in any financial loss (isolated trees or groups of trees are usually planted expressly as an amenity), there are occasions when it does.

Yet, though woodlands are primarily a timber crop from which the owner is entitled to benefit, two principles have been laid down which qualify this. First, the national interest demands that woodlands should be managed in accordance with the principles of good forestry, and second, where they

are of amenity value, the owner has a public duty to act with reasonable regard for amenity aspects. It follows that a refusal to permit felling or the imposition of conditions on operations which are either contrary to the principles of good forestry or destructive of amenity ought not to carry any compensation rights. But where there is a clash between these two principles, compensation is payable.

Thus, in a case where the principles of good forestry dictate that felling should take place, but this would result in too great a sacrifice of amenity, the owner can claim compensation for the loss which he suffers. Normally, a compromise is reached whereby the felling is deferred or phased. The commercial felling of timber is subject to a licence by the Forestry Commission, and special arrangements exist for consultation between the Commission, the central department and the local planning authority.

Planning powers go considerably further than simply enabling local authorities to preserve trees. Planning permission can be made subject to the condition that trees are planted, and local authorities themselves have power to plant trees on any land in their area. With the increasing vulnerability of trees and woodlands to urban development and the needs of modern farming, wider powers and more Exchequer aid have been provided by successive statutes. Local planning authorities are now *required* to ensure that conditions (preferably reinforced by tree preservation orders) are imposed for the protection of existing trees and for the planting of new ones.

New legislation to amend the provisions relating to tree preservation orders is to be included in forthcoming legislation, but there is no indication of when this will be.

TOURISM

Heritage is an important factor in tourism. Together with 'culture' and the countryside, heritage stimulates about two-thirds of the visits made by foreign tourists. (They certainly do not come to enjoy the weather!) In a 1993 survey of overseas visitors to

London, undertaken by the British Tourist Authority and the London Tourist Boards, 69 per cent of holiday-makers gave heritage as the main reason for visiting London (the figure for visitors from North America was 80 per cent). With domestic tourists, the most frequently mentioned main purpose of a holiday (after walking holidays) was to visit heritage attractions and sites. Additionally, one-third of domestic holiday-makers visited a heritage attraction at some point during their holiday.

The economic implications of tourism are great. Around 1.5 million people are employed in tourism. The industry contributes £30 billion to the economy: about 5 per cent of GDP and a similar proportion of export earnings. At the local level, tourism can be a major industry. In Bath, it is estimated to be worth up to £200 million a year; it supports more than 5,000 full-time jobs. Over fifty historic towns attract more than 20,000 staying visits by overseas visitors annually. (Of these, six received over 150,000: Bath, Brighton/Hove, Cambridge, Oxford, Stratford-upon-Avon and York.)

There is an abundance of such figures, and it is not necessary to labour the point. There is, however, a downside: tourism can lead to excessive wear and tear on the fabric of buildings, to congestion, to litter, and even to open hostility by residents to visitors. The generally accepted implication is that tourism has to be 'managed'. Several organisations are now devoted to this: the Historic Towns Tourism Management Group, the Heritage Cities Association (a marketing consortium) and the English Historic Towns Forum. Recently, as concern for 'vital and viable town centres' has grown, other specialist bodies have been established, such as the Association of Town Centre Management.

That tourism, as well as heritage, is a matter of importance for local planning authorities is self-evident. However, it is not a policy area which can be isolated from related ones. It is interesting to note that PPG 21 on *Tourism* refers to a long list of other relevant PPGs. This list is an eloquent testimony to the interconnectedness of planning issues; and it also points to the inherent difficulty of reconciling numerous considerations – or even giving adequate consideration to all of them. In *Our Heritage: Preserving It, Prospering from It*, the House of Commons National Heritage Committee (1994: para. 54) has complained that 'there is a serious lack of coherence about policy for the preservation of our heritage and its very important links with the tourist industry'. It pointed to the 'untidy structure' which the DNH inherited from six different departments ('or more accurately, lack of structure'). Over nine-tenths of the Department's budget is allocated to the support of thirty-four quangos. The Committee strongly argued that 'this cats-cradle of an inheritance' needed simplification. The government's response was that the DNH was a new department which was making good progress in securing a better identification of objectives, priorities and interrelationships between sectors. Indeed, the DNH was established to achieve this.

Similar issues arise at the local level though, of course, there is great variation among LPAs, partly due to their widely different tourist attractions. Some, such as Chester, have undertaken detailed studies of the city's environmental capacity. Stratford-upon-Avon was the location of the English Tourist Board's study of visitor management in historic towns. Other studies have been carried out by bodies such as the International Council on Monuments and Sites. However, the majority of these studies have focused on special sites: there are few studies of less prominent places.

Given current 'customer-oriented philosophy, and a good general awareness of business priorities in the presentation and management of properties in care' (to quote from the National Audit Office report on *Protecting and Presenting Scotland's Heritage Properties*, 1995), heritage bodies have an increased concern 'to offer more value to visitors and earn a return on the public funds required, while at the same time preserving the integrity of the properties'. However, as this quotation indicates, it is well recognised that the presentation and management of historic properties cannot be a purely commercial operation. In any case, too great a success in attracting custom could place unsustainable pressures on the very experience which the customers are seeking. As with popular countryside areas, too many visitors can be ruinous to the local environment. Nevertheless, the pressures to 'market' the heritage are clearly to be seen. Performance indicators, value for money, financial assessments, monitoring financial outcomes and the like could become dangerous threats to the heritage, even though it is recognised that enhancement of the services provided must 'avoid harm to the character and fabric of the property in each case'.

Some critics have been very outspoken on this issue. Croall, for instance, in a 1955 report published by the Calouste Gulbenkian Foundation, presents some strong arguments for greater protection of the environment against the effects of tourism. An even stronger case is made by Minhinnick (1993) who argues that 'the idea of making tourism an environmentally sustainable activity is at best an exciting pipe-dream and at worst a deceit'. This elegantly written essay is merciless in its criticisms: 'the trouble with tourism is that moderation is not part of its language'; 'local distinctiveness is erased and replaced by mediocre uniformity'. Even sharper are the sarcastic suggestions for improving tourist attractions, for example with visitor centres at the sites of pit disasters 'or even a tasteful, fee-paying Museum of the Children at Aberfan'. The stance may be regarded as an exaggerated one; but it will probably strike a sympathetic chord for some – and not only in Wales.

FURTHER READING

England

The most useful official reference sources are the PPGs and similar official publications: PPG 15 (*Planning and the Historic Environment*) and PPG 16 (*Archaeology and Planning*); DoE Circular 8/87 *Historic Buildings and Conservation Areas – Policy and Procedures*; and the DNH / Welsh Office 1996 Consultation Paper *Protecting Our Heritage*. A useful report is English Heritage (1996) *Conservation Issues in Local Plans*. Also invaluable is the annual *English Heritage Monitor*, published by the English Tourist Board. An update on English Heritage policy and programmes is given in their *Conservation Bulletin*. This is published three times a year. See also the list of English Heritage publications in the appendix on official publications (p. 369). A legal text is Suddards and Hargreaves (1995) *Listed Buildings*.

Northern Ireland, Scotland and Wales

There is little material on Northern Ireland, but two useful references are Hendry (1993) 'Conservation in Northern Ireland' in RTPI, *The Character of Conservation Areas*, vol. 2: *Supporting Information*. See also DoENI (1991) *Criteria and Standards for Listing*. On Scotland, a useful overview is given in National Audit Office (1995) *Protecting and Presenting Scotland's Heritage Properties*. See also the Scottish Office *Memorandum of Guidance on Listed Buildings and Conservation Areas 1993*. For Wales, see PPG 16 (Wales) *Archaeology and Planning*, TAN(W) 5 *Historic Buildings and Conservation Areas*, and TAN(W) 6 *Archaeology and Planning*.

Parliamentary Inquiries

There is a mammoth amount of material in three reports of the House of Commons Committees: the Environment Committee (1987) *Historic Buildings and Ancient Monuments*; the Welsh Affairs Committee (1993) *The Preservation of Historic Buildings and Monuments*; and the National Heritage Committee (1994) *Our Heritage: Preserving It, Prospering from It*.

History

An excellent, sympathetic and informative account of the history of historic preservation is given by Ross (first edition 1991; second edition 1996), *Planning and the Heritage*. Ross was former head of the Listing Branch of the DoE, and is a master of the subject.

Archaeology

On archaeology, in addition to PPG 16, see Pugh-Smith and Samuels (1993) 'PPG 16: two years on'; Redman (1990) 'Archaeology and development';

Tym & Partners (1995b) *Review of the Implementation of PPG 16: Archaeology and Planning*; English Heritage (1992) *Development Plan Policies for Archaeology*. For a well-reported case on rescue archaeology (the Rose Theatre), see Harte (1990) 'The scheduling of ancient monuments and the role of interested members of the public in environmental law'. A legal text is Pugh-Smith and Samuels (1996a) *Archaeology in Law*. A substantial, up-to-date review is given in a paper by Pugh-Smith and Samuels (1996b) 'Archaeology and planning'.

Conservation Areas

The Character of Conservation Areas (RTPI 1993) provides a broad review and (in the second volume) some useful supplementary material, including a bibliography. For a discussion of the threats to conservation areas from ill-considered alterations, see English Historic Towns Forum (1992) *Townscape in Trouble*. See also Morton (1991) 'Conservation areas – has saturation point been reached?'; Suddards and Morton (1991) 'The character of conservation areas'; Larkham and Jones (1993) 'Conservation and conservation areas in the UK: a growing problem'.

Economic Aspects

The economics of historic preservation is dealt with in Scanlon *et al.* (1994) *The Economics of Listed Build-*

ings; and Drury (1995) 'The value of conservation'. On buildings at risk, see English Heritage (1992) *Buildings at Risk*; English Heritage (1994) *Register of Buildings at Risk in Greater London*; and Davies and Keate (1995) *In the Public Interest: London's Civic Architecture at Risk*.

Tourism

The official view on the economic opportunities facing a more competitive tourist industry is expounded in DNH (1995) *Tourism: Competing with the Best*. Croall (1995) *Preserve or Destroy: Tourism and the Environment* presents a well-argued case for greater protection of the environment against the effects of tourism. Minhinnick (1993) *A Postcard Home* makes an even stronger attack on the development of tourism. (See also the references cited for Chapter 9.)

New Publications

Two recent books are Tiesdell *et al.* (1996) *Revitalising Historic Urban Quarters*, and Larkham (1996) *Conservation and the City*. See also the forthcoming *Politics and Preservation: A Policy History of the Built Heritage* by John Delafons.

9

PLANNING AND THE COUNTRYSIDE

The pace of change has quickened and much of what we most value about the rural scene seems threatened by increasing mobility, the pressures of leisure and recreation, the decline of jobs in rural industries and the demands for new jobs in businesses which once would have been found only in the towns.

Rural England (Cm 3016) 1995

THE CHANGING COUNTRYSIDE

There have been great changes in the British countryside since the early postwar policies were forged. Suburban commuter residential development, roads and transport, people seeking recreation, the changing economy, forestry, conservation, and a host of other pressures have grown beyond any expectation. The changes show no sign of abating: they never have. The British countryside has been subject to continual change: the 'natural' scenery which is now the concern of conservationists is the human-made result of earlier economic change. The changes continue, and the current ones are more far-reaching than those of the past. The most recent are those which come with agricultural surpluses and with the growth of both population and economic activity.

A major plank of postwar policy was that a prosperous agriculture would be not only of strategic economic value but also would provide the best means of preserving the countryside. Aided by the policies of the EC and by technological advances, the promotion of agricultural production has been a huge success. Unfortunately, as so often happens with policy successes, the solution of one problem gave rise to another. In place of the need for increased agricultural production is the problem of dealing with enormous surpluses and finding ways of reducing output. Matters are further complicated (throughout Europe) by more productivity increases resulting from a number of factors, including continuing technological advances (including biotechnological developments), and agricultural development in Eastern Europe (historically a major food producing region) The pressures for change in agricultural policy have been increased by mounting concern over the rural landscapes which have changed in response to newer production methods, and by growing demands for conservation and recreation.

A reversal of long-established, popular policies does not come easily; and the difficulties are increased when so many interests benefit from the subsidised regime. It is highly improbable that any political party could devise a policy that would bring about rapid and fundamental change to the traditional preoccupation with increased supply. As a result, change is taking place gradually. The 1986 Agriculture Act required agricultural ministers to maintain a balance between the interests of agriculture and wider rural and environmental interests. In particular, attention had to be given to 'the conservation and enhancement of the natural beauty and amenity of the countryside', and to 'the promotion of the enjoyment of the countryside by the public'. (More tangibly, the legislation provided for the establishment of *environmentally sensitive areas*, which are discussed later: p. 217.)

The new dimensions of policy involve issues of land management, for which the postwar planning system is inadequate. This system was designed to deal with land use, not its management; and it was largely restricted to urban land, not the countryside. It was assumed that a prosperous agriculture would by itself deal with any problems of the rural economy. One implication of this was that there appeared to be no problem about the division of central government countryside responsibilities between departments concerned with planning and those concerned with agriculture. Other divided responsibilities, particularly between countryside conservation and nature conservation, seemed more problematic: these later gave rise to several organisational changes.

NATIONAL PARKS

The demand for public access to the countryside has a long history (Eversley 1910), stretching from the early nineteenth-century fight against enclosures, James Bryce's abortive 1884 Access to Mountains Bill and the attenuated Access to Mountains Act of 1939, to the promise of the National Parks and Access to the Countryside Act of 1949: an Act which, among other things, poetically provided powers for 'preserving and enhancing natural beauty'. Many battles have been fought by voluntary bodies such as the Commons, Open Spaces and Footpaths Preservation Society and the Council for the Protection of Rural England (whose reports clearly indicate that their continued activity is still all too necessary), but they worked largely in a legislative vacuum until the Second World War. The mood engendered by the war augured a better reception for the Scott Committee's emphatic state-

ment that 'the establishment of national parks is long overdue' (Scott Report 1942: para. 178). The Scott Committee had very wide terms of reference, and for the first time an overall view was taken of questions of public rights of access to the open country, and the establishment of national parks and nature reserves within the context of a national policy for the preservation and planning of the countryside.

Government acceptance of the necessity for establishing national parks was announced in the series of debates on postwar reconstruction which took place during 1941 and 1943, and the White Paper on *The Control of Land Use* (Cmd 6537) referred to the establishment of national parks as part of a comprehensive programme of postwar reconstruction and land use planning. Not only was the principle accepted but, probably of equal importance, there was now a central government department with clear responsibility for such matters as national parks. There followed the Dower (1945) and Hobhouse (1947) reports on national parks, nature conservation, footpaths and access to the countryside, and, in 1949, the National Parks and Access to the Countryside Act which established the National Parks Commission and gave the main responsibility for the parks to local planning authorities.

The administration of the national parks has been a matter of controversy throughout their history. Dower had envisaged that there would be *ad hoc* committees with members appointed in equal numbers by the Commission and the relevant local authorities. Local representation was necessary since the well-being of the local people was to be the first consideration; but the parks were also to be *national*, and thus wider representation was essential. The lengthy arguments on this issue were eventually resolved by the 1949 Act in favour of a local authority majority, with only one-third of the members being appointed by the Secretary of State. (In line with his conception of truly national parks, Dower had proposed that the whole cost of administering them should be met by the Exchequer – an idea which was never accepted.)

Increasing pressures on the countryside have led

to a succession of policy and legislative changes. In 1968, the Countryside Act replaced the National Parks Commission with a more powerful Countryside Commission. A Countryside Commission for Scotland was established under the Countryside (Scotland) Act 1967. The Wildlife and Countryside Act 1981 strengthened the provisions for management agreements and introduced compensation for farmers whose rights were restricted (a major change in principle). The turn of the decade saw major structural changes in the organisation of agencies responsible for countryside matters, including the establishment of a separate Countryside Council for Wales (CCW), and the merging of the Countryside Commission for Scotland with the Nature Conservancy Council for Scotland as the Scottish Natural Heritage (SNH) – a reorganisation which was not followed in England. The latest change is the establishment of independent national park authorities, which have taken over the responsibilities previously exercised by local government. These are now the sole local planning authority in a national park area. In addition to the normal plans, a national park authority is required to prepare a *national park management plan*. This goes further than the scope of development plans: in addition to establishing policies, it is intended to spell out how the park is to be managed.

From their inception, the national parks have had two purposes: 'the preservation and enhancement of natural beauty', and 'encouraging the provision or improvement, for persons resorting to national parks, of facilities for the enjoyment thereof and for the enjoyment of the opportunities for open air recreation and the study of nature afforded thereby'. There is inevitably some conflict between these twin purposes, and the National Parks Review Panel (Edwards Report 1991) set up by the Countryside Commission recommended that they be reformulated to give added weight to conservation – as did the earlier Sandford Report (1974). This argument, which continues, was a major issue in the debates on the sections of the Environment Act 1995 which established independent national park authorities. Controversy centred on the need 'to promote the *quiet* enjoyment and understanding' in national parks. The final outcome is set out in the box, though a draft circular of guidance stresses that conservation takes precedence in cases of irreconcilable conflict.

THE BROADS AND THE NEW FOREST

Though proposed by the Dower Report as a national park, the Broads was rejected as such because of its deteriorated state and the anticipated cost of its management (Cherry 1975: 54). However, the need for some type of special protection continued

---

National Parks: Purposes

The Environment Act 1995 provides that the purposes of national parks shall be 'conserving and enhancing the natural beauty, wildlife and cultural heritage of the areas', and 'promoting opportunities for the understanding and enjoyment of the special qualities of those areas by the public'. If there is a conflict between these purposes, any relevant authority 'shall attach greater weight to the purpose of conserving and enhancing the natural beauty, wildlife and cultural heritage of the area'.

A national park authority, in pursuing these purposes 'shall seek to foster the economic and social well-being of local communities within the national park, but without incurring significant expenditure in doing so, and shall for that purpose cooperate with local authorities and public bodies whose functions include the promotion of economic and social development within the area of the national park'.

(Environment Act 1995)

to be debated, and the proposal surfaced again in the late 1970s. The local reaction was against this and, instead, a voluntary consortium, called the Broads Authority, was formed by the relevant public authorities (with powers and financial resources under the provisions of the Local Government Act 1972, and with 75 per cent Exchequer funding).

Discussions continued over several years among the large number of interested bodies and, in 1984, the Countryside Commission reviewed the problems of the area and the progress that had been made by the Broads Authority. Its conclusion was that, despite some achievements, the authority had not made significant improvements in water quality. Moreover, an effective framework for the integrated management of water-based and land-based recreation had not been established; and the loss of traditional grazing marsh was continuing. The outcome was the designation of the area as a body of equivalent status to a national park, but with a constitution, powers and funding designed to be appropriate to the local circumstances. A new Broads Authority (with the same name as its predecessor) was established by the Norfolk and Suffolk Broads Act 1988. The duties of the authority are extensive. It is the local planning authority and the principal unit of local government for the area. It has strong environmental responsibilities, and is required by the Act to produce a plan which has a wider remit than those required under the planning acts: it is more akin to a national park management plan.

While it took many years to devise an acceptable system for administering the Broads, it took even longer to do so for the New Forest. The New Forest was 'new' in 1079 when William the Conqueror appropriated it as his new royal hunting ground. Situated in South Hampshire, close to a large urban population, it has unique qualities of landscape and habitats, and a singular set of administrative arrangements. The Crown land, of some 27,000 hectares, is managed by the Forestry Commission. The so-called Perambulation is a wider area of about 38,000 hectares which is defined in the New Forest Act of 1964 (one of several Acts relating to the area). A larger area is the New Forest Heritage Area

which, though lacking statutory designation, was adopted in 1985 by the New Forest District Council and is identified in local development plans.

In addition to the public authorities in the area, there is a corporate body of Verderers which is responsible for managing the grazing and commoning within the forest. As well as the protection provided in these ways, much of the Crown land in the New Forest is designated as *sites of special scientific interest* (SSSIs). The southern fringe of the forest is within the South Hampshire Coast AONB, and the whole of the heritage area is within the South-West Hampshire Green Belt. It might thus appear that the New Forest is adequately (even if confusingly) protected. However, quite apart from questions of coordinating all these protectors of the forest, there is a further need to safeguard the surrounding grazing lands which are under pressure for development, and also to ensure that adequate provision is made for recreation in a manner which is in harmony with conservation requirements.

The New Forest has for long been a candidate for designation as a national park, but it has been generally agreed that, like the Broads, it has features which require special treatment. The government announced in early 1992 that the area was to be given a statutory designation, and that an *ad hoc* authority (similar to the Broads Authority) was to be established. The New Forest Committee which had been established as a non-statutory body in 1990 (with government encouragement), was to be the foundation for the new authority. It is unclear exactly what happened next, except that the decision became the victim of the political process, and it was rescinded (Tubbs 1994). The dominant theme now is cooperation between all the parties involved.

LANDSCAPE DESIGNATIONS

Both the Dower and Hobhouse Reports proposed that, in addition to national parks, certain areas of high landscape quality, scientific interest and recreational value should be subject to special protection. These areas were not considered, at that time, to require the positive management which it was assumed would characterise national parks, but 'their contribution to the wider enjoyment of the countryside is so important that special measures should be taken to preserve their natural beauty and interest'. The Hobhouse Committee proposed that such areas should be the responsibility of local planning authorities, but would receive expert assistance and financial aid from the National Parks Commission. A total of fifty-two areas, covering some 26,000 sq km was recommended, including, for example, the Breckland and much of central Wales, long stretches of the coast, the Cotswolds, most of the Downland, the Chilterns and Bodmin Moor (Cherry 1975: 55).

The 1949 Act did not contain any special provisions for the care of such areas, the power under the Planning Acts being considered adequate for the purpose. It did, however, give the Commission power to designate *areas of outstanding natural beauty* (AONBs), and provided for Exchequer grants on the same basis as for national parks. Forty-one areas have been designated covering over 20,000 sq km (7 per cent of the area of England, and 20 per cent of Wales).

Areas of outstanding natural beauty are, with some notable exceptions, generally smaller than national parks (see Figure 9.1). They are the responsibility of local planning authorities which have powers for the 'preservation and enhancement of natural beauty' similar to those of park planning authorities. There has been continuing debate on the question as to whether the designation of areas of outstanding natural beauty serves any useful function.

In addition to AONBs, there are many local authority designations designed to assist in safeguarding areas of the countryside from inappropriate development. Some of these have been given additional status through inclusion in structure and local plans. Though these are like AONBs in that they involve the application of special criteria for control in sensitive areas, they do not imply any special procedures for development control. The DoE Consultation Paper on *The Future of Development Plans*

(1986) made reference to *areas of landscape quality*, *areas of great landscape value*, *landscape conservation areas*, *coastal preservation areas* and *areas of semi-national importance*. The same paper proposed a new statutory designation, the *rural conservation area* which, it was suggested, would provide a more coherent framework. This idea, however, found little favour during the consultation process, and it was dropped. It was felt that the desired objectives could be achieved through statements of policy in development plans.

The current advice simply makes reference to 'locally devised' designations, and notes that, though they have no statutory role, they 'nevertheless serve to highlight particularly important features of the countryside that should be taken into account in planning decisions' (PPG 7: para. 3.17).

SCOTTISH DESIGNATIONS

Scotland contains large areas of beautiful unspoilt countryside and wild landscape. It has the majority of Britain's highest mountains, with nearly 300 peaks over 900 m. The northern highlands extend from the central lowlands to the north of the mainland, and are divided by the Great Glen, which stretches from Inverness to Fort William and contains Loch Ness, Loch Oich and Loch Lochy. The Grampians cover the southern highland area, and include Ben Nevis, the highest point in the British Isles (1,300 m). Scotland has the great majority of the UK islands. Its coast is over 10,000 km in length. Much of the west coast consists of fjord-type lochs, of which the longest is Loch Fyne (67 km).

Despite expectations to the contrary, there are no national parks in Scotland. Though a Scottish committee (the Ramsay Committee) recommended in 1945 the establishment of five Scottish national parks, no action followed. The reasons for this inaction were partly political and partly pragmatic (Cherry 1975). An essential element of the latter was that (with the exception of the area around Clydeside and, in particular, Loch Lomond), the

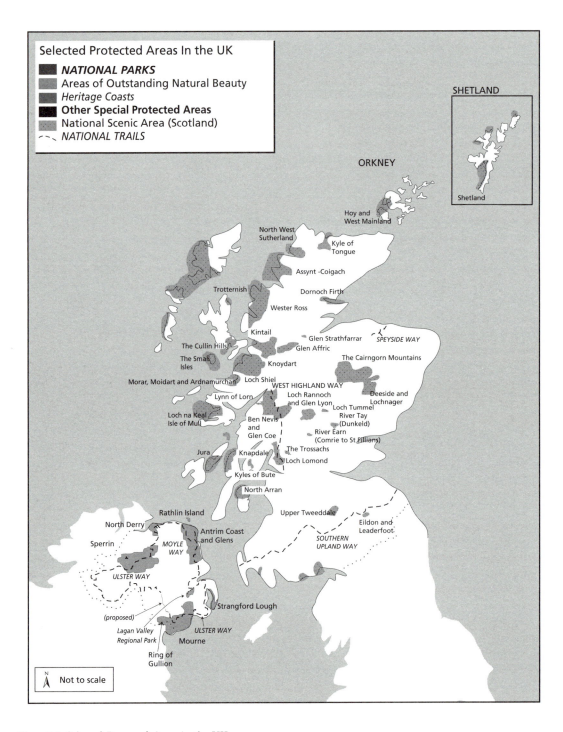

Figure 9.1 Selected Protected Areas in the UK

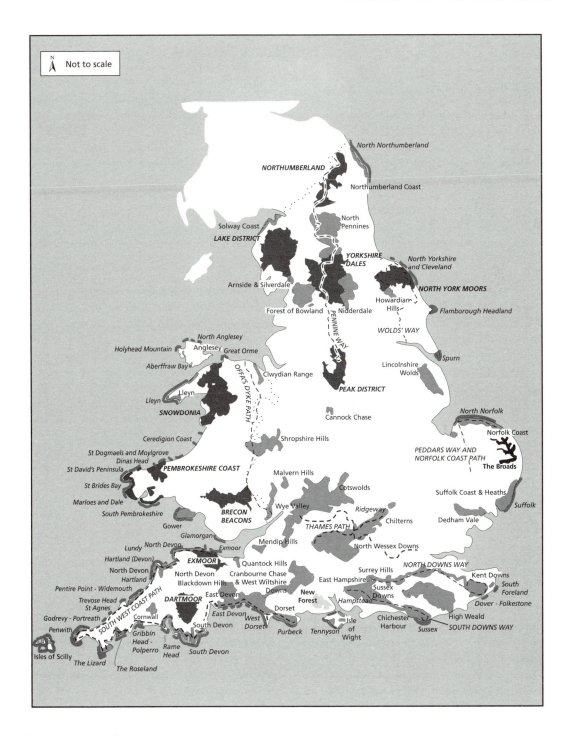

Figure 9.1 continued

pressures which were so apparent south of the border were absent.

Nevertheless, the Secretary of State used the powers of the 1947 Planning Act to issue *National Parks Direction Orders*. These required the relevant local planning authorities to submit to the Secretary of State all planning applications in the designated areas (which included Loch Lomond/Trossachs, the Cairngorms, and Ben Nevis/Glen Coe). In effect therefore, in an almost Gilbertian manner, while Scotland at this time did not have any national parks, it had an administrative system which enabled controls to be operated as if it did! But, of course, this approach was inherently negative, and it was not until the Countryside (Scotland) Act of 1967 that positive measures could be taken on a significant scale.

This Act provided for the establishment of the Countryside Commission for Scotland – recently joined with the Nature Conservancy Council for Scotland in Scottish Natural Heritage (SNH). It also enabled the establishment of regional parks and country parks. A policy framework for these was set out in the Commission's 1974 report *A Park System for Scotland*. The report also recommended the designation of national parks in Scotland, though the term *special park* was used. This has never been accepted but objectives similar to those of national parks have been achieved under other designations.

There are forty national scenic areas, four regional parks and thirty-five country parks. The *national scenic areas* are of similar status to the AONBs. They extend over an area of 1 million hectares, and include such marvellous sites as Ben Nevis and Glen Coe, Loch Lomond, the Cairngorms, and the World Heritage Site of the islands of St Kilda. Development control in these areas is the responsibility of the local planning authorities, which are required to consult with SNH for certain categories of development. As in England and Wales, there is an increasing concern for 'positive action to improve planning and land use management' in these areas, and for dealing with the erosion of footpaths. There is also a similar complaint about the lack of resources.

A *regional park* is statutorily defined simply as 'an extensive area of land, part of which is devoted to the recreational needs of the public'. The four parks are Clyde–Muirshiel, Loch Lomond, the Pentland Hills, and Fife. These parks, which cover 86,000 hectares, are primarily recreational areas, and each has a local plan which sets out management policies. Emphasis is laid on *integrated land management* schemes to ensure that public access is in harmony with other land uses. In this, they give effect to Abercrombie's green-belt philosophy, articulated in the Clyde Valley Regional Plan. He conceived these *outer scenic areas* not only as recreational areas but also as a means of protecting the rural setting of the conurbations (Smith and Wannop 1985).

Since the passing of the 1967 Act, Scottish local authorities have provided thirty-six *country parks* spread across the central belt and the north-east. The parks are 'registered' with SNH, which makes grants for capital development expenditure and also towards the cost of a ranger service. Country parks are not only of direct benefit to their eleven million annual visitors, they also have a conservation objective of 'drawing off areas that are sensitive due to productive land uses and fragile wildlife habitats'.

The Scottish legislation mirrors the English provisions in allowing planning authorities to make management agreements with private landowners. Under these agreements, of which some 500 are in operation, land uses can be modified in the interests of scenic conservation, and steps designed to promote the enjoyment of the countryside by the public.

The Natural Heritage (Scotland) Act 1991, which provided for the establishment of Scottish Natural Heritage, also introduced *natural heritage areas*. These are areas 'of outstanding value to the natural heritage of Scotland'. The Scottish Office stresses that natural heritage areas are not a substitute for national parks; they are an alternative designed to meet the particular situation in Scotland. It is intended that they will be designated for a wide range of situations in both upland and lowland Scotland. They will typically cover wide areas where there is both a landscape and a nature conservation interest, and where there is therefore a need for integrated management. Examples are the Flow Country, where there are areas of international conservation interest within a much larger area of national landscape interest, and areas of lowland Scotland where there has been considerable damage to the landscape and there is a pressing need for rejuvenation. At present there is no unified designation which deals with both nature and landscape conservation and which also provides for access and enjoyment.

Consideration is being given to the array of different designations in Scotland which (in the words of *Rural Scotland* (Cm 3041) 'has now grown to a point where there is a real risk of confusion'. Accordingly a task force has been set up to review the situation. Until its report is issued, no designations of natural heritage areas will be made.

NORTHERN IRELAND DESIGNATIONS

Northern Ireland boasts some of the finest countryside in the UK, often with special value as wildlife habitat. One reason for this is that farm and field sizes are smaller than on the mainland, and almost all the farms are owner-occupied (Glass 1994). Much of the countryside remains unspoiled, but development pressures are increasing, and there has been extensive building of isolated houses in the countryside. Progress with planning for landscape and nature conservation has been slower than in the rest of the UK. Legislation is far less developed, and there has been much criticism about the delays in designating areas needing protection and management (Dodd and Pritchard 1993).

The background to this is the social and economic conditions in the Province: these are among the poorest in the UK. GDP per head is about 78 per cent of the EU average; unemployment is twice the average; and birth rates are the highest. The whole of the Province is designated as Objective 1 for European Structural Funds, and has received substantial funding through the programmes for special rural development. Economic considerations have

therefore taken precedence over those relating to the countryside.

Criticisms of the backwardness of countryside and nature conservation led to a review on behalf of the Secretary of State by Dr Jean Balfour whose report, *A New Look at the Northern Ireland Countryside* (Northern Ireland Office 1983), confirmed the low priority given to conservation and its lack of status in the work of the DoENI. The result was the setting up of a new unit within the department and an advisory Council for Nature Conservation and the Countryside. Legislation extending nature conservation powers soon followed. Nevertheless, designations remained 'pitifully slow'. A new initiative was taken in the early 1990s, culminating in the publication in 1993 of a comprehensive *Planning Strategy for Northern Ireland*. (In the same year the Environment Committee repeated its call for the establishment of an environmental protection agency for Northern Ireland, but this was rejected by the government.)

Statutory designations are much the same as in England and Wales, and the DoENI has all powers in respect of designation and management of special areas (including those exercised by English Nature and the Countryside Commission). It can designate national parks (though it has not done so), areas of outstanding natural beauty, nature reserves, and *areas of special scientific interest* (the equivalent of the SSSI). An additional non-statutory designation is the *countryside policy area* which is employed to restrict building in the countryside.

THE COAST

A few figures underline the particular significance of the coast, and therefore of coastal planning: nowhere in the UK is more than 135 km from the sea; the coastline is 18,600 km in length, and the territorial waters extend over about one-third of a million sq km. About one-third of the coast of England and Wales is included in national parks and AONBs. Large areas of the coast are owned or protected by the National Trust. Following the *Enterprise Neptune* fund-raising appeal, the Trust protects 850 km of the coastline in England, Wales and Northern Ireland (mainly by ownership, the remainder by covenant). North of the border, the National Trust for Scotland owns large stretches of the coastline and protects further parts by way of conservation agreements. In addition, there are *marine nature reserves* (discussed later, p. 221). In spite of all this protection, the pressures on the coastline are proving increasingly difficult to cope with. Between a quarter and a third of the coastline of England and Wales is developed (PPG 20: Annex 1). Growing numbers of people are attracted to the coast for holidays, for recreation and for retirement. There are also economic pressures for major industrial development in certain parts, particularly on some estuaries (which have international importance for nature conservation).

The problem is a difficult one which cannot be satisfactorily met simply by restrictive measures: it requires a positive policy of planning for leisure provision. This has long been accepted, and the *heritage coast* designation, introduced in 1972, implies recreational provision as well as conservation. The Countryside Commission has urged that every heritage coast should have a management plan. It has also established the *Heritage Coast Forum* as 'a national body to promote the heritage coast concept and to act as a focus and liaison point for all heritage coast organisations'. This is seen as a needed addition to the activities of the Commission, whose capacity to promote all the initiatives that are necessary is limited.

By 1994, forty-five heritage coasts in England and Wales had been defined, extending laterally over 1,525 km. In Scotland, twenty-six *preferred coastal conservation zones* have been defined with a total length of 7,546 km, covering three-quarters of Scotland's mainland and islands coastlines.

The Environment Committee, in its 1992 report, *Coastal Zone Protection and Planning* has complained of the lack of coordination among the host of bodies concerned with coastal protection, planning and management. (There are over eighty Acts which deal with the regulations of activities in the coastal

zone, and as many as 240 government departments and public agencies involved in some way.) Not surprisingly, there have been suggestions that action is required to simplify, rationalise, coordinate or consolidate matters. Though an apparently obvious and sensible idea, it is remarkably difficult to see how the situation can be significantly changed; and the Environment Committee contented itself by asking for a review of legislation and responsibilities. The government response was negative. It was pointed out that, though there were many Acts relating to the coast, the same could be said about the land! Indeed, it was neither possible nor desirable to treat the coast separately from the adjoining land or from territorial and international waters. Moreover, the suggestion that the town and country planning system might be extended seaward was not persuasive, though it was agreed that 'it is now time to take this debate further'.

Two discussion papers were issued by DoE in 1993 (*Managing the Coast* and *Development Below Low Water Mark*). These drew upon the Countryside Commission's experience with Heritage Coasts and underlined the usefulness of management plans drawn up by local authorities in liaison with the relevant bodies concerned. The Countryside Commission responded that there was a need for guidance on the form and content of such plans, and for integration with the shoreline management plans for flood and coastal defence which are being promoted by the MAFF. In 1995, an organisation of maritime local authorities, the National Coasts and Estuaries Advisory Group, published a *Directory of Coastal Planning and Management Initiatives in England* (King and Bridge 1995). This was followed, in 1995, by a DoE guide to government policies affecting the coastal zones (*Policy Guidelines for the Coast*). The debate continues.

WATERWAYS

There are some 3,700 miles of waterways in Britain, of which the British Waterways Board (BWB) is responsible for about 2,000. Most of the latter are canals; the others are stretches of river navigation. The Board's waterways are divided (under the Transport Act 1968) into three categories: commercial (350 miles), which are principally for the carriage of freight; cruising (1,100 miles), which are principally for cruising, fishing and other recreational purposes; and 'remainder waterways' (500 miles). Commercial and cruising waterways have to be maintained in a navigable condition, while remainder waterways are maintained to a level consistent with public health, amenity and safety. Remainder waterways can be upgraded to cruising standard, retained or eliminated. The Board has an annual turnover (1994–5) of £85 million, of which £49 million is met from a DoE grant. A considerable proportion of expenditure (£14 million in 1994–5) is on major works, such as channel lining, embankments, aqueducts and bridges. BWB is carrying out a programme to eliminate a backlog of major works to the fabric of the canals which has accumulated during years of underfunding.

An important source of funding for improvements to the waterways is the Board's property portfolio. In conjunction with other organisations, the Board has undertaken several major urban renewal developments, such as in the development of Sheffield Basin and Paddington Basin and the comprehensive development of the Gas Street Basin in Birmingham (which has gained the city the Excellence on the Waterfront award – jointly with Boston and New York!). Since about a quarter of the total length of the waterways falls within the boundaries of the (former) metropolitan counties there is considerable potential for waterside development, (though it should be noted that most waterside property is not owned by British Waterways).

Some BWB development has attracted criticism from the Association of Pleasure Craft Operators because it has been seen as destroying the ambience of the canals and thus making them less attractive for cruising. As with many leisure pursuits, there is a problem of satisfying conflicting interests. Those who love the often closed and secretive world of much of the urban canal system may not welcome the new focal point developments, but these have proved to be extremely popular with others. Another conflict arises between the use of the waterways for leisure and their function as an aquatic habitat. Such conflicts cannot be prevented, but it is an explicit policy of the BWB 'to achieve an appropriate balance' among differing interests and uses.

The canal network has been described as a country park which is 2,000 miles long by ten yards wide; but it is much more than that: the BWB owns over 2,000 listed structures and ancient monuments and sixty-four SSSIs. It has a major programme of management and conservation in relation to these and other heritage features. Over eight million people use the waterways in the course of a year, mainly for informal recreation rather than boating. In its 1993 report on *The Waterway Environment and Development Plans*, the BWB comments that this informal recreation 'offers the greatest prospect for increased use of the waterways', and this is recognised in the Board's initiative in seeking joint study and action with local authorities and a host of other relevant agencies.

Though the Board cannot impose direct charges, related leisure facilities are income-generating: for example shops, public houses, hotels, restaurants and museums. A good example is the National Waterways Museum at Gloucester which has over 100,000 visitors a year. The Board is required by the DoE to 'run its affairs on a commercial basis' – so far as is practicable. The Board has noted that 'further provision of leisure and tourism facilities and improved access for recreation will often need to be underpinned by enabling development'. It is fortunate that national parks do not have to operate on the same basis.

PUBLIC RIGHTS OF WAY AND ACCESS

The origin of a large number of public rights of way is obscure. As a result, innumerable disputes have arisen over them. Before the 1949 Act, these disputes could be settled only on a case-by-case basis, often with the evidence of 'eldest inhabitants' playing a leading role. The unsatisfactory nature of the situation was underlined by the Scott, Dower and Hobhouse Reports, as well as by the 1948 report of the Special Committee on Footpaths and Access to the Countryside. All were agreed that a complete survey of rights of way was essential, together with the introduction of a simple procedure for resolving the legal status of rights of way which were in dispute. The 1949 Act attempted to provide for both.

This Act has been amended several times; under the current provisions, county councils have the responsibility for surveying rights of way (footpaths, bridleways and 'byways open to all traffic') and for preparing and keeping up to date what is misleadingly called a *definitive map*. The maps are supposedly conclusive evidence of the existence of rights of way but, in fact, they are not necessarily either complete or conclusive. They are incomplete because inadequate resources have been devoted to undertaking the necessary surveys; and they can be inconclusive because the map may wrongly identify a right of way. The latter is a legal matter which is not discussed here (Chesman 1991), but the former is a continuing problem of planning policy and administration.

The definitive maps show some 225,000 km of rights of way in England and Wales. Most of these are 'existing': few new recreational paths have been designated, though they are certainly needed in some parts of the country. Moreover, there has been a loss of access due to both neglect and deliberate obstruction. Each year some 1,500 formal proposals, affecting 500 km of the network, are made to change rights of way (by creation, diversion or extinguishment). Of these, about three-quarters are unopposed. The net change is negligible. It is difficult to establish what the overall effect is, though the Ramblers Association maintains that over a half of the public rights of way 'are unavailable to all but the most determined and agile person' (Blunden and Curry 1989: 135).

Another footpath problem arises from their popularity: this is the wear and tear caused by a great intensity of use. The Pennine Way in particular has suffered from this, and 'damage limitation' experiments are under way. The latest measures have included laying flagstones which are delivered by helicopter. The Pennine Way is one of the long-distance routes which now stretch over some 2,700 km. The designation of these hikers' highways has been laborious, but they have had the attention and backing of the Countryside Commission, which has official responsibility for their establishment. The Commission has set itself the target of putting the whole of the rights of way network into good order by the end of the century. As an aid to this, it has launched a *Parish Paths Partnership*. This aims to stimulate local improvement schemes through parish councils and other local groups.

In Scotland, the position in relation to rights of way and access to the countryside is different from that south of the border. The legal system is distinct, and the pressures on the countryside, until recently, have generally been fewer. There is relatively freer access to the Scottish countryside: there is 'a well-established system of mutual respect between walker and landowner' (Blunden and Curry 1989: 152). Nevertheless, this is not so in areas close to the towns, where access is severely restricted; as pressures have mounted, the inadequacy of the legal situation has become apparent. A 1994 report, *Enjoying the Countryside*, provides a comprehensive summary of the issues and an agenda for meeting the growing demands for access. It calls for additional resources, greater enterprise by local authorities in planning and providing open-air recreation in the countryside, more investment in provision for tourism, continued encouragement and support to voluntary bodies, and much more in the same vein. An agreement on access to land was signed in 1995 between SNH and farming and landowning interests, under the banner of *Scotland's Hills and Mountains: A Concordat on Access*.

The first designated long-distance footpath in Scotland was opened in 1980. This is the West Highland Way which runs for 152 km from Milngavie (a suburb of Glasgow) to Fort William at the foot of Ben Nevis. The Way is used by an estimated

90,000 walkers a year, and also injects some £3.5 million into the communities along the route. Unfortunately, this has also led to severe erosion. There has been similar erosion on sections of the 328 km Southern Upland Way which stretches from Portpatrick on the south-west coast to Cockburnspath on the east coast. Negotiations for other long-distance footpaths are at various stages of negotiation, including an extension of the Speyside Way and the development of a new route, the Great Glen Way.

Rights of way provide a structured framework for public access to the countryside, and they are of particular value to the energetic walker. However, they meet only part of the need: the number of people who enjoy a wander in the countryside is far greater than the number who hike long distances. The 1947 Hobhouse report had argued for a public right of access to all open country: among the many benefits they foresaw was the freedom to ramble across the wilder parts of the country. Their idyllic view was very much in line with the long-standing arguments for a 'right to roam' over all open country. The 1949 Act was much more circumspect, and provided for a right of access only where an access agreement was made with the owner.

The essential arguments on this issue have not changed; but circumstances have. There is now a much wider demand for access to the countryside, fostered both by increasing leisure pursuits and a huge increase in the ease of travel. But this very increase has strengthened the arguments of landowners about 'inappropriate use' of the countryside.

PROVISION FOR RECREATION

In the early postwar years, national recreation policy was largely concerned with national parks (and their Scottish shadow equivalents), areas of outstanding natural beauty and the coast. Increasingly, however, there has developed a concern for positive policy in relation to metropolitan, regional and country parks. The Countryside Act 1968 (which followed a White Paper (Cmnd 2928) with the significant title *Leisure in the Countryside*) gave additional powers to the new Countryside Commission for 'the provision and improvement of facilities for the enjoyment of the countryside', including experimental schemes to promote countryside enjoyment. At the same time, local authorities were empowered to provide *country parks*, including facilities for sailing, boating, bathing and fishing. These country parks are not for those who are seeking the solitude and grandeur of the mountains, but for the large urban populations who are 'looking for a change of environment within easy reach'. There is now a wide range of country parks, picnic sites, visitor-interpretive sites, recreation paths, interpretive trails, cycleways and similar facilities provided by local authorities and the Countryside Commission.

Contrary to popular impression, the number of visitors to the countryside has not increased significantly in recent years (as it did markedly up to the end of the 1970s). Moreover, for the most part, the impact on the environment is of manageable proportions, though traffic problems certainly can considerably reduce the 'quality of the recreational experience' (the extreme sometimes being reached in the Peak District where, on occasion, a condition of gridlock is created by the sheer volume of traffic). Though there are some areas where recreational pressures have undesirable impacts, generally leisure and tourism are less of a threat than industrial and agricultural activities. However, as the Environment Committee pointed out in their 1995 report on *The Environmental Impact of Leisure Activities*, there are difficult problems of crowding, overuse and conflict of activities in certain areas. The favourite answer is 'good management', and there is no doubt that this can help in preventing visitors 'loving to death' the beauty spots they wish to visit (to use the apt phrase adapted as the title of a report on sustainable tourism in Europe by the Federation of Nature and National Parks). So can 'countryside codes', 'visitor awareness' campaigns and such like. But more drastic measures are inevitable in the most popular locations: Dovedale attracts two million visitors a year, of whom 750,000 use the main footpath. On busy Sundays no less than two thousand people *an hour* can be crossing the river by the stepping stones. (The photograph on the cover of Jonathan Croall's 1995 book, *Preserve or Destroy: Tourism and the Environment*, is the most eloquent statement of the problem of which this is only one illustration.) In Northern Ireland, there are indications that the 'peace dividend' is already having an effect, with tourist pressures on attractive areas such as County Fermanagh.

Various schemes are being tried: from strict control of cars and the provision of public transport to a expansion of facilities in new areas. The Countryside Commission is placing increased importance on funding recreational facilities close to where people live which has the additional advantage of helping 'to reduce the number of countryside trips made by car and provides opportunities for countryside recreation for people without cars, for the young, and for those with special needs'. The Commission's annual reports record a wide range of actions on this front.

One particularly interesting initiative which started in the early 1980s and is now well established is the work of the *Groundwork Trusts*. Conceived as an *additional* resource for converting waste land to productive uses, particularly in urban fringe areas, it facilitates cooperative efforts by voluntary organisation and business, as well as public authorities. There is now a network of forty-four trusts in England and Wales. Their enterprise is wide-ranging and, in addition to providing for recreation, their work includes land reclamation, landscaping, environmental appreciation and many other activities seen as desirable and worthwhile in local communities. Examples include the development of the Taff Trail, which is a long-distance footpath and cycleway linking Cardiff and Brecon; re-creating wildlife sanctuaries and access around mining villages in east Durham; the development of the Middleton Riverside Park on a totally derelict site a few miles from Manchester; and a programme which encourages owners of industrial and commercial premises 'to stand back and take a look at the external image of their premises and then to make practical

landscape improvements'. The scheme is clearly highly adaptable to local conditions and aspirations.

Some activities are simply impossible to accommodate in popular locations: high-powered recreational vehicles and boats are not only environmentally damaging, they also destroy the pleasure that others are seeking. The Environment Committee argued that 'the principle of sustainability in leisure and recreation involves the provision of facilities of all activities, not only for the aesthetically pleasing and non-intrusive ones'. This is a curious interpretation of the concept of 'sustainability' which would not be universally accepted; but there would be more support for the Committee's proposal that sites should be selected which are 'suitable for noisy and obtrusive activities'. Whether such sites could be readily found is more problematic, even on 'derelict land or land of low amenity'. An economist would point out that this is a classic case for a charging mechanism which would enable compensation to be paid to those adversely affected.

Much of the greater willingness to provide additional opportunities and facilities for recreation in the countryside has emanated from the changed economics of agriculture. Less agricultural production and more diversified activities are desirable. Indeed, without changes in the pattern of economic activity, many rural areas will be adversely affected by the changes in agriculture. The provision of recreational facilities is a potentially lucrative business, as has been dramatically illustrated in those areas where it has been realised. The Rural Development Commission has estimated that tourism is worth £8 billion a year to the rural areas of England and generates 400,000 jobs. A study of Center Parcs holiday villages demonstrates the substantial benefits which these brought to the local areas: £4.5 million around Sherwood (Nottinghamshire) and £5.3 million around Elveden (Norfolk). The employment created could amount to well over 1,000 jobs when the Parcs are fully operational. Tourism also brings more indirect benefits to rural areas such as the survival of bus services and village shops. There is, of course, a cost to be borne for these advantages – in terms of changed character and conflicts between the interests of visitors and residents (particularly in areas which have attracted new residents). Such conflicts can be reduced by 'good management', but they are inherent in the dynamics of social and economic change.

COUNTRYSIDE GRANT PROGRAMMES

Changes in policies relating to farming have had a more tangible effect than the heated arguments of contenders for and against freer countryside access. The lower priority for food production has led to attempts to broaden the role of landowners as 'managers' of the countryside. Within this changing framework, 'access' becomes a means of diversifying the agricultural economy. Successive measures have reflected the changing priorities, and there has been increased emphasis on the role of farmers as 'stewards of the countryside' which has led to a greater concentration of funds on environmental schemes.

The Countryside Commission has been in the lead in promoting conservation and recreation as explicit objectives of agricultural policy. Its 1989 policy statement *Incentives for a New Direction in Farming* argued that the diminishing need for agricultural production provided an opportunity for 'environmentally friendly' farming. It presented a menu of incentives for farmers and landowners to provide environmental and recreational benefits. These ideas were translated into the *countryside premium*: an experimental scheme which gave incentives for land to be set aside for recreation. It was followed by *countryside stewardship*: launched in 1991, this experimental scheme provides incentives for the protection and enhancement of valued and threatened landscapes. This scheme proved to be a successful one: in its first four years, some 5,000 agreements were concluded covering 91,000 hectares. It has now been transferred (together with the hedgerow incentive scheme which provides financial incentives for the restoration of hedgerows) to the Ministry of Agriculture which operates other, similar land management schemes. In Wales, the CCW

has a parallel scheme (*Tir Cymen*). Other 'countryside' bodies also have schemes: the Forestry Commission operates a *woodland grant scheme* which includes payments for the management of woods to which the public have access; and the Ministry of Agriculture makes access payments under its *countryside access scheme*.

The 1986 Agriculture Act made provision for *environmentally sensitive areas* (ESAs) where annual grants are given by the MAFF to enable farmers to follow farming practices which will achieve conservation objectives. The introduction of ESAs marked a fundamental policy change (Bishop and Phillips 1993: 325). They provided financial support for practices which would result in environmental benefits, in contrast to the 1981 Act system which gave compensation for forgone profits.

ESAs have developed into the main plank of the ministry's countryside protection policy. There were nineteen ESAs in the original scheme introduced in 1987. By 1995, the number (in Britain) had increased to thirty-eight, extending over 31,000 sq km.

The Ministry of Agriculture also operates grant schemes for farm diversification and for farm woodlands. The former (now partly amalgamated with the EC funded *farm and conservation grant scheme*) is mainly aimed at encouraging environmentally beneficial investments. The latter (the *farm woodland scheme*) began as an experiment in 1991, but the results were disappointing and a comprehensive review led to the introduction of a new *farm woodland premium scheme* in 1992. The objectives are to enhance the farmed landscape and environment and to encourage a productive land use alternative to agriculture. Additionally, a pilot *extensification scheme for beef and sheep* provides payments to farmers who reduce output by 20 per cent and, at the same time, maintain environmental features such as hedges, grassland and moorland.

New schemes are being introduced at a bewildering rate. In addition to those already mentioned, there is the *hedgerow incentive scheme, wildlife enhancement scheme, parish paths partnership* and the Development Commission's *countryside employment programme* and *redundant building grant*.

Diversification is the theme much of current policy. In the words of PPG 7 (para 2.3): 'The priority now is to promote diversification of the rural economy so as to provide wide and varied employment opportunities for rural people, including those formerly employed in agriculture and related sectors.'

NATURE CONSERVATION

The concept of wild life sanctuaries or nature reserves is one of long standing and, indeed, antedates the modern idea of national parks. In other countries, some national parks are in fact primarily sanctuaries for the preservation of big game and other wildlife, as well as for the protection of outstanding physiological features and areas of outstanding geological interest. British national parks were somewhat different in origin, with an emphasis on the preservation of amenity and providing facilities for public access and enjoyment (though, as noted earlier, the trend has been to give increasing priority to conservation). The concept of nature conservation is primarily a scientific one concerned particularly with the management of natural sites and of vegetation and animal populations.

The Huxley Committee argued, in 1947, that there was no fundamental conflict between these two areas of interest:

> their special requirements may differ, and the case for each may be presented with too limited a vision: but since both have the same fundamental idea of conserving the rich variety of our countryside and sea-coasts and of increasing the general enjoyment and understanding of nature, their ultimate objectives are not divergent, still less antagonistic.

However, to ensure that recreational, economic and scientific interests are all fairly met presents some difficulties. Several reports dealing with the various problems were published shortly after the war (Dower Report 1945; Huxley Report 1947; Hobhouse Report 1947). The outcome was the establishment of the Nature Conservancy, constituted by a Royal Charter in March 1949 and given additional powers by the National Parks and Access to the Countryside Act. As already noted, the Conservancy has now been replaced by English Nature (EN), Scottish Natural Heritage (SNH), and the Countryside Council for Wales (CCW). In Northern Ireland, Scotland, and Wales, central responsibility for nature conservation and access to the countryside rests with a single body, but in England they remain separate (Figure 9.2).

Legislation often emerges as a response to new perceptions of problems; but sometimes legislation itself fosters such perceptions. So it was with the Wildlife and Countryside Act 1981. Introduced as a mild alternative to the Labour government's aborted Countryside Bill (and stimulated by the need to take action on several international conservation agreements), the Conservative government expected no serious trouble over the Bill. It was very mistaken: the Bill acted as a lightning rod for a host of countryside concerns that had been building up over the previous decade or so: moorland reclamation, afforestation and 'new agricultural landscapes', loss of hedgerows, damage to SSSIs, and such like. The Bill had a stormy passage through Parliament, with an incredible 2,300 proposed amendments. Though most of these failed, the Bill was considerably amended during the process. The major focus of argument (with the strong NFU and the CLA holding the line against a large but diffuse environmental lobby) was the extent to which voluntary management agreements could be sufficient to resolve conflicts of interest in the countryside. The government steadfastly maintained that neither positive inducements nor negative controls were necessary. Indeed, it was held that controls would be counterproductive in that they would arouse intense opposition from country landowners.

Of particular concern was the rate at which SSSIs were being seriously damaged; the rate at which moorland in national parks was being converted to agricultural use or afforestation; and the adverse impact of agricultural capital grants schemes both on landscape and on the social and economic well-being of upland communities. On the first issue, the

the *Countryside: Studies of Social Change in Rural Britain*; Bunce (1994) *The Countryside Ideal: Anglo-American Images of Landscape*; Cherry (1994b) *Rural Change and Planning: England and Wales in the Twentieth Century*; Cloke *et al.* (1994) *Lifestyles in Rural England*; and Newby (1985); *Green and Pleasant Land: Social Change in Rural England*.

Three current 'rural white papers' are *Rural Scotland: People, Prosperity and Partnership* (Cm 3041) (1995); *Rural England: A Nation Committed to a Living Countryside* (Cm 3016) (1995); and *A Working Countryside for Wales* (Cm 3180) (1996). See also PPG 7 *The Countryside and the Rural Economy*; Scottish Office *Rural Framework* (1992); and the highly detailed study of *Planning for Rural Diversification* by Elson *et al.* (1995a).

National Parks and Access to the Countryside

For accounts of the background to and the implementation of the 1949 Act, see Cherry (1975) *National Parks and Recreation in the Countryside* (vol. 2 of *Environmental Planning 1939–1969*), and Blunden and Curry (1989) *A People's Charter? Forty Years of the National Parks and Access to the Countryside Act 1949*. Two major reviews of national parks policies are the Sandford Report (1974) and the Edwards Report (1991). A passionate critique of the restrictions on access (and much else) is given in two books by Shoard *The Theft of the Countryside* (1980) and *This Land is Our Land* (1987)

Coastal Issues

On coastal issues, see DoE *Policy Guidelines for the Coast*; PPG 20 *Coastal Planning*; and House of Commons Environment Committee (1992) *Coastal Zone Protection and Planning*.

Waterways

There is extensive documentation on the waterways in the HC Environment Committee's report *British Waterways Board* (1989). The account in the text draws freely from this source. See also the *Government Response* to the HC report (1990).

For a study of the impact of waterways on property values, see Willis and Garrod (1993) *The Value of Waterside Properties*. The relationship between planning and waterways is dealt with in BWB (1993) *The Waterway Environment and Development Plans*.

Recreation and Tourism

Recreation in the countryside is discussed at length in HC Environment Committee, *The Environmental Impact of Leisure Activities* (1995). For a review of the impact of various recreation activities on the environment, see Sidaway (1994) *Recreation and the Natural Heritage*. A 1995 report by the Countryside Commission and others is devoted to exploring the concept of *Sustainable Rural Tourism* and the ways in which it can be translated into practice. See also PPG 21 *Tourism* (1992); Countryside Commission (1995) *Sustainable Rural Tourism*; and Segal Quince Wicksteed (1996) *The Impact of Tourism on Rural Settlements*. Other references on tourism are given in Chapter 8.

Nature Conservation

Current official guidance on nature conservation is given in PPG 9 and TAN(W) 9. Additionally, on SSSIs, see SOEnD Circular 9/1995 *Habitats and Birds Directives – Nature Conservation, etc.*, and forthcoming NPPG on *Natural Heritage*; DoENI PPS *Planning and Nature Conservation*, and Ratcliffe (1994) *Conservation in Europe: Will Britain Make the Grade?* Despite the special controls, many SSSIs have been damaged: see, for example, the National Audit Office (1994) report *Protecting and Managing Sites of Special Scientific Interest in England*.

On the issue of compensation for 'profits forgone' by farmers, see Jenkins (1990) *Future Harvest*. See also Hodge (1989) 'Compensation for nature conservation', and Hodge (1991) 'Incentive policies and the rural environment'; both these articles give useful references to other literature.

Forestry and Woodlands

Our Forests – The Way Ahead (Cm 2644) (1994) presents the conclusions from the Forestry Review. See also HC Environment Committee report *Forestry and the Environment* (1993), and the *Government Response* (1993); and HC Welsh Affairs Committee report *Forestry and Woodlands* (1994), and the *Government Response* (1994). There was also the major policy statement issued after the Rio Summit: *Sustainable Forestry: The UK Programme* (Cm 2429) (1994). An earlier statement of policy was announced in 1991: this is set out in DoE Circular 29/92, *Indicative Forestry Strategies*. See also NAO *Review of Forestry Commission Objectives and Achievements* (1986).

10

URBAN POLICIES

If local politicians cannot normally be expected to provide the visionary leadership required, neither can central government . . . Solutions imposed from the centre without the support and involvement of the local community are unlikely to succeed.

Confederation of British Industry 1988

The sets of [urban] policy instruments have continuously switched their focus, reflecting not only political priorities but also the changing interpretation of the nature of the problems. There has been a shifting target, a shifting set of priorities and a growing and fluctuating set of policies to tackle urban problems.

Robson *et al.* 1994

INTRODUCTION

The breadth of public intervention embraced by the term 'urban policies' denies any simple summary, but the quotations reflect two consistent features of urban policy over the last two decades. The first is the dominant role assumed by central government and the accompanying decline in the role and resources of local government. The second is the lack of a clear policy focus, resulting in endless experimentation with new and often disconnected initiatives. A third feature has been the growth of interest in property and infrastructure development. Since the early 1990s, there has been a welcome increase in the influence of local government, hints of a more coordinated and consistent set of programmes, and some recognition of the need to give equal attention to particular disadvantaged groups as well as to areas.

Two preliminary points need to be made about this chapter. First, the title may be somewhat mis-

leading in that a number of the policies discussed extend to rural areas (housing for example) or may have non-spatial dimensions (e.g. economic development). Nevertheless, the focus of the discussion is on urban areas in general and inner cities in particular. Second, the title is in the plural since there is no such thing as a single urban policy or a set of policies which are sufficiently cohesive as to justify the use of the word *policy* in an omnibus way.

As these points suggest, it is not a simple matter to define what are, and what are not, urban policies. (Social security and housing benefits may play a more significant role than urban aid or urban regeneration policies.) Indeed, there is an important question as to whether there is – or should be – a distinct area of urban policy. Nevertheless, there is a group of policies which are officially labelled urban and which have sufficient urban identity to justify discussing them together.

The chapter opens with a discussion of inadequate housing: the starting point of urban policy in the nineteenth century, and the major concern of policy until at least the end of the 1960s. The focus then moved from physical conditions to the social aspects of housing and, later, to areas of social need. A further shift took place in the late 1970s when economic issues were seen as the key to urban regeneration. By the mid-1980s this had become the conventional wisdom, with an accent on large-scale property development undertaken in partnership with the private sector. By the early 1990s, the value of these large projects (often modelled on the supposedly successful experience in the USA) was increasingly questioned.

INADEQUATE HOUSING

Britain has a very large legacy of old housing which is inadequate by modern standards. This results from the relatively early start of the industrial revolution in this country and the rapid, unplanned and speculative urban development which took place in the nineteenth century. (The contrast with, for example, the Scandinavian countries, whose industrial revolutions came later when wealth was greater and standards higher, is marked.) As a result, British policies in relation to clearance and redevelopment are of long standing, though it was the Greenwood Housing Act of 1930 which heralded the start of the modern slum clearance programme. Over a third of a million houses were demolished before the Second World War brought the programme to an abrupt halt.

By 1938, demolitions had reached the rate of 90,000 a year: had it not been for the war, over a million older houses would (at this rate) have been demolished by 1951. The war, however, not only delayed clearance programmes: it resulted in enforced neglect and deterioration. War damage, shortage of building resources and (of increasing importance in the period of postwar inflation) crude rent restriction policies increased the problem of old and inadequate housing.

It was not until the mid-1950s that clearance could generally be resumed, and well over two mil-

lion slum houses have been demolished since then. But the problem is still one of large dimensions. House condition surveys carried out in 1986 (in Wales) and 1991 (in the rest of the UK) showed a total of nearly 1.8 million dwellings which were below statutory minimum standards. In addition, 3.8 million were in need of urgent repairs. Some progress had been made since an earlier survey in 1981: in particular there had been a fall in the number of dwellings lacking basic amenities; but there had been no reduction in the number of dwellings below the fitness standard or in serious disrepair. This may have been related to the marked decline in slum clearance since the mid-1970s. This continued into the 1990s and, in 1993–4, the number of houses demolished or closed in Britain was less than 9,000, compared with 54,000 in 1974–5 and 90,000 in 1969.

FROM CLEARANCE TO RENEWAL

Though an improvements grants policy was introduced in 1949, it was not until the mid-1950s that it got under way. Since then, the emphasis has gradually shifted from individual house improvements, first to the improvement of streets or areas of sub-standard housing, and later to the improvement of the total environment.

Initially, it was assumed that houses could be neatly divided into two groups: according to the 1953 White Paper *Houses: The Next Step*, there were those which were unfit for human habitation and those which were essentially sound. As experience was gained the improvement philosophy broadened, and it came to be realised that there was a very wide range of housing situations related not only to the presence or otherwise of plumbing facilities and the state of repair of individual houses, but also to location, the varying socio-economic character of different neighbourhoods and the nature of the local housing market. A house lacking amenities in Chelsea was, in important ways, different from an identical house in Rochdale: the appropriate action was similarly different. Later, it was better understood that appropriate action defined in housing market terms was not necessarily equally appropriate in social terms. A middle-class invasion might restore the physical fabric and raise the quality and character of a neighbourhood, but the social costs of this were borne largely by displaced low-income families. The problem thus became redefined.

Growing concern for the environment also led to an increased awareness of the importance of the factors causing deterioration. It became clear that these are more numerous and complex than housing legislation had recognised. Through-traffic and inadequate parking provision were quickly recognised as being of physical importance. The answer, in appropriately physical terms, was the re-routing of traffic, the closure of streets and the provision of parking spaces (together with cobbled areas and the planting of trees). Most difficult of all is to assess the social function of an area, the needs it meets, and the ways in which conditions can be improved for (and in accordance with the wishes of) the people living in an area.

For a considerable time, this issue of the social function of areas was dealt with largely by ignoring it, although a strong shift towards improvement rather than clearance was heralded by the 1969 Housing Act. This increased grants for improvement and introduced *general improvement areas* (GIAs), which were envisaged as being areas of between 300 and 800 'fundamentally sound houses capable of providing good living conditions for many years to come and unlikely to be affected by known redevelopment or major planning proposals'. The enhanced grants and the GIAs made a significant contribution to the reduction in the number of unfit properties, though in some areas gentrification unexpectedly took place, reducing the amount of privately rented housing and affecting the existing communities (Wood 1991: 52).

The 1974 Housing Act, however, which represented a major reorientation of policy, brought social considerations to the fore. There was a new emphasis on comprehensive area-based strategies implementing a policy of 'gradual renewal'. The powers (and duties) conferred by the Act focused upon areas of particular housing stress. Local housing authorities were required to consider the need for dealing with these as *housing action areas* (HAAs). Though these were conceived of in terms of housing conditions, particular importance was attached to 'the concentration in the area of households likely to have special housing problems – for instance, old-age pensioners, large families, single-parent families, or families whose head is unemployed or in a low income group' (DoE Circular 13/75).

The intention was that intense activity in HAAs would significantly improve housing conditions within a period of about five years. In the event, HAA designation lasted much longer in many cities. Various additional powers were made available to local authorities within HAAs, for compulsory purchase, renewal and environmental improvement; and grant aid for renewal was targeted to HAAs.

The novel feature of HAAs was the statutory provision which made the well-being of those living in them one of its major objectives. This meant that residents had to be involved in the nature and timing of action programmes. Thus HAA implementation, in some cases, involved the setting up of residents' groups with the help and advice of the local authority, and quite extensive communication exercises to inform residents, often in several languages.

HOUSING RENEWAL AREAS

The area-based approach was retained, although substantially altered, by the Local Government and Housing Act 1989. In addition to individual income-related house renovation grants, the Act introduced *renewal areas* (RAs), which replaced GIAs and HAAs. There are also powers for local authority support for *group repair schemes* to renovate the exteriors of blocks of houses.

Detailed explanation and guidance was given in the lengthy DoE Circular 6/90. The thrust of the changes reflects a concern for a broader strategic approach, including economic and social regeneration as well as housing renewal. It involved the resumption of clearance; the use of partnerships to bring together the initiatives of local authorities, housing associations, property owners and residents; and a system of grants which were mainly both mandatory and income related. In discussing the changes, the circular heralded a significant change in policy:

> [Some authorities] have been using resources for improvement where this seems not to have been cost-effective in terms of securing the life of the stock and where there were no good social reasons why redevelopment should not have taken place. In future, therefore, authorities will be required to carry out a thorough appraisal of the various options available to them for dealing with areas of poor quality private sector housing before declaring an RA.

Thus, clearance was placed firmly back on the agenda. Moreover, it was suggested that the declaration of RAs 'may not necessarily mean tackling the worst areas first'. Rather, 'it may be better to concentrate on 'turning around' areas which have not yet reached the bottom of the spiral of decline in order to improve market confidence, and thus be able to make the maximum use of private sector inputs'.

A *neighbourhood renewal assessment* has to precede designation. This is, in effect, a plan-making and implementation programme combined, including an assessment of conditions, an estimate of the resources available and the selection of the preferred options. The procedure goes much further than the typical land use planning process to incorporate a cost-benefit analysis of different alternative policies over a thirty-year period, including the qualitative social and environmental implications. Consultation is a requirement both during and after the declaration process, with twenty-eight days given for responses to an explanatory summary of the proposals which must be delivered to every address in the area. An interesting requirement is that all those who make representations which are not accepted must be provided with a written explanation.

Renewal Areas are larger than the former HAAs and GIAs: with at least 300 properties as a minimum. More than 75 per cent of the dwellings must

be privately owned, and at least 75 per cent must be considered unfit or must qualify for grants. At least 30 per cent of the households must be in receipt of specified state benefits, thus ensuring that 'a significant proportion of residents in an RA should not be able to afford the cost of the works to their properties'.

Wood (1996) provides a summary of how the renewal areas have been implemented, drawing on the findings of two surveys undertaken by Couch and Gill (1993) and Austin (1995). Five years after the Act, in 1995, eighty RAs with 122,000 dwellings in forty-six local authorities had been designated. Their average size was 1,526, or about twice the size of GIAs and HAAs. Wood notes that this is slower progress than might have been expected and points to the variation in enthusiasm across the country. He provides three case studies of RA designation in Birmingham, a city that has 47,000 unfit privately owned dwellings and another 80,000 considered borderline. The city council has estimated that the total bill for the repair and improvement of the private housing stock in the city would be £750 million. It is perhaps not surprising therefore that Wood concludes that 6,400 properties designated in RAs in Birmingham will remain as they started (or have further deteriorated) at the end of the ten-year plan period.

Among the many other provisions of the 1989 Local Government and Housing Act was a revision of the criteria for determining whether a dwelling is unfit for human habitation. Previously, the statutory definition (in England and Wales) was related to structure, physical condition and plumbing. These nineteenth-century public health factors certainly could affect health and safety, but there was considerable scope for differing judgements, and some more modern essentials (such as hot water) were not included. The approach was altered in Scotland following the 1967 Cullingworth Report. The 1969 Housing (Scotland) Act introduced the concept of a *tolerable standard*, and this was followed in the 1989 Act for England and Wales. Dwellings which fail to meet this statutory fitness standard qualified for mandatory grants (subject to a means test) until

1996. Discretionary grants were available for minor works, such as adapting facilities to enable the elderly to remain in their homes or to adapt houses for the use of people with disabilities.

Since the renovation grants scheme was (originally) largely mandatory and therefore demand-led, some local authorities experienced growing financial difficulties. Wood quotes Morris (1994), a senior officer in Birmingham: 'In 1993 the city council was faced with a backlog of 8,000 applications for mandatory grants from private owners which, if paid, would have used up its entire renewal budget for many years to come.' This problem was one of a number which were dealt with in a 1993 consultation paper, *The Future of Private Housing Renewal Programmes*. Changes were proposed to target resources on the worst housing; to focus grant aid on the poorest households; to give local authorities more flexible powers; and to address the problems of the most vulnerable groups. These proposals found general support, and most were enacted in the Housing, Grants, Construction and Regeneration Act 1996. This replaced parts of the 1989 Act, so that grants for repairs to housing are now given at the discretion of the local authority. The rationale for this is that it enables expenditure to match the local strategy, while at the same time allowing priority to be given to those who are most in need of financial assistance. The Act also provides for grants to be made in excess of the normal maxima for altering property for the needs of people with disabilities, and also for relocation grants to enable those displaced by clearance to buy a house in the same local area.

ESTATE ACTION

In 1979, the DoE set up the *priority estates project* to explore ways in which difficult council estates could be improved. The problems of these estates varied, but all had become neglected and run down; some had been vandalised. A 1981 report *Priority Estates Project 1981: Improving Problem Council Estates* considered three experiments in their improvement, and concluded that the task involved a great deal

more than mere physical renovation: social and economic problems needed to be addressed at the same time. In 1985, the DoE established an *Urban Renewal Unit* (now called *Estate Action*) to encourage and assist local authorities to develop a range of measures to revitalise run-down estates. These measures include transfers of ownership and/or management to tenants' cooperatives or *management trusts* involving tenants; sales of tenanted estates to private trusts or developers; and sales of empty property to developers for refurbishment for sale or rent. Estate Action funding is allocated on a competitive basis. A wide range of factors is taken into account: the provisions for tenant participation; the devolution of management to tenants; increases in 'tenant mix', and right-to-buy activity; and the role of the private sector.

A research report evaluating six early Estate Action Schemes was carried out by Capita Management Consultancy (1996). In addition to the general objective of improving the quality of life of residents living in deteriorated local authority estates, there were a number of specific policy objectives: to bring empty properties quickly back into use; to improve residential quality; to reduce levels of crime and 'incivility'; to encourage participative estate-based management; to diversify tenure; and to attract private investment. Despite a large expenditure amounting (for the whole programme) to nearly £2,000 million over eight years, the reported results for the estates studied were disappointing. Indeed, the report makes very depressing reading, even though its conclusions are expressed in the guarded language which often characterises such government-sponsored research reports. A particular shortcoming was a lack of success in involving residents in the development and implementation of the schemes. More generally, the report demonstrates the difficulty of using physical plans for achieving social and economic goals.

HOUSING ACTION TRUSTS

The 1987 White Paper *Housing: The Government's Proposals* announced the creation of *housing action trusts* (HATs) to tackle the management and renewal of badly run-down housing estates. HATs are the housing equivalent of the urban development corporations, but they can be introduced only with the consent of a majority of the tenants. They are non-departmental public bodies, responsible directly to the Secretary of State for improving the physical conditions of estates by renovations and improvements to the houses and the environment. Their objectives are to secure more diverse tenure, essentially by involving private and voluntary housing agencies, and they also aim to improve economic and social conditions.

Although funding was allocated for HATs as early as 1988, there was considerable delay in getting the first ones started because of fierce opposition by affected local authorities and tenants. Rao (1990) has documented two areas in Lambeth and Sunderland that were initially identified for designation. The dilemma facing local authorities is that HATs involve a loss of control to the private sector, but are a potential source of very substantial extra funding. Tenants fear increased rents and reduced availability of housing if it is sold to private owners after improvement.

Following the embarrassing rejection of HAT designation in the first ballots, much greater care was taken with subsequent proposals to involve tenants in early discussions. The first HATs to be designated were at Waltham Forest and North Hull. In 1992, 82 per cent of the tenants of a Liverpool tower block voted in favour of the third HAT. Others have since been designated in Birmingham (Castle Vale), Tower Hamlets and Brent (Stonebridge). The HATs received £90 million funding in 1995–6 together with £6 million private finance through the PFI. They improved 486 homes, built 643 new homes, and created 2,573 training places.

The Citizen's Charter gives tenants the right to put forward proposals for HATs. Tenants' rights are further extended by the 1996 Housing Act which enables tenants on poor estates to opt out of council control by voting for transfer to a *local housing company*. This will be a non-profit social landlord providing housing at below market rates and working independently to bring in private finance.

SCOTTISH HOUSING

Scottish housing is different from that south of the border in significant ways. There is a high proportion of tenemental properties; dwellings tend to be smaller; rents are lower; and a higher proportion of the housing stock is owned by public authorities. These and other differences reflect history, economic growth and decline, local building materials and climate. Above all, Scotland faces a major problem of poor-quality tenemental housing. Despite the large amount of clearance in the postwar years, there still remains much poor quality housing in both the private and the public sectors.

The 1991 Scottish House Condition Survey, published in 1992 suggests that, because of the dominance and relatively young age of much of the public housing stock, a high proportion of housing has the basic amenities, but the condition is often poor, and requires high levels of expenditure.

Scottish housing is distinctive in important ways. First, there is the scale and character of public housing. The low council rents of the past contributed to the relatively low demand for private housing which, coupled with massive public house building programmes, gave Scotland the highest proportion of public sector housing in western Europe – and made Glasgow City Council the largest public sector landlord (McCrone 1991). (It should be noted that the Scots use the term 'house' in the English sense of 'dwelling', that is it embraces a flat or tenement.) The Right to Buy has shifted the balance somewhat between the owner-occupied and public housing sectors: owner occupation rose from 35 per cent in 1979 to 57 per cent in 1995, while public sector renting fell from 54 per cent to 30 per cent. Nevertheless, in spite of the sale of nearly a quarter of the public housing stock, the level of public renting in Scotland is still much higher than in England and Wales.

The Conservative government believed that the

large-scale public ownership of housing was at the root of much of the Scottish housing problem. The 1987 White Paper, *Housing: The Government's Proposals for Scotland* sets out the argument:

> Municipal housing was an effective means of increasing the total housing stock and of clearing the slums, but it has created problems which arose from the single mindedness with which these problems were pursued. In some areas, the system has provided good quality housing and management. In others, there are major problems of unsuitable housing types, disrepair, and management failure. In some parts of Scotland, the public landlording operations are on such a vast scale that there is an inevitable risk that they become too distant and bureaucratic to respond well to individual tenants' wishes and needs. The provision of housing in the public sector can all too easily result in inefficiencies and bureaucracy, producing queuing and lack of choice for the consumer, made worse by the existence of some housing which people do not want to live in. Poor housing conditions, just as much as tenant dissatisfaction are the result of bad management, as well as insensitive and misconceived design. Low rents in many areas have contributed to inadequate standards of management and repair, against tenants' long term interests.

The government's strong desire to reduce the public housing sector (and particularly to break up the public ownership of the large peripheral estates) has been an important background issue in Scottish housing policy for many years.

Equally distinctive has been the institutional context of Scottish housing policy. The relationship between the Scottish Office and Scottish local authorities has always been much closer than between their equivalents in England, and there has been much easier coordination. This has resulted in what Carley (1990a: 51) describes as 'a much more clearly defined and integrated housing–neighbourhood renewal policy, which covers housing and planning issues together'. The Scottish Office has tended to work with local authorities (which have a single local authority association), rather than exerting central control through such mechanisms as UDCs and HATs. Instead, there have been centrally sponsored bodies which have worked in cooperation with local government. The main housing body is Scottish Homes. Established by the Housing (Scotland) Act 1988, this incorporates the former Scottish Special Housing Association (SSHA), which dates from 1937, and the Housing Corporation in Scotland. The SSHA has been a major housing authority working alongside local authorities in the implementation of housing policies.

As in England and Wales during the 1980s, policy in relation to the older private housing stock placed more emphasis on rehabilitation than on clearance. But with much of the older tenemental properties the scope for improvement is severely restricted by the decayed fabric of the buildings, their internal layout and the high cost of alteration, as well as the practical problems of multiple ownership. Some of these difficulties have been met by the use of powers of compulsory improvement of a whole tenemental structure, and by the establishment of *ad hoc* housing associations. However, the Scottish legislation has long provided for more flexibility than the English. *Housing action areas* (which have not been superseded by renewal areas) are of three types: for demolition, for improvement, and for a combination of the two. This enables the most appropriate action to be taken according to area conditions.

Scottish practice has emphasized the involvement of residents in housing renewal. There is a requirement for a two-month period of consultation with local residents before an HAA is declared, and small community-based housing associations are established to lead renewal. This has been a particular feature of the policy of the Housing Corporation in Scotland:

> Operating with local authorities, the Corporation identified a set of adjacent Housing Action Areas. Housing associations were informed or were created to manage the rehabilitation programme in areas containing 1,000 to 2,000 units. Management committees consisting almost entirely of local residents were formed.
> (Maclennan 1989)

Given the distinctive Scottish style of implementing HAAs, there has not been the need for new legislation equivalent to the Local Government and Housing Act 1989. However, proposals for a means-tested scheme for improvement grants were issued in 1988, though no action has yet followed.

THE EMPHASIS ON AREA POLICIES

Several major elements can be identified in the development of thinking on deprived areas in Britain: inadequate physical conditions, the perception of 'large' numbers of immigrants (many of whom were born in Britain), educational disadvantage, and a multiplicity of less easily measurable social problems.

For a very long time there was a preoccupation with inadequate physical conditions (particularly in relation to plumbing). Indeed, British housing policy developed from sanitary policy (Bowley 1945), and it still remains a significant feature of it. Area policy in relation to housing was almost entirely restricted to slum clearance until the late 1960s when concepts of housing improvement widened, first to the improvement of areas of housing and then to environmental improvement. Despite a number of social surveys the policy was unashamedly physical: so much so that increasing powers were provided to *compel* reluctant owners and tenants to have improvements carried out. Not until the 1970s was attention focused on the social character and function of areas of old housing.

A further area of policy in the field of housing focused on areas of housing stress. The Milner Holland Committee (1965) looked favourably on the idea of designating the worst areas as *areas of special control* in which there would be wide powers to control sales and lettings, to acquire, demolish and rebuild property, to require and undertake improvements, and to make grants. In 1967 the Committee for Commonwealth Immigrants argued for the designation of *areas of special housing need*. Again, the crucial issue was the control of overcrowding, exorbitant rents, insanitary conditions, disrepair and the risk of fire.

These proposals were not accepted by the government, though increased powers to control multi-occupation and abuses were provided. It was left to the Plowden, Seebohm and Skeffington reports to probe more deeply into this area.

The Plowden Committee was appointed in 1963 to consider primary education *in all its aspects*. It reported in very broad terms, but underlined the complex web of factors which produced seriously disadvantaged areas (Plowden Report 1967). The web of circumstances is neatly illustrated in the following quotation which is as pertinent today as it was in 1967:

> In a neighbourhood where the jobs people do and the status they hold owe little to their education, it is natural for children as they grow older to regard school as a brief prelude to work rather than an avenue to future opportunities . . . Not surprisingly, many teachers are unwilling to work in a neighbourhood where the schools are old, where housing of the sort they want is unobtainable, and where education does not attain the standards they expect for their own children. From some neighbourhoods, urban and rural, there has been a continuing flow of the more successful young people. The loss of enterprise and skill makes things worse for those left behind. Thus, the vicious circle may turn from generation to generation and the schools play a central part in the process, both causing and suffering cumulative deprivation.

The Plowden Committee recommended a national policy of positive discrimination, the aim of which would be to make schools in the most deprived areas as good as the best in the country. Additional resources were necessary to achieve this: extra teachers and special salary increases; teachers' aides; priority for replacement and improvement in the school building programme; extra books and equipment; and expanded provision for nursery education.

The Seebohm Committee had wider terms of reference than the Plowden Committee: to review personal social services in England and Wales, and to consider what changes were desirable to secure an effective family service. Of relevance to the present discussion are the Committee's recommendations (1968) in relation to *areas of special need*. Unfortunately, the Committee did not suggest how these should be identified, in spite of a recommendation that the areas should be accorded priority in the allocation of resources.

More helpful was its reference to citizen participation, which underlined a point hardly recognised by the Skeffington Committee (1969), even though it was specifically concerned with it. It was Seebohm, not Skeffington, who clearly saw that, if area action was to be based on the wishes of the inhabitants and carried out with their participation, 'the participants may wish to pursue policies directly at variance with the ideas of the local authorities . . . Participation provides a means by which further consumer control may be exercised over professional and bureaucratic power.'

This is an issue which will be discussed in a broader context in the final chapter. Here, we need to survey briefly some of the ways in which the development of thinking on deprived areas was translated into policy.

THE URBAN PROGRAMME

Area policies in relation to housing improvement, however inadequate they may have been, were based on long experience of dealing with slum clearance and redevelopment. With other area policies there was no such base upon which to build, and both legislation and practice were hesitant and experimental. The approach, however, has remained consistently a spatial one, focusing on particular cities and areas within cities.

Legislatively, the important landmarks (though modest) were the Local Government Act 1966 (section 11) and the Local Government (Social Need) Act of 1969. These constituted the statutory basis for the *educational priority areas programme* (Halsey 1972) and the *urban aid programme*. The 1966 Act provided for grants in aid of staff costs involved in 'dealing with some of the transitional [*sic*] problems caused by the presence of Commonwealth immigrants'. The Urban Aid Programme was broader in concept; it was designed to raise the level of social services in 'localised districts' of acute social need.

The *urban aid programme* (later recast as the *urban programme*) funded mainly social schemes, but it was progressively widened in scope to embrace voluntary organisations, and to cover industrial, environmental and recreational provision. The 1977 White Paper *Policy for the Inner Cities* (Cmnd 6845) and the Inner Urban Areas Act 1978 brought about

significant changes. The new policy was 'to give additional powers to local authorities with severe inner area problems so that they may participate more effectively in the economic development of their areas' (DoE Circular 67/78). The provisions effectively designated three types of districts: seven *partnerships* in the most severely deprived large urban areas; twenty-three *programme authorities* for local authority districts with severe, but less extensive, problems; and sixteen *other designated areas* with problems of lesser severity.

It is difficult to give a coherent account of this programme since its objectives were never clearly spelled out, and its extreme flexibility gave rise to a great deal of confusion (McBride 1973). With the development of inner city partnerships and programme authorities, the position became even more confused. By 1990, the number of individual programmes had swollen to thirty-four (NAO 1990).

The *urban programme* was the major plank of the 'deprived area policy' stage. Additionally there were the *community development projects*. These produced a veritable spate of publications ranging from carefully researched analyses to neo-Marxist denunciations of the basic structural weaknesses of capitalist society, though the original aim was 'to overcome the sense of disintegration and depersonalisation felt by residents of deprived areas'. There were twelve community development projects in all: in Birmingham, Coventry, Cumbria, Glamorgan, Liverpool, Newham, Newcastle, Oldham, Paisley, Southwark, Tynemouth and West Yorkshire. Additionally, 1974 saw the introduction of a small number of *comprehensive community programmes* in areas of 'intense urban deprivation'. In the wake of these, large numbers of studies were undertaken.

The urban programme continued with an increasing emphasis on economic development. It became a valuable source of funding for the many thousands of projects and organisations that have been supported. At its height, about 10,000 projects were funded each year in the fifty-seven programme areas, costing in 1992–3 an estimated £267 million. In its later years, almost half the expenditure was devoted to economic objectives, and the rest was shared roughly equally between social and environmental objectives. The DoE estimated that this funding provided (in 1992–3) 74,000 training places; created or preserved 34,000 jobs; and helped the start-up of 788 new businesses.

POLICY FOR THE INNER CITIES AND ACTION FOR CITIES

Economic regeneration now has pride of place in urban policy. The transition has been gradual but clear. The 1977 White Paper *Policy for the Inner Cities* spoke of local authorities needing 'to stimulate investment by the private sector, by firms and by individuals, in industry, in commerce, and in housing'. The return of the Conservative government led to a review of inner-city policy which concluded that a much greater emphasis needed to be placed on the potential contribution of the private sector. One of the initiatives taken was the establishment of the *Financial Institutions Group* (FIG) led by the Secretary of State (Michael Heseltine) and staffed by twenty-five secondees from the private sector. Among the proposals of FIG, made in thirty-five unpublished reports, were recommendations for a selective employment grant; a rating and rate support grant; improvements to the business start-up scheme; better use of redundant buildings; and additional public expenditure on capital investment in the cities. Most proposals fell on stony ground, but others were acted upon. The most important proposal was for an *urban development grant* (UDG) on the lines of the American *urban development action grant*, which was introduced in 1982 (discussed below, p. 243).

Ten years later the rhetoric was much the same. After the 1987 General Election, Margaret Thatcher, then Prime Minister, announced her intention to 'do something about those inner cities'. The immediate result was the publication of a glossy brochure entitled *Action for Cities* (Cabinet Office 1988). This maintained that, though the UK had benefited during the 1980s by embracing the ethic

of enterprise, this change in attitude had not reached into the inner city. The aim of urban policy, therefore, was to establish 'a permanent climate of enterprise in the inner cities, led by industry and commerce'. This was to be achieved through creating confidence for business to grow, improving people's motivation and skills, and making inner cities safer and more attractive. Twelve specific initiatives were announced, including the extension of the urban development corporations, more direct support for private sector investment through the new city grant, and the Safer Cities Project.

At the same time, and following some encouragement from the Prime Minister, the private sector came forward with its own modest package of measures. These included *Business in the Community*, to promote business involvement in regeneration; *Investors in Industry*, to provide venture capital for inner city schemes; and *British Urban Development* (BUD), a consortium of the eleven largest construction companies who together would seek urban regeneration opportunities.

Action for Cities programmes were claimed to have involved central government expenditure of £3 billion in 1988–9 and £4 billion in 1990–1 but, in fact, little additional money was involved. Critics argued that the package merely gave the appearance that a determined effort was being made to get to grips with the problems. However debatable this might be (Lawless 1989: 155), it did indicate some reorientation of thinking on urban policy. Table 10.1 illustrates the changing patterns of funding between the main programmes of expenditure for the inner cities and regeneration.

URBAN DEVELOPMENT CORPORATIONS

The 1977 White Paper on inner cities considered the idea of using new-town style development corporations to tackle inner areas. Though such an agency 'could be expected to bring to bear single-minded management, industrial promotional expertise and experience in carrying out development',

Table 10.1 Regeneration and Inner City Expenditure and Plans, 1987–8 to 1997–8

	1987–8	1988–9	1989–90	1990–1	1991–2	1992–3	1993–4	1994–5	1995–6	1996–7 (estimate)	1997–8 (plans)
UDCs and DLR	160.2	255.0	476.7	607.2	601.8	515.0	341.2	287.1	254.9	208.1	186.9
Estates Action	75.0	140.0	190.0	180.0	267.5	348.0	357.4	372.6	313.8	256.1	177.8
Housing Action Trusts					10.1	26.5	78.1	92.0	90.0	89.0	88.9
City Challenge						72.6	240.0	233.6	241.6	224.4	143.0
Challenge Fund									125.0	265.1	483.1
City Grant	26.8	27.8	19.5	38.9	36.6	43.3	24.4[2]				
Derelict Land Grant	76.7	67.9	54.0	61.7	77.2	101.7	104.3				
English Partnerships (URA)[1]				16.8	–16.4	6.5	24.2	191.7	210.3	209.6	209.6
Manchester Regeneration Fund/Olympic bid					0.8	12.2	26.8	30.2	2.0		
Coalfield Areas Fund							2.3	2.0	0.4		
Urban Programme	245.7	224.3	222.7	225.8	237.5	236.2	166.5	67.8[3]			
Inner City Task Forces	5.2	22.9	19.9	20.9	20.5	23.6	18.0	15.7			
City Action Teams			4.0	7.7	8.4	4.6	3.4	0.2			

Sources: DoE, *Annual Report 1993*: Figure 45, p. 53; DoE, *Annual Report 1995*: Figure 43, p. 65; and DoE, *Annual Report 1996*: Figure 31, p. 50
Note: 1 Includes funding for English Estates from 1989–90 to 1993–4
2 City Grant and Derelict Land Grant became part of English Partnerships funding from November 1993 and April 1994 respectively
3 From 1995–6 urban programme and Task Force and Urban Programme funding is included in SRB. CATs were disbanded when the integrated Government Offices were set up

the then Labour government concluded that is was inappropriate for the inner cities, and that local government should be the prime agency of regeneration. The Conservative government thought differently, mainly because it had little faith in the capabilities of local government. The manifest argument, however, was that the regeneration of areas such as the London Docklands was 'in the national interest, effectively defining a broader community who would benefit from the regeneration' (Oc and Tiesdell 1991: 313).

The 1980 Local Government, Planning and Land Act made the necessary legislative provisions, and defined the role of an urban development corporation (UDC) as being:

> to secure the regeneration of its area . . . by bringing land and buildings into effective use, encouraging the development of existing and new industry and commerce, creating an attractive environment, and ensuring that housing and social facilities are available to encourage people to live and work in the area.

Though their structure and powers are based on the experience of the new town development corporations (NTDCs), the UDCs are different in several important respects. Since their task is a limited one, they were envisaged as having a relatively short life of ten years or so. Two were wound up in 1995, and all are expected to have completed their mission by 1998. Partly because of this, their designation procedure is rapid and does not provide for the consultants' reports and the public inquiry which preceded the designation of new towns. Instead, the designation of an area, known as an *urban development area* (UDA), is by way of a statutory instrument. This requires parliamentary approval under a procedure which allows a petition to be made to a Select Committee against the confirmation of the Order. No objection was sustained against the first Order, which was made in 1991 (for Merseyside), but there were several petitions against the much more controversial Docklands Order in the same year (Ledgerwood 1985; Brownill 1990).

Fourteen designations have been made, of which twelve are in England, one in Northern Ireland (Laganside) and one in Wales (Cardiff Bay); their locations are shown on Figure 10.1. Most urban development areas are not large. Indeed, in Bovaird's (1992) words, they can be seen merely as replacing local authority development programmes 'for small areas with especially severe derelict land or plant closure problems'. There are no UDCs in Scotland, since Scottish Enterprise fulfils a similar function and can operate anywhere in the country, unrestrained by the boundaries of a designated area.

UDCs have extraordinary powers of land acquisition: land may be acquired simply if it is within the designated area. Additionally, 'to reduce uncertainty', there is a provision which allows for public sector land to be transferred to the corporation by means of a *vesting order*. They also (unlike the NTDCs) usurp the local planning authority's development control functions (except in Wales), not only for determining planning applications (including their own proposals) but also for enforcement, tree preservation, conservation areas, advertisements, wasteland, and listed building controls. In short, the UDCs have very wide planning responsibilities and freedom from local authority controls. This is not accidental or incidental: it is an essential feature of their conception. Furthermore, UDCs are run by boards of directors drawn primarily from business; and they are responsible to central government through the Secretary of State.

Expenditure on UDCs rose to a peak of over £600 million in 1990–1, declined to £258 million in 1994–5, and is planned to fall to £168 million by 1997–8. The expenditure and claimed outputs of the UDCs are listed in Table 10.2. At one time, this was by far the largest share of spending on regeneration: in 1992–3 it amounted to 51 per cent of all inner city spending. This reflected the significant shift since 1984 to the centrally directed UDCs and away from locally directed expenditure. The major share of public funding, and similarly high levels of private investment, has gone to London Docklands, and within that area to major projects. Just one project, the Limehouse Link road, required the demolition of over 450 dwellings and cost

Table 10.2 Urban Development Corporations in England: Designation, Expenditure and Outputs

		Annual gross expenditure (£m)		Cumulative outputs		
		1992–3	1995–6	Jobs created	Land reclaimed (ha)	Private investment (£m)
First generation	London Docklands	293.9	129.9	63,025	709	6,084
	Merseyside	42.1	34.0	14,458	342	394
Second generation	Trafford Park	61.3	29.8	16,197	142	915
	Black Country	68.0	43.5	13,357	256	690
	Teeside	34.5	52.8	7,682	356	837
	Tyne & Wear	50.2	44.9	19,649	456	758
Third generation	Central Manchester	20.5	15.5	4,909	33	345
	Leeds	9.6	0	9,066	68	357
	Sheffield	15.9	16.4	11,342	235	553
	Bristol	20.4	8.7	4,250	56	200
Fourth generation	Birmingham Heartlands	5.0	12.2	1,773	54	107
	Plymouth	0	10.5	8	6	0
Total		621.4	398.2	165,716	2,713	11,240

Designation years: London Docklands 1981, Merseyside 1981, Trafford Park 1987, Black Country 1987, Teeside 1987, Tyne & Wear 1987, Central Manchester 1988, Leeds 1988, Sheffield 1988, Bristol 1989, Birmingham Heartlands 1992, Plymouth 1993.

Sources: DoE, Annual Reports, 1993, 1995 and 1996
Notes: 1 Leeds UDC was wound up on 31 March 1995. Bristol UDC was wound up on 31 December 1995.
2 The figures for outputs are cumulative over the lifetime of the UDCs and include all activity with the urban development area, not just those of the UDC.
3 For a critique of UDC output figures, see Shaw (1995).

£150,000 per metre, the most expensive stretch of road in the country (Brownill 1990: 139).

There are differences in style and approach among the UDCs, reflecting their time of designation, local circumstances and management. In the early years, the UDCs operated very independently of their corresponding local authorities, sometimes ignoring existing plans and thereby generating great conflict. Over recent years, there has been more cooperation between the UDCs and the local authorities, with the latter carrying out certain development control functions on an agency basis. This is most marked in the case of the LDDC: 'from being the target of considerable local abuse for its neglect of any but the commercial development of Docklands, it now openly speaks the language of community development' (N. Lewis 1992: 10).

INNER CITY TASK FORCES AND CITY ACTION TEAMS

Following Michael Heseltine's 1981 visit to Merseyside, in the wake of the Toxteth riots, the Merseyside Task Force Initiative was created. Initially, this was a task force of officials from the DoE and the then Department of Industry and Employment, established to work with local government and the private sector to find ways of strengthening the economy and improving the environment in Merseyside.

The implementation of urban policy in Merseyside, as in other cities, was characterised by a proliferation of small area schemes which, in the view of the Environment Committee, resulted in a 'complex patchwork of overlapping areas, disparate powers . . . constituting an unduly 'fragmented' system of urban governance'. Conflicts between the various authorities, and between them and central government, led to the breakdown of any genuine partnerships. This criticism was made of other areas and, as a result, action teams and task forces were used more widely from 1985.

City Action Teams (CATs) were first set up in 1985 with the aim of providing more coordination

between government departments in the implementation of central government policy in the inner cities. They were to take a broader, even regional, view of the coordination of government programmes. Each team was chaired by the regional director of one of the main departments involved: the DoE, the DTI and the Training Agency. There were eight CATs (Birmingham, Cleveland, Liverpool, London, Manchester/Salford, Tyne & Wear, Leeds/Bradford and Nottingham/Leicester/Derby). Their funding was limited, reflecting their role as coordinators rather than direct providers. At the highest in 1991–2 it was £8.4 million, which the DoE claimed helped to create or safeguard around 2,300 jobs, 2,700 training places and supported 1,000 small to medium businesses. On the other hand, the National Audit Office report *Regenerating the Inner Cities* (1990: 11) concluded that too much attention was paid to the way in which the funds were spent on particular projects and not enough on 'their effectiveness and results in coordinating programmes'. Lawless (1989: 61) argues that CATs were 'unable to devise anything that might be termed a corporate central-government strategy towards inner-city areas'. CATs were disbanded at the time of the setting up of the Government Offices for the Regions which took on the coordinating role.

Eight Inner City Task Forces were established in February 1986, and their number was increased to sixteen following the *Action for Cities* initiative in 1987, but has since been reduced to eight. Their role has been essentially one of trying to bend existing programmes and private sector investment and priorities to the inner city. The first task forces were expected to have a life of two years, but in the event their lifespan has been variable. The locations are decided on the basis of measures of urban deprivation and responses to specific problems. The task forces have the general objective of increasing the effectiveness of central government programmes in meeting the needs of the local communities. They also have resources to develop innovative approaches to the employment problems of local residents, and to support schemes that might not otherwise be covered by main programme activities.

URBAN DEVELOPMENT GRANT, URBAN REGENERATION GRANT AND CITY GRANT

The objective of the Urban Development Grant (UDG) was 'to promote the economic and physical regeneration of inner urban areas by levering private sector investment into such areas'. It was flexible in terms of the area covered, local authorities contributed 25 per cent of grant aid and the private sector contribution to a project had to be significant. The UDG was replaced by the *Urban Regeneration Grant* (URG) in 1986. In its lifetime, the UDG supported 296 projects at a cost of £136 million with a corresponding private sector investment of £555 million. This represents a leverage ratio of about 1: 4. The URG supported ten schemes at a cost of £46.5 million, with private sector investment of £208 million. Together, the government estimated that they contributed 31,966 jobs, 6,750 new homes and 1,456 acres of land brought back into use (Brunivels and Rodrigues 1989: 66). *City Grant* was launched by the *Action for Cities* initiative in 1988, and replaced several existing grants including the UDG which had been in place since 1982. City Grant has aims similar to those of the UDG, although it is paid directly to the private sector developers. It operates in the areas of the fifty-seven English programme authorities. It supports private sector projects costing more than £200,000 which would not be viable without assistance. Projects can include the provision of new or converted property for industrial or commercial development, and the erection, conversion or refurbishing of housing.

Government funding for UDG and City Grant expanded steadily, from £26.8 million in 1987–8 to £59.6 million in 1992–3, and a projected £83 million in 1995–6. The final evaluation of these grant regimes noted that they were successful in assisting private sector investment but that the grant regimes and the focus on job creation in evaluation did not 'address the basic causes of poor regional or local growth' (Price Waterhouse 1993: 63).

ENGLISH PARTNERSHIPS

The City Grant no longer exists, but its objectives, together with those of the derelict land grant and of English Estates, are now addressed by English Partnerships (the operating name of the Urban Regeneration Agency). English Estates was launched in November 1993 under the Leasehold Reform, Housing and Urban Development Act with the objective of promoting the regeneration of areas of need through the reclamation or redevelopment of land and buildings. Its work is conducted through six regional offices and is largely concerned with the reuse of vacant, derelict and contaminated land and the provision of floorspace for industry, commercial and leisure activities, and housing.

English Partnerships is funded through a ring-fenced element of the Single Regeneration Budget (SRB). An investment fund has been created which can be used with more flexibility than previous grant regimes to provide advice; to take a stake in joint ventures with the private sector and others; to provide loans or loan guarantees; and to support direct development. There are three programmes (amounts allocated for 1995–6 are given in brackets): the partnership investment programme for joint venture investments with the private sector and others (£105 million); the land reclamation programme primarily for local authority-led land reclamation (£28 million); and the direct development programme for providing serviced sites and commercial premises on its own account (£38 million).

In the first two categories, English Partnership funding complements that of other programmes. It is expected to contribute to wider strategies, and to work within the planning framework established by regional guidance and development plans. Thus, collaboration, especially with local authorities, is an essential feature of its work.

English Partnerships is a 'roving agency' to facilitate property-led urban regeneration: it is not tied to designated areas. However, it has some priority areas which include areas identified under the European Union ERDF regulations as Objective 1, 2 and 5b, the coalfield closure areas, inner cities

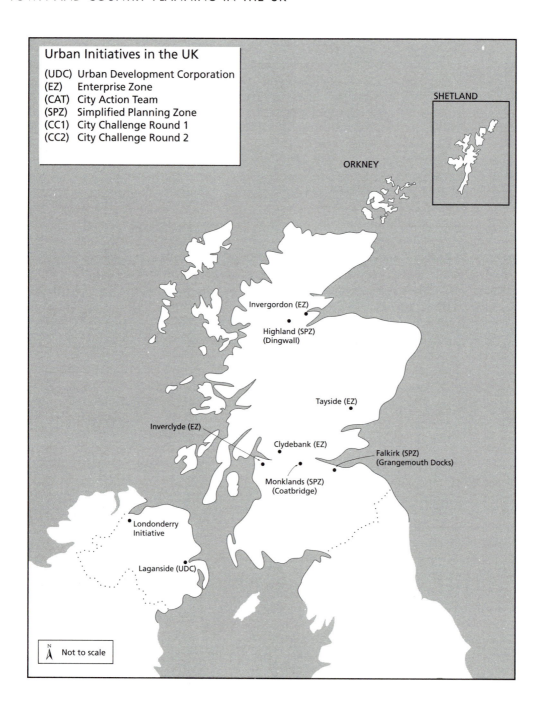

Figure 10.1 Urban Policy Initiatives in the UK

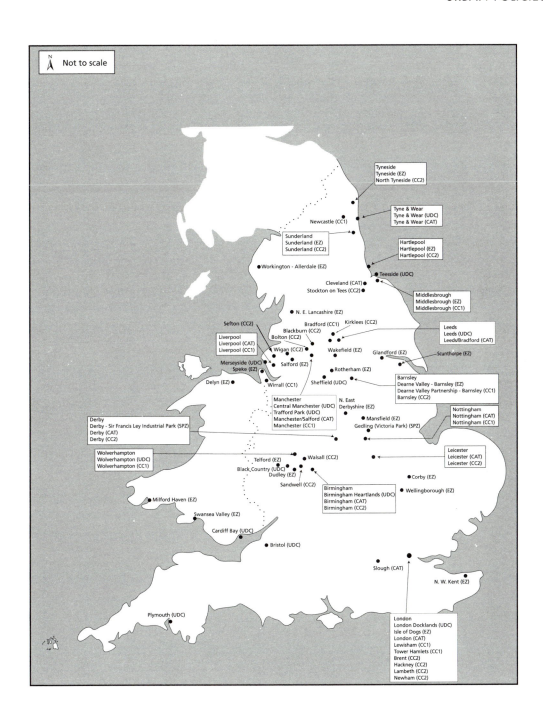

Figure 10.1 continued

(especially City Challenge areas) and other assisted areas. This is a long list and it has led one critic to comment that 'as a list of *priority* areas it is meaningless' (C. Jones 1996: 204). It has been asked not to duplicate the work of the Rural Development Commission, and generally will not be involved in the areas of UDCs.

CITY CHALLENGE

A major switch in other funding mechanisms was announced in May 1991, in the form of *City Challenge*. This marked a significant change in policy. Though the emphasis on land and property development remained, City Challenge implied a recognition that this should be more closely linked to the needs of local communities. It was also intended to encourage a long-term perspective on change; that it should integrate the work of different programmes and agencies; that this would require some effort in establishing appropriate new institutional arrangements or partnerships; and that the local authority should be the initiating body. However, City Challenge runs parallel with the UDCs and other centrally controlled initiatives, and the most dramatic immediate impact has been on the Urban Programme, from which substantial resources have been transferred. Nevertheless, it is still seen as a marked and unexpected turnaround in the government's approach, due at least in part to the views of the then Secretary of State, Michael Heseltine, who believed the urban programme to be moribund (Gahagan 1992: 3).

The explicit objectives of City Challenge emphasise physical change through land and property development, but these are intended to be linked with the provision of opportunities for disadvantaged residents. The aim is to provide major impetus to area improvement, leading to self-sustaining economic regeneration. In the words of a DoE *Annual Report* (1993: 55), the 'winners' of the competitive bidding process seek 'to provide major impetus to area improvement, leading to self-sustaining economic regeneration'.

Only fifteen authorities were invited to bid for funds in the first round, eleven of which were selected to start on their programmes in 1992. Bradford, the Dearne Valley Partnership (Barnsley, Doncaster and Rotherham), Lewisham, Liverpool, Manchester, Middlesbrough, Newcastle, Nottingham, Tower Hamlets, Wirral and Wolverhampton. A second round of City Challenge bidding was opened up to all fifty-seven urban programme areas, and a further twenty were chosen to develop programmes in 1993 (Barnsley – the only authority to win in both rounds, Birmingham, Blackburn, Bolton, Brent, Derby, Hackney, Hartlepool, Kensington & Chelsea, Kirklees, Lambeth, Leicester, Newham, North Tyneside, Sandwell, Sefton, Stockton-on-Tees, Sunderland, Walsall and Wigan).

Most projects are in inner city locations, but a few are on the urban fringe. The Dearne Valley is unique in being 20 square miles in area, covering a number of smaller settlements in the South Yorkshire conurbation. City Challenge encourages an integrated approach, with a focus on property development but cutting across a range of topic areas, including economic development, housing, training, environmental improvements, social programmes relating to such matters as crime, and equal opportunities.

The competition prize appears substantial: £7.5 million for each area for each of five years of funding. The total central government expenditure (including £19 million a year earmarked for City Challenge from the Housing Corporation) amounted to £242 million in 1995–6. But this was not new money: City Challenge is a different approach to spending rather than an allocation of new funds. Spending for those authorities which win a City Challenge bid can begin only after an action programme is agreed. Projects within the action programme effectively continue to be funded through existing programmes, with similar, if simplified, procedures for making and considering applications. The private sector is expected to play a significant role, and its involvement needs to be demonstrated before projects are agreed. This has forced a quickening in the trend towards partnership arrangements and their institutional management structures:

What is new about City Challenge is that it requires councils to obtain money for inner city regeneration on the basis of a highly politicised competitive bidding process which has no objective relationship to need or even ability to deliver. The rules of the competition and the whole ethos behind the process has prioritised skills, alliances and attitudes not traditionally associated with either local or central government! It has also required from both councillors and officers a flexibility and intensity of activity which has not previously been associated with urban regeneration work.

(de Groot 1992: 197)

Local communities are seen to be important partners, but their place in the management structures is variable. Sometimes they are represented at the policy-making level, while in others they are involved only as consultees or in the detailed implementation of projects. Tower Hamlets has a *regeneration corporation* which is a limited company in which the local authority has only a minority holding. Other board members are drawn from 'the private sector, the community, the voluntary sector and the other statutory agencies' (Forsyth 1992: 20). Such a commitment may be feasible in an authority such as Tower Hamlets where partnerships and targeted action plans have been evolving over several years, but elsewhere the great speed needed to respond to invitations to bid (only six weeks in the first round) has worked against much meaningful community participation in the early stages (Bell 1993). Nevertheless, it is the impetus that City Challenge gives in bringing together different sectors and creating new and positive relationships between them, shifting attitudes and mainstream investment, that offers the greatest potential.

Whilst local authorities play the primary role in coordination, leadership and contact with central government, an essential feature of City Challenge, at least in theory, is a devolution of both control and responsibility to the 'partnership'. But the partnership boards have no statutory status or powers and cannot receive money directly from government: their funds are channelled through existing local government structures, sometimes requiring the same committee and council approval as mainstream spending programmes. There is thus considerable potential for conflict between the lead authorities and partnership boards.

THE SINGLE REGENERATION BUDGET

In response to criticisms of the fragmented nature of programmes for regeneration, and building on the competitive approach of City Challenge, came the Single Regeneration Budget (SRB). This is intended to promote economic, social and physical regeneration through a more flexible funding mechanism. The SRB was introduced in 1994 to promote integrated strategy, partnership and competition. It is administered at the regional level, although the centre has had considerable influence on writing the rules, and ministers have the final say on recommendations made by the regions. It brings together twenty previously separate funding programmes from five government departments though, in fact, nearly nine tenths of the funding is from seven DoE programmes.

Funding amounted to £1.46 billion in total in 1994–5, fell to £1.36 billion in 1995–6, and is projected at £1.32 in 1996–7. These are big figures, but the SRB does not provide any additional funding; indeed, there has been a small decline (Hogwood 1995). A large proportion of the funds are earmarked for English Partnerships, City Challenge and the UDCs and HATs: these, of course, are central government, not local authority, initiatives.

Local authorities must bid for SRB, and the competitive funding model of City Challenge was used as a basis. Bids must be 'strategic': this means that they must establish links to other investment plans, such as single programming documents for European Structural Funds, and economic development strategies. They can include projects needing longer-term funding up to seven years. Of the total 469 bids in the first round, 200 were funded. Of 329 bids in the second round, 172 were funded to begin in the year 1996–7.

The bidding process is conducted in two stages, the first requiring the submission of outline plans describing the partners, the principal outputs, the funding sought and relationship to other funding sources. The bids must include a long list of anticipated outputs in terms of jobs created or safeguarded, number of people trained, new business start-ups and hectares of land to be improved.

The first bidding round was evaluated by the HC Select Committee on the Environment in 1995, where the concept was generally supported subject to a number of recommendations to reduce bureaucracy, improve consistency and increase the involvement of voluntary and community groups. The Committee also noted the need for the regional offices to take account of existing regional guidance and development plans. Other early evaluations of the new approach note extensive 'inter and intra-organisational bargaining'; a skewed investment pattern due to a few very large bids; and 'the domination of employment and economic development related output measures' (S. Hall 1995). Crucially, less money is available, with cuts in real terms of 47 per cent over the first three years of SRB (Hogwood 1995: 20). The investments through SRB are relatively minor compared with the main programmes of participating departments (Hill and Barlow 1995).

SCOTTISH URBAN POLICIES

Scottish urban policies have evolved differently from those in England and Wales, with a long history of partnerships in economic policy (Boyle 1993). There are several reasons for this, including the different urban history of Scotland, its distinctive governmental organisation, and the close relationship that exists between local and central government. Glasgow and its extreme urban problems have been highly significant in determining the course of urban policy. Of particular note was the establishment of the Scottish Development Agency (SDA) in 1975, and its role in the *Glasgow Eastern Area Renewal* (GEAR) project.

The rationale for the SDA was that economic regeneration required measures to promote local growth and overseas investment, and that local authorities did not have the managerial expertise and dynamism needed for the task. But, unlike the case with the later UDCs in England, importance was attached to retaining the role of local authorities as well as the many other agencies involved:

> It was felt that this was more likely to be effective, that it was not appropriate simply to override local democracy if cooperation could be obtained and that, given the important functions local authorities would retain in any event, cooperation between willing partners, rather than a provoked antagonism, was the best approach.
>
> (McCrone 1991: 926)

The GEAR project had economic, environmental and social objectives. Despite initial teething problems associated with a lack of clarity about the scope of the project and consequent delays, the physical impact on the area was massive. Economic revitalisation, however, proved much more elusive, as was underlined in an unpublished 1988 report by PIEDA (McCrone 1991: 927). Though up to 2,000 jobs were retained or created, the benefits to the resident disadvantaged population were limited.

The experience of the GEAR project was used subsequently in tackling disadvantaged areas facing the additional problem of a severe local employment crisis resulting from the closure of a dominant firm: the steel-making plant at Glengarnock in North Ayrshire, the Singer sewing machine factory in Clydebank and the Leyland plant at Bathgate. The accomplishments were impressive, in terms of job creation, the provision of new infrastructure and widespread environmental improvement. This was largely attributed to the establishment of a clear strategy; the integrated nature of the approach, with environmental recovery, job creation and retraining all working together; and the active participation of local authorities in the scheme (in the knowledge that, if they did not participate, the scheme would not go ahead) (McCrone 1991: 929).

Also benefiting from experience, the SDA assumed a coordination role in later projects in Leith, Dundee, Motherwell, Monklands and Inverclyde. Most of these projects involved 'issues that were socially related in that the areas suffered from decay and deprivation of long standing', but they all included an objective of economic regeneration. Moreover, the SDA, the local authorities and other bodies committed themselves in advance to a project agreement which spelled out who was to do what, where and when. A more ambitious scheme was undertaken jointly by SDA and the City of Glasgow in the old Merchant City area, with the aid of housing improvement grants and the SDA's *local enterprise grants for urban projects* (LEG-UP). The outcome has been judged a success, with the transformation of a large area. Together with other initiatives in the City, there has been an about-turn in both the City's morale and outside perceptions of Glasgow as an attractive place for investment. The extent of the change was epitomised by the designation in 1990 of Glasgow as European City of Culture.

The Scottish Office statements *New Life for Urban Scotland* (1988) and *Urban Scotland into the 90s: New Life Two Years On* (1990) reviewed the achievements and the priorities for further urban action. Much attention is devoted to the large housing estates on the periphery of Scottish cities. Those of Paisley (Ferguslie Park), Glasgow (Castlemilk), Edinburgh (Wester Hailes) and Dundee (Whitfield) have undergone regeneration through partnerships established by the Scottish Office. These involve the local authorities, the local community organisations, the Scottish Office, Scottish Enterprise and other agencies operating in the areas.

There is a justified emphasis on the physical improvement of these monuments to old problems (all were built by local authorities to cater for their horrendous postwar housing shortages), but it is now accepted that such physical improvements must be carried out as part of an integrated strategy of economic, social, educational and training measures. Above all, there has to be sensitive concern for local needs and a commitment to the involvement of local people in the regeneration process.

It is this which is the major advance made in Scottish urban policy. To put the matter in the lowest terms, 'experience has shown that unless tenants can be encouraged to take more responsibility for their living conditions through being involved

more closely in management, the areas will quickly deteriorate again' (McCrone 1991: 936). Participation is thus seen to be good economics as well as good politics. The other major ingredient is the crucial importance given to training: a problem on which it is hoped that greater success will be achieved by the formation, in 1991, of Scottish Enterprise and Highlands and Islands Enterprise. These are effectively mergers of the Training Agency with the Scottish Development Agency and the Highlands and Islands Development Board. The legislation lists the functions of Scottish Enterprise as being to develop the Scottish economy; to enhance the skills of the workforce; to promote the efficiency and competitiveness of Scottish industry; and to improve the environment. Merger effectively brings these activities under one umbrella.

Further developments in Scottish urban policy took place in 1993 with the publication by the Scottish Office of *Progress in Partnership* and, in 1995, with proposals for change in *Programme for Partnership* and *Partnership in the Regeneration of Urban Scotland*. As in England and Wales, a competitive system is proposed to allocate funds to partnership projects across the country. This would replace the current urban programme which, like its counterpart in England, supports a very large number of small projects: 1,142 in 1995–6 (230 of which were approved in that year). The main share of urban regeneration resources would be ring-fenced for *priority partnership areas* designated by the Scottish Office in response to proposals from local authority-led partnerships, covering the whole of their area. However, at the time of writing there is a concurrent proposal for further devolution of power to the new unitary councils. This could result in a much reduced role for the Scottish Office in centrally allocating specific urban regeneration funding.

EMPLOYMENT, TRAINING AND ENTERPRISE AGENCIES

Employment is, of course, one of the principal economic considerations in town and country planning, yet it cannot be said that employment policies have ever been successfully integrated with physical planning policies. This is due, at least in part, to organisational separatism and the fact that local authorities have little responsibility for employment and training policies. Under the Conservative administration, even this small responsibility waned.

Employment and training policy and services have been a function largely of central government. These grew in number and administrative complexity until a major reorganisation was made following the 1990 White Paper *Employment in the 1990s*. There are now networks of eighty-two *training and enterprise councils* (TECs) in England and Wales, and twenty-three *local enterprise companies* (LECs) in Scotland. This new system is aimed at providing value for money through a privatised organisation, linked, in this case, to the additional objectives of localism and decentralisation (Bennett 1990).

Copied from the USA (where TECs are known as *private industry councils*), the underlying rationale is that if local businesses take a central place in guiding the support programmes for employment and training, the result will be a better response to local employment needs, a more business-like mode of operation, and increased leverage of private sector support for training (N. Lewis 1992).

The TECs have taken over youth and employment training together with business support and growth initiatives, again with the objective of tailoring these to local conditions. Their activities include maintaining a knowledge base of local labour markets and training needs and provision; providing information to employers, employees and the unemployed; encouraging employers to meet training needs; and directly providing or commissioning training for employees or the self-employed. All this (to the extent that it is actually done) is decided, in the manner of private board meetings, behind closed doors, although there is evidence of increasing attention to public relations if not true accountability (Hart *et al.* 1996). As with other current urban policies, accountability has been sacrificed for the elusive goal of efficiency. N. Lewis (1992:

41) is not alone in concluding that 'the emergence of TECs can scarcely be seen as an articulate and coherent response to the problems of the inner cities'.

NORTHERN IRELAND

Urban policy in Northern Ireland has followed a similar pattern to that in the rest of the UK, with early emphasis on social problems giving way to a concentration on property-led regeneration and, in recent years, a shift of attention to place marketing, public–private partnerships and community development (Berry and McGreal 1995). However, the special circumstances, notably the violence and subsequent central government control of policy and implementation, have been important in shaping the problem and responses. Urban conditions and unfairness in employment and housing allocations were important factors in the start of 'the troubles' from 1969, and terrorism has subsequently accentuated the difficulties of tackling them.

Urban problems are concentrated in Belfast and Londonderry, although there is a separate *Community Regeneration and Improvement Programme* for smaller towns. The urban policy budget in Northern Ireland amounts to more than £34 million (1993–4) but to this must be added the considerable funding that comes by virtue of the designation of the whole of the Province as an Objective 1 region for the purposes of ERDF, and the significant inward cash flows for employment and housing investment through mainstream funding programmes.

Urban renewal and all other housing functions are undertaken by a single authority: the Northern Ireland Housing Executive. It has followed policies similar to those in England with the sale of public housing to sitting tenants, the progressive renewal of existing stock (where great improvements have been made), and only very limited new building. Urban policy in the most deprived areas of Belfast is implemented through nine Belfast Action Teams which coordinate the activities of different departments as did the CATs in England (PA Cambridge Economic Consultants 1992). A separate Londonderry Initiative was established in 1988. *Making Belfast Work* is an urban regeneration programme for the thirty-two most deprived wards in Belfast and promotes business growth and training. Both the Belfast Action Teams and the *Making Belfast Work* programme promote community participation in their schemes. A third programme, the Community Economic Regeneration Scheme enables communities in the most severely depressed areas to bid for substantial capital investment (property development) to promote business and employment. There is one UDC – the Laganside Corporation – set up in 1989 following closure of shipbuilding yards.

EVALUATION OF URBAN POLICY

One remarkable change in government policy over the last decade or so has been the embracing of research as an essential part of the policy-making process. There has been a huge expansion of research studies, as the list of official publications in the appendix demonstrates. This is, at first sight, surprising, since governments typically find it easier to express their goals in broad terms which encompass the widest possible range of outcomes. Modern management techniques, however, have brought to the fore the formulation of specific objectives which can be monitored and evaluated. This has made the easy option of vagueness unacceptable in principle; but practice is a different matter, as is illustrated by the emphasis laid on undefined 'regeneration'. Much of the recently commissioned research is limited to the evaluation of specific programmes, as the NAO report (1990: 3.19) pointed out: 'most of the [evaluation] measures so far developed are intermediate measures related to activities rather than effectiveness'.

These 'intermediate measures' include a plethora of numbers relating to matters such as training places, jobs, visitors, roads and reclaimed land. Nevertheless, many of the projects have produced conclusions that have wider import, and some of them have clearly demonstrated that urban issues are so complex that it is very difficult to devise effective policies to deal with them. Indeed, there

is an extraordinary difficulty at the outset: how to define clearly what the objectives of policy are. The problem is very familiar to policy analysts (Rittel and Webber 1973), but it has to be constantly tackled anew by policy-makers. A good example (and this discussion has to be illustrative rather than comprehensive) is the apparently simple matter of increasing employment. The problem is one of finding a satisfactory definition (or even concept) of the term 'new employment'. The effectiveness of policy, as measured by the cost per job created, is greatly affected by the definition used: in the Aston University evaluation of the UDG programme, the cost per new job ranged from £3,100 to £18,300 on different definitions (Public Sector Management Research Unit, 1988a).

Other complications arise when account it taken of the 'life' of new jobs created: many of those created in the course of regional development programmes later disappeared (Hughes 1991). Moreover, there may be a lag in the growth of jobs which might be difficult to take into account. Again, policy may be directed not to the objective of short-term job creation, but towards increasing the long-term competitiveness of the area in a changing national and world economy. This poses obvious difficulties of evaluation. Going further, if the aim of policy is wealth creation (a term that was popular for a short time in the mid-1980s), any thought of evaluation becomes mind-boggling. How is wealth to be defined (particularly in these environmentally conscious days)? Is the object to raise the average level of wealth, or the level of wealth of the poorest? Such questions quickly banished the term from general use.

Despite all the conceptual and practical difficulties, researchers have been able to draw some important conclusions from their evaluations of urban policy. For example, many of the jobs that have been 'created' would have arisen without any intervention. Indeed, if the projects that have been evaluated are typical, 'such programmes are unlikely to make more than a modest contribution to the economic regeneration of the inner cities' (Martin 1989: 638). On reflection, this is perhaps unsurprising. In a complex, interdependent society (and, increasingly, an interdependent world) 'local' issues are elusive. In Kirby's (1985: 216) words, 'we cannot attempt to understand the complexities of local economic affairs *in situ'*. Much research corroborates this view.

Another issue in which difficulties of assessment abound (and in which myths live on) is that of the impact of property-led development. It is frequently assumed that property development will somehow or other stimulate economic growth. Indeed, this is a central plank of government policy (typically concentrated in small areas):

> Property development offers the potential to achieve visible results, changing the appearance if not the underlying characteristics of places . . . Precisely how property development is intended to bring about the economic revival of urban areas has not been officially articulated. In practice, it is often simply assumed that private sector property development is synonymous with economic development, or that there is an inevitable one-way process leading from physical to economic regeneration and community prosperity.
>
> (Turok 1992: 363)

Though 'the links between property development and economic regeneration are universally poorly understood, there has been little detailed research on the subject' (Turok 1992). Such research as has been undertaken offers no clear conclusions, though recent studies 'suggest that access to markets, management abilities, and the availability of finance are more important than buildings' and 'levels of investment in product development and production technology, together with differences in the way human resources are managed, are most significant' (ibid.).

The experience of the Scottish new towns underlines the importance of factors other than property (such as the availability of a skilled labour force, ready finance and an attractive environment). The experience of the SDA in Glasgow showed that, though the provision of premises attracted some firms to the area, this was 'at the expense of other parts of the city, and most jobs were filled by inward commuters anyway' (ibid.: 372). Moreover, property development can present its own problems: this

became clear in 1990, when rental values began to fall as the economy dipped. The slowing down of property investment in 1990 turned into a spectacular collapse, with catastrophic effects on the construction industry and local economies.

UDCs above all have failed to consider the relationship between the local economy and property development. This is especially so in terms of the variable nature of local conditions which affect the value of property development as an economic development tool. In some cases, urban policy can have a detrimental effect on local economic activity as, for example, when the precarious position of small local firms is challenged with competition from outside the locality. Urban policy has been characterised by short-term thinking, centred on getting the best return from particular sites (CLES 1992b). It is instructive to note here that there has been little systematic evaluation of the UDCs comparable to that done for the enterprise zones. (Perhaps it is considered sacrilegious to investigate the government's leading policy prescription?) Independent assessments have been highly critical, though they have been made from very different perspectives.

Although social, community and employment objectives were secondary for urban policy in the 1980s, the assumption was that property-led urban regeneration produces a trickle-down of benefits for the local disadvantaged community as the local economy improves. There is little to support this view.

The rationale for the concentration of urban policy on property development and the private sector is not simply that this creates more physical development and infrastructure. It is argued that it is cost-effective in that longstanding revitalisation of the local economy is stimulated by involving and encouraging renewed private investment; that benefits eventually 'trickle down' to local communities in the form of jobs, facilities and quality of life; and, more recently, that such an approach generates an 'enterprise culture' and results in 'moral regeneration':

> In shorthand terms, this policy could be labelled 'privatism': the attracting into the Inner City of private developers, whose activities can in turn demonstrate

that regeneration is taking place. Such regeneration should be tangibly evident, in the form of 'cranes on the horizon'; but also extend to the intangibles – the putting to rout of dependency.

(Edwards and Deakin, 1992: 362)

Conclusions from the most comprehensive evaluation of urban policy recently undertaken suggest other policy priorities would have greater impact (Robson *et al.* 1994). Given the difficulties of evaluation, it is not surprising that the researchers describe as daunting the task of providing a sweeping overview of urban policy impact. They underline as the main difficulties the interlocking nature of urban policy, continual change in programmes and the difficulty of identifying a single unambiguous set of objectives against which to measure progress.

In response, the researchers used a variety of techniques, first measuring socio-economic outcomes (or change), including unemployment, small-firm creation, house price change and migration of 25–34 year olds; second, a survey of residents in areas targeted by urban policy; and third, discussions with experts including civil servants, business people, local authority officers, quangos and community workers. The evaluation concentrates on the fifty-seven urban priority areas and case studies and interviews in three conurbations: Greater Manchester, Merseyside and Tyne & Wear.

Some key points emerge from the wealth of data. Although targeting resources on the most deprived areas is a key objective, in practice it has been weak. Furthermore, cuts in mainstream funding (particularly revenue support grant and the housing investment programme) have fallen more heavily on some of the most deprived areas, leaving them even worse off per capita. Not surprisingly, therefore, there is increasing concentration of the most disadvantaged in the urban programme areas in comparison with other authorities. Within the areas themselves there is an increasing concentration of unemployment in the worst areas. Thus there is growing polarisation within the conurbations, with the benefits of targeting being felt most by the surrounding areas.

Generally, job creation and falling unemployment is positively correlated with urban policy spending, and there have been similar relative (if not absolute) improvements in socio-economic conditions. Funding has also had a positive impact on residents' perceptions of their area.

The conclusion made in a summary of the findings by Bradford and Robson (1995) is that the research 'suggests some limited success for government policy', particularly in smaller cities and the outer districts of conurbations. Where well-coordinated multi-agency approaches have been taken, some policy instruments have worked well:

> On the other hand, the amount of money going into urban policy is minuscule compared to the size of the problems which are being tackled . . . many of the poorer areas are not improving or at least not nearly as much as the better-off areas within the districts. The property-led developments have not produced any trickling down effects. The economic and environmental emphases of policies have not tackled these areas of social disadvantage, where the place loyalty of the local residents suggests that there is plenty of human capital in which to invest effectively.

The report identifies five principal conclusions for future policy, which are summarised in the box (p. 252). There is some indication that the breadth of opinion on the need for change has had some effect over recent years. The introduction of the SRB and integrated government offices for the regions is about providing more coordination and consistency from different departments. City Challenge has taken a longer-term view, gives a leading role to local authorities and seeks to incorporate (although not without some difficulty) local communities. But it is generally agreed that much more change is needed, in both policy themes and the context in which it is formulated and implemented. In a review of the prospects for change, Lawless (1996) points to the need for a context which encourages local strategy building rather than centrally directed *ad hoc* responses. This, in turn, calls for a new form of urban governance which is stronger at both the regional and local community levels, and a more politically mature programme: 'Extraordinary as it may seem, and admittedly with the important exception of jobs and employment, problems which

Assessing the Impact of Urban Policy: Conclusions for Future Urban Policy

There are clear indications of the importance of creating effective coalitions of 'actors' within localities and that these are most likely to result from . . . mechanisms which encourage or require long-term collaborative partnerships.

Local authorities – in their newly emergent roles as enablers and facilitators – need to be given greater opportunities to play a significant part in such coalitions.

Local communities equally need to be given opportunities to play roles in such coalitions. The evidence of increasing polarisation suggests the need for specific resources to address the scope for community capacity-building within deprived areas.

There remains a need to improve the coherence of programmes both across and within govern-

ment departments. This requires a greater emphasis on the identification of strategic objectives which can guide departmental priorities. Area targeting has played an important part in those cases where separate programmes have been successfully linked so as to create additionality, thereby suggesting the value of giving even greater emphasis to area based approaches . . .

An important part of such coherence must derive from less ambiguity in the targeting of resources. There is a strong argument for the development of an urban budget which might be administered at regional level so as to reflect the varying constraints and opportunities across different regions, and to improve coordination across programmes and departments.

(Robson *et al.* 1994)

most affect the urban disadvantaged have received minimal attention.'

URBAN PROBLEMS AND POLICY

Underlying the transition from a physical policy concerned with deteriorated housing to a social policy in aid of deprived areas and then to an economic policy for strengthening the base for local growth there is the dramatic change which has taken place in the character of inner cities. There has been a relentless decline of population and employment in the urban areas generally, and particularly in the inner areas. (The other side of the coin is the growth in smaller towns and rural areas.) The economy of the older industrial British cities has been transformed: their manufacturing base has been eroded, and there has been little in the way of expanding tertiary industries to take its place. Peter Hall (1981: 138) has suggested that it may well be that 'certain kinds of area no longer represent an appropriate milieu for the expanding kinds of economic activity'. There are many possible reasons for

this, ranging from climate and infrastructure to the local political culture. Cameron (1990: 486) goes further: 'The aggregate decline of many major British cities is inevitable and indeed desirable. Probably a growing percentage of British consumers and producers will seek locations for living and producing outside such cities.'

Currently, explicit inner city policies are restricted to selected areas. It is not clear whether this is for experimental reasons (we are not sure *how* to go about solving the problem) or, more simply, for financial reasons (there is not enough money to go round). But it is clear that to attempt to explain, or even describe, the inner city problem is to fall into a major trap. There is no such thing as *the* inner city, still less a conceivable set of plans for it. The historian has the luxury of looking backwards and attempting to explain what happened why, when and where (Checkland 1981), but those concerned with current issues have no such privilege. They have to wrestle with the pressing, and perhaps misleadingly articulated, problems of today.

Cities (and still less inner cities) are not islands: they are part of the wider dynamic of change. Traditionally, urban populations have improved their conditions by moving out to suburban locations and, in the process, have made room for newcomers, who have chosen the city centres because of the opportunities which they have provided. The situation is now more complex: there are few newcomers; there have been changes in the economic structure of cities (and the regions of which they form a part); in the ease of mobility (which greatly increased the area of job markets); in the difficulties of access to good-quality housing (considerably exacerbated by housing policies which have strangled the private provision of rental housing and made inner city owner-occupation prohibitively expensive for the majority); in new laws of settlement (operated through the council housing system and through bureaucratic and political stances in relation to 'homelessness'); and, more generally, in the character of the postindustrial city.

These are but some of the changes which have transformed the functions of the inner cities. When other factors such as education, urban roads (and parking), the impact of public and private forces on 'community', the spread of urban blight and dereliction, the concentration of families with children in high-rise flats (thereby raising 'child densities' to unmanageable levels), teenage unemployment, and so forth are added, the list becomes frightening to contemplate. And overshadowing such features are fears of racial (or sectarian) violence, of a general run-down in the quality of life and in the concentration of self-exacerbating social problems.

It is questionable whether there is any useful validity in the notion of an inner city economy as a spatially and economically discrete entity (Edwards and Deakin 1992). There are important interrelationships between the local, national and international economy. Major new employment-generating uses will frequently be determined by national and international corporate decisions. They do not grow out of the inner city. Employment uses do not have to be located in the inner city to serve those residents. The direct benefits of urban regeneration in economic development are therefore questionable.

The basic problem with current urban policies is that they assume that the issue to be tackled lies in a particular part of particular cities: but 'inner city' problems is a misleading abstraction. Adapting a passage from Marc Fried (1969) (who was commenting upon the concept of poverty), 'the inner city [is] an empirical category, not a conceptual entity, and it represents congeries of unrelated problems'. The problems posed in the inner city 'are not readily accessible for study or resolution in the name' of the inner city.

Progress will not be made by 'comprehensive action' but by identifying priority fields in which effort should be concentrated. (This, of course, is precisely what the Conservative government did, even if its choice of focus was debatable.) Most of the problems identified in inner cities are matters of national policy relating to all areas. (Thus, though poverty is undoubtedly a problem which arises in inner cities, most of the residents are not in poverty: and most poverty is not in inner areas.)

The arguments *against* inner city, or indeed any area-based, policy are strong (Townsend 1976). To the extent that the problems relate to the deprived, it makes more sense to channel assistance to them directly, irrespective of where they live. Only to the extent that the problems are locationally concentrated should remedies focus on specific locations – as in the case of *renewal areas*.

None of this is to deny the importance of tackling directly those problems of decay and disadvantage which are all too apparent in many inner areas. Nor is there any argument against the desirability of attempting better organisation of services at local levels, or improved coordination both within and between agencies. What is crucial is to identify the forces which have created the problems and to establish means of stemming or redirecting them.

This certainly means a reorganisation of urban policy and implementation processes. Currently, both suffer from the multiplicity of agencies, the project-led nature of policy, the absence of any strategic view (or even the machinery for preparing one), and the secrecy with which the many *ad hoc* agencies surround themselves. The only agency which over-

comes these problems is the much-maligned local
government. But local authorities need a framework
of central policy and finance geared to assist them 'to
improve the process of adjustment to new economic
structure and to avoid welfare losses by those who
suffer during the process' (Cameron 1990: 486).

FURTHER READING

General

There is a wide and expanding literature on urban
policies. General textbooks include the earlier
reviews by Lawless (1989) *Britain's Inner Cities*; Robson (1988) *Those Inner Cities*; and MacGregor and
Pimlott (1990) *Tackling the Inner Cities*. These are
now joined by Blackman (1995) *Urban Policy in
Practice*, which gives a contemporary review and
analysis of the impact of recent change in the management of local government services on urban policy; and Atkinson and Moon (1994) *Urban Policy in
Britain*, which provides a historical review and evaluation. See also Turok and Shutt (eds) (1994),
Urban Policy into the 21st Century (special issue of
Local Economy).

Brief summaries of current programmes, their
objectives and expenditure are given in the *Annual
Reports* of the relevant government ministries. The
DoE report is particularly useful. Wilmott (1994)
Urban Trends 2 analyses changes in the most
deprived parts of Britain.

Housing Renewal

There are only a few recent texts on housing
renewal. Carley (1990b) *Housing and Neighbourhood
Renewal* is a principal source, and see also Couch
(1990) *Urban Renewal: Theory and Practice*. Balchin
(1995) *Housing Policy* and Malpass (1996) 'The unravelling of housing policy in Britain' places renewal
within its wider context. The Joseph Rowntree
Foundation sponsors a considerable body of research
on urban renewal and housing improvement. Of
particular interest is *Made to Last: Creating Sustain-*

able Neighbourhood and Estate Regeneration (Fordham
1995). Summaries of research findings are available
free from The Homestead, 40 Water End, York
YO3 6LP.

History of Urban Policies

Early approaches to the urban problem (before the
focus on economic development) are summarised in
Edwards and Batley (1978) *The Politics of Positive
Discrimination*, and Hambleton *et al.* (1980) *Inner
Cities: Management and Resources*. There are many
commentaries on urban policy in the Thatcher years,
notably Lawless (1991) 'Urban policy in the
Thatcher decade'; Parkinson (1989) 'The Thatcher
government's urban policy'; Haughton and Lawless
(1992) *Policies and Potential: Recasting British Urban
and Regional Policies*; and Hambleton (1993) 'Issues
for urban policy in the 1990s'. A special edition of
Planning Practice and Research (vol. 10, nos 3/4) is
devoted on urban policy. Healey *et al.* (1995a)
Managing Cities explores the links between local
problems and global economic forces.

Urban Development Corporations

There is a large number of writings on urban development corporations, predominantly of a highly
critical nature. These include Audit Commission
(1989) *Urban Regeneration and Economic Development:
The Local Government Dimension*; Brownill (1990)
Developing London's Docklands; N. Lewis (1992) *Inner
City Regeneration*; National Audit Office (1988)
Urban Development Corporations and (1993) *Regenerating the Inner Cities*; and Thornley (1993) *Urban Planning under Thatcherism* (ch. 8). Imrie and Thomas
(1993) *British Urban Policy and the Urban Development
Corporations* contains case studies from eight UDCs.

Task Forces and City Action Teams

The setting up of task forces and city action teams is
discussed in HC Environment Committee, *The Problems of Management of Urban Renewal* (1983). See also

Tibbs (1991) 'The objectives and implementation of the Bristol Inner City Project'.

UDG, URG, City Grant and English Partnerships

A DoE-sponsored study is Price Waterhouse (1993) *Evaluation of Urban Development Grant, Urban Regeneration Grant and City Grant.* Earlier evaluations were made by Jacobs (1985) 'Urban development grant'; S. Martin (1989) 'New jobs in the inner city'; and Public Sector Management Research Unit (1988a) *An Evaluation of the Urban Development Grant Programme.* An early review of English Partnerships is given by C. Jones (1996) 'Property-led local economic development policies'.

City Challenge and Competitive Bidding

Among the many useful studies are Bailey and Barker (1992) *City Challenge and Local Regeneration Partnerships*; de Groot (1992) 'City Challenge: competing in the urban regeneration game'; Davoudi and Healey (1994) *Perceptions of City Challenge Processes*; Oatley (1995) 'Competitive urban policy and the regeneration game'; Oatley and Lambert (1995) 'Evaluating competitive urban policy'. On the community view of City Challenge, see Bell (1993) *Key Trends in Communities and Community Development*, and McFarlane (1993a and 1993b) *Community Involvement in City Challenge* (2 vols). *Laxton's Guide to Single Regeneration Budgets* (1996) is intended as a handbook of 'good bidding practice'; it contains case studies of successful bids.

Single Regeneration Budget

Early evaluation of SRB is provided by the HC Environment Committee report (1995) on the *Single Regeneration Budget*, and Mawson (1995) *The Single Regeneration Budget.* See also Hogwood (1995) *The Integrated Regional Offices and the Single Regeneration Budget*, and Hill and Barlow (1995) 'Single regeneration budget'.

Training and TECs

Accounts of the performance of the TECs vary dramatically. A critical review of their performance was published in 1994 (Coopers & Lybrand) in their final report on the *Employment Department Baseline Follow-up Studies.* The findings are assessed by M. Jones (1996) 'TEC policy failure'. In contrast, see the glowing account of the success of one TEC by Groves (1992) 'A chairman's view of a TEC'. A different view is expressed in CLES (1992a) *Reforming the TECs: Towards a Strategy.* See also Haughton (1990) 'Targeting jobs to local people: the British urban policy experience'; Bovaird (1992) 'Local economic development and the city'; and M.R. Jones (1995) 'Training and enterprise councils: a continued search for local flexibility'.

Scotland, Wales and Northern Ireland

Many of the references cited above cover practice across different parts of the UK. For specific examination of urban policy in Scotland, see McCrone (1991) 'Urban renewal: the Scottish experience', and Boyle (1993) 'Changing partners: the experience of urban economic policy in West Central Scotland'. The thinking behind the creation of Scottish Enterprise and the Highlands and Islands Development Board are explained in the 1988 White Paper: *Scottish Enterprise: A New Approach to Training and Enterprise Creation* (Cm 534). For a critical review of this change, see Danson *et al.* (1989) 'Rural Scotland and the rise of Scottish Enterprise', and Hayton (1993) 'Scottish Enterprise: a challenge to local land use planning'. A research report by McAllister (1996) reviews *Partnership in the Regeneration of Urban Scotland.* See also Scottish Office: *Progress in Partnership: A Consultation Paper on the Future of Urban Regeneration Policy in Scotland.*

For reviews of urban policy in Wales, see Alden and Romaya (1994) 'The challenge of urban regeneration in Wales'. See also Welsh Office Paper *Programme for the Valleys: Building on Success* (1993), and a more critical account by Morgan (1995) *Rebuilding the Valleys; The Urban Challenge in South Wales.*

Berry and McGreal (1995a) 'Community and inter-agency structures in the regeneration of inner-city Belfast' give a perspective of urban policy in Northern Ireland.

Evaluation

Many of the references mentioned so far include some evaluation of particular programmes or projects but there is a body of literature which takes evaluation of urban policy as its central theme. The principal source is Robson *et al.* (1994) *Assessing the Impact of Urban Policy*. The National Audit Office has also reviewed urban policy in *Regenerating the Inner Cities* (1990). A broader perspective, including consideration of the methodological problems of evaluation, is given in the edited text by Hambleton and Thomas (1995) *Urban Policy Evaluation*.

On the contribution of property development to urban regeneration, see Healey *et al.* (1992a) *Rebuilding the City: Property-led Urban Regeneration*; Turok (1992) 'Property-led urban regeneration'; and Bovaird *et al.* (1991a) 'Improved performance in local economic development'.

The concept of partnership and its progress in urban policy is recorded in Bailey *et al.* (1995) *Partnership Agencies in British Urban Policy* which also contains a number of case studies. See also Whitney and Haughton (1990) 'Structures for development partnerships in the 1990s: practice in West Yorkshire'; Lawless (1994) 'Partnerships in urban regeneration in the UK'; Hastings and McArthur (1995) 'A comparative assessment of government approaches to partnership with the local community'; Nevin and Shiner (1995) 'Community regeneration and empowerment'; and Littlewood and Whitney (1996) 'Re-forming urban regeneration?'.

11

TRANSPORT PLANNING

What nobler agent has culture or civilisation than the great open road made beautiful and safe for continually flowing traffic, a harmonious part of a great whole life?

Frank Lloyd Wright, 1963

Development plans should aim to reduce the need to travel, especially by car.

PPG 13, 1994

MOBILITY AND ACCESSIBILITY

Transport is many things: it is a means of getting from one place to another. It includes a range of very different forms of travel – walking, cycling, travelling by car, bus or train, or flying. A journey to work on the London Underground is very different from a country holiday tour. Except perhaps for the latter type of journey, transport is unlike other goods in that it is a means to an end: it is not an end in itself. Indeed, much transport is an impediment to the enjoyment of something else. It is a means of providing access. Mobility is not important of itself: its importance is in providing access. Yet the debate on transport often forgets this elementary point and focuses on mobility: faster roads, faster trains and more frequent buses. The advantage of focusing on accessibility rather than mobility is that it opens up the possibility of alternative means: changing land use relationships, for example. As an advert on a condominium tower above a Toronto metro station neatly pointed out, 'If you lived here, you would be home now'. The greater the accessibility, the less the need for 'transport'. Thus, transport planning is much more than the building of

roads, even though this does not always appear to be the case. It should involve a consideration of the relationship between different land uses, and between land uses and transport feasibilities, as well as the relationships between different transport modes and their relative effectiveness in meeting economic, financial, social and environmental goals.

There is nothing profound in these observations but, until recently, transport policy appeared to deny their validity. Roads have formed the major focus of policy. It is therefore fitting that we start by considering road traffic.

THE GROWTH OF TRAFFIC

Between 1950 and 1960, the number of vehicles on the roads of Britain more than doubled, from 4.0 million to 8.5 million (Table 11.1). The number more than doubled again by the end of 1980, to 19.2 million. In the following decade, there was a further increase of a quarter. The figure for 1994 was 25.2 million. The most dramatic increase was in cars, from 2.0 million in 1950 to 20.5 million in 1994. The proportion of households owning a car

Table 11.1 Number of Vehicles, Great Britain, 1950–94

	Cars ('000)	All vehicles ('000)
1950	1,979	3,970
1960	4,900	8,512
1970	9,971	13,548
1980	14,660	19,199
1990	19,742	24,673
1994	20,479	25,231

Source: DoT, *Transport Statistics Great Britain, 1995*, Table 9.5

has grown from about one-third at the end of the 1950s to over two-thirds in the mid-1990s (Table 11.2). In terms of total road traffic (measured in billion vehicle kilometres), the increase has been from 46.5 in 1949 to 309.7 in 1985, and 421.9 in 1994. Despite a massive road building programme, including some 3,200 km of motorway, the increase in the length of the road network has been far less than the increase in traffic. There were 365,000 km of road in 1994: an increase of less than a quarter over the 1947 total of 295,000 km (which was only fractionally greater than the 1909 figure of 282,000). The consequence, of course, has been that roads have become far more crowded: the 'average daily flow' on all roads increased by nearly a half between 1981 and 1993.

The increase in traffic shows little sign of abating, and a major increase is currently forecast. Traffic forecasts are, of course, only estimates and no more reliable than weather forecasts – less so in fact. (Forecasts based on other forecasts are particularly

suspect: the traffic forecast is based mainly on assumed economic growth and, of course, on an absence of serious impediments to car ownership and use.) The Buchanan Report of 1963 referred to the prospect (a carefully chosen word) of 27 million vehicles by 1980, whereas the actual figure in that year was 19 million. Since then, forecasts of population growth have been drastically reduced, and this was a main factor in the reduction of later traffic forecasts. However, the 1984 forecast proved to be far too low, and a revised forecast was made in 1989 (DoT, *National Road Traffic Forecasts, Great Britain, 1989*). This (NRTF) reflected more optimistic views on the future rate of economic growth and the level of fuel prices (both of which, it should be noted, could change dramatically). The 1989 forecast, which is the latest, is for an increase in total traffic of between 27 per cent and 47 per cent by 2000, and between 83 per cent and 142 per cent by 2025 (Table 11.3). The increase in car traffic is forecast at between 29 per cent and 49 per cent by 2000, and between 82 per cent and 134 per cent by 2025.

With the emphasis which is so often placed on increases in cars and traffic, it is easy to forget that a third of households do not have a car. The proportion without a car is higher (40 per cent) in the north and in Scotland. It is also higher for the economically inactive (55 per cent) and for unskilled manual workers (45 per cent).

The car ownership forecasts have two components: car ownership and car use. Car ownership is still increasing. Indeed, the 'saturation' level has not yet been reached in any country (not even in the USA). It is therefore not easy to guess what this level

Table 11.2 Proportion of Households with Regular Use of a Car, Great Britain, 1951–93

	No car	1 car	2 cars	3 cars	Households (mn)
1951	86	13	1	—	14.5
1970	48	45	6	1	18.2
1980	41	44	13	2	19.9
1990	33	44	19	4	21.9
1993	32	45	19	4	22.9

Source: DoT, *Transport Statistics, Great Britain, 1995*, Table 9.2

Table 11.3 Forecasts of Increase in Traffic, Great Britain, 1988–2025 (%)[1]

| | *2000* | | *2025* | |
	Low	*High*	*Low*	*High*
Cars	29	49	82	134
Light goods	26	46	101	215
Heavy goods	17	31	67	141
Buses	0	0	0	0
All traffic	27	47	83	142

Source: DoT, *National Road Traffic Forecasts, Great Britain, 1989*
Note: 1 Per cent increase from base year 1988

may be, and, of course, it may well differ among countries. For the purpose of forecasting, the saturation level is assumed to occur when 90 per cent of the driving age-group (17–74) own a car. This works out at 650 cars per thousand people. Official forecasts are, therefore, estimates of how quickly the saturation level is reached. Factors taken into account include income, the cost of buying and running cars (including the after-tax price of cars and petrol), the availability and quality of public transport, and changes in attitudes to car ownership. Clearly, these are not simple matters to evaluate and, in practice, the greatest weight is given to income.

Car use is also difficult to predict. It fell during the period 1973–6 when GDP fell and real fuel prices rose, but there was no fall when similar conditions occurred during 1979–82. Use of second and third cars is not lower than the use of first cars: in fact it is higher.

Of the vehicles on the road, four-fifths are cars. Most of the remainder are goods vehicles. The forecasts for these are calculated separately for light vans and heavy goods vehicles. Light goods traffic is forecast to increase by between 101 and 215 per cent by 2025. Heavy goods traffic is more problematic. Previous forecasts were proved to be far too low (the high forecast for the period 1982–7 was 4 per cent; the actual was 22 per cent). In fact, there are no obvious trends on which to base a forecast. The Channel Tunnel may divert some heavy goods traffic to rail, but the impact on domestic road movement is not expected to be significant. The

Table 11.4 Road Traffic, Great Britain, 1949–94 (bn vehicle km)

	All motor vehicles	*Cars*	*Goods vehicles*
1949	46.5	20.3	19.0
1960	112.3	68.0	30.4
1970	200.5	155.0	37.9
1980	271.9	215.0	45.7
1990	335.9	410.8	64.8
1994	345.1	421.9	67.9

Source: DoT, *Transport Statistics, Great Britain, 1995*, Table 9.4

Single Market may increase international road transport, but it is difficult to predict how great this might be. For these and other reasons, the DoT forecast is based on a simple ratio of tonne km to GDP. This gives an increase of between 67 and 141 per cent by 2025.

Buses and coaches account for only 1 per cent of total vehicle miles. Though there has been some increase in recent years ('possibly reflecting the effects of deregulation', according to NRTF), there is no way of determining whether this will continue. Consequently, the forecast assumed no change in this mode of travel.

Since the forecasts related to road use, they did not deal with rail traffic. Passenger traffic had increased from a total of 30,300 million km in 1980 to 34,100 million km in 1990 (though it has fallen since then). Freight traffic by rail had declined (while freight traffic by road had increased

Table 11.5 Passenger Transport by Mode, Great Britain, 1952–94 (bn passenger km)

	Total	Bus	Car	Pedal cycle	All road	Rail	Air
1952	219	93	65	23	180	39	0.2
1970	403	60	301	4	365	36	2.0
1994	689	43	600	5	648	35	5.5

Source: DoT, *Transport Statistics, Great Britain, 1995*, Table 9.1

Table 11.6 Freight Transport: Goods Moved, Great Britain, 1952–94 (bn tonne km)

	Road	Rail	Water	Pipeline	Total
1952	31	37	20	0.2	88
1970	85	25	23	3.0	136
1994	144	13	52	12.0	221

Source: DoT, *Transport Statistics, Great Britain, 1995*, Table 9.3

greatly). Well over half of freight (measured in freight tonne miles) went by road, and the official view is that, since road and rail serve mainly different markets, there is little scope for transferring road freight to the railways.

The huge increase in car traffic and the relative decline in bus and rail travel (both in the past and in the forecasted future) does not signify a massive transfer from public to private transport (Table 11.5). On the contrary, the figures show that most of the increase in car usage is newly generated traffic. Though the issue has not been subject to research (incredible though this seems) some of the increase must have resulted from the dispersed pattern of activities and the increased separation of home and work. This itself has been facilitated by road improvements, thus illustrating the impact of 'transport supply' on demand. Moreover, since this new traffic is based on dispersal, it may be very difficult to change it to a public transport mode. Though Britain is far from being as car-dependent as the USA, much new development is of this American character.

TRANSPORT POLICIES

Public policy on transport has a long history (Barker and Savage 1974), but postwar policy began with a plan for a network of new trunk roads (which was not implemented) and a plan for the nationalisation of road haulage and the railways (which was). Much energy was dissipated in the nationalisation and denationalisation processes, and more attention was paid to ownership and control than to transport policy. Experience with the centralised and, later, the decentralised British Railways left a legacy of unease about railway spending in the Transport Department which persisted (Truelove 1992: 4; Kay and Evans 1992: 34). With road haulage, the role of government since denationalisation has been largely restricted to safety controls, though there has been acrimonious argument over axle weights (or, in popular parlance, juggernauts). Bus services have been particularly affected by conflicting political philosophies. Indeed, fights over fares policy were a significant factor in the Conservative government's decision to abolish the GLC and the MCCs.

Cycling and walking get relatively little attention, though their importance is increasingly being acknowledged. But the major focus of transport policy has always been on roads, and only in recent years has it become generally accepted that they have to be considered within a wider framework. The starting point for any discussion of this must be the Buchanan Report.

THE BUCHANAN REPORT 1963

It is traffic in towns which forcibly demonstrates that the motor car is a 'mixed blessing', to borrow the title of an earlier book by Buchanan (1958). As a highly convenient means of personal transport it cannot, other things being equal, be bettered. But its mass use restricts its benefits to car users, imposes severe penalties (in congestion, pollution and reduction of public transport) on non-motorists, involves huge expenditure on roads, and at worst plays havoc with the urban environment.

A major landmark in the development of thought in this field was the 1963 Buchanan Report. This eloquent survey surmounted the administrative separatism which had prevented the comprehensive coordination of the planning and location of buildings on the one hand, and the planning and management of traffic on the other. With due acknowledgement to the necessarily crude nature of the methods and assumptions used, the report proposed as a basic principle the canalisation of larger traffic movements on to properly designed networks that would service areas within which environments suitable for a civilised urban life could be developed. The two main ideas here were for primary road networks and environmental areas:

> There must be areas of good environment – urban rooms – where people can live, work, shop, look about and move around on foot in reasonable freedom from the hazards of motor traffic, and there must be a complementary network of roads – urban corridors – for effecting the primary distribution of traffic to the environmental areas.

The simplicity of this concept is in stark contrast to the complexity and huge cost of its application. But what of the alternatives? Buchanan stressed that the general lesson was unavoidable: 'if the scale of road works and reconstruction seems frightening, then a lesser scale will suffice *provided there is less traffic*'. The accompanying report of the Steering Committee argued that the scope for deliberate limitations on the use of vehicles in towns would be almost impossible to enforce, even if a car-owning electorate were prepared to accept such limitations

in principle. Not all would agree and, as traffic has grown, the practical possibilities of the various forms of control have assumed an increased significance.

The great danger, in Buchanan's view, lay in the temptation to seek a middle course between massive investment in replanning and curtailing the use of vehicles 'by trying to cope with a steadily increasing volume of traffic by means of minor alterations resulting in the end in the worst of both worlds: poor traffic access and a grievously eroded environment'. (This, of course, is precisely what has happened.)

An improvement of public transport is no answer to these problems, though it must be an essential part of an overall plan. Indeed, it is quite impossible to dispense with public transport. The implication is that there must be a planned coordination between transport systems, particularly with regard to journeys to work in concentrated centres. On this, Buchanan recommended that transportation plans should be included as part of the statutory development plans. This was accepted and passed into legislation by the 1968 Town and Country Planning Act.

TRAFFIC PLANNING MACHINERY

Traffic policies and planning have evolved (and continue to evolve) over a long period of time. Initially, the main, if not the only, relevant matter was a road network plan. There have always been differing views on this. During the 1930s, there was rivalry between the highway engineers championing a 2,800-mile motorway system and the county surveyors favouring a more realistic 1,000-mile network – which became the basis for the motorway building programme (Kay and Evans 1992: 18). But there was little interest in plans for transport as a whole.

A major change came with the Labour government's 1967 White Paper *Public Transport and Traffic* (Cmnd 3481). This heralded a new approach to transport planning: 'our major towns and cities can only be made to work effectively and to provide a decent environment for living by giving a new dynamic role to public transport as well as expanding facilities for private cars'. Since local authorities were responsible for 'planning' they were obviously the appropriate authorities for transport. All forms of transport, it was argued, needed to be planned together in a coordinated way. However, in major urban areas existing local governments were too numerous and too small. Moreover, the traffic situation was so bad and was deteriorating so rapidly that reorganisation could not await general legislation on local government. Thus some kind of *ad hoc* system was necessary.

This led to the establishment of Passenger Transport Authorities (PTAs) (under powers provided by the 1968 Transport Act) in Greater Manchester, Merseyside, West Midlands, Tyneside and Greater Glasgow. The original intention had been to give these PTAs wide powers to coordinate different forms of transport but, as a result of local government objections, they were limited to public transport (including local rail services).

Subsequent local government reorganisation gave PTA status to all the English metropolitan county councils and the Strathclyde Regional Council. (Following the abolition of the metropolitan county councils, the PTAs were resuscitated as *ad hoc* authorities.) County councils in the non-metropolitan counties were given parallel duties in relation to 'a coordinated and efficient system of public transport'. There was corresponding provision in the Scottish Local Government Act for the regional and islands authorities. In London, the Greater London Council was the PTA (until 1984 when London Regional Transport was established).

A major feature of this new organisation was that it facilitated the preparation of comprehensive public transport plans, as well as providing a mechanism for channelling financial support not only to roads but also to public transport services. This was particularly attractive to a number of Labour councils which wished to strengthen, and indeed favour, public transport. Unfortunately, a period of financial stringency followed, and this, together with the

fragmented nature of the grant system (despite original aims of 'integration'), killed the rational basis of the new system (Skelcher 1985). Nevertheless, transport policies and programmes (TPPs) survived in an attenuated form, and they still operate. Their main concern has been with roads (Land Use Consultants 1993), but in recent years there has been a noticeable shift in emphasis to a more balanced approach. A 'package approach' now encourages local authorities to submit package bids covering investment proposals for both roads and public transport, together with a supporting comprehensive transport strategy (Cook and Davis 1993). DoT's guidance to local authorities in 1995 outlines the information which is required in TPPs:

> Each TPP should contain a statement of the authority's policies for improving its transport infrastructure, supported by a description of the transport needs of its areas, diagrams of the main traffic flows, and notes on prominent industries or activities giving rise to that traffic. This statement should include an outline of the authority's overall transport strategy . . . It should include an account of their policies for public transport, cycling, walking, and for encouraging interchange between different forms of transport . . .
> (DoT, *Transport Policies and Programme Submission for 1996–97*: para. 93)

The complexities of Transport Supplementary Grant (TSG) need not be discussed here. (In any case, they are subject to change.) It is, however, necessary to indicate its broad outlines and the policy which it embodies. The grant is termed 'supplementary' since it is additional to the general Exchequer grants paid to local authorities. Originally, TSG was paid on a proportion of capital and current transport expenditure. In 1985, however, TSG was restricted (partly to cut bus subsidies) to capital expenditure on roads and traffic regulation. Thus, in financial terms, there was not even a pretence of comprehensive transport planning. Under the package approach, local authorities are free to allocate their expenditure on either roads or public transport.

Unlike development plans, TPPs are not statutory documents; nor are they subject to any formal inquiry procedures. Nevertheless, they obviously must be closely related to development plans. In the words of the DoT: 'in view of the vital relationship between land use and transport it should now be automatic for authorities to explain in their TPPs how their proposals fit into the local structure and development plans'.

From 1992, policy guidance has promoted closer integration of development plans and transport policy. Structure plans must include proposals for major changes to the primary route network, and may be discussed at the examination in public. Local plans will set out the detailed alignment of roads, and may add proposals for non-strategic changes. Objections to road alignments are heard at the local plan inquiry. Plans should also include policies for traffic management, including the coordination of public transport services.

However, development plans are not an adequate vehicle for a comprehensive local transport policy. An example of what this might entail is well illustrated by a plan prepared by the Central Regional Council. This Scottish local authority had been spending almost £5 million a year on new transport schemes, of which over 80 per cent had been devoted to new roads and town centre car parks. Only 5 per cent had been targeted at public transport and helping cyclists and pedestrians. In 1993, the county council adopted a policy of redirecting investment 'towards facilities which provide alternatives to the car'. The objectives were: to reduce road accident casualties; to promote public transport; to control traffic growth; and to meet the needs of those who do not use cars. To achieve this, expenditure plans were recast (see Table 11.7).

The policy was elaborated in plans for the region as a whole and for its constituent towns. It was translated into a statutory *alteration* to the structure plan, which was approved by the Secretary of State in 1994. In the first two years, significant progress was made, including the building of a new railway station; the introduction of 'easy boarder' buses, dial-a-journey and taxi card schemes; and various safety and traffic calming measures. More ambitious proposals, however, were made difficult (or even impossible) to achieve by the reorganisation of local government, the preliminary stages of railway privatisation, and the fragmentation of transportation responsibility. Strategic planning for transportation planning (or any type of planning) is problematic when several agencies are involved: the chances are that their objectives will clash.

PUBLIC TRANSPORT PLANNING

The Labour government of 1974–9 laid emphasis on the development of public transport, and the Transport Act 1978 was intended 'to provide for the planning and development of public passenger transport services in the counties of England and Wales'. All non-metropolitan counties were required by the Act 'to develop policies which will promote the provision of a coordinated and efficient system of public passenger transport to meet the country's needs . . . and to prepare and publish a passenger transport plan'. This plan was to have a five-year time-scale and be revised each year.

This statutory requirement for a further plan (that is, in addition to the TPP and the structure plan) may have been more symbolic than substantive. Be that as it may, it was part of a belief strongly held by the Labour government of the time that there were serious deficiencies in transport planning and policy. Part of this, it was felt, could be met by additional

Table 11.7 Central Region, Scotland: Transportation Strategy

	Target %	Actual 1994–5 %	Planned 1995–9 %
New roads	32	33	32
Improvements for pedestrians	18	30	19
Traffic calming	16	6	13
Bus facilities	10	5	12
Rail facilities	8	14	6
Cycling facilities	4	1	4
Car parking (mainly park and ride)	4	0.2	5
Other schemes	7	12	8

statutory plans (and a range of specific provisions relating, for example, to concessionary fares, community bus services and car-sharing). But matters went deeper, and a better basis was needed for coordinated pricing and investment decisions.

That there was also public concern on some transport issues was apparent from opposition to specific highway proposals – particularly at Airedale, Winchester and Archway. Though the anti-road lobbies took up an extreme stance (eloquently justified by Tyme 1978), there was clearly a more broadly based lack of confidence in the system by which highway needs and routes were assessed. In addition to the new Act, therefore, a more effective and flexible approach to transport planning was promised, with more systematic and open public participation in policy formulation, annual White Papers and, in the longer run, policies geared 'to decrease our absolute dependence on transport and the length and number of some of our journeys'. An independent assessment of the government's methods of appraising road schemes and forecasting needs was established, and a review of highway inquiry procedures was set up. (This is discussed later, pp. 266–9.)

The Conservative government, elected in 1979, had a very different approach to public transport: it strongly believed that it should be subject to the discipline of market forces (or, to use different terminology, consumer preferences). The main initial focus of debate was on the subsidisation of fares. The issue could be argued at length, but the 1982 White Paper *Public Transport Subsidy in Cities* (Cmnd 8735) concluded, in succinct terms, that there was a need for 'legislation which will provide for a reasonable, stable and lawful subsidy regime'. What this meant was that central government would take unto itself greater powers to control local government in its transport policies. Legislative effect to the proposals was provided by the Transport Act 1983.

Another White Paper, *Buses* (Cmnd 9300), published in 1984, argued that it was inappropriate for county councils to coordinate public transport: 'it is for passengers to demonstrate what they want and for operators to respond'. In this way, greater effi-

ciency would be secured. The White Paper estimated that there was the potential for a reduction of up to 30 per cent in the cost of public operators. Many of the cost comparisons presented were made in terms of costs per vehicle mile and thus ignored the wide variation in traffic operating conditions. There were many similar debatable statistical inferences (for example that subsidy payments leak substantially into higher operating costs). A major policy issue, of course, was that of cross-subsidy: the practice of using surpluses from some parts of the system to offset deficits in others. The government argued against this on two main grounds. First, users of good routes were being penalised 'by being made to pay excessive fares in relation to the cost of providing the service they use'. Second:

> It leaves to operators for decision, matters which should not be so left. Services which the market does not provide and which therefore need subsidy if they are nevertheless to continue, should get that subsidy only by decision of elected representatives after proper testing that they constitute good value for public money and are within the resources available to them.

These and other arguments are more complex than they appear. Estimating cross-subsidy is difficult with the particular costing conventions used, and revenue attribution is problematic, particularly with return journeys and linked trips. Moreover, bus travellers are buying a transport *network*: availability of services at unpopular hours and in areas of low demand are parts of a package.

RAPID TRANSIT

It is central government that has the main power in relation to rapid transit systems – almost inevitably so, given their huge cost. Financial procedures virtually ruled out new rapid transit systems until 1989. Until that year, central government grants (under section 56 of the Transport Act 1968) had to be justified mainly on the grounds of time savings to existing passengers and benefits arising from generated trips. Since then, it has been a requirement of grant that, where practicable, users should pay for the benefits they obtain. Subsidies are payable for the benefits accruing to non-users (for example, through reductions in road congestion). Rapid transit systems tend to be very expensive (particularly if they use a fixed rail), and they have been out of favour for the last quarter of a century. Schemes had earlier been approved for several cities, including Glasgow, Tyne & Wear, Merseyside and London (the Jubilee Line), but many more were shelved. Increased road congestion (and prospects of much more in the future), the model of the London Docklands Light Railway (promoted as part of the Docklands renewal strategy), and increasing experience of foreign systems has reawakened political interest in rapid transit.

The HC Transport Committee's 1991 report (*Urban Public Transport: The Light Rail Option*) noted that there were some forty urban areas with proposals for rapid transit. (The report listed twenty-two of the schemes which were well advanced: their total cost was estimated at £2,700 million.) Underlining the change in the government's attitude to these schemes (which, though costly, are less so than alternative road works), the Minister of State for Transport is quoted by the Committee as saying that, within ten years, 'based on the track record [*sic*] so far, we may be looking perhaps at up to a dozen schemes in operation'. The Committee concluded: 'it is clear that, even if all the current plans are not realised, light rail represents a significant trend rather than a flash in the pan'.

Britain does not compare well with other European countries on rapid transit, and this may in part be due to the fact that public transport generally is expected to cover a large proportion of its operating costs. In a useful 1992 'state of the art review', Walmsley and Perrett note that the 'even before deregulation, bus services in the major cities typically achieved revenue-cost ratios of 70 per cent, and only in the most highly subsidised were the ratios as low as the systems abroad'. This makes the outlook for rapid transit in Britain less certain, though there is some comfort in the fact that most foreign public transport networks have improved their 'revenue-operating cost ratio'. There is, how-

ever, the added difficulty in Britain that any rapid transit system would find itself in competition with deregulated bus services. On this, Walmsley and Perrett comment that 'it could be argued that if a bus system is able to compete successfully, the case for rapid transit must be weak anyway, especially as it will inevitably cost a great deal more than a bus system'. As for promoting private enterprise, the authors have some pithy things to say. In particular, private development is likely to be stimulated only if there is a comprehensive planning framework:

> There are many different types of planning regimes. The least successful are those without any positive development powers, for such authorities can do little to make things happen. The most successful in terms of rapid transit will be those that have similar powers (and money) to those of the current development corporations. There also need to be supporting policies to limit development elsewhere.

Though the report is from the government Transport Research Laboratory, and it naturally has the usual disclaimers, its publication is significant nevertheless. Even the most rigid governments find that changing electoral attitudes demand flexibility in policy.

ROAD POLICIES IN THE 1980s

Whatever reservations the Conservative government had about public transport, it had no doubts on the economic and social value of roads. During most of the 1980s however, its concern for reducing public expenditure took priority. The Thatcher government's White Papers on the trunk road system all stressed the importance of roads for economic growth, but 'national economic recovery' demanded a close rein on public expenditure. Within a programme smaller than that of the previous Labour government, the top priority was for 'roads which aid economic recovery and development' (foremost among which was the M25). Other priorities were for environmental improvement (by the building of bypasses), road maintenance ('preserving the investment already made'), and improved road safety. It was expected that the balance of the programme was likely to change as the major inter-urban routes were completed. Increasingly (so it was thought) the emphasis would shift to schemes which were required to deal with specific local problems. This perception was dramatically altered by the 1989 traffic forecasts, and a White Paper of that year, *Roads for Prosperity* (Cm 693), announced a massive increase in road building.

Between 1980 and 1990, the road network increased by 18,400 km. By 1989, investment in trunk roads was nearly 60 per cent higher in real terms than ten years earlier. The policies continued to follow earlier ones in emphasising the importance of roads to economic growth (despite little evidence on the matter). The M25 is now the most heavily used road in Britain. Far more traffic uses it than was forecast (though at the planning stage, the Transport Department's forecast was fiercely attacked as being too high). This 'immediate success' (to use the Department's eccentric description) led to a major *M25 Action Plan* (1990). This included increased lighting, improved signalling and widening the road to dual four lanes, with a later possibility of five lanes in certain sections. The 1989 *Roads for Prosperity* and the 1990 *Trunk Roads, England: Into the 1990s* set out a programme which consisted mainly of widening existing inter-urban roads (including some 600 miles of motorway). Until the reversal of road building policies in 1994, this involved a doubling of the cost of the forward road programme to around £12 billion. This 'massive' investment (as it was advertised by the government) was of course a response to the 1989 traffic forecasts, although the government stressed that the forecasts were not really forecasts at all:

> They are in no sense a target or an option; they are an estimate of the increase in demand as increased prosperity brings more commercial activity and gives more people the opportunity to travel, and to travel more frequently and for longer distances.
>
> (DoT, *Trunk Roads, England*)

But if the NRTF is not to be used as a basis for policy, what is the alternative? The forecasts were based on the assumption that the demand for roads would be met with an appropriate supply, that there would be no significant policy of traffic restraint, and that attitudes towards motoring (and its cost) would not change. Though the official stance on such issues has been a coy one, they raise important questions which are now at the forefront of debate (Goodwin *et al.* 1991; FoE 1991b).

Increasing concern about the impact of specific road construction schemes, coupled with a more general concern about traffic congestion, has resulted in a near-paralysis of policy. This has the tangible advantage of assisting in the restraint of public expenditure, which was further helped by the 1994 roads review. Before discussing this, however, it is useful to examine how the need for roads has been approached by the Department of Transport. Successive governments have discovered that 'need' (whether for roads, health services or houses) is an elusive concept, and many have tried to find ways of giving it an objective basis. Not only is this appealingly rational: it also changes the nature of the debate. Argument can be settled by recourse to the 'facts'. In a democracy, such nonsense encounters stiff resistance. The history of assessing the need for roads is a good illustration of this.

ASSESSING THE NEED FOR ROADS

Trunk Road Assessment (1977)

An independent assessment (the 1977 Leitch Report) of the methods used for assessing the need for roads was highly critical. The conventional methodologies were judged to be essentially 'extrapolatory', 'insensitive to policy changes', and partly self-fulfilling. Public concern about road planning was shown to be well founded. There were, however, no easy solutions: indeed, the issues were inherently complex. The way forward lay in a more balanced appraisal process, 'ongoing monitoring arrangements' and more openness – with no attempt 'to disguise the uncertainties inherent in the whole process'.

The Labour government's response was positive, and its 1978 White Paper, *Report on the Review of Highway Inquiry Procedures* (Cmnd 7133), represented a marked change in approach. National policies were to be set out for parliamentary debate in White Papers: these would 'also serve as an authoritative background against which local issues can be examined at public inquiries into particular road schemes'. It was hoped that this would avoid the confusion at local inquiries between national policies and their application in specific areas. It was pointed out, however, that this would work only if the methods of assessing national needs (what the Leitch Committee termed 'a highly esoteric evaluation process') were acceptable. Since these methods could not be properly examined at local inquiries (or, indeed, by Parliament), they were to be subject to 'rigorous examination' by the independent Standing Advisory Committee on Trunk Road Assessment (SACTRA). The Committee's report was published in 1979, under the title *Trunk Road Proposals: A Comprehensive Framework for Appraisal*.

This SACTRA report examined the techniques used to evaluate the economic value of proposed road schemes. The Department's system of cost-benefit analysis (the COBA programme) was criticised for the narrowness of its approach, and certain changes followed. For example, instead of using only one traffic forecast, high and low levels were introduced.

Urban Road Appraisal (1984)

The 1979 SACTRA report dealt with inter-urban roads; in 1984 it was given the task of assessing the traffic, environmental, economic and other effects of road improvements within urban areas. Some urgency for a review was added by the abolition of the GLC and the metropolitan county councils: this gave the Secretary of State, the London boroughs and the metropolitan district councils new responsibilities for tackling the transport problems of major urban areas. The SACTRA report, *Urban Road Appraisal*, together with the government response, was published in 1986.

Much of the report sets out recommendations as principles rather than as detailed prescriptions, and it thus seems to have more than a fair share of platitudes. As a result, though the government accepted many of these, they 'can only be applied once detailed guidance on their application has been prepared'. Moreover, 'where further development or research is needed, the detailed guidance required to implement them will take some time to prepare, and progress must be subject to the availability of resources'. The Committee could hardly have found this a very helpful response!

There were four broad areas which the Committee saw as contributing to the nature and extent of change needed:

- concern about the way in which the Department develops, assesses and justifies major schemes, particularly in urban areas;
- expectations about public involvement;
- the importance of integration between transport proposals and broader land use planning and environmental considerations; and
- the complexity of the planning, assessment and decision-making processes, especially in their application to urban areas.

Interestingly (in view of the common contention that public participation adds to delay), the Committee maintained that one reason why many road schemes took so long to bring to completion was that 'the opportunity to debate their justification comes too late in the procedural chain'. This was accepted, as was a recommendation that national and local objectives should be treated separately 'so that conflicts and common denominators can be readily seen'. However, a recommendation that there should be a two-stage assessment and public inquiry process for the larger and more complex schemes 'presented difficulties'. Though it was regarded as 'a constructive proposal', the government was 'not convinced of the practicality of separating the examination of policy options from consideration of detailed design and local issues'.

The *Urban Road Appraisal* report also echoed the

widespread unease about the methods used to assess the economic value of road schemes, but its recommendation was couched in such broad terms that the government had no difficulty in side-stepping it by maintaining that it reflected existing practice.

Environmental Impact (1992)

The issues, which successive governments might have hoped would be settled by the various inquiries, refused to disappear: in fact, they became more problematic as public attention widened to encompass more and more matters which had not traditionally been regarded as pertaining to roads. Above all, concern has grown enormously about the environmental effects of roads and, indeed, of all forms of traffic. Inevitably, further inquiries were commissioned including another SACTRA Report (*Assessing the Environmental Impact of Road Schemes*), which was published together with the government's response in 1992.

By this date, the arguments about the limitations of COBA had intensified. Not only were environmental considerations now at the forefront of the debate (particularly after the shock of the 1989 traffic forecasts), but it was being argued that any sharp distinction between economic and environmental impacts was false (Pearce *et al.* 1989). Earlier reports had advised that environmental benefits and costs should not be evaluated in money terms but should be subject to 'professional judgement'. The rationale for this is a simple one: there is no acceptable way to estimate the 'value' of a cathedral, a marvellous view or other such 'non-economic goods' – though this has not stopped economists from trying (Schofield 1987). But this leads to a host of mind-boggling questions. What is the value of land which is safeguarded from development? Is it the 'economic' value for development, or the lower 'social' value which is determined by planning controls? Which value should be used in evaluating alternative routes for a road? Other questions are equally baffling: if environmental factors are important, which should be taken into account and which should be ignored? (The SACTRA report has a long list of local, regional, national and global factors.)

How are the cumulative effects of a multiplicity of apparently unimportant decisions to be dealt with? Will future increases in traffic increase environmental damage, or will technological innovations more than offset these? The range and number of questions seem endless. No wonder that cost-benefit analysis is having a hard time!

The report deals at length with such issues, and it emphasises the importance of clear policy objectives and 'strategic' and long-term effects. These, of course, are precisely the policy issues with which the political process has difficulty, though the *Government's Response* is eloquent in listing good intentions, including a continuing research programme. The problem will never be 'solved' – it isn't that sort of problem. Instead, it falls into that class of which Rittel and Webber (1973) have neatly termed 'wicked problems': problems that are unique to a specific place and time, which defy definitive formulation and which can only be 'resolved' by political judgement.

Do New Roads Generate Traffic? (1994)

A major issue in the debate on forecasting traffic needs is the extent to which new roads actually generate extra traffic. Certainly, their immediate effect on pre-existing roads can be dramatic, but this may be short-lived. Traffic seems to increase faster than new roads can be built. American studies have argued that, typically, the creation of new road space is eventually (and it may be sooner rather than later) taken up by increased traffic. Where does this traffic come from? Downs (1992) has put forward an elegant explanation in his theory of 'triple convergence'. This is based on the simple fact that since every driver seeks the easiest route, the cumulative result is a convergence on that route. If it then becomes overcrowded, some drivers will switch to an alternative route which has become relatively less crowded. These switches continue until there is an equilibrium situation (which, like any human equilibrium, is not stable – conditions constantly change). On this theory, building a new road or expanding an existing one will have a 'triple convergence'. First, motorists will switch from other routes to the new one ('spatial convergence'); second, some motorists who avoided the peak hours will travel at the more convenient peak hour ('time convergence'); third, travellers who had used public transit will switch to driving since the new road now makes the journey faster ('modal convergence').

The eventual outcome depends upon the total amount of traffic (actual and potential) in relation to the available roads. If the increase in traffic stimulated by the new road is modest, there will be an observable benefit for all. Though peak-hour traffic may be congested, this is simply because so many drivers are travelling at the time which is most convenient to them. (There may, however, be a loss to transit passengers if the 'modal convergence' leads to a reduction in service.)

On this argument, new roads can generate traffic by diversion from public transport. Are there other ways in which extra traffic can be generated? Intuitively, it would seem obvious that there are: the more congested and difficult a road journey is, the more likely it is that a potential traveller will seek an alternative. Conversely, ease of road journeys must generate increased trips. A more sophisticated argument which supports the logic of induced traffic lies in the DoT methodology for traffic forecasting and road scheme evaluation. In the words of the SACTRA report (1994: ii):

> First, it is well established in the literature and in the methods used to produce the National Road Traffic Forecasts that the total amount of travel undertaken by vehicle users is somewhat responsive to the level of fuel prices. Secondly, it is also well established, and indeed is the cornerstone of the Department's road scheme evaluation procedures, that people attach value to savings in time spent travelling, and that these values can be represented in money terms with a reasonable degree of accuracy. It follows from these two propositions that travellers must, as a matter of logic, be assumed to respond to reductions in travel time brought about by road improvements by travelling more or further.

Of course, if this were self-evident it would not have taken SACTRA over 200 pages to discuss it;

nor would the Department of Transport have been so resistant to it. Indeed, the matter is a complex one, mainly because it is difficult to establish cause and effect over time on matters where there are many variables. However, the DoT has accepted the thrust of the SACTRA report, even though 'clear evidence' is lacking. A Guidance Note (*Guidance on Induced Traffic*) has been issued, and the Department's on-going research programme has been augmented.

There is a wider point: new roads not only induce traffic, they also encourage car ownership and use. As car use increases, other methods of transport are used less and, as a result, standards of service fall (thereby further increasing the attraction of car use). Moreover, road building and ease of car use has major impacts on the location of new developments (of all kinds – housing, employment, shopping and leisure). Many new locations are car dependent, and therefore may increase the demand for road travel and hence the need for more road construction. In this cumulative way roads certainly generate more traffic. However, it is extremely difficult to forecast patterns of land use, travel and the interactions between land use and transport.

TRUNK ROADS REVIEW OF 1994

The early 1990s witnessed increasing recognition of the impossibility of catering for a continuation in the growth of road traffic, though acceptable alternative policies seemed elusive. A bumper crop of reports in 1994 (including the SACTRA report discussed in the previous section) provided conflicting advice and a massive excuse for further delays in taking positive decisions. However, the *Trunk Roads in England: 1994 Review* (and its 1995 successor *Managing the Trunk Road Programme*) did signal a significant shift in road-building policy. It detailed a reduced road programme (announced in the previous year): a total of forty-nine road schemes were withdrawn completely and many others were postponed. Four programmed new routes were abandoned: M12–M25 (Chelmsford); A5–M11

(Stansted); a new motorway to the south and west of Preston; and the M55–A585 (near Blackpool). These were in addition to the motorway links between the M56 and M62 (Manchester) and the M1–M62 (Yorkshire) which had been abandoned earlier 'because of the difficulty in finding an environmentally acceptable route'.

This is one of the first steps in a major reorientation of transport policy. Though 'it is no part of this government's policies to tell people when and how to travel . . . we must be aware of the consequences if people continue to exercise their choices as they are at present. There is no realistic possibility of simply halting traffic growth . . . The government's policy for sustainable development is to strike the right balance between securing economic development, protecting the environment, and sustaining future quality of life . . . ' Central to the change in policy is a search for 'alternative solutions'. The alternatives identified are more easily stated than implemented, but clearly there has been a sea change in the rhetoric of policy. In the meantime, the road programme continues on a reduced scale. The policies it embodies are set out in the *Review*:

- the majority of resources should be devoted to improving sections of existing key routes likely to experience congestion in the near future, primarily by adding lanes to existing motorways;
- priority should be given to providing urgently needed bypasses of towns and villages;
- proposals for building new trunk routes, particularly those which go through open countryside should be reduced still further;
- the programme of major urban road improvements should be a very limited one, in the light of continuing substantial investment in urban transport initiatives;
- some schemes with particular environmental disadvantages or those not likely to be progressed in the foreseeable future should be withdrawn in order to reduce blight and uncertainty.

Where new roads are to be provided, the private sector is being encouraged to supply them through privately funded *design, build, finance and operate* (DBFO) schemes. The government forecasted a £1 billion private investment in some thirty-seven road schemes in the three years up to 1998–9, though this looks increasingly optimistic in view of the poor response to the PFI (outlined in Chapter 3). The first privately funded toll road will be the Birmingham Northern Relief Road, which offers an alternative to the overcrowded (untolled) M6 in the West Midlands.

EUROPEAN UNION TRANSPORT POLICY

A common transport policy, furthering the free movement of people and goods, has been an objective of the EU since the Treaty of Rome. However, little progress was made on this until the mid-1980s when the European Parliament challenged the Council of Ministers in the European Court for failing to meet its transport obligations. The Commission's 1992 White Paper *The Future Development of the Common Transport Policy* emphasised the positive role that a coordinated transport policy could play in promoting economic growth through the creation of the single market and what is described as 'sustainable mobility'.

Until the 1990s, policy has been directed primarily at measures to deregulate cross-frontier movements and to increase competition in the transport sectors. A more explicit spatial dimension has since been added with the identification of the Trans-European Networks (see box). There has also been a shift in emphasis on the contribution that policy can make to reducing the impact of pollution on the environment. There are three key themes: the promotion of intermodal connections (which entails the identification of nodes where changeovers can take place); the integration of national transport networks and policies; and the reduction in travel demand by ensuring that the price of using roads accurately reflects the true costs.

Implementation of policy relating to new infra-

Trans-European Networks

The European Commission's White Paper *Growth, Competitiveness and Employment* emphasised the importance of creating EU and pan-European infrastructure networks. This was followed by firm commitments in the Maastricht Treaty. The objective is to increase the integration of existing networks for transport, telecommunications and energy as a means of improving the competitiveness of the European economy. Priority projects were identified, including the high-speed rail links Paris–Brussels–Cologne–Amsterdam–London and Cork–Dublin–Belfast–Larne–Stranraer. The route for the Channel Tunnel Rail Link (and the builders and operators, London and Continental Railways) was announced in 1996. Other priority routes extend the TGV from France into Italy and Germany, and across southern Europe, and also the Øresund fixed link between Denmark and Sweden. Some 30,000 km of new and upgraded high-speed rail track and 12,000 km of motorways are planned by 2020. Concern has been voiced about the effects of this major investment on increasing the amount of travel, on the attractiveness of the metropolitan nodes and on the relative disadvantage of peripheral areas.

structure is the responsibility of Member States (though the EU has made substantial contributions to major projects, including the improvement of links between the UK and Ireland). The most recent directions in European policy are introduced in the 1996 Green Paper *The Citizens' Network: Fulfilling the Potential of Public Passenger Transport in Europe*. This responds to concerns that the Commission's road building plans do not square up with its commitment to sustainable development. Transport Commissioner Neil Kinnock has shifted the emphasis from the supply of infrastructure (especially roads) to managing the demand for travel. Trans-European Networks remain important, but there is increased emphasis on local connections to public transport feeder services. This change is evident also in EU environmental and energy policies which will increasingly affect transport. Thus the Fifth Action Programme on the Environment notes the need to improve planning to reduce transport requirements, to further the integration of public transport, and to promote specific policy instruments such as road pricing and the increase of parking charges in urban centres.

CYCLING

There is a feast to be had from cycling statistics. The world has twice as many bicycles (around 800 million) as cars. Global production of cycles outnumbers car production by three to one. These figures, of course, include the less-developed countries where cycles play a very significant role in transport. But the British figures are higher than one might expect. There are over 13 million cycles, and over a third of all British households have at least one. About 11 million people use their cycles at least once a year; in an average week about 3.6 million are used. Over a million people use a bicycle as their main means of transport to work. These seemingly impressive figures have to been viewed in context. Only 2 per cent of total trips are made by cycle. The total distance travelled by cycles is between 5 and 6 million km a year, compared with around 350 million km by cars. On the other hand, a greater distance is travelled by cycle than by bus and train combined.

Despite this apparent abundance of statistics, it is difficult to obtain an accurate picture of cycle use (and still less of any potential increase). There is, in fact, a shortage of data: the National Travel Survey, for instance, ignores journeys of less than a mile, yet these account for over one-third of all personal journeys (representing about one-third of all cycle journeys); traffic counts are usually taken on main roads that are less used by cyclists. Despite the inadequacy of statistics, it is clear that there are great variations in the use of cycles in different developed countries (that is, ignoring the vastly different economic conditions in lesser developed countries). While cycling in the UK accounts for less than 2 per cent of trips (and is declining), the proportions are 10 per cent in Sweden, 11 per cent in Germany, 15 per cent in Switzerland and 18 per cent in Denmark (DoT, *National Cycling Strategy*, 1996: 7).

It is on the basis of figures such as these that it is argued that there is a significant potential for an increase in cycling, with benefits all round: 'Cycling offers a widely accessible, convenient and environmentally-friendly means of making local journeys, especially in urban and suburban areas. And it is a healthy, enjoyable, economic and efficient means of travelling' (ibid.: 2).

Clearly there is much to be said for such a healthy and efficient mode of transport; but there are problems as well. Cyclists are exposed and vulnerable, and it is simply not possible to provide them with their own tracks on all roads. The DoT has, in the past, expressed great concern about cycle accidents – or, to be more precise, accidents to cyclists. In evidence to the HC Transport Committee (*Cycling*, 1991: 54) the Head of the Traffic Policy Division warned that 'we need to be very careful about artificially boosting cycling as a means of transport'.

The Department of Transport's policy initially was that, though the cycle accident rate was of great concern to central government, most cycling was on local roads, and it was therefore considered that local authorities were best able to deal with the problems according to their local situations. However, the 1977 White Paper, *Transport Policy*, which took a more positive approach to cycling, accepted the need for the provision of facilities for cyclists, and there followed increased study and promotion of experimental schemes. The Department introduced an *innovatory projects budget* which provided for contributions to the costs of local authority experimental cycling schemes. The success of the early schemes

led to the *Statement of Cycling Policy* issued by the DoT in 1982 which extended the scope of financial assistance to *networks* of cycle routes. In fact, however, apart from Nottingham, the supported schemes consisted of a single route with connecting spurs. In all, there have been around eighty schemes supported under this programme. The general assessment is that they have been reasonably successful, though there has not been the large increase in new cyclists which some optimists envisaged. Instead, existing cyclists have transferred to the more suitable routes provided. Other local authority schemes have included the conversion of disused railway lines to cycleways and, in London, the fifty projects planned and carried out by the Greater London Council's *Cycling Project Team* (McClintock 1992: 24).

Cycling rose higher on the political agenda in 1994 when the DoT published its *June 1994 Cycling Statement*. However, this rather unconvincingly extolled the virtues of cycling, and exhorted local authorities to do more to facilitate and encourage it as healthy, environmentally friendly, economical and efficient means of transport for local journeys. A more tangible boost came with the Millennium Commission's grant of £43 million towards the 6,000–mile National Cycle Network. Much of the route already exists, and the organising charity, Sustrans, reports that there is good prospect of the additional £140 million being raised from local authorities, voluntary organisations and donations. The benefits of the scheme have been prominently advertised: '6,000 miles of high quality cycle and pathways within two miles of 21 million people will be provided, linking towns and cities from Dover to Inverness to Belfast' (Sustrans 1996). Half the network will be entirely free of vehicular traffic. Sustrans estimates that there will be more than 100 million journeys a year on the network. Perhaps the greater long-term benefit, however, will be an improvement in the image of the cycle as a means of transport.

It was, however, the *National Cycling Strategy* (DoT 1996) which marked a genuine change in governmental attitudes to cycling. This claims to represent 'a major breakthrough in transport thinking in the UK'. The target is to double the number of trips by cycle by 2002, and quadruple the number by 2012. It is not clear where these figures come from, but they exhibit an eagerness which to date has been restricted to cycling enthusiasts. The *Strategy* covers a wide range of relevant issues, including appropriate planning measures, safety, provision of parking for cycles (and accommodation for them on public transport), integrating cycling with traffic management, cycle security and the 'communication programme' needed to change attitudes to cycling.

The new DoT approach is almost euphoric by contrast with traditional attitudes:

> The *Strategy* will generate a culture change for cycling. More people want to cycle, especially for local trips. With safer condition on the road a 'critical mass' of

cyclists will be encouraged . . . Cycling has a bright future, contributing significant benefits to the nation.
>
> (Ibid.: 2)

Whether the necessary investment will be forthcoming to translate these hopes into reality is not at all clear.

TRAFFIC CALMING

Traffic calming is an expressive term which, though used in different ways, essentially refers to measures for reducing the harmful effects of motor traffic. In its limited sense, it refers to speed reductions, parking restrictions, pedestrianisation schemes and such like. In a wider sense, it is synonymous with overall traffic policy, including car taxation and land-use

---

The Bypass Demonstration Project

Traffic calming is particularly appropriate after the completion of a bypass. Though the town may feel that the bypass has solved their local problems, in fact it can bring new problems in its wake. The old route will have all the features of a heavily trafficked route: it will bear all the marks of a road which has been adapted (and perhaps mutilated) to accommodate high levels of traffic. Quite apart from the poor appearance of the place, the reduced traffic will facilitate higher speeds and paradoxically, new traffic hazards may appear. Also (the final irony), though traffic will initially decrease significantly, it can soon build up again.

To explore how these problems can be dealt with, the DoT in conjunction with local authorities mounted a Bypass Demonstration Project (announced in the White Paper *This Common Inheritance*, and completed in 1995). A major object of this was to demonstrate how the benefits of a bypass can be enhanced by an overall improvement scheme. Six towns (Berkhamsted, Dalton in Furness, Market

Harborough, Petersfield, Wadebridge, and Whitchurch) were selected and, with some financial support from the DoT, major traffic calming and other 'town enhancement works' were undertaken.

The report on the study revealed the problems and opportunities. The removal of through traffic allows a radical change in the street space. Inevitably the benefits are not equally shared, but 'the six project towns have shown how pedestrians, visitors, cyclists, disabled people, and civic uses in general can benefit substantially, whilst still maintaining vehicular access in a traffic-calmed environment'. Such schemes demand a great deal of professional input, coordination of effort, public involvement and cost: the expenditure in the six towns ranged between £1 million and £2 million. The benefits are striking, and can amount to a transformation of the area.

(DoT, *Better Places through Bypasses: Report of the Bypass Demonstration Project*, 1995)

measures designed to reduce the need for car journeys. Advocates of traffic calming can make some telling points in its support. For instance, a 50 kph speed limit (about 30 mph) in residential areas is 'acknowledged in many European countries' to be 'far too high'; at speeds of 30 kph or below additional road space is created since cars can travel closer together because they need less stopping distance; if traffic calming is restricted to a few streets, its benefits are reduced: traffic simply redistributes itself to neighbouring streets. Complete exclusion of traffic can have a dramatic impact on town centres; it has been widely adopted on grounds of safety, amenity and increased turnover for shops in pedestrianised streets – though the economic benefits are far from certain.

Hass-Klau *et al.* (1992) have noted that the British approach is focused on the reduction of accidents, whereas the objectives in other European countries are typically wider and embrace environmental improvement and urban regeneration. Their book contains detailed technical descriptions of well-established methods such as speed bumps, chicanes (kinks in a road to slow down traffic), pinch points, as well as some lesser-known techniques. It also presents an assessment of traffic calming experience in Germany, the Netherlands, Denmark and Sweden. It describes and comments on some forty British traffic calming schemes. (The term 'traffic calming' was introduced by Dr Hass-Klau as the translation of the German term *verkehrsberuhigung*.)

The traditional approach has been to segregate traffic and pedestrians: with reductions in traffic speed they can both be accommodated, but with the pedestrian instead of the car being master. The Devon County Council calming guidelines (1991) describe nineteen different measures 'to moderate driver behaviour and to exploit the potential for safety and environmental improvement'.

A major shortcoming of many traffic calming schemes is that they are essentially local in concept and operation. Rather than being parts of a comprehensive transport policy, they are typically reactions to vocal residents. As a result, the effect of calming in some areas is to move the problem elsewhere.

Indeed, Banister (1994a: 212) has suggested that 'positive responses from those living in the traffic-calmed area are more than outweighed by anger from those living in adjacent areas where traffic levels (and accidents) have increased'. It is unfortunate that calming does not simply reduce the total amount of traffic; perhaps with proper planning, it can (Tolley 1990). The same point arises in relation to parking policy, which is arguably the simplest and most effective method of reducing private car use.

Many of the measures now discussed under the heading of traffic calming have more traditionally been known as traffic management, though the concern is now with wide environmental and amenity issues as well as with traffic flow. This is becoming an increasingly sophisticated area of policy. Additional legislation is a testament to the importance now attached to it: the Traffic Calming Act 1992 extends the statutory provisions for 'the carrying out on highways of works affecting the movement of vehicular and other traffic for the purposes of promoting safety and of preserving or improving the environment'.

TRAFFIC MANAGEMENT IN LONDON

The Road Traffic Act 1991 provided a new legislative framework for traffic management for London. The Act empowers the Secretary of State to designate a network of priority routes (commonly known as *red routes* because of their distinctive red road markings and signs) which are subject to special parking and other traffic controls. They are aimed at reducing traffic congestion and improving traffic conditions on main routes, particularly for buses, without encouraging additional car commuting into central London. A pilot scheme in 1991 proved successful: overall journey times improved by 25 per cent, bus journey times were reduced by more than 10 per cent, and reliability increased by one-third. More people used buses; and road casualties fell significantly. As a result, a permanent scheme was introduced in 1992: 500 km of red routes have been designated throughout the capital. This network covers all trunk roads in London as well as local roads which are of strategic importance.

The planning, coordination and implementation, maintenance and monitoring of traffic management on the network is the responsibility of an *ad hoc* (non-departmental) body, the Traffic Director for London. This was also established by the 1991 Act. The network plan forms the framework for detailed local plans which are the responsibility of the London local authorities.

Central to this new system is a package of traffic management schemes and a reform of on-street parking in Greater London. Traffic management measures include increased priority for buses, improved pedestrian crossings, enforcement of parking regulations, and encouragement to cyclists to use roads other than red routes except where separate cycle tracks can be provided. A range of traffic calming measures are being implemented on side roads which might be affected by the red route traffic. These regulate speed and deter motorists from using side roads as 'rat runs'. The DoT recommends the introduction of 20 mph zones to discourage through traffic. In such zones, there is the additional advantage that the road hump regulations are far more relaxed: for example, warning signs are not required. Doubts about the legality of traffic calming measures have been settled by the Traffic Calming Act 1992 which provides for the making of regulations governing them.

In addition to other parking restrictions, the 1991 Act provided for a new system of *special parking areas*. In these areas, which are designated by London local authorities, parking contraventions are no longer criminal offences and traffic wardens are replaced by local authority *parking attendants*. The powers of these attendants include issuing *penalty charge notices* and authorising wheel clamping and the removal of vehicles. A *joint parking committee*, appointed by the London local authorities, has the duty of setting certain parking charges and appointing *parking adjudicators* who have a comparable role to magistrates' courts under the former system of criminal parking controls. (The new system is technically known as decriminalised parking enforcement.)

These and similar provisions amount to an elaborate new system for traffic management in London. (Local authorities outside London can apply to the DoT for the powers to be extended to their areas). They constitute a new, and perhaps final, attempt to make regulatory controls work. If they prove inadequate, some form of road pricing will probably be necessary. This has been resisted up to now, partly because of technical problems, but mainly because of the fear of a political backlash. The fear is that the public would be opposed to such a measure or, at least, that the public is 'not yet ready' for it. This issue of public attitudes is worth examining further.

PUBLIC ATTITUDES

Measures such as those outlined above would have been unacceptable without public support for stronger controls. This is a crucial factor in transport policy. It is also one that changes over time, particularly as the impacts of increased traffic are experienced. Yet there is a very real problem in reconciling private and public interests. Each car owner regards congestion problems as being created by *other* motorists; the individual's contribution to the total is negligible. This zero marginal cost for the individual imposes high costs on the collectivity of users, but car users have no incentive to economise in their use of road space: to them it is a free good.

The car can be more than a means of transport. It can be an extension of a driver's personality, a symbol of affluence or power, an object to be loved as well as used. The 'love affair' with the car is, however, under strain: mass ownership and use has made it less appealing than it was (Goodwin *et al.* 1991: 144). Whether the disenchantment has gone far enough to warrant more-penal methods of controlling its use is the basic political question. Recent surveys are helpful in showing the nature of public opinion and the scope that might exist for radical changes in policy.

In a 'poll of polls', P.M. Jones (1991b) summarised the major findings of ten surveys carried out between 1988 and 1990. First, there is no doubt about the general realisation of the seriousness of the problem of traffic congestion, but most people would be reluctant to reduce significantly their dependence on car use: despite the congestion, it still remains the favoured form of transport among car owners. Nevertheless, there seems to be an increasing willingness to consider switching to public transport for some types of journey *if public transport were better*. In looking at public transport as an alternative to the car, quality of service is much more important than the level of fares. (But the converse seems to be the case for car use: higher petrol prices are more significant than a doubling of journey times.) There is support for an increase in the range and quality of alternatives (or supplements) to car use and for more effective parking restrictions, but little backing for road pricing. Londoners, however, are more favourably inclined towards road pricing, and Jones suggests that 'the government is being unduly cautious in not proposing a road pricing scheme for London'.

Attitude surveys, of course, are not necessarily a good guide for policy-makers. A measure may be popular but ineffective: 'indeed its popularity may lie precisely in its ineffectiveness and lack of impact on car-based life styles. Some drivers may support better public transport, for example, because they believe that *other* drivers will use it, and so clear the roads for them.'

In another study commissioned by the Oxford Transport Studies Unit (not covered by Jones's review), Cullinane (1992) noted the extent of car-dependence: about a half of households in the survey perceived a car to be essential to their lifestyle and a further 13 per cent would not want to be without one. She also noted, however, that a quarter of households did not have a car and had no intention of getting one. The overall conclusion was that car dependence was increasing, and that, if things are allowed to continue as they are, it would become increasingly difficult to persuade owners to reduce their car usage.

Cullinane's study included that idiosyncratic British institution for the privileged: the company car. Company-assisted motoring increased dramatically during the 1970s, to the point where 60 per cent of new car registrations are attributable to companies and the self-employed (TEST 1984). The government has recognised the importance of the issue in reducing the tax benefits of company cars, though critics, including the Royal Commission on Environmental Pollution (Cm 2674, 1994: ch. 7) have argued that the changes do not go far enough. The government disagrees (*Transport: The Way Forward* 1996: 52).

Although a company car was perceived to be a necessity by three-quarters of those who had one, Cullinane suggests that 'there is some reason to believe that they are in some sense an unnecessary addition to the national park of cars'. Over a quarter of these drivers say that they would not replace their company car with a private one if it was withdrawn from them, and a third said they would reduce their mileage if they no longer had a company car.

Reflecting on the survey as a whole, Cullinane concludes that congestion seems likely to increase:

> and that there will be some voluntary reduction in traffic as the problems intensify. However, the level of attachment of most people to their car is such that it will take some positive action from outside to force any real reduction in traffic, and this positive action will have the most impact if it hits people's purses.

A number of studies have pointed in the same direction. Cars are highly valued by those who can afford them, but the problems of congestion are becoming increasingly burdensome. There is support for better public transport and, though no massive changeover by car users is to be expected, 'all changes take place at the margin'. Goodwin *et al.* (1991: 147) suggest that 'policies that would have been very difficult to implement successfully only a few years ago, may now be more popular'. Though it is very apparent that motorists do not like the idea of road pricing, there is good evidence that they would accept it (reluctantly) if it were part of a package which provided them with some offsetting benefits, particularly in the form of good public transport. This is perhaps the most important finding of research both in Britain and abroad. Cervero (1990) has reviewed North American studies of transit pricing and concludes:

> For the most part, riders are insensitive to changes in either fare levels, structures, or forms of payments, though this varies considerably among user groups and operating environments. Since riders are approximately twice as sensitive to changes in travel time as they are to changes in fares, a compelling argument can be made for operating more premium quality transit services at higher prices. Such programs could be supplemented by vouchers and concessionary programs to reduce the burden on low-income users.

Goodwin (1989) underlines the point in his 'rule of three'; this useful article deals succinctly with a difficulty about road pricing which is frequently not addressed by its advocates. Though it is simple to demonstrate that road pricing increases total economic efficiency (since traffic will use the priced road only when the benefits are greater than the costs), it is also true that those who no longer use the road are losers. To make road pricing equitable and attractive, it is necessary to provide benefits to non-users; and this can be done with the revenues that are collected from pricing. Goodwin divides these benefits into three groups: environmental improvement; traffic attracted to the less congested, faster, priced road; and the increased speed of all traffic on the priced road. The revenue raised by road pricing can be allocated in a similar three-fold way: to general purposes; to new roads; and to the improvement of public transport. Such an approach is attractive since it can satisfy a range of interests – which road pricing by itself does not.

One further issue is selected for mention here: the effect of localised road pricing on land use. It is apparent that, if pricing were restricted to certain areas (e.g. within the area bounded by the M25) this would provide an incentive to relocation outside the priced area. It could also lead to a more dispersed pattern of origins and destinations. An immediate response is that this is precisely what

can be expected to happen with a high degree of congestion in a central urban area. But the implications are also the same in either case: first, transport policies should not be divorced from land use planning policies. Transport is the linkage between land uses, and both should be planned in a coordinated way (to the extent that this is possible). At the least, policies in each area should take account of the context of, and the implications for, the other. An important land use planning objective ought to be to minimise the need to travel (as indicated in PPG 12). Second, there is a case for road pricing to be extended over the whole road system, not merely the 'congested' parts (Hillman 1992: 228). If this were politically feasible, road taxes might replace some or all current transport taxes. This, of course, raises some much wider issues than are being discussed here.

Even the most ardent supporter of road pricing admits that there are many unknowns in the matter. Theoretical studies may be suggestive, but many of the issues are empirical – or at least need empirical testing. Unfortunately, as May (1986: 120) has pointed out, it is one of the dilemmas of research on traffic restraint that, although empirical evidence is needed, this is difficult to obtain since governmental authorities are unwilling to experiment without adequate predictions. This is a case where both academic and political considerations call for more research. The bibliography of research reports seems set to expand.

At the time of writing, transport policy in the UK is in the doldrums. The clear and explicit policy of catering for the motor car by an extensive road building programme has been replaced by uncertainties and doubts. It seems unlikely that these will be readily settled. Though there is a general agreement about the need for coordination of transport planning with land use planning, it is unclear how effective this could be – or even what it means.

Impact of Transport Policies in Five Cities

There is a wealth of experience on the impact of different types of transport policy on road traffic, but there is a bewildering range of possibilities. An indication is given by a report from the Transport Research Laboratory (Dasgupta *et al.* 1994). This investigated the effects of various policies on urban congestion in five cities (Leeds, Sheffield, Derby, Bristol and Reading). It was found that halving public transport fares increased bus use by between 7 per cent and 20 per cent, but the proportionate effect on car use was slight: only 1 to 2 per cent. Other options examined included raising fuel costs by 50 per cent, doubling parking charges, halving the number of parking places, and applying a cordon charge of £2 in the peak and £1 in the off-peak period. The latter two measures had the greatest effect: they reduced car use in the central areas by about one-fifth (and increased it in the outer area by between 3 and 5 per cent). Different types of policies have different effects, but they also vary among cities and between peak and off-peak periods. The study concludes that, when interpreting the results, it is important to take into account the complicated interrelationships among 'modal transfer', redistribution, changes in vehicle-km and changes in trends.

(Dasgupta *et al.* 1994)

FURTHER READING

The main sources of transport statistics are DoT *Transport Statistics for Great Britain*, and ONS *Social Trends*. Both are annual publications.

The Buchanan Report (*Traffic in Towns*, 1963) is historically a landmark, but see also his earlier (1958) *Mixed Blessing: The Motor Car in Britain*. Truelove (1992) *Decision Making in Transport Planning* provides an excellent overview of transport policies and politics. Banister (1994a) *Transport Planning in the UK, USA, and Europe* has a broader canvas, with some comparative analysis; it also has a useful bibliography. See also *Transport, the Environment and Sustainable Development*, jointly edited by Banister and Button (1993). Other general works on transport include Bruton (third edition 1985) *Introduction to Transportation Planning*. Downs (1992) takes a stimulating and iconoclastic approach in his *Stuck in Traffic*.

An influential report by Goodwin *et al.* was published in 1991: *Transport: the New Realism*. See also Goodwin (1995) *The End of Hierarchy? A New Perspective on Managing the Road Network*. The report of the Royal Commission on Environmental Pollution (1990) *Transport and the Environment*, and the various reports of the Standing Advisory Committee on Trunk Road Assessment (SACTRA) are referred to in the chapter.

Road and Congestion Charges

The amount of writing on charging road users is in striking contrast to the amount of action. The number and character of government and parliamentary reports on the issues is testament to the political difficulties involved.

The Smeed Report (1964) is an early milestone, but it was not until the end of the 1980s that it became clear that government had to take some hard decisions (or avoid them, and take a decision by default). At first, attention was more focused on attracting private sector involvement in the provision of roads (*New Roads by New Means: Bringing in Private Finance* (Cm 698), 1989). Attention soon switched to more pressing matters of congestion with the 1993 Green Paper *Paying for Better Motorways*, which was followed by a HC Transport

Committee (1994) report *Charging for the Use of Motorways*, and the governmental response (1994) *Government Observations*. The Transport Committee launched a wider inquiry later that year: (1995) *Urban Road Pricing*. The government has commissioned a number of studies in this and related fields: see, for instance, MVA Consultancy (1995) *The London Congestion Charging Research Programme: Principal Findings*. Outside Parliament, a succession of reports emerged from various interest groups such as the Chartered Institute of Transport and the London Boroughs Association (see below). A succinct survey of the field is provided by Lewis (1994) *Road Pricing: Theory and Practice*.

See also Chartered Institute of Transport (1990) *Paying for Progress: A Report on Congestion and Road Use Charges*, and (1992) *Paying for Progress: A Report on Congestion and Road Use Charges: Supplementary Report*; Grieco and Jones (1994) 'A change in the policy climate? Current European perspectives on road pricing'; Jones (1991b) 'UK public attitudes to urban traffic problems and possible countermeasures: a poll of polls'; London Boroughs Association (1990) *Road Pricing for London*; Nevin and Abbie (1993) 'What price roads? Practical issues in the introduction of road-user charges in historic cities in the UK'; Newberry (1990) 'Pricing and congestion: economic principles relevant to pricing roads'.

Freight Transport

Dealing specifically with freight transport are the Armitage Report (1980); Wardroper (1981) *Juggernaut*; National Audit Office (1997) *Regulation of Heavy Lorries*; Plowden and Buchan (1995) *A New Framework for Freight Transport*; Royal Commission on Environmental Pollution (1994) *Transport and the*
Environment (Cm 2674) (ch. 10); DoT (1996) *Transport: The Way Forward* (Cm 3234) (ch. 15).

Traffic Calming and Management

DoE (1995) *PPG 13: A Guide to Better Practice: Reducing the Need to Travel through Land Use and Transport Planning*; DoT (1995) *Better Places Through Bypasses: Report of the Bypass Demonstration Project*; Hass-Klau (1990) *The Pedestrian and City Traffic*; and Hass-Klau *et al.* (1992) *Civilised Streets: A Guide to Traffic Calming*.

Details of the traffic management system for London are set out in DoT (1992) *Traffic in London: Traffic Management and Parking Guidance* (Local Authority Circular 5/92). On the extension of special parking areas to local authorities outside London, see DoT (1995) *Guidance on Decriminalised Parking Enforcement outside London* (Local Authority Circular 1/95). The government's *Transport Strategy for London* was published in 1996. In *Speed Control and Transport Policy* (1996) Plowden and Hillman explore the potential for speed limits as a policy instrument.

Cycling

DoT (1996) *The National Cycling Strategy* is the most important official document published in this field: it is obtainable free from the DoT. See also McClintock (1992) *The Bicycle and City Traffic*, and the HC Transport Committee's (1991) evidence on *Cycling*. A list of official reports from the DoT and other government departments and agencies is given in the appendix on *Official Publications*.

12

PLANNING, THE PROFESSION AND THE PUBLIC

Publicity is the greatest and most effective check against arbitrary action.

Margaret Thatcher 1990

Democracy means that citizens have a significant influence over what happens, have equitable rights to exercise influence, and are entitled to know why policies have been adopted and that action taken is in line with these policies.

Patsy Healey, 1991b

PLANNING AND POLITICS

Throughout the 1950s and 1960s, planning proposals were generally presented to the public as a *fait accompli*, and only rarely were they given a thorough public discussion (Davies 1972; Dennis 1972). Though there was machinery for objections and appeals, this quasi-judicial process was devised only for specified uses by a restricted range of interested parties. As will be shown later in this chapter, it was modified in its operation in response to increasing public pressures, but the general attitude to this system was (and remains) less favourable than to the normal judicial system with which it is frequently, though inappropriately, compared.

The lack of concern for public participation was a result partly of the political consensus of the postwar period and partly of the trust that was accorded to 'experts' – which, by definition, included professionals. The time was perceived to be one of rapidly expanding scientific achievement, and the methods that had made such progress in the physical sciences

were thought to be transferable to the problems of social and political organisation (Hague 1984). This, together with the advent of new social security, health and other social and public services, led to a rapid growth in the professions and the bureaucracies in which they worked. Town planners, though having identity problems which took many years to settle, had a good public image: they were to be the builders of the Better Britain which was to be won now that the military battle was over. In the same spirit as the established professions, they sought to establish a strong, scientific and objective knowledge base. Armed with the right techniques in manipulating the environment, they were to address the physical problems of the nation and, at least by implication, the underlying social and economic forces.

In retrospect, the approach implied a depoliticising of issues which were later appreciated to be of intense public concern. This was further obscured by professional techniques and language which the public could not be expected to understand (Glass

1959). Of course, planners were not alone in this: on the contrary, they simply took the same stance as other 'disabling professions' – to use Illich's term (1977). At the time, however, the lack of political debate and participation was not widely recognised as a problem. Professionals were perceived as acting in everyone's interest – the general public interest.

It was in the 1960s that these ideas were effectively challenged in the UK, closely following experience in the USA. (See, for example, Broady 1968 on the UK and Gans 1968 on the USA.) By this time, the political consensus had broken down, and there was widespread dissatisfaction both with the lack of access to decision-making within government and to the way in which benefits were being distributed. Though it claimed to serve the public interest, the planning system began to be seen as an important agent in the distribution of resources – frequently with regressive effects (Pickvance 1982). The idea of an objective, neutral planning system was increasingly recognised to be false.

In particular, the physical bias of the planning system had failed to address social and economic problems: perhaps it even made them worse. There was growing concern for a new type of 'social planning' which would seek to redress the imbalance in the access to goods, services, opportunities and power. To achieve this, some saw the need for 'advocacy planning', which would provide experts to work directly with disadvantaged groups.

This critique has had consequences for planning practice of greater permanence than that achieved by the intellectual arguments themselves. Changes were made in the statutory planning procedures, and consultation and participation gradually became an important feature of the planning process. The more open system was generally acclaimed by the profession, though there were critics who questioned underlying assumptions. Neo-Marxists drew attention to the more fundamental divisions of power in the political and economic structure of capitalist society. In practice also, extensive participation exercises produced only limited gains, and some advocate planners were amongst the first to reject the approach for its weaknesses. The critics argued that,

like all 'agents of the state', planners operate within a 'structural straitjacket' and, irrespective of their own values, will inevitably serve the very interests which they are supposed to control. This was supported by research findings which demonstrated that planning had operated systematically in the interests of property owners. There was also substantial theoretical work concerned with the role played by the planning system in the interests of capital.

These critiques were powerful but, by their very nature, they could offer little guidance to planners working in a professional, politically controlled system. Indeed, how could they respond to allegations that a fundamental purpose of planning in society is the legitimation of the existing order? If participation merely supports a charade of power-sharing, leaving entrenched interests secure, what alternatives do planners have? They are, of course, used to unfair criticism for things that go wrong: the postwar housing estates that were built as quickly as possible (and with few resources left over for 'amenities'); the motorways that were belatedly built to cope with the great increase in traffic congestion, but which destroyed the social and physical fabric of towns; and the participation processes that raised hopes which were dashed by the outcomes. In Ambrose's words (1986), planning was the scapegoat.

More radical approaches have emerged. Popular planning aims

> to democratise decision making away from the state bureaucrats or company managers to include the workforce as a whole or people who live in a particular area . . . empowering groups and individuals to take control over decisions which affect their lives, and therefore to become active agents of change.
>
> (Montgomery and Thornley 1988: 5)

Not surprisingly, the few examples of such practice are to be found in the left-wing strongholds of the former GLC and a small number of the other metropolitan district councils, and with only limited success. Decentralisation of decision-making has only been effected (and perhaps only *can* be effected) on a very limited basis.

Interestingly, popular planning approaches have

generally not been innovative in their methods of establishing community needs. Whilst the language is one of empowerment, they have generally involved the usual mixture of meetings, publicity and leaflets. Few community development projects have managed to achieve a real strengthening of a community's social resources and management capabilities to allow for devolution of real power to local people; rather, they have generally focused on improving the resource base of communities – an admirable, but different objective (Thomas 1995).

THE PLANNING PROFESSION

In 1996, there were 17,220 members of the Royal Town Planning Institute, of whom 13,630 were corporate members; three-quarters were in the public sector (with about two-thirds in local government), and about a fifth were in private consultancies and the development industry. Half of the local government members are in the non-metropolitan district councils (Nadin and Jones 1990: 18).

A 1993 survey of planners employed in local government shows that only about half of the professional staff employed in planning are members of the Institute (LGMB and RTPI 1993). The others are either professionals who are not eligible for membership, including members of other professions, such as architects who are working principally on planning matters, or students working for qualifications that lead to membership. If it were assumed that a similar balance of Institute members exists in the rest of the planning field, in the private and voluntary sectors and in education, the total 'professional body' in its widest sense is more than 35,000 strong. In addition, there are support staff who, in 1991, totalled over 8,000 in local government and an equal number elsewhere. In total, therefore, there is a planning workforce of some 50,000.

The overwhelming majority of the membership is young, white and male. Women and ethnic minorities are under-represented. Even in the younger age groups, women account for only about 30 per cent of members. At the higher levels, women are very poorly represented. For example, in the local government survey, only 1.6 per cent of chief planning officers are women. By contrast, women occupy two-fifths of junior posts and over four-fifths of part-time posts.

In a 1988 survey of Institute members, only seventy UK-based respondents fell outside the category of 'white European', the highest proportion of these being in the London boroughs. Not surprisingly, the dominant activities undertaken by planners are in development control and development planning, though they are engaged in a very wide range of other jobs.

The planning profession is thus well established in Britain, and the RTPI is easily the largest professional planning body in Europe. How this position has been reached in the years since 1913 when the Institute was established is worthy of a note of explanation. Under the Royal Charter, the Institute is responsible for the education of prospective planners, and has the power 'to devise standards of knowledge and skills for persons seeking corporate membership of the Chartered Institute . . . '. The Institute held its own examinations for membership until 1992 but, from the earliest days, specific courses were set up to train planners, the first being at Liverpool University in 1909 followed by University College London in 1914 (Batey 1993; Collins 1989). By 1945, there were nine courses in town and country planning, all of which were postgraduate. Student numbers in town and country planning increased sharply to 959 enrolments on fifty-four separate courses in 1978 (Thomas 1990b, 1990c). By 1981, the number of courses had increased to fifty-seven, including twenty at the undergraduate level.

Despite a spirited defence by the planning schools and the RTPI, some courses were forced to cease recruitment during the 1980s following a government-inspired review of *Manpower Requirements for Physical Planning* (Amos *et al.* 1982). However, the impact of the closures was reduced because of increased intakes in the other schools. In 1988,

766 students were recruited onto thirty-one courses. There was a dramatic increase in student numbers in the following years. This was, in part, a result of the 1988 Education Act which promoted much wider participation in higher education generally. Additionally, the boom years at the end of the 1980s increased intakes of graduates with other qualifications into postgraduate planning courses. In the year 1995–6, 988 students enrolled on planning courses in the twenty-three recognised schools and the Joint Distance Learning Course. The total student body was 3,195.

The Institute visits accredited schools at least once every five years and publishes guidelines setting out policy on the education of planners, together with a core curriculum. Students completing accredited courses are, after two years' practical experience, eligible for corporate membership of the Institute. Despite pessimistic forecasts in the 1980s, the demand for planning graduates has increased significantly over the last fifteen years, but demand has not kept up with increasing supply. At the time of writing, there is a shortage of opportunities for planning graduates and concerns about the large number of accredited courses.

INTERESTS IN PLANNING

'Interests' in planning are usually thought of in terms of the organisations, groups and individuals who are actively engaged in the planning arena: they are identified by their participation in the land development and planning processes. These include land and property owners, developers, special interest groups, national government and its agencies, and local authorities themselves (in both their land-owning and regulatory capacities). Some organisations have become particularly skilled in presenting their views at both national and local levels. The House Builders' Federation, for example, is an important national organisation that regularly presents evidence at public inquiries in support of the interests of the housebuilding industry. It has a special place in the planning process by virtue of

its role in the land availability studies discussed in Chapter 6. Voluntary organisations operate in a variety of ways, from lobbying to education, and from active participation in planning processes to the ownership and management of protected lands. Membership of such organisations has increased significantly. Between 1971 and 1994 membership of the Royal Society for the Protection of Birds increased from 98,000 to 870,000; that of the National Trust from 278,000 to 2.2 million; and the relative newcomer, Friends of the Earth, increased from a mere 1,000 to 112,000 (*Social Trends* 1996, Table 11.4).

National voluntary organisations obviously do not command the resources available to commercial interests, but they do employ experts and they can be very effective in promoting their causes. Some of them have become increasingly sophisticated and can appropriately be described as 'major elites' (Goldsmith 1980). There are innumerable 'minor elites' of small groups who become involved in an *ad hoc* way with particular issues. The evolution of participation and consultation in planning has favoured these self-defined elites, and given them a relatively privileged position in the planning process. It has long been appreciated that such groups are not necessarily representative of anything wider than the interests of their active supporters. Many people are not able or willing to take the time to engage in 'participation', and some groups who have a clear stake in planning outcomes (such as home buyers or job seekers) are too diffuse to have become effective participators, and 'rarely if ever emerge as definable actors in the development process' (Healey *et al.* 1988: ch. 7). In the following sections, discussion focuses on three 'interests' that permeate almost all planning activity – race, gender and disability – yet which are rarely represented in disputes over particular development projects.

RACE AND PLANNING

Questions of equal racial opportunities have figured prominently on the town planning agenda over the last twenty years, though with questionable impact on practice. The Race Relations Act 1976 places a duty on all local authorities to eliminate unlawful racial discrimination in their activities and to promote 'good relations between persons of different racial groups'. A joint working party of the RTPI and the Commission for Racial Equality (CRE) reported in 1983 on *Planning for a Multi-racial Britain*. This was a frank assessment of the inadequacies of the then current thinking on race and planning. It is perhaps only a little less relevant today, and its recommendations apply to all planners, wherever they work.

The working party's deliberations were spurred by the deteriorating race relations in Britain's cities, and by numerous reports calling for more action from central government and other bodies (Scarman Report 1981). These and other studies demonstrated that:

> Black people in Britain share all the problems of social malaise and multiple deprivation with their white neighbours. In addition, however, they experience the additional difficulties of racial discrimination and disadvantage, because of the colour of their skin.
>
> (CRE 1982: 9)

The RTPI/CRE report argued that, despite its record of innovation in participation, the profession had failed to address the issue of race. Indeed, the RTPI's own 1982 report on participation omitted any explicit consideration of the racial dimension. Ethnic minorities were mentioned only once, with reference to the need to provide interpretation at public meetings. *Planning for a Multi-racial Britain* sought to improve sensitivity to racial issues 'and show how planning practice can be modified to avoid racial discrimination, promote equality of opportunity, and improve race relations for the benefit of the whole community'. It identified three elements in the racial dimension of planning:

1 reviewing the impact of current policies, practices and procedures upon different racial groups, with a view to ascertaining actual or potential racial discrimination;

2 building racial distinctions into surveys, analyses and monitoring with a view to identifying the special needs of different racial groups; allowing the impact of policies to be assessed, and providing a basis for any appropriate positive action;

3 positive action in planning policies, procedures, standards and decision-making, partly by directing positive non-racial policies and actions towards groups containing high proportions of black people, and partly by taking special steps to ensure that black people have equal access to the benefits offered by town planning.

The report also considered how the profession could be made more representative in terms of its membership, and the implications of this for the education of planners. It became an important guide for committed practitioners and teachers, and formed the basis for other more detailed guidance, foremost of which was the GLC's *Race and Planning Guidelines*. Its publication was squeezed in just before abolition of the GLC (when perhaps the most vigorous supporter of equal opportunities was lost).

The GLC report, however, has subsequently provided a rich source of material for guiding good practice in the London boroughs and elsewhere. Its recommendations illustrate the inadequacy of the 'colour-blind' approach identified as commonplace in the RTPI/CRE report. Such an approach, though holding that all people should be treated equally, is insensitive to the cultural traditions and needs of particular ethnic groups. Unintentional discrimination can therefore result. It is perhaps surprising, therefore, that Thomas and Krishnarayan (1993) are able to note the almost complete absence of reference to racial matters in major planning reports such as the Audit Commission's *Building in Quality* and the DoE's *Development Plans: A Good Practice Guide*. Despite evidence of innovative practice in a few places (Best and Bowser 1986), planning has made only a weak contribution to challenging racial discrimination and disadvantage. Thomas and Krishnarayan (1994) explain this with reference to the 'socially conservative' nature of planning: the tendency to focus on technical problems of land use management rather than radical social goals which has been strongly reinforced by professionalisation, the narrow interpretation of the objectives of planning by government and the courts, and the reproduction of existing social and spatial divisions in planning policy. The result is that the outcomes of planning intervention 'tend to reflect existing patterns of social and economic disadvantage', a comment which applies equally to the issues of gender and disability.

PLANNING AND GENDER

The absence of policy explicitly related to the concerns of women has attracted little attention until recently. One of the many reasons for this is the inadequacy of the classic texts on social theory and urban studies. In reviewing these, Greed (1993: 233) writes:

> Women appear in studies of working class communities as a variety of oversimplified stereotypes, based on observing them as mono-dimensional residents tied to the area rather than as people with jobs, interests and aspirations beyond its boundaries. Young and Willmott give emphasis to women in their study, but their fondness for seeing them in the role of 'Mum', as virtually tea machines, and almost as wall paper to the main action of life, is open to question.

Greed has elsewhere (1994) described the male domination of the profession as 'only a temporary intermission'. Women were primary contributors to the social movement which promoted town planning at the turn of the century. She argues that the professionalisation of planning, its institutionalisation within the government structure and the limited access to qualifying courses has 'kept most women out'. The gender bias reflects, and in part perpetuates, the patriarchal structure of British society, and continues to influence the recruitment and education of planners.

An illustration of the gender bias is the way in which some issues are defined in land use terms, whereas others are labelled 'social'. Thus, the provision of sporting facilities and the open space standards applied to them are routinely regarded as legitimate land use matters. These predominantly male activities can be contrasted with crèches which are commonly regarded as social issues, even though they 'may have major implications for central area office development' (Greed 1993: 237).

Awareness of gender issues grew strongly in the 1980s, and was reflected in important and influential reports from the GLC (1986) and the RTPI (1987, 1988). It is less clear whether such efforts have had a significant impact. One difficulty (in addition to the power of traditional attitudes and ways of thinking and perceiving) is the general lack of explicit social policies in plans. There has long been argument about the extent to which land use plans should incorporate social issues, and the DoE has been generally opposed to them. A survey of development plans found that a number of local authorities had had some success in incorporating the special needs of women in their development plans, even weathering DoE scrutiny (L. Davies 1993). But this has not been taken up widely, and where relevant policies have been included they have often been deleted following requests from the DoE after plan scrutiny, or in response to inspectors' reports after the inquiry. Davies (1996) notes the removal of a policy from the Hammersmith and Fulham UDP which required sheltered lockable spaces for baby buggies in new large-scale shopping developments.

Nevertheless, the Development Plan Regulations provide some leeway in that they require local authorities 'to have regard to social and economic considerations'. Furthermore, although PPG 12 does not explicitly refer to 'women' or 'gender', it does advise that:

> Authorities will wish to consider the relationship of planning policies and proposals to social needs and problems, including their likely impact on different groups in the population, such as ethnic minorities, religious groups, elderly and disabled people, single parent families, students, and disadvantaged and deprived people in inner urban areas.

Concern for the needs of women is generally less articulated than those of race or disability, especially outside London where some authorities simply 'didn't consider this an issue' (Davies 1996). Even where they are, 'child care' is the predominant 'women's' issue'; reflecting the assumption of women's primary role as carers.

As Greed (1994: 176) argues, to take on the gender issue would require a more fundamental rethinking of planning policy and action:

> Planning for women inevitably means altering planning for men and may be seen as an all-inclusive approach which affects everyone and every aspect of urban form and development, and not as a limited, marginal little policy adoption for a small group in society.

PLANNING AND PEOPLE WITH DISABILITIES

People with disabilities have received more attention from the planning system. The 1990 Planning Act requires local authorities to draw to the attention of planning applicants the need to consider the requirements of the Chronically Sick and Disabled Persons Act 1970, and the Building Regulations require 'access provision' in new buildings (extended in 1992 to include access for people with sensory impairments). Disability in this sense has a broad definition including (as the RTPI's Practice Advice Note 3 points out) those who suffer breathlessness or pain, who need to walk with a stick, are partially sighted, have difficulty in gripping because of arthritis, or are pregnant. Access policies can also generally make life easier for parents with children and the elderly. As Gilroy (1993: 24–5) notes, most people are or will be physically impaired at some time during their life. That impairment or disability can be turned into a major handicap by an unaccommodating environment.

Despite this explicit concern, the needs of people with disabilities have been considered largely in terms of the building design rather than of planning. The main emphasis is on access requirements. Important though this is, it leads to an overly simplistic stereotyping of the problems faced by individuals with disabilities. Thomas (1992: 25) gives a strong critique of current attitudes:

> The 'regs' can become a checklist which defines the needs of disabled people, ignoring, indeed disallowing, the possibility that individual professionals dealing with particular cases need to learn from the experience of disabled people themselves. The British legislation which relates specifically to planning with its references to practicality and reasonableness, reinforces a strand in planners' professional ideologies which emphasises the role of the planner in reaching optimum solutions in situations involving competing needs or interests. Thus might a fundamental right to an independent and dignified life be reduced to an 'interest' to be balanced against the 'requirements' of conservation or aesthetics.

Davies (1996) concludes that policies related to access for the disabled (and to women's special needs) are now finding their way into plans, especially in the London boroughs. She is, however, critical of central government advice which, although often only very recently updated, fails to give the necessary impetus. The DoE *Good Practice Guide*, for example, is described as 'woefully lacking'. She also observes that, whereas negotiations between applicants and planners have often strengthened plans on these matters, further objection by other interests and scrutiny by the DoE later in the process have resulted in a 'watering down of policies'. Above all, policies apply only to new building: much of our environment (including the Portland Place headquarters of the RTPI and universities – as Greed ruefully notes: 1996: 233) presents major access problems for many people.

PUBLIC PARTICIPATION

Concern for, and even interest in, public participation has not been a particularly obvious strength of British local government. With little experience to build on, it was perhaps inevitable that a committee should be appointed 'to consult and report on the best methods, including publicity, of securing the participation of the public at the formative stage in the making of development plans for their area'. The committee was set up under the chairmanship of Arthur Skeffington (a junior minister at the Ministry of Housing and Local Government), and its report was published in 1969.

The Skeffington Report made a number of rather obvious recommendations which did not carry the issues much further. (For example, it pointed to the need for information, publicity and participation.) The mundane nature of many of the recommendations is testimony to the distance which British local government had to go to make citizen participation a reality.

Unfortunately, the report did not discuss many of the crucial issues, though passing references suggest that the committee was aware of some of them. For instance, it is rightly stated that planning is only one service, 'and it would be unreasonable to expect the public to see it as an entity in itself'. The report continues: 'public participation would be little more than an artificial abstraction if it becomes identified solely with planning procedures rather than with the broadest interests of people'. This has major implications for the internal organisation and management of local authorities (which are not discussed). So have the proposals for the appointment of 'community development officers . . . to secure the involvement of those people who do not join organisations' and for 'community forums' which would 'provide local organisations with the opportunity to discuss collectively planning and other issues of importance to the area'.

What was conspicuously lacking in this debate on public participation was an awareness of its political implications. The Skeffington Report noted that it was feared that a community forum might become the centre of political opposition: but the only comment made was: 'we hope that would not happen; it seems unlikely that it would, as most local groups are not party political in their membership'. The issue is not, however, one of *party* politics: it is one of local concerns, pressures and interests. Public participation implies a transfer of some power from local councils to groups of electors, but it was not Skeffington but the Seebohm Committee

(1968) which highlighted the tension between participation and traditional representative democracy.

An essential ingredient of effective public participation is a concern on the part of elected members and professional staffs to make participation a reality. This cannot be effective unless it is organised, but this, of course, is one of the fundamental difficulties. Though a large number of people may feel vaguely disturbed in general about the operation of the planning machine (and particularly upset when they are individually affected), it is only a minority who are prepared to do anything other than grumble. The minority may be growing but, as far as can be seen, public participation will always be restricted: 'the activity of responsible social criticism is not congenial to more than a minority' (Broady 1963). Similar conclusions have been drawn more recently after comparing practice in the UK with that in Sweden and France, where there is equal public reluctance to engage in plan-making as opposed to consideration of site-specific proposals (Barlow 1995). Edmundson (1993) has argued that the switch to a plan-led system is unlikely to spur greater general public involvement in plan-making (although it has generated much more interest from certain interests) and attention will continue to be focused on development control.

LOCAL CHOICE

Chris Patten is often credited for an about-turn in policy on local participation during his short term of office at the Department of the Environment in 1989. In fact, his predecessor, Nicholas Ridley, had already begun the shift to more local decision-making. It was during his time as Environment Secretary that a clearer national planning framework was instituted through PPGs, that local authorities were strongly encouraged to produce more plans, and that warnings were given that developers would have to bear costs if they pursued applications contrary to up-to-date statutory plans. On coming into office, Patten accelerated the rate of change. As a result of concerted opposition, much of it from within the ranks of the Conservative Party, he reversed the judgment on the proposal for a new settlement at Foxley Wood in Hampshire. His statement, under the banner *Planning and Local Choice*, made much of the opportunity for local communities to make their own decisions:

> While there is undoubtedly a continuing need for more houses, there are choices about the way we meet that need. One of the functions of the planning system is to help us identify those choices, and make them sensibly. What is more, many of the important choices are decisions which can and should be made locally, to reflect the values which local communities place on their surroundings. If the planning system works properly at the local level, there is less need for the central government decision taking – by me or by my Inspectors – which can so easily appear to the local community to attach too little weight to their views.
> (Patten 1989)

This was not a signal to local authorities that they could respond as they wished to local demands. Local autonomy was to be exercised only where it was within parameters laid down by the centre in national and regional policy statements. Plans and development control decisions still needed to be 'realistic about the overall level of provision'. Nevertheless, the statement marks an important milestone in the relationship between central and local decision-making. It was a clear move away from the previous line of argument, so closely linked to Nicholas Ridley's term of office, that local concerns would have to be set aside for a presumption in favour of new development.

The most recent and important developments in public participation are arising from the promotion of sustainable development through the Local Agenda 21 (LA21) initiatives of local authorities (as described in Chapter 7). The local authority is recognised as the principal actor in implementing global objectives for sustainable development at the local level and participation is central to the formulation of strategies. LA21 embraces a far wider range of concerns than planning, including, for example, green accounting and purchasing, energy conservation and recycling, but always at its heart is the objective of raising awareness and engaging citizens

in promoting good environmental practice. LA21 has become an important stimulus for promoting more creative ways of engaging local people in policy-making, some of which are finding their way into the separate arrangements for consultation on planning matters.

ACCESS TO INFORMATION

Information is power; so it not surprising that undemocratic societies guard it jealously. It is surprising, however, that secrecy is so prevalent in Britain. The fiasco over the Crossman Diaries revealed the absurdities in striking detail. British secrecy is a legacy of old styles of government. These persist in many ways, and the elitist origins are still apparent. Government is carried on in an elitist atmosphere of determining what actions are in the public interest, as viewed through the eyes of government. Participation is limited and information restricted.

Ironically, though traditional attitudes still prevail at the centre, it is the central government which has forced local government to provide greater 'freedom of information'. A major milestone in this is the Local Government (Access to Information) Act 1985, which imposes a duty on local authorities and other bodies such as PTAs 'to publish information . . . about the discharge of their functions and other matters'. The underlying principle of this Act is that all meetings of local authorities, their committees and sub-committees should be open to the press and the public. There is a power to exclude only in narrowly defined circumstances. The Act also provides for public access to agendas, reports and minutes of all meetings, and opens for public inspection certain background papers which relate to the subject matter of reports to council, committee or sub-committee meetings.

A review of the Act after ten years of operation (Steele 1995) was generally positive, finding that it had established 'minimum standards of openness and accountability'; that the provisions were exceeded by four out of five councils; and that this had been done at little extra cost. However, most local authorities do little more than make information *available*: few take positive steps to make it truly *accessible*.

A major new influence on improving accessibility is the EU, particularly with its 1990 *Directive on Freedom of Access to Information on the Environment* (Birtles 1991) This was a product of the commitment in the EC's Fourth Action Programme to enable groups and individuals to take a more effective part in protecting and promoting their interests. The Directive was brought into force in the UK by the Environmental Information Regulations 1992. The objective of the Directive is 'to ensure freedom of access to, and dissemination of, information on the environment held by public authorities and to set out the basic terms and conditions on which such information should be made available'. This has potentially significant implications for a very wide range of bodies, including central and local government departments, environmental and conservation agencies.

The regulations define environmental information as 'the state of any water or air, the state of any flora or fauna, the state of any soil, or the state of any natural site or other land', together with activities or measures adversely affecting or designed to protect these states. Organisations affected are given two months to respond to any requests for information, and there are only limited provisions for refusal to supply information, for example, where it is incomplete or subject to legal proceedings. An important provision in the regulations is the requirement for relevant organisations to produce a list of all information sources and for this to be made publicly available. The Directive, however, does not address the two most important issues: 'knowing that there is someone to ask, and knowing that there is something to ask for' (Clabon and Chance 1992: 25). Once an issue is identified, there is a major problem of *navigation* around complex information systems (Moxen *et al.* 1995).

In order to encourage an active approach by governments, the Union Treaty also enables any citizen to make a complaint to the Commission if an EU provision is not being applied. Thus the EC is cast

in the role of an ombudsman for the environment (Krämer 1991). Additional rights to 'environmental information' arise from other Directives such as that on environmental assessment. This gives the general public the right of access to any statement made; and 'people affected' have additional rights to express their views before implementation.

The use that is to be made of these rights to environmental information remains to be seen. Some evidence is available from research concerning the use of the public registers on the discharge of trade or sewage effluent under the Control of Pollution Act 1974 – a major anti-secrecy piece of legislation. From an analysis of enquiries made to three water authorities, Burton (1989: 193) concludes that 'interest to date has been disappointing, both in terms of the number of enquiries made and the range of people making those enquiries'. There are several difficulties facing the searcher for information: it must be known to exist; it must be accessible; and it must not involve unreasonable cost. Neither the water authorities nor the DoE made any significant effort on these counts. All these problems can be met, but a positive attitude, some imagination and modest resources are required: 'The bare legal requirement to maintain registers is not itself sufficient to increase public awareness of their existence and encourage their use' (ibid.: 206).

The cost issue is worth further note in respect of planning documents. Central government publications are now highly priced (presumably at market levels), as a result of which their cost is prohibitive to ordinary citizens. There is a striking contrast with many US government and European Union publications which the public can obtain free of charge. A good example has been set by the Scottish Office; it announced at the end of 1992 that it had decided to issue its planning guidance (NPPGs, Circulars and PANs) free of charge, although retaining the discretion to make a charge for PANs where appropriate. This followed the responses to a Consultation Paper: 'most consultees opposed imposing a charge for better quality publications, arguing that information on government policy should be freely available' (SoED *Planning Bulletin 6*, 1992).

With local government, a bigger problem is the paucity of publications, but the cost of planning documents tends to be very high. Taken almost at random, a list of plans published in 1996, which appeared in *Planning*, included the *Waltham Forest Unitary Development Plan* (£40); the *West Wiltshire District Local Plan* (£40); the *Bracknell Forest Borough Local Plan* (£20); and the *Epsom and Ewell District Plan* (£50 plus postage). No doubt, these expensive reports are intended for a small high-price market, but high quality of plan production is not important in relation to the needs for public information and participation. A number of authorities have printed plans in newspaper or poster format. These are models well worth copying, as is the policy of North Kesteven District Council whose plan is free, and the London Boroughs of Tower Hamlets and Barnet who gave plans away free to residents.

Another means of providing information is on the Internet. Not much use has yet been made of this, though central government has greatly expanded its information services recently. At the local authority level, the London Borough of Wandsworth has its planning applications available on the Internet, complete with scanned drawings. There will be considerable scope for more such enterprising ideas as its use spreads.

THE CITIZEN'S CHARTER

The Citizen's Charter applies six principles: setting standards and monitoring performance against them; the provision of full and accurate information about how services are run; the creation of greater choice between service providers; courtesy and helpfulness in service provision; making sure things are put right where they go wrong; and value for money. A central aim is to furnish more information to the public about the performance of service providers. In this, it is closely linked to the extension of performance indicators, first established in 1982, emerging from the Audit Commission's Report on *Building in Quality*. Although often credited to the Citizen's Charter initiative of the Prime Minister,

John Major, 'quality review' has been a steadily growing component of public, private and voluntary sector management practice.

The general concept has been taken up with enthusiasm by many organisations, and in 1996 there were forty Charters for the major public service providers alone. This can be multiplied many times over for the Charters emerging from local authorities and other smaller organisations. Amongst the main government public services for transport, for example, there is the *Travellers Charter*, the *Passenger's Charter*; the *Bus Passenger's Charter*; the *Rail User's Charter*; the *London Underground Customer's Charter* and the *Road User's Charter*.

Planning is well represented with the *Development Control Charter* and the *Planning Charter* for the entire service. There is also the *Local Environment Charter*, which addresses the right of access to environmental information held by public authorities; the right to participate in decision-making on environmental issues; and the right to seek remedies in the event of shortcomings in environmental services. (Note that the environmental charters, like many others, appear in different forms for the four countries of the UK.) Local authorities, agencies, private companies and even universities have prepared their own customer charters. Local authority planning charters are based on the model provided by the Planning Forum with local variations.

It is not clear if these charters are making any difference to the relationship between the public and the planning and environmental protection systems. On this, there are two important questions to consider. The first is the long-standing debate on the problems of measuring quality in planning. Quality is an elusive concept, but some guidance and encouragement is provided by the British Standards Institution (1991). This has been one of the factors responsible for much of the increasing interest within management circles about quality, service delivery and customers' experience. (Interest has been prompted also by the government's drive to introduce competition in the public sector.) The British Standard lays emphasis on extending control to the wide range of factors that might affect the quality of service for the customer. These include management, personnel, training, organisational objectives and resources. The thrust of the charters, however, is to focus on easily measurable facets of service delivery. In some cases, it is reduced to rather shallow measurements and the setting of targets for responding to inquiries, making decisions and the like. The value of these is limited, and their impact will depend on how challenging they are.

The second question is the extent to which the charters extend the rights of the citizen in planning matters. The clear answer is 'very little'. The charters spell out existing rights and, in so far as this helps to increase understanding, they are to be welcomed. But the relationship between citizen and government is not fundamentally changed: indeed, it is defined in a one-dimensional way: the citizen as consumer. Thus, attention is focused on aspects of information provision, consultation, complaint and redress, equal access and even empowerment. But 'the ultimate power of decision must remain with those who have acquired legitimacy thanks to the ballot box' (LGMB 1992). It is difficult to see how it could be otherwise in the provision of the planning 'service'. By its very nature, planning deals with an array of competing interests, and it involves difficult questions concerning the distribution of costs and benefits.

The RTPI has argued that the role of interest groups and the existing provisions for managing public participation in the planning system should be recognised. Indeed, it has been suggested that the statutory consultation in the planning system has provided a 'de facto Citizen's Charter for many years'. This is discussed further in the next section.

DEVELOPMENT PLANS

A major landmark in the growth of public participation was the 1968 Planning Act (and its Scottish equivalent of 1969) which made public participation a statutory requirement in the preparation of development plans.

The main stimulus for this came, not from the grass-roots, but from central government. Under the old development plan system, the Department was becoming crippled by what a former permanent secretary called 'a crushing burden of casework'. The concept of ministerial responsibility was clearly shown to be inapplicable over the total field of development plan approval and planning appeals. Not only was much of this work inappropriate to a central government department, its sheer weight prevented central government from fulfilling its essential function of establishing major planning policies. A new system was therefore required which would remove much of the detailed work of planning, including approval of local plans, from central to local government. This was provided by the 1968 Act, discussed in Chapter 4. The 1991 Act further extended the powers of local authorities for the adoption of plans to include structure plans as well as local plans.

The procedures for the adoption of plans ensure that local authorities stay within the parameters set down by central government and generally act responsibly. The procedural safeguards do this, however, at some cost of time and resources. The problem for government has been to balance the controls over local discretion with the need for a locally responsive and efficient development plan system. During the 1970s and 1980s, many local authorities avoided preparing statutory development plans, in part because they believed the costs of taking a plan through the formal procedures outweighed any benefits (Bruton and Nicholson 1983). Benefits were limited because the plan was only one consideration in development control, and might be overturned with relative ease. As a result, the legitimacy which plans provided to decision-making was limited. The failure of local authorities to keep plans up to date exacerbated this.

The perverse outcome of all this was an over-reliance on informal policy (or in some cases no policy at all), ad hoc decision-making, and consequently much less accountability. Much of the debate over 'the future of development plans' has turned on this issue. From the local authorities' point of view, it was argued that in order to facilitate the production of plans, the procedures should be streamlined, for example by removing the requirement for a public inquiry. But local authorities in general had demonstrated little commitment to plan production.

It is against this background that plans have been made mandatory and their significance in development control increased. Furthermore, central government is now playing a more active role in setting limits to local discretion and in determining the content of plans.

The full procedure for the adoption of development plans is set out in Chapter 4. The legislation has always provided only the bare bones of the system. Consultation and participation have in practice been much more than adherence to formal procedures. This is now even more the case following the Planning and Compensation Act 1991. The general effect of the changes made by this Act has been to reduce the emphasis on public participation and consultation in the statutory procedure, whilst increasing the rights of those who have made formal objections later in the process. Local authorities now have discretion to decide the appropriate publicity for individual plans. The result of this is a rather illusory reduction in the procedural steps required by statute, whilst the imperative for local authorities to involve and consult local communities in plan preparation is in reality no less than before. In contrast, the rights of formal objection after deposit of the plan have been extended by including the provision for objections where the local authority has not accepted the Inspector's or Panel's recommendations.

In Scotland, advice on consultation follows similar lines, but lays greater stress on the benefits to be gained from early consultation with a wide range of interests.

In general, however, there is no doubt that over the last twenty years there has been a shift in emphasis away from early participation in the plan-making process, towards formal opportunities for objecting to plan policies in the post-deposit stages. This has been reinforced by the increasing pressure to produce plans within a reasonable time-

scale, as well as by the sheer size and complexity of plans. Inevitably, this means that the plan preparation process will focus more on the concerns of those whose interests are most directly affected and who also have the inclination, skills and resources to participate. Whilst the procedural safeguards are in principle open to all, it is the better organised and well financed groups who are able to make most use of them.

PLANNING APPLICATIONS

Around half of a million applications for planning permission are made each year to local planning authorities in Britain, of which over four-fifths are granted. This enormous spate of applications involves great strains on the local planning machinery which, generally speaking, is not adequately staffed to deal with them and at the same time undertake the necessary work involved in preparing and reviewing development plans. Yet full consideration by local planning staffs is needed if planning committees are to have the requisite information on which to base their decisions.

The importance of this is underlined by the fact that planning committees often have remarkably little time during a meeting in which to come to a decision. Agendas for meetings tend to be long: an average of five to six minutes for consideration of each application is nothing unusual, and in some cases the time spent on an application may be much less. It cannot, therefore, be surprising that in a large proportion of cases (in the bigger authorities at least) the recommendations of the planning officer are approved *pro forma*. This may, of course, result from a harmonious relationship between councillors and officers, although there have been a number of well publicised cases where elected members have consistently acted against the recommendations of officers, as discussed below (pp. 301–2). Nevertheless, the general point still holds that members and planning officers are hard-pressed to cope with the constant flood of applications. This workload and the lack of need for discussion on many applications led the DoE to consider the formal delegation of decisions to officers. The 1968 provisions allowed local authorities to delegate decisions on a wide range of matters, although this is entirely discretionary and there are wide variations in practice.

Several important implications follow from this. First, and most obvious, is the danger that decisions will be given which are 'wrong' – that is, they do not accord with planning objectives. Second, good relationships with the public in general and unsuccessful applicants in particular are difficult to attain: there is simply not sufficient time. Third, this lack of time corroborates the view of many (unsuccessful) applicants that their case has never had adequate consideration: a view which is further supported by the manner in which refusals are commonly worded. Phrases such as 'detrimental to amenity' or 'not in accordance with the development plan' or similar, mean little or nothing to the individual applicant, who suspects that his or her application has been considered in general terls rather than in the particular detail which would naturally seem important in that particular case. And the applicant may be right: understaffed and overworked planning departments cannot give each case the individual attention which is desirable.

This, of course, is not the whole picture. For instance, individuals who may not wholly agree with a general planning principle will tend to see it in a different light when it is applied to their own applications:

> The man [*sic*] who has his home in one part of a green belt and owns what an estate agent would call 'fully ripe building land' in another part, is as vociferous in relying on green belt principles to oppose building near his home as he is in denouncing the extreme and ridiculous lengths to which those principles have been carried when he is refused planning permission on his other land, and frequently seems to achieve this without any conscious hypocrisy.
>
> (Grove 1963: 130)

This natural human failing is encouraged by the curious compromise situation which currently exists in relation to the control of land. On the one hand, it is accepted in principle (and law) that there is no right to develop land, unless the development is publicly acceptable (as determined by a political instead of a financial decision). On the other hand, though the allocation of land to particular uses is determined by a public decision, the motives for private development are financial, and the financial profits which result from the development constitute private gain. This unhappy circumstance (which is discussed at length in Chapter 6) involves a clash of principles which the unsuccessful applicant for planning permission experiences in a particularly sharp manner. It follows that local planning officials may have a peculiarly difficult task in explaining to landowners why, for example, their particular fields need to be 'protected from development'.

Nevertheless, the success which attends this unenviable task does differ markedly among different local authorities. The question is not simply one of the great variations in potential land values in different parts of the country or in the relative adequacy of planning staffs. Though these are important factors, there remains the less easily documented question of attitudes towards the public. Some local authorities make a great effort to assist and explain matters to an applicant, while others give the impression of a bureaucratic machine which displays little patience or kindness towards the individual applicant who does not understand 'planning procedures'.

PLANNING APPEALS

An unsuccessful applicant for planning permission can, of course, appeal to the Secretary of State and, as already noted, a large number do so. Each case is considered by the department on its merits. This allows a great deal of flexibility, and permits cases of individual hardship to be sympathetically treated. At the same time, however, it can make the planning system seem arbitrary, at least to the unsuccessful appellant. Although broad policies are set out in such publications as the DoE's *Planning Policy Guidance Notes*, the general view in the central departments is that a reliance on precedent could easily give rise to undesirable rigidities.

Other issues relevant to this view are the flexibility of the development plan, the wide area of discretion legally allowed to the planners in the operation of planning controls, and the very restricted jurisdiction of the courts. All these necessitate a judicial function for the department. However, this function is only quasi-judicial: decisions are taken not on the basis of legal rules as in a court of law or in accordance with case-law, but on a judgment as to what course of action is, in the particular circumstances and in the context of ministerial policy, desirable, reasonable and equitable. By its very nature this must be elusive, and the unsuccessful appellant may well feel justified in believing that the dice are loaded. The very fact that public inquiries on planning appeals are heard by ministerial 'Inspectors' (and probably in the town hall of the authority whose decision is being appealed) does not make for confidence in a fair and objective hearing.

Of course, part of the expressed dissatisfaction comes from those who are compelled to forgo private gain for the sake of communal benefit: the criticisms are not really of procedures, and they are not likely to be assuaged by administrative reforms or good public relations. Fundamentally, they are criticisms of the public control of land use – in particular, if not in principle.

THIRD-PARTY INTERESTS AND NEIGHBOUR NOTIFICATION

The rights of third parties – those affected by planning decisions but with no legal 'interest' in the land subject to decision – were highlighted in the so-called Chalk Pit case (Griffith and Street 1964). This, in brief, concerned an application to 'develop' certain land in Essex by digging chalk. On being refused planning permission, the applicants appealed to the minister, and a local inquiry was held. The Inspector recommended dismissal partly because of the impact on the neighbouring property

of a Major Buxton. The minister disagreed and allowed the appeal.

Major Buxton then appealed to the High Court, partly on the ground that in rejecting his Inspector's findings of fact, the minister had relied on further information supplied by the Minister of Agriculture without giving the objectors any opportunity to correct or comment upon it. But Major Buxton now found that he had no legal right of appeal to the courts: indeed, he apparently had no legal right to appear at the inquiry. (He only had what the judge thought to be a 'very sensible' administrative privilege.) In short, Major Buxton was a 'third party': he was in no legal sense a 'person aggrieved'. Yet clearly in the wider sense of the phrase Major Buxton was very much aggrieved, and at first sight he had a moral right to object and to have his objection carefully weighed. But should the machinery of town and country planning be used for this purpose by an individual? Before the town and country planning legislation, any landowner could develop his land as he liked, provided he did not infringe the common law which was designed more to protect the right to develop rather than to restrain it. However, as the judge stressed, the planning legislation was designed 'to restrict development for the benefit of the public at large and not to confer new rights on any individual member of the public'.

This, of course, is the essential point. It is the job of the local planning authority to assess the public advantage or disadvantage of a proposed development, subject to a review by the Secretary of State if those with a legal interest in the land in question object. Third parties cannot usurp these government functions. Nevertheless, it might be generally agreed that those affected by planning decisions should have the right to make representations for consideration by a planning committee. The present position is that third parties have an administrative privilege to appear at an appeal inquiry, but generally no similar privilege in relation to a planning application (Hough 1992; Sharman 1985).

Of course, those who object to a proposed development always have the traditional recourse to the political process, but this is of no relevance if a proposal is not known. In some areas, a small breed of energetic 'application watchers' will regularly consult the *Register of Planning Applications*, but these do not provide a general means of alerting the general public to applications (Sharman 1985).

There have been wider arrangements for publicity in Scotland and Northern Ireland for some years, particularly for 'neighbour notification'. In the Scottish system, notification is the responsibility of the applicant, who certifies to the local authority that neighbours, as well as owners and lessees, have been notified. This can be problematic for the applicant, and can lead to false certification (whether inadvertent or deliberate).

In England and Wales, there is a similar requirement for notification of owners and other interests in land which is the subject of a planning application, but this does not necessarily extend to neighbours. Provision for notification of neighbours is dealt with under the provisions for publicity. Local authorities have the responsibility of deciding, on a case-by-case basis, what type of publicity to require. Major developments, as defined in the GDPO, require *either* site notices or neighbour notification *and* a newspaper advertisement. A survey of London authorities found that direct contact by letter was most effective, although the letters used by about a third of authorities were written in a way which meant they were unlikely to be understood by two out of three people (Edmundson 1993: 13). The requirements of the GDPO have been described as 'overkill' and unnecessarily expensive, especially in the need for newspaper advertisements which have questionable effect (Harrison 1994).

This system was introduced in 1992 and replaces the former requirements in relation to specified 'bad neighbour' developments (such as sewage works, dance halls and zoos). Thus, the statutory requirements for publicity have been reduced rather than extended. Moreover, in Circular 15/92, the DoE stresses that obligations to publicise applications should not jeopardise the target of deciding 80 per cent of applications within a period of eight weeks.

Speed is thus apparently of higher priority than public participation.

In Northern Ireland, a non-statutory system requires the Planning Service to notify neighbours, but their identification (through presentation of a list of 'notifiable interests') is undertaken by the applicant. The merits of the different approaches have been compared, with the Northern Ireland system coming out on top.

The Scottish system has been found to be burdensome to both the applicant and the local authority (Brand and Thompson 1982; Thompson 1985). The procedure is a constant source of complaint, with about half the complaints to the Ombudsman relating to notification. The onus to make the notification lies clearly with the applicant, but the Ombudsman has found that 'the public expect the authority to shoulder some responsibility for the proper implementation' (Renton 1992: 43), and there must be some sympathy for this view.

A more recent study of the Scottish system (Edinburgh School of Planning and Housing *et al.* 1995) found that policing the notification efforts of applicants would only increase the workload for councils, and the effectiveness could only be increased by transferring responsibility for notification to them. Overall, it advocated the Northern Ireland approach in which responsibilities are divided.

If planning permission is granted and it later emerges that a notifiable neighbour has not been notified, the local authority cannot revoke the planning permission. The only recourse open to the third party is a private action against the applicant, but it has to be shown that failure to notify was carried out 'knowingly and with deceitful intention' (Berry *et al.* 1988: 806).

THE CHANGING NATURE OF PUBLIC INQUIRIES

Public inquiries into major planning appeals and called-in planning applications have had a stormy passage for many years, particularly those held in connection with highways and major developments such as Stansted, Windscale and Sizewell. Similar difficulties are now being experienced with local plan inquiries, which, following the 1991 changes to the development plan regime, are bigger and more keenly contested affairs.

The planning inquiry is a microcosm of the land use planning system, and it reflects many of its competing positions and underlying conflicts of interest. It is perhaps in the inquiry where the clash of ideologies is most easily seen. McAuslan (1980: 72) has used the example of road inquiries:

> The disenchantment with public inquiries into road proposals is only the most public and publicised manifestation of a general disenchantment with the system of land use planning, to which the conflict of ideologies within and over the use of the law is an important contributor. This conflict is heightened in public inquiries into road proposals because the issues of substance give rise to such sharp divisions of opinion, and because participators are making such explicit use of the inquiry for the promotion of alternative policies and versions of the public interest. What this use of the inquiry has shown is that the reforms introduced as a result of the Franks Report twenty years ago based on the principle of openness, fairness and impartiality, and concentrating on procedures did not change (perhaps were not designed to change) the overriding purpose of the public local inquiry which was and is to advance the administration's version of the public interest.

Using McAuslan's terminology, this is a triumph of 'the public interest ideology' over 'the ideology of public participation'. The important point here is that the inquiry is not an extension of public participation, but 'a limited and carefully controlled and confined discussion of specific proposals . . . inimical to the kind of wide ranging discussion that participators are demanding'. This applies equally to major planning inquiries and examinations in public (discussed in the following section).

A difficulty with many inquiries is determining where the boundaries of discussion are to be drawn: there is always the danger that argument will spill over into a broader policy framework. It is common at inquiries into particular matters for the most general questions of policy to arise. This is hardly surprising, since typically the development being debated is, in fact, the application of one or more policies to a particular situation: this readily offers the opportunity for questioning whether the policy is intended to apply to the case at issue – or whether it should. Even wider issues arise, such as the desirability of supporting a particular way of generating nuclear power, or the need for more roads, or the role of the planning system in providing affordable housing. Pressure groups which, for example, may be opposed to the building of new roads or out-of-town shopping centres anywhere, irrespective of the merits (or otherwise) of particular projects, will want to use the inquiry as a platform on which to make their wider case.

This raises the question as to whether the provisions for national policy debate are adequate. It makes sense, of course, to argue that Parliament should be the arena for the national policy debate and the local authority for debate on local policies. It also seems reasonable to maintain that it is quite inappropriate for major issues of principle to be raised when they are simply being applied locally. But issues are not so easily packaged: for instance, some fall between national and local levels and need consideration at a regional tier of government which does not exist; some site-specific proposals raise acute issues of policy which have not been settled or adequately discussed; and sometimes government may avert proper discussion because of the complexity and sensitivity of the issues involved.

There have been several proposals for the funding of third parties at major public inquiries, though they differ on the form that this should take – and the difficulties to which it could give rise. The government has taken the narrow approach that most objectors participate in public inquiries to defend their own interests, and that there is no reason why this should be financed out of public funds. In Canada, however, the Berger Commission on the Mackenzie Valley Pipeline Inquiry arranged for a funding programme which cost nearly $2 million, for 'those groups that had an interest that ought to be represented, but whose means would not allow it'. The federal government has an 'intervenor funding programme' which was used in the

Beaufort Sea environmental assessment review (Cullingworth 1987). On the basis of Canadian experience, Purdue and Kemp (1985: 685) have advocated some limited state funding on the basis that some objectors 'genuinely contribute to the wider understanding of the issues involved'.

EXAMINATIONS IN PUBLIC

In the case of development plan inquiries, the separation of broad strategic policy and detailed site-specific issues has been widely, if not completely, accepted. Whilst delegating the adoption of local plans to local authorities, the 1968 Act confirmed that the public local inquiry would continue to be used for all development plans, and that objectors would maintain the statutory right to have their objections heard. The maintenance of the rights of those affected to obtain an independent hearing was thought to be particularly important, given concerns raised in debate about the empowerment of local authorities to adopt their own plans (Bridges 1979). But this argument quickly lost ground with the realisation of the practical consequences, evidenced in the inquiry into the Greater London Development Plan (GLDP). This inquiry considered 28,000 objections over twenty-two months in the years from 1970 to 1972. The GLDP was not typical of the emerging notion of a structure plan and contained many detailed proposals, but the experience led to support for the introduction, in 1972, of the *examination in public* (EIP), a major departure from former practice. This involves a panel which considers only those matters which are selected for discussion. Objectors have no statutory right to be heard, 'effectively relieving the Secretary of State of any duty to inquire, in public, into objections made to a submitted development plan' (Dunlop 1976: 9). These changes produced a system which is the opposite of that intended by the Planning Advisory Group (PAG). The introduction of the EIP made the procedure for structure plans 'almost entirely administrative in character, being governed at almost every stage by discretion' exer-

cised, until 1992, by central government (Bridges 1979: 246).

The rationale for EIPs, however, is far more than the negative one of avoiding lengthy, time-consuming and quasi-judicial public inquiries: it is related essentially to the basic purpose and character of the plan. A structure plan does not set out detailed proposals and, therefore, does not show how individual properties will be affected. It deals with broad policy issues such as the future level and distribution of population and employment, transport strategies, any major inconsistencies within the plan, and issues on which there is unsettled controversy.

An EIP is a required part of the adoption procedure for structure plans, unless the Secretary of State decides otherwise. The issues selected are placed on deposit prior to the examination, together with a list of those selected to participate. There will usually be a preliminary meeting where the procedure and agenda can be discussed. This meeting, like the examination itself, is carried out by a panel with an independent chairman, supported by a panel secretary. The chairman has the discretion, both before and during an examination, to invite additional participants in addition to those selected by the authority, and to adjust the form of the proceedings if he or she considers this necessary. Though the Act made provision for regulations governing the conduct of the EIP, none have been made: instead the DoE published a Code of Practice which is now a joint code with that for local plan inquiries. This stresses that the proceedings should be organised so as to promote 'intensive discussion without formality'. For example, 'objections to the general policies and proposals of the plan may be heard by means of a round-table session of the inquiry, chaired by the Inspector'. It is not necessary for participants to be professionally represented, and they should not be made 'to feel at a disadvantage if they are not' (PPG 12 Annex A).

The experience of EIPs has sometimes been quite different from this ideal, with excessively formal hearings involving senior counsel acting for local authorities and others. The proceedings certainly do not lend themselves to involvement by ordinary members of the public, and in fact this is not encouraged at any stage in the structure planning process. The main participants tend to be representative interest groups, notably the House Builders' Federation and the Council for the Protection of Rural England, as well as local authorities.

This procedure clearly gives rise to difficulties, particularly since there is likely to be considerable criticism by any objectors who are excluded from participation in the examination. This is likely to become even more sensitive in England and Wales where, since 1992, the planning authority itself is responsible for adopting the structure plan, holding and paying for the EIP, and selecting issues and participants. In this connection, it is important to stress that the examination is envisaged as only one part of the process by which the plan is adopted (or in Scotland, approved by the Secretary of State). Of crucial importance in this process is the extent to which effective citizen participation has taken place in the preparation of the plan, but here, as noted above, it can be argued that the 'safeguards' have been weakened.

LOCAL AND UNITARY DEVELOPMENT PLAN INQUIRIES

The precise role of the local plan inquiry has long been a subject of debate. The procedure is a long-standing feature of British government administration, with its origin in the Parliamentary Private Bill Procedure which provided an opportunity for objections to government proposals to be heard by a Parliamentary Committee (Wraith and Lamb 1971). The procedure has grown as much by accident as design because it has been successively amended to take into account changes elsewhere in the system. Its twofold purpose has continued: to gather information for government and to provide a route for individual redress. Like appeal inquiries, local plan inquiries involve the same balancing of private and public interests through a procedure which, although essentially administrative, has many of the hallmarks of judicial courtroom practice. How-

ever, the essential nature of the planning procedure is administrative. Final decisions are taken by government, at either the central of the local level. In making these decisions, the administration may take into account matters not discussed at the inquiry. As the Franks Report (1957: para. 272) noted, the process 'must allow for the exercise of a wide discretion in the balancing of public and private interests'. The legitimacy of the decisions rests with the political accountability of the decision-maker (Parliament or the local council) rather than on the weighing and testing of evidence as in a court of law.

Whilst the Secretary of State has ultimate discretion, the procedure also attempts to safeguard the rights of the individual citizen. Some aspects of the inquiry procedure are much more akin to a judicial process. Objectors have a statutory right to appear, and the evidence is tested through a process of adversarial questioning before an independent party. There is inherent ambiguity in a system which has as its main objective the gathering of evidence to assist in the making of a governmental decision, whilst at the same time operating in the manner of a judicial hearing (Wraith and Lamb 1971). The essential dilemma of this quasi-judicial process was described in the Franks Report (1956: paras 273–5):

> If the administrative view is dominant the public enquiry cannot play its full part in the total process, and there is a danger that the rights and interests of the individual citizen affected will not be sufficiently protected . . . If the judicial view is dominant there is a danger that people will regard the person before whom they state their case as a kind of judge provisionally deciding the matter, subject to an appeal to the minister.

The difficulties have been increased by the 1991 reforms. In confirming the role of districts as the responsible authority for adopting local plans and extending the same responsibility to counties for structure plans, the administrative role of the inquiry is reinforced. In introducing the provision which allows for objections where the local authority does not accept the recommendations of an Inspector or panel, the changes lend weight to the judicial role. Thus, the central questions have changed little over the years following the Franks Committee, not least because, whilst identifying the ambiguity in objectives, the report found in favour of neither, and fell back instead on the need for balance and the need for the inquiry to be conducted with 'openness, fairness and impartiality'.

The three principles of 'openness, fairness and impartiality' have guided Inspectors with some success, and the courts have played only a small part in the planning process. Nevertheless, each of the three principles requires some qualification. The Franks Report itself recognised that impartiality needed to be qualified since in some circumstances central government was both a party to the debate (perhaps putting forward proposals) and at the same time the decision-maker. How, in this situation, can the procedure be impartial? This is the major complicating factor for local plan inquiries, EIPs and major call-in inquiries. Here, one of the parties to the dispute will make the final decision, giving at least the appearance of being the judge and jury in its own court.

The openness and fairness of the inquiry also need to be qualified. First, there is widespread misunderstanding of the procedure, especially the respective roles of Inspector and local authority. The adversarial nature of the inquiry, with the Inspector playing a passive role while objectors and the local authority exchange evidence and questions has important implications for the way in which the agenda is structured; and it limits potential outcomes. In his case study of the Belfast Urban Areas Plan inquiry, Blackman (1991a) points out how an adversarial hearing focuses attention on the evidence brought forward to support the position of particular interests. The inquiry becomes moulded into a battle over which interest should prevail; and this precludes debate about alternative and potentially shared solutions, which may be in a 'common or generalisable social interest'.

All this has now to be considered in the context of the 'plan-led system'. More emphasis on statutory plans means that more development interests, neighbouring authorities and service providers will be concerned to influence the content of plans and

thus the outcome of inquiries. Despite the recession, the number of objections to plans has increased dramatically over recent years, due largely to an increase in the length of inquiries (from just over two weeks in 1988 to eight weeks in 1993). The full duration from the opening of the inquiry to receipt of the Inspector's report was forty-nine weeks in 1994–5. The average number of objections in the same period was 1,400.

Many have questioned the system's capacity to cope with this burden, particularly since the biggest test for contentious plans may be later in the modifications stages. One response to the increased scale of work at inquiries (and the burden on Inspectors) has been to emphasise the potential that pre-deposit consultation and post-deposit negotiation might have in reducing the number of objections. Local authorities are being encouraged to come forward with 'suggested changes' agreed with objectors before the inquiry. The proposed changes have to be widely advertised before the Inspector deals with the issue, so that the likely response can be assessed.

In response to widespread concerns about the mounting costs of inquiry work the DoE commissioned research on the *Efficiency and Effectiveness of Local Plan Inquiries* (Steel *et al.* 1995). This concluded that the crucial problems lay with local authorities. Many authorities were insufficiently prepared for the scale of the plan inquiry task and furthermore paid only passing attention to the advice that was then offered. Many were unable or unwilling to explore the possibility of compromise (or provide information and clarify uncertainties) with objectors. Very few offered the support to objectors that would have reduced the effort needed from inspectors to clarify objections at the inquiry stage. In the main, local authorities took a very defensive attitude at the inquiry stage and were not well prepared. The Planning Inspectorate has now published new *Guidance for Local Planning Authorities* (1996) which brings the advice into line with the new demands of inquiries.

In contrast to local authorities, Inspectors were judged to perform well, and were highly regarded by participants. There was, however, some concern over inconsistencies in procedure which flowed from the wide discretion given to them. Furthermore, when problems arose in the procedure, Inspectors had few powers and resources to put them right. The researchers' conclusion was that inquiry procedure rules were needed to provide Inspectors with additional powers of sanction. This has not been acted upon. Instead, following a consultation process (DoE 1994: *Improving the Local Plan Process*), the Secretary of State has made some minor (but important) revisions to the regulations and Code of Conduct.

Although not the subject of detailed research, it is obvious that procedural problems are often compounded by the complex nature of the plans being examined. It is perhaps surprising, therefore, that attention is still primarily focused on the procedures. The DoE, the Town and Country Planning Association and the Local Authority Associations are all (in 1996) undertaking reviews of development planning which are mostly focused on procedural improvements. The form and content of development plans receives much less attention, although its basic design is about as old as the Mini.

All this demonstrates the important difference between participation before deposit of the plan and objections at the inquiry after deposit. The opportunity to object formally to a plan is not a general extension of 'participation'. It merely provides a limited opportunity for individuals and groups to air particular grievances. But, at this stage, the local authority will be committed to the plan, and any significant modifications will be effectively a challenge to the overall strategy.

MALADMINISTRATION, THE OMBUDSMAN AND PROBITY

Most legislation is based on the assumption that the organs of government will operate efficiently and fairly. This is not always the case but, even if it were, provision has to be made for investigating complaints by citizens who feel aggrieved by some action (or inaction). As modern post-industrial society becomes more complex, and as the rights of electors and consumers are viewed as important, pressures for additional means of protest, appeal and restitution grow.

At the parliamentary level, the case for an ombudsman was reluctantly conceded by the government, and a Parliamentary Commissioner for Administration was appointed in 1967. The Commissioner is an independent statutory official whose function is to investigate complaints of maladministration referred to him through Members of Parliament (Gregory and Pearson 1992). Powers of investigation extend over all central government departments, and there is an important right of access to all departmental papers.

Only a small fraction of the Parliamentary Commissioner's cases relate to planning matters and, of course, the concern is with administrative procedures, not with the merits of planning decisions. The Commissioner's reports give full but anonymised texts of reports of selected cases which have been investigated. Illustrative cases include a complaint that the Secretary of State for the Environment failed to understand the grounds on which a request had been made for intervention; a complaint that following a motorway inquiry, the Inspector called for further evidence from the DoE on which objectors were not given the opportunity to cross-examine; and a complaint by a group of local residents that an appeal decision to allow a gypsy caravan site paid little heed to local residents' objections, ignored important relevant facts and was taken on the basis of inconsistent attitudes. In all these cases, the Parliamentary Commissioner concluded that the complaint could not be upheld. This is not always the case, however; and the Commissioner's subsequent criticisms have led to changes in internal administrative procedures in the DoE.

The cases in which the Commissioner does find 'maladministration' are often of extraordinary complexity, if not real confusion. Indeed, complexity and confusion can be major factors in the failures in communication and the misunderstandings which result in 'maladministration'.

The popularity of the Parliamentary Commissioner led to pressures for the establishment of a similar institution for local government. In the mid-1970s, *Commissioners for Local Administration* were set up for England, Scotland, and Wales. With good sense, they recently decided that they should be known as the *Local Government Ombudsmen* (using the Swedish word, where the office dates back to 1809: Renton 1991). This is not only their popular name but also makes explicit that their responsibilities are confined almost entirely to local government (though it is hardly gender-sensitive!). The Ombudsmen also deal with police authorities and UDCs.

A high proportion of complaints concern planning matters: in 1994–5, about one-third in Wales and a quarter in England, but only 18 per cent in Scotland. As with the Parliamentary Commissioner, complaints have to be referred via an elected member (although in exceptional circumstances the Ombudsman can accept a complaint direct), a requirement on which there is considerable controversy. There is also concern about the situation which arises when a local authority refuses to 'remedy' a case in which maladministration or injustice is found by a Commissioner. To date, however, only limited legislative changes have been made. These include a power for local authorities to incur expenditure to remedy injustice without specific authorisation by the Secretary of State; a requirement that local authorities must notify the Ombudsman of action taken in response to an adverse report; a power for the Ombudsman to publish in a local newspaper a statement concerning cases in which a local authority has refused to comply with the Ombudsman's recommendations; and a new responsibility for the Ombudsmen to provide local authorities with advice on good practice, based on the experience of their investigations.

Not all have taken kindly to the 'interference' of the local ombudsman, and the annual reports (while noting with satisfaction a general improvement in the handling of complaints by local authorities) often name authorities which have refused to remedy cases of maladministration and personal injustice.

The irony is that the Ombudsman is wholly funded by contributions from local authorities.

The Ombudsmen have constantly noted that aggrieved objectors to planning permissions (who account for about half of all cases) have little or no redress – unlike the aggrieved applicant who can appeal to the Secretary of State. The Ombudsmen have no power to deal with the merits of planning decisions, but they have difficulty in explaining the difference between a planning decision which constitutes maladministration and one which is simply disputed (Macpherson 1987). It is, of course, not surprising that, for example, unsatisfied neighbours should take the view that their objections have been ignored, even though a local authority will usually have considered the objections and rejected them. It would be extremely difficult to substantiate a claim that the objections had been disregarded. In the meantime, the Ombudsmen can take pride in the impact they have had on encouraging local authorities to improve their planning procedures and to go beyond minimum statutory requirements.

Despite improvements to procedure in many local authorities, there have recently been well-publicised cases of extreme maladministration. Although very small in number, such cases raise more general concerns about the integrity of officers and the probity of councillors. These concerns have assumed greater significance against the background of wider national debate on 'standards in public life'.

The best-known case is that of North Cornwall District Council, though the London Borough of Brent, Bassetlaw and Warwick District Councils have also been subject to similar inquiries. In North Cornwall, complaints were first taken up by the Ombudsman, followed by the district auditor, the police and Channel 4 television. Finally, the DoE set up an official enquiry (Lees 1993) which unequivocally condemned the local councillors for granting permissions to local people for development in the open countryside contrary to planning policy.

The 1992 Local Government Act and the Code of Conduct for Councillors establish the principle that councillors should not take part in proceedings if they have a direct or indirect pecuniary interest in the issue under discussion. But such a simple distinction does not cover the many ways that councillors and officers can be influenced or can themselves influence decisions. At the heart of the issue is lobbying. Applicants and objectors lobby councillors (and sometimes officers). Councillors may lobby colleagues (although not taking part in the decision themselves); committee chairpersons can put pressure on officers; and so on. Whether or not such practices constitute improper activity is not always easy to discern.

IN CONCLUSION

The planning scene has been dominated for many years by a veritable orgy of institutional change. Though all this was intended as a means of facilitating better planning, it is possible that it has had the opposite effect of restraining the development of policies appropriate to changing conditions and perceptions. If the filing cabinets are being constantly moved, it is difficult to bring their contents up to date. Furthermore, some of the institutional changes (even if promising in the longer run) may have added to the confusion over the role of 'town and country planning' in relation to regional and national economic planning, to the management of the economy, to the increasingly strident demands for environmental protection, to the place of public participation in the planning process, and to even more intractable issues such as 'the energy question', the distribution of incomes and 'access to opportunity'. It is, however, a nice question as to whether a more stable institutional structure would have facilitated the formulation of more appropriate and effective policies in the context of the baffling economic and social problems of the time.

What does seem clear is that the faith in the efficacy of institutional change was misplaced. The reorganisation of local government (the term 'reform' has significantly fallen out of use) seems to have created as many problems as it solved, and another reorganisation is now being put into place.

As the wag said, 'When in doubt reorganise'; or, in the words of Matthew Arnold, 'faith in machinery is our besetting danger'.

The basic problems lie deeper: they relate to the functions, scope and practicability of 'town and country planning'. The Greater London Development Plan Inquiry was perhaps the most dramatic illustration of the fact that many of the crucial issues with which 'planning' is concerned do not fall within the responsibility or competence of the planning authority, or even within that of local government – jobs and incomes being the two most obvious ones. Hence central government wrestles with the political pressures to which problems in such areas give rise, though typically with disappointing results.

From a cynical viewpoint, much effort is wasted at both local and central levels in attempting to control the uncontrollable. The proclamations of politicians are given a credibility which is unwarranted. It also has unfortunate consequences, since the illusion that problems can be 'solved' turns easily into a delusion, and constant failure debases the political process and breeds cynicism. Illich (1977: 11) noted that the mid-twentieth century was 'the age of the disabling professions': an age when people had 'problems', experts had 'solutions', and scientists measured such imponderables as 'abilities' and 'needs'. The words have more than a suggestion of truth, though Illich was surely wrong in asserting that the age which he caricatures was ending.

More positively, there has been wider discussion of the limits, role and purpose of planning. A greater understanding of the operations of government has been provided by writings ranging from the Crossman Diaries to academic studies and reflections such as those of Heclo and Wildavsky, King and Solesbury (to mention but a few). More remarkable were the clear signs of some fundamental rethinking within the planning profession itself, which was heralded by the RTPI 1976 discussion paper on *Planning and the Future*. Many more such thoughtful and thought-provoking papers have followed.

It would be interesting to speculate why this untypically deep questioning started when it did. Perhaps it was a sign of the coming of age of planning. Two factors were of particular importance: an awakening of concern for making government more responsive (what was inadequately termed 'public participation') and a sea change in the economy.

The first started with protests against unwanted developments, big and small, particular and generalised. Some of these protests led to gargantuan 'inquiries' – of which Roskill and Windscale were the epitome. Others were more modest and localised, but also much more numerous. With hindsight, the most important were those which in reality were protests not simply against a particular development (though that was the manifest objective), but against the policies which these represented. Typically, these were not the responsibility of planners, but of other professions and, above all, of politicians who forged the policies. Politicians, at both central and local levels, perceived problems (understandably) in the terms in which they were presented. Problems labelled as housing shortages, road congestion, slum clearance and redevelopment portrayed the obvious solutions: build houses quickly; build more roads; clear the slums; redevelop the worn out parts of the inner city. The political responses were to 'solve' these clearly articulated problems. But policies involve choices and, again with hindsight, some of the choices had undesirable results: more houses involved high densities and few amenities; new roads increased the attraction of private transport and the decline of public alternatives; slum clearance destroyed communities; and so on.

The perceived 'failures' of planning – high-rise development, difficult-to-let council housing schemes, urban motorways, inner city decline and the like – added to the mounting concern about the role and character of planning. Whether, or to what extent, these were 'failures' and, if so, the degree to which 'planning' was to blame are questions which were seldom raised, let alone answered in their historical context. But they were seen to symbolise the inadequacy of planning.

An alternative interpretation would lay emphasis on the growth of real public participation. Public participation is not a subsidiary process which can be held in check: once it begins to work effectively it transforms the nature of the planning process. On occasion, it can get 'completely out of control', as it did in some well-publicised highway inquiries. Though disruptive, these led to a major reappraisal of both highway inquiry procedures and highway planning. Here the point is that the lesson was learned: it had become apparent that participation could work.

The professional acceptance of public participation (though by no means unanimous) was a remarkable feature of the 1970s. That it came first in planning, but not in other fields such as education or health, may be related to the transformed nature of planning education and the changed character of the 'intake' to the profession (Cherry 1974; Centre for Environmental Studies 1973). Indeed, it may be that it is this which above all explains the new humility, the introspective questioning and the new intellectualism which was so marked a feature of the time. In this respect, planners departed from the norms of professionalism, though not without internal strife.

The profession's commitment to public participation continued into the 1980s, despite an increasingly hostile political framework. The growth of a participatory ethic, however, may have been of lesser importance than the impact of economic change. A new humility grew in response to a gradual realisation that changes in the economy were structural rather then cyclical. Policies based on the assumption that the task in hand was to channel the forces of economic growth were increasingly perceived to be misplaced. Planning was no longer to be preoccupied with controls over the location of growth: it was to be remoulded to assist in the actual promotion of growth.

Certainly, the 1980s saw a remarkable change in the political scene. A new and clear political philosophy emerged: the objective of planning now is to facilitate enterprise with the minimum of constraints. Planning controls have been reduced, the

GLC and the metropolitan county councils have been abolished, and statutory requirements for public participation in the preparation of plans have been reduced. Local government has been increasingly bypassed in favour of *ad hoc* bodies designed to promote private sector involvement.

Much of this dramatic change stems from the explicit political stance of the central government, and the belief that, somehow or other, planning itself was part of the problem – an attitude encapsulated by the remark that jobs were being locked up in the filing cabinets of planners. But there are also some deeper undercurrents. Above all, the poor state of the economy has raised the importance of the *promotion* of development in contrast to its *control*. Successive public expenditure crises have also taken their toll. The problems of urban decay are seemingly of a growing intractability: or at least it looks as if traditional planning approaches are inadequate.

More widely, there is increased confusion about the role of planning. The promises held out by the introduction of structure planning failed to materialise. It appeared to be no more effective, speedy, flexible or satisfying than the system it was designed to replace. Whether the latest changes will prove to be more effective remains to be seen. When the needs for economic growth and for good planning clash, the former is likely to win.

More subtly, the fact that local planning powers can now be exercised (and to some extent defined) by local authorities with lesser ministerial supervision widens the questions relevant to 'what is planning?'; and these questions are underlined by the range of initiatives promoted by the central government in relation to regeneration and the harnessing of market forces.

It is always difficult to see current events in perspective, and there is abundant scope for debating whether the changes that have taken place are fundamental or not. More likely they will be overtaken by new problems or by the redefinition of old problems which cannot readily be foreseen.

FURTHER READING

Planning and Politics

This is an enormous topic and many of the central texts on planning address it. Some recommendations for required reading on this topic are Ambrose (1986) *Whatever Happened to Planning?*; Reade (1987) *British Town and Country Planning*; Blowers (1980) *The Limits of Power*; Low (1991) *Planning, Politics and the State*; Thornley (1993) *Urban Planning Under Thatcherism*. See also the references for Chapter 1.

The Profession

The intellectual development of the planning profession is explained by Healey (1985) 'The professionalisation of planning in Britain'. See also Taylor (1992) 'Professional ethics in town planning'; Evans (1993) 'Why we no longer need a planning profession'.

A new RTPI report is to be published in 1997 on the relationship between councillors and officers following research undertaken by Oxford Brookes University.

Race and Planning

In addition to RTPI (1993) *Ethnic Minorities and the Planning System*, the earlier RTPI/CRE (1983) *Planning for a Multi-racial Britain* should be consulted. See also Thomas and Krishnarayan (1994) 'Race, disadvantage and policy processes in British planning'.

Women and Planning

Two recent textbooks provide a comprehensive analysis of theory and practice. Greed (1994) *Women and Planning* is a mine of interesting examples and a guide to reading. Little (1994) *Gender, Planning and the Policy Process* links the issue to wider debates. Other key sources are Sandercock and Forsyth (1992) 'A gender agenda: new directions for planning theory'; Gilroy (1993) *Good Practices in Equal Opportunities*. See also RTPI (1988) *Planning for Choice and Opportunity* and (1995) Practice Advice Note no. 12 *Planning for Women*; Little (1994b) 'Women's initiatives in town planning in England'; Davies (1996) *Equality and planning: gender and disability*.

People with Disabilities

The British Standards Institution publish the *Code of Practice for Access for the Disabled to Buildings* (BS 5810) and *Code of Practice for Design for the Convenience of Disabled People* (BS 5619). The DoE provides advice in Development Control Policy Note 16 (1985) which has been supplemented by the RTPI's Practice Advice Note: *Access for Disabled People* (1988) and (1985) *Access Policies in Local Plans*.

Participation

Early writings include Dennis (1970) *People and Planning*; Davies (1972) *The Evangelistic Bureaucrat: A Study of a Planning Exercise in Newcastle-upon-Tyne*; Wates (1977) *The Battle for Tolmers Square*. Recent titles are Barlow (1995) *Public Participation in Urban Development*; Colenutt and Cutten (1994) 'Community empowerment in vogue or vain'; and Reeves (1995) 'Developing effective public consultation'.

The DoE (1995) study *Community Involvement in Planning and Development Processes* reconsiders some classic principles in the very different context of the 1990s. It also contains a list of other key references and an interesting classification of involvement methods and practices.

For international comparisons, see Barlow (1995) *Public Participation in Urban Development: The European Experience*.

Access to Information

On the general topic the starting point should be Steele (1995) *Public Access to Information*. On access to environmental information, see European Environmental Bureau (1994) *Your Rights under European Union Environment Legislation*. Procedure for making a complaint to the Commission and petitions to the European Parliament is given in the CPRE (1992a) *Campaigners' Guide to Using EC Environmental Law*. Another useful guide is Moxen *et al.* (1995) *Accessing Environmental Information in Scotland*.

The relative ease of actually getting at planning information is considered in CPRE (1994d) *Public Access to Planning Documents*, Ramblers' Association (1993) *Open Door Planning*, and Planning Aid for Scotland: *Access to Planning Information*. The RTPI's (1991) report on *Planning – Is it a Service and How Can it be Effective?* provides data on the information practices of local authorities.

Public Inquiries

The DoE study on *The Efficiency and Effectiveness of Local Plan Inquiries* provides the most up-to-date review of inquiry practice for plans; it also contains an extensive bibliography. The Planning Inspectorate has a number of useful publications: (1996) *Guide to Local Planning Authorities*; its annual *Business and Corporate Plan* (which gives statistics and some interpretation of trends); and the recently launched *Inspectorate Journal* (which will no doubt prove a good source over coming years).

For a recent explanation of the EIP in practice, see Phelps (1995) 'Structure plans: the conduct and conventions of examinations in public'.

BIBLIOGRAPHY

Abel-Smith, B. and Townsend, P. (1965) *The Poor and the Poorest*, London: Bell

Abercrombie, P. (1945) *Greater London Plan*, London: HMSO

Adair, A. S., Berry, J. N., and McGreal, W. S. (1991) 'Land availability, housing demand and the property market', *Journal of Property Research* 8: 59–69

Adams, B., Okely, J., Morgan, D., and Smith, D. (1976) *Gypsies and Government Policy in Britain*, London: Heinemann

Adams, D. (1992) 'The role of landowners in the preparation of statutory local plans', *Town Planning Review* 63: 297–323

Adams, D. (1994) *Urban Planning and the Development Process*, London: UCL Press

Adams, D. (1996) 'The use of compulsory purchase under planning legislation', *Journal of Planning and Environment Law* 1996: 275–85

Adams, D. and Pawson, G. P. (1991) *Representation and Influence in Local Planning*, Manchester: Department of Planning and Landscape, University of Manchester

Adams, D., Russell, L., and Taylor-Russell, C. (1994) *Land for Industrial Development*, London: Spon

Adams, J. G. U. (1990) 'Car ownership forecasting: pull the ladder up, or climb back down?', *Traffic Engineering and Control* 31: 136–41

Adams, J. G. U. (1995) *Risk*, London: UCL Press

Adams, W. M. (1986) *Nature's Place: Conservation Sites and Countryside Change*, London: Allen & Unwin

Advisory Committee on Business and the Environment (annual) *Progress Report*, London: Department of Trade and Industry

Advisory Committee on Business and the Environment (1993) *Report of the Financial Sector Working Group*, London: Department of Trade and Industry

Advisory Committee on the Safe Transport of Radioactive Materials (1988) *The Transport of Low Level Radioactive Waste in the UK*, London: HMSO

Advisory Council on Science and Technology (1990) *The Enterprise Challenge: Overcoming Barriers to Growth in Small Firms*, London: HMSO

Albrechts, L. (1991) 'Changing roles and positions of planners', *Urban Studies* 28: 123–37

Albrechts, L., Moulaert, F., Roberts, P., and Swyngedouw, E. (eds) (1989) *Regional Policy at the Crossroads: European Perspectives*, London: Jessica Kingsley

Alden, J. (1992) 'Strategic planning guidance in Wales', in Minay, C. L. W. 'Developing regional planning guidance in England and Wales: a review symposium', *Town Planning Review* 63: 415–34

Alden, J. and Boland, P. (eds) (1996) *Regional Development Strategies: A European Perspective*, London: Jessica Kingsley

Alden, J. and Offord, C. (1996) 'Regional planning guidance', in Tewdwr-Jones, M. (ed.) *British Planning Policy in Transition: Planning in the 1990s*, London: UCL Press

Alden, J. and Romaya, S. (1994) 'The challenge of urban regeneration in Wales: principles, policies and practice', *Town Planning Review* 65: 435–61

Aldous, T. (1989) *Inner City Urban Regeneration and Good Design*, London: HMSO

Aldridge, H. R. (1915) *The Case for Town Planning*, London: National Housing and Town Planning Council

Aldridge, M. (1979) *The British New Towns*, London: Routledge & Kegan Paul

Aldridge, M. (1996) 'Only demi-paradise? Women in garden cities and new towns', *Planning Perspectives* 11: 23–39

Aldridge, M. and Brotherton, C. J. (1988) 'Being a programme authority: is it worthwhile?', *Journal of Social Policy* 16: 349–69

Alexander, A. (1982) *Local Government in Britain since Reorganisation*, London: Allen & Unwin

Alexander, E. R. (1981) 'If planning isn't everything, maybe it's something', *Town Planning Review* 52: 131–42

Alexander, E. R. (1992) *Approaches to Planning: Introducing Current Planning Theories, Concepts and Issues* (second edition), Philadelphia: Gordon & Breach

Allaby, M. (1994) *Macmillan Dictionary of the Environment*, London: Macmillan

Allen, J. and Hamnett, C. (eds) (1991) *Housing and Labour Markets: Building the Connections*, London: Unwin Hyman

Allen, K. (1986) *Regional Incentives and the Investment Decision of the Firm* (DTI), London: HMSO

Allen, R. (1995) 'Policy and grassroots action – a vital mix', in S. Whitaker, *First Steps: Local Agenda 21 in Practice*, London: HMSO

Allison, L. (1986) 'What is urban planning for?', *Town Planning Review* 57: 5–16

Allmendinger, P. (1996a) 'Twilight zones', *Planning Week* 4 (29), 18 July 1996: 14–15

Allmendinger, P. (1996b) *Thatcherism and Simplified Planning*

Zones: An Implementation Perspective, Oxford Planning Monographs vol. 2, no. 1., Oxford: Oxford Brookes University

Alonso, W. (1971) 'Beyond the inter-disciplinary approach to planning', *Journal of the American Institute of Planners* 37: 169–73 (reprinted in J. B. Cullingworth (ed.) *Planning for Change* (vol. 3 of *Problems of an Urban Society*), London: Allen & Unwin, 1973)

Alterman, R. and Cars, G. (eds) (1991) *Neighbourhood Regeneration: An International Evaluation*, London: Mansell

Alterman, R., Harris, D., and Hill, M. (1984) 'The impact of public participation on planning: the case of the Derbyshire structure plan', *Town Planning Review* 55: 177–96

Ambrose, P. (1974) *The Quiet Revolution: Social Change in a Sussex Village 1871–1971*, London: Chatto & Windus

Ambrose, P. (1986) *Whatever Happened to Planning?*, London: Methuen

Ambrose, P. (1994) *Urban Process and Power*, London: Routledge

Ambrose, P. and Colenutt, B. (1975) *The Property Machine*, Harmondsworth: Penguin

Amery, C. and Cruickshank, D. (1975) *The Rape of Britain*, London: Paul Elek

Amin, A. and Tomaney, J. (1991) 'Creating an enterprise culture in the North East? The impact of urban and regional policies of the 1980s', *Regional Studies* 25: 479–87

Amos, F. J. C., Davies, D., Groves, R., and Niner, P. (1982) *Manpower Requirements for Physical Planning*, Birmingham: Institute of Local Government Studies

Amundson, C. (1993) 'Sustainable aims and objectives: a planning framework', *Town and Country Planning* 62: 20–2

Anderson, M. A. (1981) 'Planning policies and development control in the Sussex Downs AONB', *Town Planning Review* 52: 5–25

Anderson, M. A. (1990) 'Areas of outstanding natural beauty and the 1949 National Parks Act', *Town Planning Review* 61: 311–39

Anderson, W. P., Kanaroglou, P. S., and Miller, E. J. (1996) 'Urban form, energy and the environment', *Urban Studies* 33: 7–35

Anon (1956) 'Ye olde English green belt', *Journal of the Town Planning Institute* 42: 68–9

Archbishop of Canterbury's Commission on Urban Priority Areas (1985) *Faith in the City: A Call for Action by Church and Nation*, London: Church House Publishing

Archbishop of Canterbury's Commission on Urban Priority Areas (1990) *Living Faith in the City: A Progress Report*, London: General Synod of the Church of England

Archbishops' Commission on Rural Areas (1990) *Faith in the Countryside*, Stoneleigh Park, Warwickshire: ACORA Publishing

Armitage Report (1980) *Lorries, People and the Environment* (Report of the Inquiry), London: HMSO

Armstrong, J. (1985) 'The Sizewell inquiry', *Journal of Planning and Environment Law* 1985: 686–9

Arnell, N. W., Jenkins, A., and George, D. G. (1994) *Implications of Climate Change for the NRA* (National Rivers Authority), London: HMSO

Arnstein, S. R. (1969) 'A ladder of citizen participation', *Journal of the American Institute of Planners* 35: 216–24

Arton Wilson Report (1959) *Caravans as Homes*, Cmnd 872, London: HMSO

Arup Economic Consultants (1990) *Mineral Policies in Development Plans* (Department of the Environment), London: HMSO

Arup Economic Consultants (1991) *Simplified Planning Zones: Progress and Procedures* (Department of the Environment), London: HMSO

Arup Economics and Planning (1994) *Assessment of the Effectiveness of Derelict Land Grant in Reclaiming Land for Development* (Department of the Environment), London: HMSO

Arup Economics and Planning (1995a) *Coastal Superquarries: Options for Wharf Facilities on the Lower Thames* (Department of the Environment), London: HMSO

Arup Economics and Planning (1995b) *Derelict Land Prevention and the Planning System* (Department of the Environment), London: HMSO

Ashby, E. (1978) *Reconciling Man with the Environment*, Stanford, CA: Stanford University Press

Ashby, E. and Anderson, M. (1981) *The Politics of Clean Air*, Oxford: Clarendon Press

Ashford, D. E. (ed.) (1979) *The Politics of Urban Resources*, Chicago: Maaroufa

Ashworth, W. (1954) *The Genesis of Modern British Town Planning*, London: Routledge & Kegan Paul

Assembly of Welsh Counties (1992) *Strategic Planning Guidance in Wales*, Mold: Clwyd County Council

Association of Conservation Officers (1992) *Listed Buildings Repair Notices*, Brighton: The Association

Association of County Archaeological Officers (1993) *Archaeological Heritage*, London: Association of County Councils

Association of County Councils (1992) *National Parks: The New Partnership*, London: ACC

Association of County Councils (1993) *The Enabling Authority and County Government*, London: ACC

Association of London Government (1996a) *Paying the Wrong Fare: Affordable Public Transport*, London: ALG

Association of London Government (1996b) *Red Routes: Do They Work?*, London: ALG

Association of National Park Officers (1988) *National Parks: Environmentally Favoured Areas? A Proposal for a New Agricultural Policy to Achieve the Aims of National Park Designation*, Bovey Tracey, Devon: The Association

Association of Town Centre Management (1992) *Working for the Future of our Towns and Cities*, London: ATCM

Association of Town Centre Management (1994) *The Effectiveness of Town Centre Management*, London: ATCM

Astrop, A. (1993) *The Trend in Rural Bus Services since Deregulation*, Crowthorne, Berks.: Transport Research Laboratory

Atkinson, R. and Moon, G. (1994) *Urban Policy in Britain: The City, The State and the Market*, London: Macmillan

Audit Commission: see appendix on *Official Publications*

Austin, J. (1995) 'Renewal areas: the research findings' London: Austin, Mayhead

Automobile Association (1995a) *Transport and the Environment: The AA's Response to the Royal Commission on Environmental Pollution*, London: AA

Automobile Association (1995b) *Shopmobility: Good for People and Towns*, London: AA

Bailey, N. and Barker, A. (eds) (1992) *City Challenge and Local Regeneration Partnerships: Conference Proceedings*, London: Polytechnic of Central London

Bailey, N., Barker, A., and MacDonald, K. (1995) *Partnership Agencies in British Urban Policy*, London: UCL Press

Bailey, S. J. (1990) 'Charges for local infrastructure', *Town Planning Review* 61: 427–53

Bain, C., Dodd, A., and Pritchard, D. (1990) *RSPB Planscan: A Study of Development Plans in England and Wales*, Sandy, Beds.: Royal Society for the Protection of Birds

Baker, S., Kousis, M., Richardson, D., and Young, S. (1996) *The Politics of Sustainable Development*, London: Routledge

Balchin, P. (1995) *Housing Policy: An Introduction*, London: Routledge

Baldock, D., Cox, G., Lowe, P., and Winter, M. (1990) 'Environmentally sensitive areas: incrementalism or reform?', *Journal of Rural Studies* 6: 143–62

Baldock, D., Bishop, K., Mitchell, K., and Phillips, A. (1996) *Growing Greener: Sustainable Agriculture in the UK*, London: CPRE

Ball, M. (1983) *Housing Policy and Economic Power*, London: Methuen

Ball, R. M. (1989) 'Vacant industrial premises and local development: a survey, analysis, and policy assessment of the problems in Stoke-on-Trent', *Land Development Studies* 6: 105–28

Ball, R. M. (1995) *Local Authorities and Regional Policy in the UK: Attitudes, Representations and the Local Economy*, London: Paul Chapman

Ball, S. and Bell, S. (1994) *Environmental Law* (second edition), London: Blackstone

Banham, R., Barker, P., Hall, P., and Price, C. (1969) 'Non-plan: an experiment in freedom', *New Society* 26: 435–43

Banister, D. (1994a) *Transport Planning in the UK, USA and Europe*, London: Spon

Banister, D. (1994b) 'Reducing the need to travel through planning', *Town Planning Review* 65: 349–54

Banister, D. (ed.) (1995) *Transport and Urban Development*, London: Spon

Banister, D. and Button, K. (eds) (1993) *Transport, the Environment and Sustainable Development*, London: Spon

Banister, D. and Watson, S. (1994) *Energy Use in Transport and City Structure*, London: Planning and Development Research Centre, University College London

Banister, D., Capello, R., and Nijkamp, P. (eds) (1995) *European Transport and Communications Networks*, Chichester, W. Sussex: John Wiley

Bannon, M. J., Nowlan K.I., Hendry, J., and Mawhinney, K. (1989) *Planning: The Irish Experience 1920–1988*, Dublin: Wolfhound Press

Barker, T. C. and Savage, C. I. (1974) *An Economic History of Transport*, London: Hutchinson

Barkham, J. P., MacGuire, F. A. S., and Jones, S. J. (1992) *Sea-level Rise and the UK*, London: Friends of the Earth

Barkham, R., Gudgin, G., Hart, M., and Harvey, E. (1996) *The Determinants of Small Firm Growth* (Regional Studies Association), London: Jessica Kingsley

Barlow, J. (1982) 'Planning practice, housing supply and migration', in Champion, A. G. and Fielding, A. J. (eds) *Migration Processes and Patterns*, vol. 1: *Research Progress and Prospects*, London: Belhaven

Barlow, J. (1986) 'Landowners, property ownership, and the rural locality', *International Journal of Urban and Regional Research* 10: 309–29

Barlow, J. (1988) 'The politics of land into the 1990s: landowners, developers, and farmers in lowland Britain', *Policy and Politics* 16: 111–21

Barlow, J. (1995) *Public Participation in Urban Development: The European Experience*, London: Policy Studies Institute

Barlow, J. and Chambers, D. (1992) *Planning Agreements and Affordable Housing Provision*, Brighton: Centre for Urban and Regional Research, University of Sussex

Barlow, J. and Duncan, S. (1992) 'Markets, states and housing provision: four European growth regions compared', *Progress in Planning* 38, 2: 93–177

Barlow, J. and Duncan, S. (1994) *Success and Failure in House Provision: European Systems Compared*, Oxford: Elsevier Science

Barlow, J. and Gann, D. (1993) *Offices into Flats*, York: Joseph Rowntree Foundation

Barlow, J. and Gann, D. (1995) 'Flexible planning and flexible buildings: reusing redundant office space', *Journal of Urban Affairs* 17: 263–76

Barlow, J. and King, A. (1992) 'The state, the market, and competitive strategy: the housebuilding industry in the United Kingdom, France, and Sweden', *Environment and Planning A* 24: 381–400

Barlow, J., Cocks, R., and Parker, M. (1994a) *Planning for Affordable Housing* (Department of the Environment), London: HMSO

Barlow, J., Cocks, R., and Parker, M. (1994b) 'Delivering affordable housing: law, economics and planning policy', *Land Use Policy* 11: 181–94

Barlow Report (1940) *Report of the Royal Commission on the Distribution of the Industrial Population*, Cmd 6153, London: HMSO

Barnekov, T., Boyle, R., and Rich, D. (1989) *Privatism and Urban Policy in Britain and the United States*, Oxford: Oxford University Press

Barnekov, T., Hart, D., and Benfer, W. (1990) *US Experience in Evaluating Urban Regeneration* (Department of the Environment), London: HMSO

Barrett, S. and Fudge, C. (eds) (1981) *Policy and Action*, London: Methuen

Barrett, S. and Healey, P. (eds) (1985) *Land Policy: Problems and Alternatives*, Aldershot, Hants.: Avebury

Barrett, S. and Whitting, G. (1983) *Local Authorities and Land Supply*, Bristol: School for Advanced Urban Studies, University of Bristol

Barrett, S., Stewart, M., and Underwood, J. (1978) *The Land Market and Development Process*, Bristol: School for Advanced Urban Studies, University of Bristol

Barton, H., Davis, G., and Guise, R. (1995) *Sustainable Settlements: A Guide for Planners, Designers and Developers*, Bristol: University of the West of England; and Luton, Beds.: Local Government Management Board

Bassett, K. (1993) 'Urban cultural strategies and urban regeneration: a case study and critique', *Environment and Planning A* 25: 1773–88

Bassett, K. (1996) 'Partnerships, business elites and urban politics: new forms of governance in an English city?', *Urban Studies* 33: 539–55

Bateman, M. (1985) *Office Development: A Geographical Analysis*, New York: St Martin's Press

Batey, P. (1985) 'Postgraduate planning education in Britain', *Town Planning Review* 56: 407–20

Batey, P. (1993) 'Planning education as it was', *The Planner* 79, 4 (April 1993): 25–6

Batho Report (1990) *Report of the Noise Review Working Party*, London: HMSO

Batley, R. (1989) 'London Docklands: an analysis of power relations between UDCs and local government', *Public Administration* 67: 167–87

Batley, R. and Stoker, G. (eds) (1991) *Local Government in Europe: Trends and Development*, London: Macmillan

Batty, M. (1984) 'Urban policies in the 1980s: a review of the OECD proposals for managing urban change', *Town Planning Review* 55: 489–98

Batty, M. (1990) 'How can we best respond to changing fashions in urban and regional planning?', *Environment and Planning B: Planning and Design* 17: 1–7

Baxter, J. D. (1990) *State Security, Privacy and Information*, Hemel Hempstead, Herts.: Harvester Wheatsheaf

Bean, D. (ed.) (1996) *Law Reform for All*, London: Blackstone

Beckerman, W. (1974) *In Defence of Economic Growth*, London: Jonathan Cape

Beckerman, W. (1990) *Pricing for Pollution: Market Pricing, Government Regulation, Environmental Policy* (second edition), London: Institute of Economic Affairs

Beckerman, W. (1995) *Small is Stupid: Blowing the Whistle on the Greens*, London: Duckworth

Bedfordshire County Council (1996) *A Step by Step Guide to Environmental Appraisal*, Sandy, Beds.: RSPB

Beesley, M. E. and Kain, J. F. (1964) 'Urban form, car ownership and public policy: an appraisal of *Traffic in Towns*', *Urban Studies* 1: 174–203

Begg, H. M. and Pollock, S. H. A. (1991) 'Development plans in Scotland since 1975', *Scottish Geographical Magazine* 107: 4–11

Begg, I. (1991) 'High technology location and the urban areas of Great Britain: development in the 1980s', *Urban Studies* 28: 961–81

Bell, J. L. (1993) *Key Trends in Communities and Community Development* London: Community Development Foundation

Bell, S. (1992) *Out of Order. The 1987 Use Classes Order: Problems and Proposals*, London: London Boroughs Association

Bendixson, T. (1989) *Transport in the Nineties: The Shaping of Europe*, London: Royal Institution of Chartered Surveyors

Benington, J. (1994) *Local Democracy and the European Union: The Impact of Europeanisation on Local Governance*, London: Commission for Local Democracy

Bennett, R. and Errington, A. (1995) 'Training and the small rural business', *Planning Practice and Research* 10: 45–54

Bennett, R. J. (1990) 'Training and enterprise councils (TECs) and vocational education and training', *Regional Studies* 24: 65–82

Benson, J. F. and Willis, K. G. (1992) *Valuing Informal Recreation on the Forestry Commission Estate* (Forestry Commission Bulletin 104), London: HMSO

Bentham, C. G. (1985) 'Which areas have the worst urban problems?', *Urban Studies* 22: 119–31

Benyon, J. (ed) (1984) *Scarman and After: Essays Reflecting on Lord Scarman's Report, the Riots, and their Aftermath*, Oxford: Pergamon

Berry, J. and McGreal, S. (1995a) 'Community and inter-agency structures in the regeneration of inner-city Belfast', *Town Planning Review* 66: 129–42

Berry, J. and McGreal, S. (eds) (1995b) *European Cities, Planning Systems and Property Markets*, London: Spon

Berry, J. N., Fitzsimmons, D.F., and McGreal, W.S. (1988) 'Neighbour notification: the Scottish and Northern Ireland models', *Journal of Planning and Environment Law* 1988: 804–8

Berry, J., McGreal, W. S., and Deddis, W. G. (1993) *Urban Regeneration: Property Investment and Development*, London: Spon

Bertuglia, C. S., Clarke, G. P., and Wilson, A. G. (eds) (1994) *Modelling the City: Performance, Policy and Planning*, London: Routledge

Best, R. (1981) *Land Use and Living Space*, London: Methuen

Best, J. and Bowser, L. (1986) 'A people's plan for central Newham', *Planner* 27, 11: 21–5

Beveridge Report (1942) *Social Insurance and Allied Services*, Cmd 6404, London: HMSO

Bibby, P. R. and Shepherd, J. W. (1991) *Rates of Urbanisation in England 1981–2001* (Department of the Environment), London: HMSO

Bibby, P. R. and Shepherd, J. W. (1993) *Housing Land Availability: The Analysis of PS3 Statistics on Land with Outstanding Planning Permission* (Department of the Environment), London: HMSO

Birch, A. H. (1993) *The British System of Government* (ninth edition), London: Routledge

Birkenshaw, P. (1990) *Government and Information*, London: Butterworths

Birtles, W. (1991) 'The European directive on freedom of access to information on the environment', *Journal of Planning and Environment Law* 1991: 607–10

Bishop, J. (1994) 'Planning for better rural design', *Planning Practice and Research* 9: 259–70

Bishop, K. (1992) 'Assessing the benefits of community forests: an evaluation of the recreational use benefits of two urban fringe woodlands', *Journal of Environmental Planning and Management* 35: 63–76

Bishop, K. and Hooper, A. (1991) *Planning for Social Housing*, London: National Housing Forum (Association of District Councils)

Bishop, K. D. and Phillips, A. C. (1993) 'Seven steps to market: the development of the market-led approach to countryside conservation and recreation', *Journal of Rural Studies* 9: 315–38

Bishop, K. D., Phillips, A. C., and Warren, L. (1995a) 'Protected for ever? Factors shaping the future of protected areas policy', *Land Use Policy* 12: 291–305

Bishop, K. D., Phillips, A. C., and Warren, L. (1995b) 'Protected areas in the United Kingdom: time for new thinking', *Regional Studies* 29: 192–201

Blackaby, D. H. and Manning, D. N. (1990) 'The north-south divide . . . ', *Papers of the Regional Science Association* 69: 43–65

Blackhall, J. C. (1993) *The Performance of Simplified Planning Zones*, Newcastle upon Tyne: Department of Town and Country Planning, University of Newcastle

Blackhall, J. C. (1994) 'Simplified planning zones (SPZs) or simply political zeal?' *Journal of Planning and Environment Law* 1994: 117–23

Blackman, T. (1985) 'Disasters that link Ulster and the North East', *Town and Country Planning* 54: 18–20

Blackman, T. (1991a) *Planning Belfast: A Case Study of Public Policy and Community Action*, Aldershot, Hants.: Avebury

Blackman, T. (1991b) 'People-sensitive planning: communication, property and social action', *Planning Practice and Research* 6, 3: 11–15

Blackman, T. (1995) *Urban Policy in Practice*, London: Routledge

Blaker, G. (1995) 'Gypsy law', *Journal of Planning and Environment Law* 1995: 191–6

Blowers, A. (1980) *The Limits of Power: The Politics of Local Planning Policy*, Oxford: Pergamon

Blowers, A. (1983) 'Master of fate or victim of circumstance: the exercise of corporate power in environmental policy-making', *Policy and Politics* 11: 375–91

Blowers, A. (1984) *Something in the Air: Corporate Power and the Environment*, London: Harper & Row

Blowers, A. (1986) 'Environmental politics and policy in the 1980s: a changing challenge', *Policy and Politics* 14: 11–18

Blowers, A. (1987) 'Transition or transformation? Environmental policy under Thatcher', *Public Administration* 65: 277–94

Blowers, A. (ed) (1993) *Planning for a Sustainable Environment: A Report by the Town and Country Planning Association*, London, Earthscan

Blunden, J. and Curry, N. (1988) *A Future for our Countryside?*, Oxford: Basil Blackwell

Blunden, J. and Curry, N. (1989) *A People's Charter? Forty Years of the National Parks and Access to the Countryside Act 1949*, London: HMSO

Body, R. (1982) *Agriculture: The Triumph and the Shame*, London: Maurice Temple Smith

Bond, M. (1992) *Nuclear Juggernaut: The Transport of Radioactive Materials*, London: Earthscan

Bongers, P. (1990) *Local Government and 1992*, Harlow, Essex: Longman

Bonnel, P. (1995) 'Urban car policy in Europe', *Transport Policy* 2: 83–95

Bonyhandy, T. (1987) *Law and the Countryside: The Rights of the Public*, Oxford: Professional Books

Booker, C. and Green, C. L. (1973) *Goodbye London: An Illustrated Guide to Threatened Buildings*, London: Fontana

Booth, P. (1996) *Controlling Development: Certainty and Discretion in Europe, the USA, and Hong Kong*, London: UCL Press

Booth, P. and Beer, A. R. (1983) 'Development control and design quality', *Town Planning Review* 54: 265–84 and 383–404

Borchardt, K. (1995) *European Integration: The Origins and Growth of the European Community* (fourth edition), Luxembourg: Office for the Official Publications of the European Communities

Borins, S. F. (1988) 'Electronic road pricing: An idea whose time may never come', *Transportation Research A* 22A: 37–44

Borraz, O. *et al.* (1994) *Local Leadership and Decision Study of France, Germany, the United States and Britain*, London: Local Government Chronicle / Joseph Rowntree Foundation

Boucher, S. and Whatmore, S. (1990) *Planning Gain and Conservation: A Literature Review*, Reading: Department of Geography, University of Reading

Boucher, S. and Whatmore, S. (1993) 'Green gains? Planning by agreement and nature conservation', *Journal of Environmental Planning and Management* 36: 33–50

Bourne, F. (1992) *Enforcement of Planning Control* (second edition), London: Sweet and Maxwell

Bovaird, T. (1992) 'Local economic development and the city', *Urban Studies* 29: 343–68

Bovaird, T., Gregory, D., Martin, S., Pearce, G., and Tricker, M.

(1990a) *Evaluation of the Rural Development Programme Process*, London: HMSO

Bovaird, T., Tricker, M., Martin, S. J., Gregory, D. G., and Pearce, G. R. (1990b) *An Evaluation of the Rural Development Programme Process*, London: HMSO

Bovaird, T., Gregory, D., and Martin, S. (1991a) 'Improved performance in local economic development: a warm embrace or an artful sidestep?', *Public Administration* 69: 103–19

Bovaird, T., Tricker, M., Hems, L., and Martin, S. (1991b) *Constraints on the Growth of Small Firms: A Report on a Survey of Small Firms* (Aston Business School), London: HMSO

Bowers, J. (1990a) *Economics of the Environment: The Conservationists' Response to the Pearce Report*, Newbury: British Association of Nature Conservationists

Bowers, J. (ed) (1990b) *Agriculture and Rural Land Use*, Swindon, Wilts.: Economic and Social Research Council

Bowers, J. (1992) 'The economics of planning gain: a reappraisal, *Urban Studies* 29: 1329–39

Bowers, J. (1995) 'Sustainability, agriculture, and agricultural policy', *Environment and Planning A* 27: 1231–43

Bowers, J. K. and Cheshire, P. C. (1983) *Agriculture, the Countryside and Land Use: An Economic Critique*, London: Methuen

Bowley, M. (1945) *Housing and the State 1919–1944*, London: Allen & Unwin

Bowman, J. C. (1992) 'Improving the quality of our water: the role of regulation by the National Rivers Authority', *Public Administration* 70: 565–75

Boydell, P. and Lewis, M. (1989) 'Applications to the High Court for the review of planning decisions', *Journal of Planning and Environment Law* 1989: 146–56

Boyle, R. (ed) (1985) 'Leveraging urban development: a comparison of urban policy directions and programme impact in the United States and Britain', *Policy and Politics* 13: 175–210

Boyle, R. (1993) 'Changing partners: the experience of urban economic policy in West Central Scotland, 1980–90', *Urban Studies* 30: 309–24

Boyne, G., Jordan, G., and McVicar, M. (1995) *Local Government Reform: A Review of the Process in Scotland and Wales*, London: Local Government Chronicle Communications in association with the Joseph Rowntree Foundation

Boynton, J. (1986) 'Judicial review of administrative decisions: a background paper', *Public Administration* 64: 147–61

Bradford, M. and Robson, B. (1995) 'An evaluation of urban policy', in Hambleton, R. and Thomas, H. (eds) (1995) *Urban Policy Evaluation: Challenge and Change*, London: Paul Chapman

Bramley, G. (1989) *Land Supply, Planning, and Private Housebuilding*, Bristol: School for Advanced Urban Studies, University of Bristol

Bramley, G. (1993) 'The impact of land use planning and tax subsidies on the supply and price of housing in Britain', *Urban Studies* 30: 5–30

Bramley, G. and Smart, G. (1995) *Rural Incomes and Housing Affordability*, Salisbury, Wilts.: Rural Development Commission

Bramley, G. and Watkins, C. (1995) *Circular Projections: Household Growth, Housing Development, and the Household Projections*, London: CPRE

Bramley, G. and Watkins, C. (1996) *Steering the Housing Market: New Building and the Changing Planning System*, Bristol: Policy Press

Bramley, G., Bartlett, W., and Lambert, C. (1995) *Planning, the Market and Private House-building*, London: UCL Press

Brand, C. M. and Thompson, B. (1982) 'Third parties and development control: a better deal for Scottish neighbours?', *Journal of Planning and Environment Law* 1982: 743–63

Bray, J. (1992) *The Rush for Roads: A Road Programme for Economic Recovery. A Report by Movement Transport Consultancy for ALARM UK and Transport 2000*, London: Transport 2000

Breheny, M. J. (1983) 'A practical view of planning theory', *Environment and Planning B: Planning and Design* 10: 101–15

Breheny, M. J. (1989) 'Chalkface to coalface: a review of the academic–practice interface', *Environment and Planning B: Planning and Design* 16: 451–68

Breheny, M. J. (1991) 'The renaissance of strategic planning', *Environment and Planning B: Planning and Design* 18: 233–49

Breheny, M. J. (ed) (1992) *Sustainable Development and Urban Form*, London: Pion

Breheny, M. J. and Congdon, P. (eds) (1989) *Growth and Change in a Core Region: The Case of South East England*, London: Pion

Breheny, M. J. and Hall, P. (eds) (1996) *The People – Where Will They Go? National Report of the TCPA Regional Inquiry into Housing Need and Provision in England*, London: TCPA

Breheny, M. J. and Hooper, A. (1985) *Rationality in Planning: Critical Essays on the Role of Rationality in Urban and Regional Planning*, London: Pion

Breheny, M., Gent, T., and Lock, D. (1993) *Alternative Development Patterns: New Settlements* (Department of the Environment), London: HMSO

Brenan, J. (1994) 'PPG 16 and the restructuring of archaeological practice in Britain', *Planning Practice and Research* 9: 395–405

Bridge, G. (1993) *Gentrification, Class and Residence,* Bristol: School for Advanced Urban Studies

Bridges, L. (1979) 'The structure plan examination in public as an instrument of intergovernmental decision making', *Urban Law and Practice* 2: 241–64

Briggs, A. (1952) *History of Birmingham* (2 vols), Oxford: Oxford University Press

Brindley, T., Rydin, Y., and Stoker, G. (1996) *Remaking Planning* (second edition), London: Routledge

British Road Federation (annual) *Basic Road Statistics*, London: BRF

British Waterways Board (1993) *The Waterway Environment and Development Plans*, Watford, Herts.: BWB

Britton, D. (ed) (1990) *Agriculture in Britain: Changing Pressures and Policies*, Wallingford, Berks.: CAB International

Broady, M. (1968) *Planning for People*, London: Bedford Square Press

Bromley, M. P. (1990) *Countryside Management*, London: Spon

Bromley, R. and Thomas, C. (eds) (1993) *Retail Change: Contemporary Issues*, London: UCL Press

Brooke, C. (1996) *Natural Conditions: A Review of Planning Conditions and Nature Conservation*, Sandy, Beds.: RSPB

Brooke, R. (1989) *Managing the Enabling Authority*, Harlow, Essex: Longman

Broome, J. (1992) *Counting the Cost of Global Warming*, Cambridge: White Horse Press

Brotherton, D. I. (1989) 'The evolution and implications of mineral planning policy in the national parks of England and Wales', *Environment and Planning A* 21: 1229–40

Brotherton, D. I. (1992a) 'On the control of development by planning authorities', *Environment and Planning B: Planning and Design* 19: 465–78

Brotherton, D. I. (1992b) 'On the quantity and quality of planning applications', *Environment and Planning B: Planning and Design* 19: 337–57

Brownill, S. (1990) *Developing London's Docklands: Another Great Planning Disaster?*, London: Paul Chapman

Brundtland Report (1987) *Our Common Future* (World Commission on Environment and Development), Oxford: Oxford University Press

Brunivells, P. and Rodrigues, D. (1989) *Investing in Enterprise: A Comprehensive Guide to Inner City Regeneration and Urban Renewal*, Oxford: Basil Blackwell

Brunskill, I. (1989) *The Regeneration Game: A Regional Approach to Regional Policy*, London: Institute for Public Policy Research

Bruton, M. J. (1980) 'PAG revisited', *Town Planning Review* 48: 134–44

Bruton, M. J. (1985) *Introduction to Transport Planning*, London: UCL Press

Bruton, M. J. and Nicholson, D. (1983) 'Non-statutory plans and supplementary planning guidance', *Journal of Planning and Environment Law* 1983: 432–43

Bruton, M. J. and Nicholson, D. J. (1985) 'Supplementary planning guidance and local plans', *Journal of Planning and Environment Law* 1985: 837–44

Bruton, M. J. and Nicholson, D. J. (1987) *Local Planning in Practice*, London: Hutchinson

Bryant, B. (1995) *Twyford Down: Roads, Campaigning and Environmental Law*, London: Spon

Buchanan, C. D. (1958) *Mixed Blessing: The Motor Car in Britain*, London: Leonard Hill

Buchanan, C. D. (1963) *Traffic in Towns* (Buchanan Report), London: HMSO

Budd, L. and Whimster, S. (eds) (1992) *Global Finance and Urban Living: A Study of Metropolitan Change*, London: Routledge

Building Design Partnership (BDP) (1996) *London's Urban Environment: Planning for Quality* (Government Office for London), London: HMSO

Building Research Establishment (BRE) (1994) *The Noise Climate Around Our Homes*, Watford, Herts.: BRE

Bunce, M. (1994) *The Countryside Ideal: Anglo-American Images of Landscape*, London: Routledge

Bunnell, G. (1995) 'Planning gain in theory and practice: negotiation of agreements in Cambridgeshire', *Progress in Planning* 44, 1: 1–113

Burbridge, V. (1990) *Review of Information on Rural Issues* (Central Research Unit Papers), Edinburgh: Scottish Office

Burchell, R. W. and Listokin, D. (eds) (1982) *Energy and Land Use*, New Brunswick, NJ: Center for Urban Policy Research, Rutgers University

Burchell, R. W. and Sternlieb, G. (eds) (1978) *Planning Theory in the 1980s*, New Brunswick, NJ: Center for Urban Policy Research, Rutgers University

Burke, T. and Shackleton, J. R. (1996) *Trouble in Store: UK Retailing in the 1990s*, London: Institute of Economic Affairs

Burns, D., Hambleton, R., and Hoggett, P. (1994) *The Politics of Decentralisation: Revitalising Local Democracy*, London: Macmillan

Burton, T. P. (1989) 'Access to environmental information: the UK experience of water registers', *Journal of Environmental Law* 1: 192–208

Burton, T. P. (1992) 'The protection of rural England', *Planning Practice and Research* 7, 1 (Spring 1992): 37–40

Business in the Community (1990) *Leadership in the Community: A Blueprint for Business Involvement in the 1990s*, London: Business in the Community

Business in the Community (1992) *A Measure of Commitment: Guidelines for Measuring Environmental Performance*, London: Business in the Community

Byrne, D. (1989) *Beyond the Inner City*, Milton Keynes, Bucks.: Open University Press

Byrne, S. (1989) *Planning Gain: An Overview. A Discussion Paper*, London: Royal Town Planning Institute

Byrne, T. (1992) *Local Government in Britain* (fifth edition), London: Penguin

Cairncross, F. (1993) *Costing the Earth* (second edition), London: Business Books/Economist Books

Cairncross, F. (1995) *Green Inc: Guide to Business and the Environment*, London: Earthscan

Calder, N., Cavanagh, S., Eckstein, C., Palmer, J., and Stell, A. (1993) *Women and Development Plans*, Newcastle upon Tyne: Department of Town and Country Planning, University of Newcastle

Callies, D. L. and Grant, M. (1991) 'Paying for growth and planning gain: an Anglo-American comparison of development conditions, impact fees and development agreements', *Urban Lawyer* 23: 221–48

Cameron, G. C. (1990) 'First steps in urban policy evaluation in the United Kingdom', *Urban Studies* 27: 475–95

Cameron, G. C., Monk, S., and Pearce, B. J. (1988) *Vacant Urban Land: A Literature Review*, London: Department of the Environment

Campbell, M. (ed)(1990) *Local Economic Policy*, London: Cassell

Campbell, S. and Fainstein, S. S. (1996) *Readings in Planning Theory*, Oxford: Blackwell

Capita Management Consultancy (1996) *An Evaluation of Six Early Estate Action Schemes* (Department of the Environment), London: HMSO

Capner, G. (1994) 'Green belt and rural economy', in Wood, M. (ed) *Planning Icons: Myth and Practice* (Planning Law Conference, *Journal of Planning and Environment Law*), London: Sweet & Maxwell

Card, R. and Ward, R.. (1996) 'Access to the countryside: the impact of the Criminal Justice and Public Order Act 1994', *Journal of Planning and Environment Law* 1996: 447–62

Carley, M. (1990a) 'Neighbourhood renewal in Glasgow: policy and practice', *Housing Review* 39: 49–51

Carley, M. (1990b) *Housing and Neighbourhood Renewal: Britain's New Urban Challenge*, London: Policy Studies Institute

Carley, M. (1991) 'Business in urban regeneration partnerships: a case study in Birmingham', *Local Economy* 6: 100–15

Carmichael, P. (1992) 'Is Scotland different? Local government policy under Mrs Thatcher', *Local Government Policy Making* 18, 5: 25–32

Carnie, J. K. (1996) *The Safer Cities Programme in Scotland* (Scottish Office, Central Research Unit), Edinburgh: HMSO

Carnwath, R. (1991) 'The planning lawyer and the environment', *Journal of Environmental Law* 3: 57–67

Carnwath, R. (1992) 'Environmental enforcement: the need for a specialist court', *Journal of Planning and Environment Law* 1992: 799–808

Carnwath Report (1989) *Enforcing Planning Control: Report by Robert Carnwath QC*, London: HMSO

Carson, R. (1951) *The Sea Around Us*, New York: Oxford University Press

Carson, R. (1962) *Silent Spring*, Harmondsworth: Penguin edition (1965)

Carter, C. and John, P. (1992) *A New Accord: Promoting Constructive Relations Between Central and Local Government*, York: Joseph Rowntree Foundation

Carter, N., Brown, T., and Abbott, T. (1991) *The Relationship Between Expenditure-based Plans and Development Plans*, Leicester: School of the Built Environment, Leicester Polytechnic

Carter, N., Oxley, M., and Golland, A. (1996) 'Towards the market: Dutch physical planning in a UK perspective', *Planning Practice and Research* 11: 49–60

Castells, M. (1991) *The Informational City: Economic Restructuring and Urban Development*, Oxford: Blackwell

Catalano, A. (1983) *A Review of Enterprise Zones*, London: CES

Central Housing Advisory Committee (1967) *The Needs of New Communities*, London: HMSO

Central Office of Information (COI) (1996) *Britain 1996: Official Handbook*, London: HMSO

Centre for Environmental Studies (1970) *Observations on the Greater London Development Plan*, London: CES (reprinted in Cullingworth, J. B. (ed) *Planning for Change* (vol. 3 of *Problems of an Urban Society*), London: Allen & Unwin, 1973)

Centre for Environmental Studies (1973) 'Education for planning', *Progress in Planning* 1: 1–100

Centre for Local Economic Strategies (1990) *Inner City Regeneration: A Local Authority Perspective*, Manchester: CLES

Centre for Local Economic Strategies (1991) *City Centres, City Cultures: The Role of the Arts in the Revitalisation of Towns and Cities*, Manchester: CLES

Centre for Local Economic Strategies (1992a) *Reforming the TECs: Towards a Strategy. Final Report of the CLES TEC/LEC Monitoring Project*, Manchester: CLES

Centre for Local Economic Strategies (1992b) *Social Regeneration: Directions for Urban Policy in the 1990s*, Manchester: CLES

Cervero, R. (1990) 'Transit pricing research: a review and synthesis', *Transportation* 17: 117–39

Cervero, R. (1995) 'Planned communities, self-containment and commuting: a cross-national perspective', *Urban Studies* 32: 1135–1161

Chadwick, G. F. (1978) *A Systems View of Planning: Towards a Theory of the Urban and Regional Planning Process*, Oxford: Pergamon

Champion, A. G. (1989) *Counterurbanisation: The Changing Pace and Nature of Population Deconcentration*, London: Edward Arnold

Champion, A. G. and Fielding, A. J. (eds) (1982) *Migration Processes and Patterns*, vol. 1: *Research Progress and Prospects*, London: Belhaven

Champion, A. G. and Green, A. E. (1992) 'Local economic performance in Britain during the late 1980s: the results of the third booming towns study', *Environment and Planning A* 24: 243–72

Champion, T. and Watkins, C. (1991) *People in the Countryside: Studies of Social Change in Rural Britain*, London: Paul Chapman

Chandler, J. A. (1996) *Local Government Today* (second edition), Manchester: Manchester University Press

Chapman, D. and Larkham, P. (1992) *Discovering the Art of*

Relationship, Birmingham: Faculty of the Built Environment, Birmingham Polytechnic

Charles, HRH The Prince of Wales (1989) *A Vision of Britain. A Personal View of Architecture*, London: Doubleday

Chartered Institute of Environmental Health (1995) *Travellers and Gypsies: An Alternative Strategy*, London: CIEH

Chartered Institute of Transport (1990) *Paying for Progress: A Report on Congestion and Road Use Charges*, London: CIT

Chartered Institute of Transport (1991) *London's Transport: The Way Ahead*, London: CIT

Chartered Institute of Transport (1992) *Paying for Progress: A Report on Congestion and Road Use Charges.: Supplementary Report*, London: CIT

Checkland, S. G. (1981) *The Upas Tree: Glasgow 1875–1975 and after 1975–1980*, Glasgow: University of Glasgow Press

Cherry, G. E. (1974) *The Evolution of British Town Planning*, London: Leonard Hill

Cherry, G. E. (1975) *National Parks and Access to the Countryside: Environmental Planning 1939–1969*, vol. 2, London: HMSO

Cherry, G. E. (1981) *Pioneers in British Planning*, London: Architectural Press

Cherry, G. E. (1982) *The Politics of Town Planning*, London: Longman

Cherry, G. E. (1988) *People and Plans: The Shaping of Urban Britain in the Nineteenth and Twentieth Centuries*, London: Edward Arnold

Cherry, G. E. (1994a) *Birmingham: A Study in Geography, History and Planning*, Chichester, W. Sussex: John Wiley

Cherry, G. E. (1994b) *Rural Change and Planning: England and Wales in the Twentieth Century*, London: Spon

Cherry, G. E. (1996) *Town Planning in Britain since 1900: The Rise and Fall of the Planning Ideal*, Oxford: Blackwell

Cherry, G. E. and Penny, L. (1990) *Holford: A Study in Architecture, Planning and Civic Design*, London: Spon

Cherry, G. E. and Rogers, A. (1996) *Rural Change and Planning: England and Wales in the Twentieth Century*, London: Spon

Cheshire, P. and Sheppard, S. (1989) 'British planning policy and access to housing: some empirical estimates', *Urban Studies* 26: 469–85

Cheshire, P., D'Arcy, E., and Giussani, B. (1991) *Local, Regional and National Government in Britain: A Dreadful Warning*, Reading: Department of Economics, University of Reading

Cheshire County Council (1995) *Cheshire County Structure Plan: New Thoughts for the Next Century*, Chester: Cheshire County Environmental Planning Service

Chesman, G. R. (1991) 'Local authorities and the review of the definitive map under the Countryside and Wildlife Act 1981', *Journal of Planning and Environment Law* 1991: 611–14

Chisholm, M. (1995) *Britain on the Edge of Europe*, London: Routledge

Christensen, T. (1979) *Neighbourhood Survival*, Dorchester: Prism Press

Chubb, R. N. (1988) *Urban Land Markets in the United Kingdom* (Department of the Environment), London: HMSO

Church, A. (1988) 'Urban regeneration in London Docklands: a five-year review', *Environment and Planning C: Government and Policy* 6: 187–208

Churchill, R., Warren, L. M., and Gibson, J. (eds) (1991) *Law, Policy and the Environment*, Oxford: Blackwell

Civic Trust (1991) *Audit of the Environment*, London: The Trust

Clabon, S. and Chance, C. (1992) 'Legal profile: freedom of access to environmental information', *European Environment* 2, 3: 24–5

Clark, A., Lee, M., and Moore, R. (1996) *Landslide Investigation and Management in Great Britain* (a support document for PPG 14) (Department of the Environment), London: HMSO

Clark, D. M. (1992) *Rural Social Housing: Supply and Trends. A 1992 Survey of Affordable New Homes*, Cirencester, Glos.: Association of Community Councils in Rural England

Clark, G., Darrall, J., Grove-White, R., Macnaghten, P., and Urry, J. (1994) *Leisure Landscapes: Leisure, Culture and the English Countryside. Challenges and Conflicts*, London: Council for the Protection of Rural England

Clark, M., Smith, D., and Blowers, A. (1992) *Waste Location: Spatial Aspects of Waste Management, Hazards, and Disposal*, London: Routledge

Clawson, M. and Hall, P. (1973) *Land Planning and Urban Growth*, Baltimore MD: Johns Hopkins University Press

Clayton, A. M. H. and Radcliffe, N. J. (1996) *Sustainability: A Systems Approach*, London: Earthscan

CLES: *see* Centre for Local Economic Strategies

Cloke, P. (ed) (1987) *Rural Planning: Policy into Action?*, London: Paul Chapman

Cloke, P. (ed) (1988) *Policies and Plans for Rural People*, London: Allen & Unwin

Cloke, P. (ed) (1992) *Policy and Change in Thatcher's Britain*, Oxford: Pergamon

Cloke, P. (1993) 'On problems and solutions: the reproduction of problems for rural communities in Britain during the 1980s', *Journal of Rural Studies* 9: 113–21

Cloke, P. and Little, J. (1990) *The Rural State: Limits to Planning in Rural Society*, Oxford: Clarendon Press

Cloke, P., Doel, M., Matless, D., Phillips, M., and Thrift, N. (1994a) *Writing the Rural: Five Cultural Geographies*, London: Paul Chapman

Cloke, P., Milbourne, P. and Thomas, C. (1994b) *Lifestyles in Rural England*, Salisbury, Wilts.: Rural Development Commission

Clotworthy, J. and Harris, N. (1996) 'Planning policy implications of local government reorganization', in Tewdwr-Jones, M. (ed) *British Planning Policy in Transition: Planning in the 1990s*, London: UCL Press

Coates, D. (1994) 'Wanted? More land, enquire within'. *House Builder* April: 29–30

Coccossis, H. and Nijkamp, P. (eds) (1995) *Sustainable Tourism Development*, Aldershot, Hants.: Avebury

Cochrane, A. (1991) 'The changing state of local government: restructuring for the 1990s', *Public Administration* 69: 281–302

Cochrane, A. (1993) *Whatever Happened to Local Government?*, Buckingham: Open University Press

Cochrane, A. and Clarke, A. (1990) 'Local enterprise boards: the short history of a radical initiative ', *Public Administration* 68: 315–36

Cocks, R. (1991) 'First responses to the new "breach of condition" notice', *Journal of Planning and Environment Law* 1991: 409–18

Cole, I. and Smith, Y. (1996) *From Estate Action to Estate Agreement: Regeneration and Change on the Bell Farm Estate, York*, Bristol: Policy Press

Cole, J. and Cole, F. (1993) *The Geography of the European Community*, London: Routledge

Coleman, A. (1990) *Utopia on Trial* (second edition), London: Hilary Shipman

Colenutt, B. and Cutten, A. (1994) 'Community empowerment in vogue or vain?', *Local Economy* 9: 236–50

Collar, N. A. (1994) *Green's Concise Scots Law: Planning*, Edinburgh: W. Green

Collins, L. (1996) 'Recycling and the environmental debate: a question of social conscience or scientific reason?' *Journal of Environmental Planning and Management* 39: 333–55

Collins, M. P. (1989) 'A review of 75 years of planning education at UCL' *Planner* 75, 6: 18–22

Collins, M. P. and McConnell, S. (1988) *The Use of Local Plans for Effective Town and Country Planning: Report of a Research Project Funded by the Nuffield Foundation*, London: Bartlett School of Architecture and Planning, University College London

Commission for Racial Equality (1982) *Local Government and Racial Equality*, London: CRE

Commission on Social Justice (1994) *Social Justice: Strategies for National Renewal*, London: Vintage

Confederation of British Industry (1988) *Initiatives Beyond Charity: Report of the CBI Task Force on Business and Urban Regeneration*, London: CBI

Confederation of British Industry (1995) *Missing Links: Setting National Transport Priorities*, London: CBI

Confederation of British Industry (1996) *Winning Ways: Developing the UK Transport Network*, London: CBI

Consortium Developments (1985) *New Country Towns*, London: Consortium Developments

Cook, A. J. and Davis, A. L. (1993) *Package Approach Funding: A Survey of English Highway Authorities*, London: Friends of the Earth

Cooke, P. and Morgan, K. (1993) 'The network paradigm: new departures in corporate and regional development', *Environment and Planning D* 11: 543–64

Coombes, M., Raybould, S., and Wong, C. (1992) *Developing Indicators to Assess the Potential for Urban Regeneration* (Department of the Environment), London: HMSO

Coombes, T., Fidler, P., and Hathaway, A. (1992) 'South West regional planning review: towards a regional strategy', in Minay, C. L. W., 'Developing regional guidance in England and Wales: a review symposium', *Town Planning Review* 63: 415–34

Coon, A. (1988) 'Local plan provision: the record to date and prospects for the future', *Planner* 74, 5: 17–20

Coon, A. (1989) 'An assessment of Scottish development planning', *Planning Outlook* 32, 2: 77–85

Cooper, D. E. and Palmer, J. A. (eds) (1992) *The Environment in Question*, London: Routledge

Coopers & Lybrand (1985) *Land Use Planning and the Housing Market: An Assessment*, London: Coopers & Lybrand

Coopers & Lybrand (1987) *Land Use Planning and Indicators of Housing Demand*, London: Coopers & Lybrand

Coopers & Lybrand (1994) *Employment Department Baseline Follow-up Studies: Final Overall Report to the Employment Department*, London: Coopers & Lybrand

Cornford, J. and Gillespie, A. (1992) 'The coming of the wired city? The recent development of cable in Britain', *Town Planning Review* 63: 243–64

Costonis, J. J. (1989) *Icons and Aliens: Law, Aesthetics, and Environmental Change*, Urbana-Champaign: University of Illinois Press

Couch, C. (1990) *Urban Renewal: Theory and Practice*, London: Macmillan

Couch, C. and Gill, N. (1993) *Renewal Areas: A Review of Progress*, Working Paper 119, Bristol: Bristol University School for Advanced Urban Studies

Council for the Protection of Rural England (1988) *Welcome Homes: Housing Supply from Unallocated Land*, London: CPRE

Council for the Protection of Rural England (1990a) *From White Paper to Green Future*, London: CPRE

Council for the Protection of Rural England (1990b) *Our Finest Landscapes: CPRE Submission to the National Parks Review Panel*, London: CPRE

Council for the Protection of Rural England (1990c) *Future Harvest: The Economics of Farming and the Environment. Proposals for Action* (Jenkins, T. N.), London: CPRE and WWF

Council for the Protection of Rural England (1990d) *Planning Control over Farmland*, London: CPRE

Council for the Protection of Rural England (1991) *Energy Conscious Planning* (Owens, S.), London: CPRE

Council for the Protection of Rural England (1992a) *Campaigners Guide to Using EC Environmental Law* (Macrory, R.), London: CPRE

Council for the Protection of Rural England (1992b) *Campaigners Guide to Local Plans* (Green Balance), London: CPRE

Council for the Protection of Rural England (1992c) *Our Common Home: Housing Development and the South East's Environment*, London: CPRE

Council for the Protection of Rural England (1992d) *Transport and the Environment: CPRE's Submission to the Royal Commission on Environmental Pollution*, London: CPRE

Council for the Protection of Rural England (1992e) *Where Motor Car is Master: How the Department of Transport Became Bewitched by Roads* (Kay, P. and Evans, P.), London: CPRE

Council for the Protection of Rural England (1992f) *The Lost Land: Land Use Change in England 1945–1990* (Sinclair, G.), London: CPRE

Council for the Protection of Rural England (1993a) *Index of National Planning Policies*, London: CPRE

Council for the Protection of Rural England (1993b) *Preparing for the Future: A Response by the CPRE to the Consultation Paper on the UK Strategy for Sustainable Development*, London: CPRE

Council for the Protection of Rural England (1993c) *Water for Life: Strategies for Sustainable Water Resource Management*, London: CPRE

Council for the Protection of Rural England (1994a) *Leisure Landscapes: Leisure, Culture and the English Countryside. Challenges and Conflicts* (Clark, G., Darrall, J., Grove-White, R., Macnaghten, P., and Urry, J.), London: CPRE

Council for the Protection of Rural England (1994b) *Environmental Policy Omissions in Development Plans*, London: CPRE

Council for the Protection of Rural England (1994c) *The Housing Numbers Game*, London: CPRE

Council for the Protection of Rural England (1994d) *Public Access to Planning Documents*, London: CPRE

Council for the Protection of Rural England (1994e) *Coal Mining and Colliery Spoil Disposal: Revision of MPG 3*, London: CPRE

Council for the Protection of Rural England (1995a) *Renewable*

Energy in the UK: Financing Options for the Future (Mitchell, C.), London: CPRE

Council for the Protection of Rural England (1995b) *The End of Hierarchy: A New Perspective on Managing the Road Network* (Goodwin, P.), London: CPRE

Council for the Protection of Rural England (1995c) *Circular Projections: Household Growth, Housing Development, and the Household Projections* (Bramley, G. and Watkins, C.), London: CPRE

Council for the Protection of Rural England (1996a) *The Campaigners' Guide to Minerals*, London: CPRE

Council for the Protection of Rural England (1996b) *Growing Greener: Sustainable Agriculture in the UK* (Baldock, D., Bishop, K., Mitchell, K., and Phillips, A.), London: CPRE

Council of Europe (1989) *European Campaign for the Countryside: Conclusion and Declarations*, Strasbourg: The Council

Council of Europe (1991) *The Bern Convention of Nature Conservation*, Strasbourg: The Council

Council of Europe (1992) *The European Urban Charter*, Strasbourg: Council of Europe Standing Conference of Local and Regional Authorities of Europe

Countryside Commission: *see* appendix on *Official Publications*

Countryside Council for Wales: *see* appendix on *Official Publications*

County Planning Officers Society (annual) *Opencast Coalmining Statistics*, CPOS (published annually by Durham County Council)

County Planning Officers Society (1991) *Regional Guidance and Regional Planning Conferences {Progress Report}*, CPOS

County Planning Officers Society (1993) *Planning for Sustainability*, CPOS, Winchester: County Planning Department, Hampshire County Council

County Planning Officers Society, Metropolitan Planning Officers Society, and District Planning Officers Society (1992) *Planning in the Urban Fringe: Final Report of the Joint Special Advisory Group*, CPOS, Middlesbrough: Department of Environment, Development and Transportation, Cleveland County Council

Cowan, R. (1995) *Planning Aid Handbook*, London: Royal Town Planning Institute

Cox, A. (1984) *Adversary Politics and Land*, Cambridge: Cambridge University Press

Cox, G., Lowe, P., and Winter, M. (1986) *Agriculture, People and Policies*, London: Allen & Unwin

Cox, G., Lowe, P., and Winter, M. (1990) *The Voluntary Principle in Conservation*, Chichester, W Sussex: Packard

Crabtree, J. R. and Chalmers, N. A. (1994) 'Economic evaluation of policy instruments for conservation: standard payments and capital grants', *Land Use Policy* 11: 94–106

Crawford, C. (1989) 'Profitability and its role in planning decisions', *Journal of Environmental Law* 1: 221–44

Cripps, J. (1977) *Accommodation for Gypsies: A Report on the Working of the Caravan Sites Act 1968*, London: HMSO

Croall, J. (1995) *Preserve or Destroy: Tourism and the Environment*, London: Calouste Gulbenkian Foundation

Cross, D. (1992) 'Regional planning guidance for East Anglia', in Minay, C. L. W., 'Developing regional guidance in England and Wales: a review symposium', *Town Planning Review* 63: 415–34

Cross, D. and Bristow, R. (eds) (1983) *English Structure Planning*, London: Pion

Crouch, C. and Marquand, D. (eds) (1989) *The New Centrism: Britain out of Step with Europe?*, Oxford: Blackwell

Crow, S. (1996) 'Lessons from *Bryan*', *Journal of Planning and Environment Law* 1996: 359–69

Cuddy, M. and Hollingsworth, M. (1985) 'The review process in land availability studies: bargaining positions for builders and planners', in Barrett, S. and Healey, P. (eds) *Land Policy: Problems and Alternatives*, Aldershot, Hants.: Avebury

Cullinane, S. (1992) 'Attitudes towards the car in the UK: some implications for policies on congestion and the environment', *Transportation Research* 26A: 291–301

Cullingworth, J. B. (1960a) 'Household formation in England and Wales', *Town Planning Review* 31: 5–26

Cullingworth, J. B. (1960b) *Housing Needs and Planning Policy: A Restatement of the Problems of Housing Need and 'Overspill' in England and Wales*, London: Routledge & Kegan Paul

Cullingworth, J. B. (ed) (1973) *Planning for Change* (vol. 3 of *Problems of an Urban Society*), London: Allen & Unwin

Cullingworth, J. B. (1975) *Reconstruction and Land Use Planning 1939–1947: Environmental Planning 1939–1969*, vol. 1, London: HMSO

Cullingworth, J. B. (1979) *New Towns Policy: Environmental Planning 1939–1969*, vol. 3, London: HMSO

Cullingworth, J. B. (1981) *Land Values, Compensation and Betterment: Environmental Planning 1939–1969*, vol. 4, London: HMSO

Cullingworth, J. B. (1987) *Urban and Regional Planning in Canada*, New Brunswick, NJ: Transaction

Cullingworth, J. B. (ed) (1990) *Energy, Land, and Public Policy* (Energy Policy Studies, vol. 5), New Brunswick, NJ: Transaction

Cullingworth, J. B. (1993) *The Political Culture of Planning: American Land Use Planning in Comparative Perspective*, New York and London: Routledge

Cullingworth, J. B. (1997) *Planning in the USA*, New York: Routledge

Cullingworth, J. B. and Karn, V. A. (1968) *The Ownership and Management of Housing in the New Towns*, London: HMSO

Cullingworth Report (1967) *Scotland's Older Houses* (Report of the Scottish Housing Advisory Committee, Subcommittee on Unfit Housing in Scotland), Edinburgh: HMSO

Curry, N. (1992a) 'Nature conservation, countryside strategies and strategic planning', *Journal of Environmental Management* 35: 79–91

Curry, N. (1992b) 'Controlling development in the national parks of England and Wales', *Town Planning Review* 63: 107–21

Curry, N. (1993) 'Negotiating gains for nature conservation in planning practice', *Planning Practice and Research* 8, 2: 10–15

Curry, N. (1994) *Countryside Recreation, Access and Land Use Planning*, London: Spon

Curry, N. and Peck, C. (1993) 'Planning on presumption: strategic planning for countryside recreation in England and Wales', *Land Use Policy* 10: 140–50

Cyclists Public Affairs Group (1995) *Investing in the Cycling Revolution*, Godalming, Surrey: Cyclists Touring Club

Cyclists Touring Club (1993) *Cycle Policies in Britain: The 1993 CTC Survey*, Godalming, Surrey: CTC

Dabinett, G. and Lawless, P. (1994) 'Urban transport investment and regeneration: researching the impact of South Yorkshire Supertram', *Planning Practice and Research* 9: 407–14

Dales, J. H. (1968) *Pollution, Property and Prices*, Toronto: University of Toronto Press

Dalziel, M. and Rowan-Robinson, J. (1986) 'Resurrecting the two price system for land', *Journal of Planning and Environment Law* 1986: 409–15

Damer, S. and Hague, C. (1971) 'Public participation in planning: evolution and problems', *Town Planning Review* 42: 217–24

Damesick, P. J. (1986) 'The M25: a new geography of development?', *Geographical Journal* 152: 155–60

Daniels, S. (1993) *Fields of Vision: Landscape Imagery and National Identity in England and the United States*, Oxford: Polity Press

Danson, M. W., Lloyd, M. G., and Newlands, D. (1989) 'Rural Scotland and the rise of Scottish Enterprise', *Planning Practice and Research* 4, 3: 13–17

Danson, M. W., Lloyd, M. G., and Newlands, D. (1992) *The Role of Regional Development Agencies in Economic Regeneration*, London: Jessica Kingsley

Darley, G., Hall, P., and Lock, D. (1991) *Tomorrow's New Communities*, York: Joseph Rowntree Foundation

Dasgupta, M., Oldfield, R., Sharman, K., and Webster, V. (1994) *Impact of Transport Policies in Five Cities*, Crowthorne, Berks.: Transport Research Laboratory

Davidoff, P. and Reiner, T. A. (1962) 'A choice theory of planning', *Journal of the American Institute of Planners* 28: 103–15 (reprinted in Faludi, A., *A Reader in Planning Theory*, Oxford: Pergamon, 1973)

Davies, C., Pritchard, D., and Austin, L. (1992) *Planscan Scotland: A Study of Development Plans in Scotland*, Sandy, Beds.: Royal Society for the Protection of Birds

Davies, H. W. E. (1992) 'Britain 2000: the impact of Europe for planning and practice', *Planner: TCPSS Proceedings* 78, 21: 21–2

Davies, H. W. E. (1994) 'Towards a European planning system?', *Planning Practice and Research* 9: 63–9

Davies, H. W. E. (1996) 'Planning and the European question', in Tewdwr-Jones, M. (ed), *British Planning Policy in Transition: Planning in the 1990s*, London: UCL Press

Davies, H. W. E. and Gosling, J. A. (1994) *The Impact of the European Community on Land Use Planning in the United Kingdom*, London: Royal Town Planning Institute

Davies, H. W. E., Edwards, D., and Rowley, A. R. (1984) 'The relevance of development control', *Town Planning Review* 51: 5–24

Davies, H. W. E., Edwards, D., and Rowley, A. R. (1986a) *The Relationship between Development Plans, Development Control, and Appeals*, Reading: Department of Land Management and Development, University of Reading

Davies, H. W. E., Edwards, D., Roberts, C., Rosborough, L., and Sales, R. (1986b) *The Relationship between Development Plans and Appeals*, Reading: Department of Land Management and Development, University of Reading

Davies, H. W. E., Rowley, A. R., Edwards, D., Blom-Cooper, A., Roberts, C., Rosborough, L., and Tilley, R. (1986c) *The Relationship between Development Plans and Development Control*, Reading: Department of Land Management and Development, University of Reading

Davies, H. W. E., Edwards, D., and Rowley, A. R. (1989a) *The Approval of Reserved Matters following Outline Planning Permission* (Department of the Environment), London: HMSO

Davies, H. W. E., Hooper, A. J., and Edwards, D. (1989b) *Planning Control in Western Europe*, London: HMSO

Davies, J. G. (1972) *The Evangelistic Bureaucrat: A Study of a Planning Exercise in Newcastle-upon-Tyne*, London: Tavistock

Davies, L. (1996) 'Equality and planning: race' and 'Equality and planning: gender and disability', in Greed, C. (ed), *Implementing Town Planning: The Role of Town Planning in the Development Process*, Harlow, Essex: Longman

Davies, P. and Keate, D. (1995) *In the Public Interest: London's Civic Architecture at Risk*, London: English Heritage

Davies, R. (1995) *Retail Planning Policies in Western Europe*, London: Routledge

Davies, R. L. and Campion, A. G. (1983) *The Future of the City Centre* (Institute of British Geographers) London: Academic Press

Davis, K. C. (1969) *Discretionary Justice: A Preliminary Inquiry*, Baton Rouge: Louisiana State University Press

Davis, M. (1990) *City of Quartz: Excavating the Future in Los Angeles*, London: Verso

Davison, I. (1990) *Good Design in Housing: A Discussion Paper*, London: House Builders Federation/Royal Institute of British Architects

Davison, R. C. (1938) *British Unemployment Policy: The Modern Phase*, London: Longman

Davoudi, S. (1995) 'City challenge: the three-way partnership', *Planning Practice and Research* 10: 333–44

Davoudi, S. and Healey, P. (1994) *Perceptions of City Challenge Policy Processes: The Newcastle Case*, Newcastle upon Tyne: Department of Town and Country Planning, University of Newcastle upon Tyne

Davoudi, S. and Healey, P. (1995) 'City challenge: sustainable process or temporary gesture?', *Environment and Planning C: Government and Policy* 13: 79–95

Dawson, D. (1985) 'Economic change and the changing role of local government', in Lochlin, M., Gelfand, M. D., and Young, K. (eds), *Half a Century of Municipal Decline*, London: Allen & Unwin

Dawson, J. A. (1994) *Review of Retailing Trends, with Particular Reference to Scotland*, Edinburgh: Scottish Office Central Research Unit Papers

Dawson, J. and Walker, C. (1990) 'Mitigating the social costs of private development: the experience of linkage programmes in the United States', *Town Planning Review* 61: 157–70

Deakin, N. and Edwards, J. (1993) *The Enterprise Culture and the Inner Cities*, London: Routledge

Dear, M. and Scott, A. J. (eds) (1981) *Urbanization and Urban Planning in a Capitalist Society*, London: Methuen

Debenham, Tewson & Chinnocks (1988) *Planning Gain: Community Benefit or Commercial Bribe?*, London: Debenham, Tewson & Chinnocks

de Groot, L. (1992) 'City challenge: competing in the urban regeneration game', *Local Economy* 7: 196–209

Deimann, S. (1994) *Your Rights under European Union Environment Legislation*, Brussels: European Environmental Bureau

Delafons, J. (1990a) *Development Impact Fees and Other Devices*, Berkeley, CA: Institute of Urban and Regional Development, University of California at Berkeley

Delafons, J. (1990b) *Aesthetic Control: A Report on Methods Used in the USA to Control the Design of Buildings*, Berkeley, CA: Institute of Urban and Regional Development, University of California at Berkeley

Dennington, V. N. and Chadwick, M. J. (1983) 'Derelict land and waste land: Britain's neglected land resource', *Journal of Environmental Management* 16: 229–39

Denington Report (1966) *Our Older Homes: A Call for Action* (Central Housing Advisory Committee), London: HMSO

Dennis, N. (1970) *People and Planning*, London: Faber & Faber

Dennis, N. (1972) *Public Participation and Planners' Blight*, London: Faber & Faber

Department of the Environment: *see* appendix on *Official Publications*

Department of National Heritage (annual) *Annual Report* (HMSO)

Department of National Heritage (1996) *Protecting Our Heritage: A Consultation Paper on the Built Heritage of England and Wales*

Department of Transport: *see* appendix on *Official Publications*

Derby City Council (1989) *Sir Francis Ley Industrial Park: Simplified Planning Zoning. The First Twelve Months*, Derby: The Council

Derby City Council (1991) *Sir Francis Ley Industrial Park: Simplified Planning Zoning. The First Three Years*, Derby: The Council

Derthick, M. (1972) *New Towns In-town: Why a Federal Program Failed*, Washington, DC: Urban Institute

De Soissons, M. (1988) *Welwyn Garden City*, Cambridge: Publications for Companies

Devon County Council (1991) *Traffic Calming Guidelines*, Exeter: The Council

Diamond, D. (1979) 'The uses of strategic planning: the example of the National Planning Guidelines in Scotland', *Town Planning Review* 50: 18–25

Diamond, D. and Spence, N. (1989) *Infrastructure and Industrial Costs in British Industry*, London: HMSO

Dicken, P. (1992) *Global Shift: The Internationalization of Economic Activity*, London: Paul Chapman

Diefendorf, J. (1990) *Rebuilding Europe's Blitzed Cities*, Basingstoke, Hants.: Macmillan

Distributive Trades Economic Development Committee (1988) *The Future of the High Street*, London: NEDO

District Planning Officers Society (1992) *Affordable Housing*, DPOS

Dobry Report (1974a) *Control of Demolition*, London: HMSO

Dobry Report (1974b) *Review of the Development Control System: Interim Report*, London: HMSO

Dobry Report (1975) *Review of the Development Control System: Final Report*, London: HMSO

Dobry, G., Hart, G., Robinson, P., and Williams, A. (1996) *Blundell and Dobry: Planning Applications, Appeals and Proceedings*, London: Sweet & Maxwell

Dobson, A. (1990) *Green Political Thought: An Introduction*, London: HarperCollins

Dobson, A. (1991) *A Green Reader*, London: Andre Deutsch

Dobson, A. (1995) 'No environmentalism without democratisation', *Town and Country Planning* 64: 322–3

Docklands Consultative Committee (1991) *Ten Years of Docklands: How the Cake Was Cut*, London: DCC

Docklands Consultative Committee (1992) *All That Glitters is Not Gold: A Critical Assessment of Canary Wharf*, London: DCC

Dodd, A. M. and Pritchard, D. E. (1993) *RSPB Planscan Northern Ireland: A Study of Development Plans in Northern Ireland*, Sandy, Beds.: Royal Society for the Protection of Birds

Donnison, D. and Middleton, A. (eds) (1987) *Regenerating the Inner City: Glasgow's Experience*, London: Routledge

Doogan, K. (1996) 'Labour mobility and the changing housing market', *Urban Studies* 33: 199–221

Dorling, D. and Atkins, D. (1995) *Population Density, Change and Concentration in Great Britain 1971, 1981, and 1991*, Studies on Medical and Population Subjects 58, Office of Population Censuses and Surveys, London: HMSO

Dower Report (1945) *National Parks in England and Wales*, Cmd 6628, London: HMSO

Dowling, J. A. (1995) *Northern Ireland Planning Law*, Dublin: Gill & Macmillan

Downs, A. (1992) *Stuck in Traffic: Coping with Peak-hour Traffic Congestion*, Washington, DC: Brookings Institution/Lincoln Institute of Land Policy

Draper, P. (1977) *Creation of the Department of the Environment: A Study of the Merger of Three Departments to Form the Department of the Environment*, Civil Service Studies 4, London: HMSO

Drewry, G. and Butcher, T. (1991) *The Civil Service Today* (second edition), Oxford: Blackwell

Drivers Jonas (1992) *Retail Impact Assessment Methodologies*, Edinburgh: Scottish Office Central Research Unit Papers

Drury, P. (1995) 'The value of conservation', *Conservation Bulletin* (English Heritage), July: 20

Dubben, N. and Sayce, S. (1991) *Property Portfolio Management*, London: Routledge

Dunlop, J. (1976) 'The examination in public of structure plans: an emerging procedure', *Journal of Planning and Environment Law* 1976: 8–17

Dunmore, K. (1992) *Planning for Affordable Housing*, London: Institute of Housing

Durman, M. and Harrison, M. (1996) *Bournville 1895–1914*, Birmingham: Article Press, Department of Art, University of Central England

Duxbury, R. M. C. (1996) *Telling and Duxbury's Planning Law and Procedure* (tenth edition), London: Butterworths

Dwyer, J. and Hodge, I. (1996) *Countryside in Trust: Land Management by Conservation, Recreation and Organisation*, Chichester, W. Sussex: John Wiley

Dynes, M. and Walker, D. (1995) *The Times Guide to the New British State: The Government Machine in the 1990s*, London: Times Books

Earp, J. H., Headicar, P., Banister, D., and Curtis, C. (1995) *Reducing the Need to Travel: Some Thoughts on PPG 13*, Oxford: School of Planning, Oxford Brookes University

Easteal, M. (1995) 'A thoroughly modern review', *Local Government Chronicle*, 22 September: 14–15

ECOTEC Research and Consulting (1990) *Dynamics of the Rural Economy*, London: DoE

ECOTEC Research and Consulting (1993) *Review of UK Environmental Expenditure: A Final Report to the Department of the Environment*, London: HMSO

Edinburgh School of Planning and Housing, Edinburgh College of Art, and Heriot-Watt University, and Peter P. C. Allan (Chartered Town Planning Consultants) (1995) *Review of Neighbour Notification*, Edinburgh: Scottish Office Central Research Unit

Edmundson, T. (1993) 'Public participation in development control', *Town and Country Planning Summer School Proceedings*, London: RTPI, 59–62

Edwards, J. (1987) *Positive Discrimination, Social Justice and Social Policy: Moral Scrutiny of a Policy Practice*, London: Tavistock

Edwards, J. (1990) 'What is needed from public policy?', in Healey, P. and Nabarro, R. (eds) *Land and Property Development in a Changing Society*, Aldershot, Hants.: Gower

Edwards, J. and Batley, R. (1978) *The Politics of Positive Discrimination*, London: Tavistock

Edwards, J. and Deakin, N. (1992) 'Privatism and partnership in urban regeneration', *Public Administration* 70: 359–68

Edwards Report (1991) *Fit for the Future: Report of the National Parks Review Panel*, Cheltenham, Glos.: Countryside Commission

Ekins, P. (1986) *The Living Economy: A New Economics in the Making*, London: Routledge

Elcock, H. (1994) *Local Government*, London: Routledge

Elkin, S. H. (1974) *Politics and Land Use Planning: The London Experience*, Cambridge: Cambridge University Press

Elkin, T., McLaren, D., and Hillman, M. (1991) *Reviving the City: Towards Sustainable Development*, London: Policy Studies Institute

Elsom, D. (1996) *Smog Alert: Managing Urban Air Quality*, London: Earthscan

Elson, M. J. (1986) *Green Belts: Conflict Mediation in the Urban Fringe*, London: Heinemann

Elson, M. J. (1990) *Negotiating the Future: Planning Gain in the 1990s*, Gloucester: ARC Ltd

Elson, M. J. and Ford, A. (1994) 'Green belts and very special circumstances', *Journal of Planning and Environment Law* 1994: 594–601

Elson, M. J., Walker, S., and Macdonald, R. (1993) *The Effectiveness of Green Belts* (Department of the Environment), London: HMSO

Elson, M. J., Macdonald, R., and Steenberg, C. in association with Broom, G. (1995a) *Planning for Rural Diversification* (Department of the Environment), London: HMSO

Elson, M. J., Steenberg, C., and Wilkinson, J. (1995b) *Planning for Rural Diversification: A Good Practice Guide* (Department of the Environment), London: HMSO

Elson, M. J., Steenberg, C., and Mendham, N. (1996) *Green Belts and Affordable Housing: Can We Have Both?*, Bristol: Policy Press

Emms, J. E. (1980) 'The Community Land Act: a requiem', *Journal of Planning and Environment Law* 1980: 78–86

English Heritage: *see* appendix on *Official Publications*

English Historic Towns Forum (1992) *Townscape in Trouble: Conservation Areas. The Case for Change*, Bath: The Forum

English Nature: *see* appendix on *Official Publications*

English Tourist Board (annual) *English Heritage Monitor*, London: ETB

English Tourist Board (1991) *Tourism and the Environment: Maintaining the Balance* (prepared in conjunction with the Employment Department Group), London: ETB

English Tourist Board and Civic Trust (1993) *Turning the Tide: A Heritage and Environment Strategy for a Seaside Resort*, London: ETB

Ennis, F. (1994) 'Planning obligations in development plans', *Land Use Policy* 11: 195–207

Environ (1994) *Parking Provision in City Centres: A Lot to be Desired?*, Leicester: Environ

Environ (1995) *Indicators of Sustainable Development in Leicester*, Leicester: Environ

Environment Agency: *see* National Rivers Authority, in appendix on *Official Publications*

Environmental Resources Ltd (1992) *Economic Instruments and Recovery of Resources from Waste* (Department of the Environment), London: HMSO

Erikson, R. A. and Syms, P. M. (1986) 'The effects of enterprise zones on local property markets', *Regional Studies* 20: 1–4

Esher, L. (1981) *A Broken Wave: The Rebuilding of England 1940–1980*, London: Allen Lane (Penguin edition 1983)

Essex County Council (1973) *Design Guide for Residential Areas*, Chelmsford: The Council

Etzioni, A. (1967) 'Mixed-scanning: a "third" approach to decision-making', *Public Administration Review*, December 1967 (reprinted in Faludi, A., *A Reader in Planning Theory*, Oxford: Pergamon, 1973

European Conference of Ministers of Transport (1995) *Urban Travel and Sustainable Development*, Paris: OECD.

European Environmental Bureau (1994) *Your Rights Under European Union Environmental Legislations*, Brussels: EEB

Evans, A. W. (1988) *No Room! No Room! The Costs of the British Town and Country Planning System*, London: Institute of Economic Affairs

Evans, A. W. (1991) 'Rabbit hutches on postage stamps: planning, development and political economy', *Urban Studies* 28: 853–70

Evans, A. W. (1995) 'The property market: ninety per cent efficient?' *Urban Studies* 32: 5–29

Evans, A. W. (1996) 'The impact of land use planning and tax subsidies on the supply and price of housing in Britain: a comment', *Urban Studies* 33: 581–5

Evans, A. W. and Eversley, D. (eds) (1980) *The Inner City: Employment and Industry*, London: Heinemann

Evans, B. (1993) 'Why we no longer need a town planning profession', *Planning Practice and Research* 8: 9–15

Evans, B. (1995) *Experts and Environmental Planning*, Aldershot, Hants.: Avebury

Evans, D. (1992) *A History of Nature Conservation*, London: Routledge

Eve, G. and Department of Land Economy, University of Cambridge (1992) *The Relationship between House Prices and Land Supply* (Department of the Environment), London: HMSO

Eve, Grimley J. R. (1992) *Use of Planning Agreements*, London: HMSO

Eve, G. (Gerald Eve Research) (1995) *Whither the High Street?*, London: Gerald Eve Research

Everest, D. A. (1990) 'The provision of expert advice to government on environmental matters: the role of advisory committees', *Science and Public Affairs* 4: 17–40

Eversley, D. E. C. (1973) *The Planner in Society*, London: Faber & Faber

Eversley, Lord (1910) *Commons, Forests and Footpaths: The Story of the Battle during the Past Forty-five Years for Public Rights over the Commons, Forests and Footpaths of England and Wales*, London: Cassell

Fainstein, S. S. (1994) *The City Builders: Property, Politics, and Planning in London and New York*, Oxford: Blackwell

Fainstein, S. S. and Campbell, S. (1996) *Readings in Urban Theory*, Oxford: Blackwell

Fainstein, S. S., Gordon, I., and Harloe, M. (eds) (1992) *Divided*

Cities: New York and London in the Contemporary World, Oxford: Blackwell

Faludi, A. (1973) *A Reader in Planning Theory*, Oxford: Pergamon

Faludi, A. (1987) *A Decision-centred View of Environmental Planning*, Oxford: Pergamon

Fergusson, A. (1973) *The Sack of Bath: A Record and an Indictment*, Salisbury, Wilts.: Compton Russell

Ferris, J. (1972) *Participation in Urban Planning: The Barnsbury Case. A Study of Environmental Improvement in London*, London: Bell

Fielden, G. B. R., Wickens, A. H., and Yates, I. R. (1994) *Passenger Transport after 2000 AD*, London: Spon (for the Royal Society)

Fielder, S. (1986) *Monitoring the Operation of the Planning System: The National Dimension*, Reading: Department of Land Management and Development, University of Reading

Fielding, T. and Halford, S. (1990) *Patterns and Processes of Urban Change in the United Kingdom* (Department of the Environment Reviews of Urban Research), London: HMSO

Fischer, F. and Forester, J. (eds) (1993) *The Argumentative Turn in Policy Analysis and Planning*, London: UCL Press

Fitzsimmons, D. S. M. (1995) 'Removal of agricultural occupancy conditions: the Northern Ireland controversy', *Journal of Planning and Environment Law* 670–8

Flowers Report (1981) *Coal and the Environment*, London: HMSO

Flynn, A. and Marsden, T. K. (1995) 'Rural change, regulation, and sustainability', *Environment and Planning A* 27: 1180–92

Flynn, A., Leach, S., and Vielba, C. (1985) *Abolition or Reform? The GLC and the Metropolitan County Councils*, London: Allen & Unwin

Foley, P. (1992) 'Local economic policy and job creation: a review of evaluation studies', *Urban Studies* 29: 557–98

Fontaine, P. (1995) *Europe in Ten Lessons*, (second edition), Luxembourg: Office for the Official Publications of the European Communities

Fordham, G. (1995) *Made to Last: Creating Sustainable Neighbourhood and Estate Regeneration*, York: Joseph Rowntree Foundation

Fordham, R. (1989) 'Planning gain: towards its codification', *Journal of Planning and Environment Law* 1989: 577–84

Fordham, R. (1990) 'Planning consultancy: can it serve the public interest?', *Public Administration* 68: 243–8

Forester, J. (1980) 'Critical theory and planning practice', *Journal of the American Planning Association* 46: 275–86

Forester, J. (1982) 'Planning in the face of power', *Journal of the American Planning Association* 48: 67–80

Forman, C. (1989) *Spitalfields: A Battle for Land*, London: Hilary Shipman

Forshaw, J. H. and Abercrombie, P. (1943) *County of London Plan*, London: Macmillan.

Forsyth, J. (1992) 'Tower Hamlets: setting up a regeneration corporation in Bethnal Green', in Bailey, N. and Barker, A. (eds) *City Challenge and Local Regeneration Partnerships: Conference Proceedings*, London: Polytechnic of Central London

Fothergill, S. and Gudgin, G. (1982) *Unequal Growth: Urban and Regional Change in the UK*, London: Heinemann

Fothergill, S. and Guy, N. (1990) *Retreat from the Regions: Corporate Change and the Closure of Factories*, London: Jessica Kingsley/ Regional Studies Association

Fothergill, S., Kitson, M., and Monk, S. (1985) *Urban Industrial Change: The Causes of the Urban–Rural Contrast in Manufacturing*

Employment Change (Department of the Environment), London: HMSO

Fothergill, S., Kitson, M., and Perry, M. (1987) *Property and Industrial Development*, London: Hutchinson

Fox, M. and Turner, S. (1994) 'Northern Ireland', in Freshfields Environment Group (eds) *Tolley's Environmental Handbook*, Croydon: Tolley

Franks Report (1957) *Report of the Committee on Administrative Tribunals and Enquiries*, Cmnd 218, London: HMSO

Freestone, D. (1991) 'European Community environmental policy and law', in Churchill, R., Warren, L. M., and Gibson, J. (eds) *Law, Policy and the Environment*, Oxford: Blackwell

Freilich, R. H. and Bushek, D. W. (eds) (1995) *Exactions, Impact Fees and Dedications*, Chicago: American Bar Association

Fried, M. (1969) 'Social differences in mental health', in Kosa, J., Antonovsky, A., and Zola, I. K., *Poverty and Health: A Sociological Analysis*, Cambridge, MA: Harvard University Press

Friends of the Earth (1989) *Action for People: A Critical Appraisal of Government Inner City Policy*, London: FoE

Friends of the Earth (1991a) *Air Quality and Health*, London: FoE

Friends of the Earth (1991b) *Local Responses to 1989 Traffic Forecasts*, London: FoE

Friends of the Earth (1992) *Less Traffic, Better Towns*, London: FoE

Friends of the Earth (1994a) *Planning for the Planet: Sustainable Development Strategies for Local and Strategic Plans*, London: FoE

Friends of the Earth (1994b) *Working Future: Jobs and the Environment*, London: FoE

Frost, M. and Spence, N. (1993) 'Global city characteristics and Central London's employment', *Urban Studies* 30: 547–58

Froud, J. (1994) 'The impact of ESAs on lowland farming', *Land Use Policy* 11: 107–18

Fry, G. K. (1984) 'The attack on the civil service and the response of the insiders', *Parliamentary Affairs* 37: 353–63

Fudge, C., Lambert, C., and Underwood, J. (1982) Local plans: approaches, preparation and adoption', *Planner* 68 (March/ April): 52–3

Fudge, C., Lambert, C., Underwood, J. and Healey, P. (1983) *Speed, Economy and Effectiveness in Local Plan Preparation and Adoption*, Final Report of the DoE Research Project, School for Advanced Urban Studies Occasional Paper no. 11, University of Bristol

Fulton Report (1968) *Report of the Committee on the Civil Service*, Cmnd 3638, London: HMSO

Gahagan, M. (1992) 'City challenge: a solution to regeneration through partnership?', in Bailey, N. and Barker, A. (eds) *City Challenge and Local Regeneration Partnerships: Conference Proceedings*, London: Polytechnic of Central London

Gans, H. J. (1968) *People and Plans: Essays on Urban Problems*, New York: Basic Books

Gans, H. J. (1991) *People, Plans and Policies: Essays on Poverty, Racism and Other National Urban Problems*, New York: Columbia University Press

Gardner, B. (1996) *Farming for the Future: Policies, Production and Trade*, London: Routledge

Garside, P.L. and Hebbert, M. (eds) (1989) *British Regionalism 1900–2000*, London: Mansell

Gatenby, I. and Williams, C. (1992) 'Section 54A: the legal and practical implications', *Journal of Planning and Environment Law* 1992: 110–20

Geddes, M. (1995) *Poverty, Excluded Communities and Local Democracy*, London: Commission for Local Democracy

Geddes, P. (1915) *Cities in Evolution*, London: Benn

Gentleman, H. (1993) *Counting Travellers in Scotland*, Edinburgh: Scottish Office

Gentleman, H. and Swift, S. (1971) *Scotland's Travelling People*, Edinburgh: HMSO

Geraghty, P. J. (1992) 'Environmental assessment and the application of an expert systems approach', *Town Planning Review* 63: 123–42

Gibbs, D. (1994) 'Towards the sustainable city', *Town Planning Review* 65: 99–109

Gibbs, D., Longhurst, J., and Braithwaite, C. (1996) 'Moving towards sustainable development? Integrating economic development and the environment in local authorities', *Journal of Environmental Planning and Management* 39: 317–32

Gibson, J. (1991) 'The integration of pollution control', in Churchill, R., Warren, L. M., and Gibson, J. (eds) *Law, Policy and the Environment*, Oxford: Blackwell

Gibson, M. S. and Langstaff, M. J. (1982) *An Introduction to Urban Renewal*, London: Hutchinson

Gilbert, A. and Healey, P. (1985) *The Political Economy of Land: Urban Development in an Oil Economy*, Aldershot, Hants.: Gower

Gilg, A. W. (ed) (1983) *Countryside Planning Yearbook 1983*, Norwich: Geo Books

Gilg, A. W. (ed) (1988) *The International Year Book of Rural Planning*, London: Elsevier

Gilg, A. W. (ed) (1991) *Progress in Rural Policy and Planning*, vol. 1: 1991, London: Belhaven

Gilg, A. W. (ed) (1992) *Progress in Rural Policy and Planning*, vol. 2: 1992, London: Belhaven

Gilg, A. W. (ed) (1994) *Progress in Rural Policy and Planning*, Chichester, W. Sussex: John Wiley

Gilg, A. W. (ed) (1995) *Progress in Rural Policy and Planning*, Chichester, W. Sussex: John Wiley

Gillett, E. (1983) *Investment in the Environment: Planning and Transport Policies in Scotland*, Aberdeen: Aberdeen University Press

Gillingwater, D. (1992) 'Regional strategy for the East Midlands', in Minay, C. L. W. 'Developing regional guidance in England and Wales: a review symposium', *Town Planning Review* 63: 415–34

Gilpin, A. (1995) *Environmental Impact Assessment: Cutting Edge for the Twenty-first Century*, Cambridge: Cambridge University Press

Gilpin, A. (1996) *Dictionary of Environmental and Sustainable Development*, Chichester, W. Sussex: John Wiley

Gilroy, R. with Marvin, S. (1993) *Good Practices in Equal Opportunities*, Aldershot, Hants.: Avebury

Ginsburg, L. (1956) 'Green belts in the Bible', *Journal of the Town Planning Institute* 42: 129–30

Glaister, S. and Mulley, C. (1983) *Public Control of the British Bus Industry*, Aldershot, Hants.: Gower

Glass, R. (1959) 'The evaluation of planning: some sociological considerations', in Faludi, A. (1973) *A Reader in Planning Theory*, Oxford: Pergamon

Glass, W. D. (1994) 'Regional rural planning policies', *Town and Country Planning Summer School 1994: Proceedings*: 21–4

Glasson, B. and Booth, P. (1992) 'Negotiation and delay in the development control process', *Town Planning Review* 63: 63–78

Glasson, J. (1992a) *An Introduction to Regional Planning*, London: UCL Press

Glasson, J. (1992b) 'The fall and rise of regional planning in the economically advanced nations', *Urban Studies* 29: 505–31

Glasson, J. (1993) 'The fall and rise of regional planning in the economically advanced nations', in Paddison, R., Money, J., and Lever, B. (eds) *International Perspectives in Urban Studies*, London: Jessica Kingsley

Glasson, J., Therivel, T., and Chadwick, A. (1994) *Introduction to Environmental Impact Assessment*, London: UCL Press

Glasson, J., Godfrey, K., and Goodey, B. with Absalom, H. and Borg, J (1995) *Toward Visitor Impact Management: Visitor Impacts, Carrying Capacity and Management Responses in Europe's Historic Towns and Cities*, Aldershot, Hants.: Avebury

GMA Planning, P-E International, and Jacques & Lewis (1993) *Integrated Planning and Granting of Permits in the EC* (DoE Planning Research Programme), London: HMSO

Goldsmith, F. B. and Warren, S. A. (eds) (1993) *Conservation in Progress*, Chichester, W. Sussex: John Wiley

Goldsmith, M. (1980) *Politics, Planning and the City*, London: Hutchinson

Goldsmith, M. and Newton, K. (1986) 'Central – local government relations: a bibliographic summary of the ESCR research initiative', *Public Administration* 64: 102–8.

Gomez-Ibanez, J. A. and Meyer, J. R. (1993) *Going Private: The International Experience with Transport Privatization*, Washington, DC: Brookings Institution

Goodchild, R. N. and Munton, R. J. C. (1985) *Development and the Landowner: An Analysis of the British Experience*, London: Allen & Unwin

Goodin, R. E. (1992) *Green Political Theory*, Oxford: Polity Press

Goodwin, P. B. (1989) '*The rule of three*: a possible solution to the political problem of competing objectives for road pricing', *Traffic Engineering and Control* 30: 495–7

Goodwin, P. B. (1992) 'A review of new demand elasticities with special reference to short and long run effects of price changes', *Journal of Transport Economics and Policy* 26: 155–69

Goodwin, P. B. (1995) *The End of Hierarchy? A New Perspective on Managing the Road Network*, London: Council for the Protection of Rural England

Goodwin, P., Hallett, S., Kenny, F., and Stokes, G. (1991) *Transport: The New Realism*, Oxford: Transport Studies Unit, University of Oxford

Gore, T. and Nicholson, D. (1991) 'Models of the land development process: a critical review', *Environment and Planning A* 23: 705–30

Gosling Report (1968) *Report of the Footpaths Committee*, London: HMSO

Gowers Report (1950) *Houses of Outstanding Historic and Architectural Interest*, London: HMSO

Graham, S. and Marvin, S. (1996) *Telecommunications and the City: Electronic Spaces, Urban Places*, London: Routledge

Graham, T. (1991) 'The interpretation of planning permissions and a matter of principle'', *Journal of Planning and Environment Law* 1991: 104–12

Graham, T. (1993) 'Presumptions', *Journal of Planning and Environment Law* 1993: 423–8

Graham, T. (1996) 'Contaminated land investigations: how will they work under PPG 23?' *Journal of Planning and Environment Law* 1996: 547–53

Grant, J. S. (1987) 'Government agencies and the Highlands since 1945', *Scottish Geographical Magazine* 103: 95–9

Grant, M. (1979) 'Britain's Community Land Act: a post mortem, *Urban Law and Policy* 2: 359–73

Grant, M. (1982) *Urban Planning Law* (Supplement 1990), London: Sweet & Maxwell

Grant, M. (1986) 'Planning and land taxation', *Journal of Planning and Environment Law* 1986: 92–106

Grant, M. (1992) 'Planning law and the British planning system', *Town Planning Review* 63: 3–12

Grant, M. (1996) *Permitted Development* (second edition), London: Sweet & Maxwell

Grant, M. (ed) (1997) *Encyclopedia of Planning Law and Practice* (6 vols), London: Sweet & Maxwell (looseleaf; updated regularly) [A supplementary updating *Monthly Bulletin* is issued to subscribers]

Gray, C. (1994) *Government Beyond the Centre: Sub-national Politics in Britain*, London: Macmillan

Gray, T. S. (1995) *UK Environmental Policy in the 1990s*, London: Macmillan

Grayson, L. (1990) *Green Belt, Green Fields and the Urban Fringe: The Pressure on Land in the 1980s: A Guide to Sources*, London: London Research Centre

Greater London Council (GLC) (1986) *Race and Planning Guidelines*, London: GLC

Greed, C. (1993) *Introducing Town Planning*, Harlow, Essex: Longman

Greed, C. (1994) *Women and Planning: Creating Gendered Realities*, London: Routledge

Greed, C. (ed) (1996a) *Implementing Town Planning: The Role of Town Planning in the Development Process*, Harlow, Essex: Longman

Greed, C. (ed) (1996b) *Investigating Town Planning: Changing Perspectives and Agendas*, Harlow, Essex: Longman

Green, A., Hasluck, C., Owen, D., and Winnett, C. (1993) *Local Unemployment Change in Britain: Leaps and Lags in the Response to National Economic Cycles*, London: Jessica Kingsley

Green, H., Thomas, M., Iles, N., and Down, D. (1996) *Housing in England 1994/95: A Report of the 1994/95 Survey of English Housing Carried Out by the Social Survey Division of ONS on Behalf of the Department of the Environment*, London: HMSO

Green, R. E. (ed) (1991) *Enterprise Zones: New Directions in Economic Development*, Newbury Park, CA: Sage

Green Balance (1992) *Campaigners' Guide to Local Plans*, London: Council for the Protection of Rural England

Greenhalgh, L., Worpole, K., and Grove-White, R. (1996) *People, Parks and Cities: A Guide to Current Good Practice in Urban Parks* (Department of the Environment), London: HMSO

Greer, A. (1996) *Rural Politics in Northern Ireland: Policy Networks and Agricultural Development since Partition*, Aldershot, Hants.: Avebury

Greer, P. (1992) 'The next steps initiative: an examination of the agency framework documents', *Public Administration* 70: 89–98

Gregory, R. and Pearson, J. (1992) 'The parliamentary ombudsman after twenty-five years', *Public Administration* 70: 469–98

Grieco, M. and Jones, P. M. (1994) 'A change in the policy climate? Current European perspectives on road pricing', *Urban Studies* 31: 1517–32

Griffith, J. A. G. and Street, H. (1964) *A Casebook on Administrative Law*, London: Pitman

Griffiths, D. (1996) *Thatcherism and Territorial Politics: A Welsh Case Study*, Aldershot, Hants.: Avebury

Griffiths, R. (1993) 'The politics of cultural policy in urban regeneration strategies', *Policy and Politics* 21: 39–46

Grove, G.A. (1963) 'Planning and the appellant', *Journal of the Town Planning Institute* 49: 128–33

Groves, P. (1992) 'A chairman's view of a TEC', *Local Government Policy Making* 19, 1: 9–17

Gunther, W. D. and Leathers, C. G. (1987) 'British enterprise zones: a critical assessment', *Review of Regional Studies* 17: 1–2

Guy, C. (1994) *The Retail Development Process*, London: Routledge

Guy, S. and Marvin, S. (1995) *Planning for Water: Space, Time and the Social Organisation of Natural Resources*, Newcastle upon Tyne: Department of Town and Country Planning, University of Newcastle upon Tyne

Gyford, J. (1990) 'The enabling authority: a third model', *Local Government Studies* 17, 1: 1–4

Gyford, J. (1991) *Citizens, Consumers and Councils: Local Government and the Public*, London: Macmillan

Gyford, J., Leach, S., and Game, C. (1989) *The Changing Politics of Local Government*, London: Unwin Hyman

Haar, C. M. (1951) *Land Planning in a Free Society*, Cambridge, MA: Harvard University Press

Hackett, P. (1995) *Conservation and the Consumer: Measuring Environmental Concern*, London: Routledge

Hadjilambrinos, C. J. (1993) *Energy regimes and the development of the European Community*, Ph.D. dissertation, University of Delaware, Newark, DE: College of Urban Affairs and Public Policy, University of Delaware

Hagman, D.G. and Misczynski, D.J. (1978) *Windfalls for Wipeouts: Land Value Capture and Compensation*, St Paul, MN: West

Hague, C. (1984) *The Development of Planning Thought: A Critical Perspective*, London: Hutchinson

Hague, C. (ed) (1996) *Planning and Markets*, Aldershot, Hants.: Avebury

Haigh, N. (1990) *EEC Environmental Policy* (second edition), Harlow, Essex: Longman

Haigh, N. and Irwin, F. (eds) (1990) *Integrated Pollution Control in Europe and North America*, Bonn: Institute for European Environmental Policy; and Washington, DC: Conservation Foundation

Halcrow Fox & Associates, and Birkbeck College, University of London (1986) *Investigating Population Change in Small to Medium-sized Areas*, London: Department of the Environment

Hall, A. C. (1990) 'Generating urban design objectives for local areas: a methodology and case study application to Chelmsford, Essex', *Town Planning Review* 61: 287–309

Hall, A. C. (1996) *Design Control: Towards a New Approach*, Oxford: Heinemann

Hall, D. (1989) 'The case for new settlements', *Town and Country Planning* 58: 111–14

Hall, P. (1980) *Great Planning Disasters*, London: Weidenfeld & Nicolson

Hall, P. (1981a) *The Enterprise Zone Concept: British Origins, American Adaptations*, Berkeley, CA: Institute of Urban and Regional Development, University of California at Berkeley

Hall, P. (ed) (1981b) *The Inner City in Context: The Final Report of the Social Science Research Council Inner Cities Working Party*, London: Heinemann (reprinted 1986, Aldershot, Hants.: Gower)

Hall, P. (1982) 'Enterprise zones: a justification', *International Journal of Urban and Regional Research* 6: 415–21

Hall, P. (1983) 'The Anglo-American connection: rival rationalities in planning theory and practice, 1955–1980', *Environment and Planning B: Planning and Design* 10: 41–6

Hall, P. (1988) *Cities of Tomorrow*, Oxford: Blackwell

Hall, P. (1989) *London 2001*, London: Unwin Hyman

Hall, P. (1991) 'The British enterprise zones', in Green, R. E. (ed) *Enterprise Zones: New Directions in Economic Development*, Newbury Park, CA: Sage

Hall, P. (1992) *Urban and Regional Planning* (third edition), London: Routledge

Hall, P. (1993) 'Planning in the 1990s: an international agenda', *European Planning Studies* 1: 3–12

Hall, P. and Hass-Klau, C. (1985) *Can Rail Save the City?*, Aldershot, Hants.: Gower

Hall, P. and Markussen, A. (eds) (1985) *Silicon Landscapes*, London: Allen & Unwin

Hall, P., Gracey, H., Drewett, R., and Thomas, R. (1973) *The Containment of Urban England*, London: Allen & Unwin

Hall, P., Breheny, M., McQuaid, R., and Hart, D. (1987) *Western Sunrise: Genesis and Growth of Britain's Major High Tech Corridor*, London: Unwin Hyman

Hall, S. (1995) 'The SRB: Taking stock', *Planning week* 27 April: 16–17

Hall, T. (ed) (1991) *Planning and Urban Growth in the Nordic Countries*, London: Spon

Hallett, G. (ed) (1977) *Housing and Land Policies in West Germany and Britain*, London: Macmillan

Hallsworth, A. G. (1992) *The New Geography of Consumer Spending*, London: Frances Pinter

Hallsworth, A. G. (1994) 'Decentralization of retailing in Britain: the breaking of the third wave', *Professional Geographer* 46: 296–307

Halsey, A. J. (ed) (1972) *Educational Priority*, vol. 1: *EPA Problems and Policies*, London: HMSO

Ham, C. and Hill, M. (1993) *The Policy Process in the Modern Capitalist State*, (second edition), London: Harvester Wheatsheaf

Hambleton, R. (1986) *Rethinking Policy Making*, Bristol: School for Advanced Urban Studies, University of Bristol

Hambleton, R. (1990) *Urban Government in the 1990s: Lessons from the USA*, Bristol: School for Advanced Urban Studies, University of Bristol

Hambleton, R. (1991) 'The regeneration of U.S. and British cities', *Local Government Studies* 17, 5: 53–69

Hambleton, R. (1993) 'Issues for urban policy in the 1990s', *Town Planning Review* 64: 313–23

Hambleton, R. and Hoggett, P. (1990) *Beyond Excellence: Quality Government in the 1990s*, Bristol: School for Advanced Urban Studies, University of Bristol

Hambleton, R. and Taylor, R. (eds) (1993) *People in Cities: A Transatlantic Policy Exchange*, Bristol: School for Advanced Urban Studies, University of Bristol

Hambleton, R. and Thomas, H. (eds) (1995) *Urban Policy Evaluation: Challenge and Change*, London: Paul Chapman

Hambleton, R., Stewart, J., and Underwood, J. (1980) *Inner Cities: Management and Resources*, Bristol: School for Advanced Urban Studies, University of Bristol

Hamnett, C. and Randolph, B. (1991) *Cities, Housing and Profits: Flat Break-up and the Decline of Private Renting*, London: UCL Press

Hansen, A. (ed) (1993) *The Mass Media and Environmental Issues*, Chichester, W. Sussex: Belhaven Press

Harding, A. (1991) 'The rise of urban growth coalitions, UK-style?', *Environment and Planning C: Government and Policy* 9: 295–317

Harding, A., Evans, R., Parkinson, M., and Garside, P. (1996) *Regional Government in Britain: An Economic Solution?* (Joseph Rowntree Foundation), Bristol: Policy Press

Hardy, D. (1991a) *From Garden Cities to New Towns: Campaigning for Town and Country Planning*, vol. 1: 1899–1946, London: Spon

Hardy, D. (1991b) *From New Towns to Green Politics: Campaigning for Town and Country Planning*, vol. 2: 1899–1946, London: Spon

Hardy, S., Hart, M., Albrechts, L., and Katos, A. (1995) *An Enlarged Europe: Regions in Competition?*, London: Regional Studies Association and Jessica Kingsley.

Harloe, M. (1995) *The People's Home? Social Rented Housing in Europe*, Oxford: Blackwell

Harman, R. (1995) *New Directions: A Manual of European Best Practice in Transport Planning*, London: Transport 2000

Harrap, P. (1993) *Charging for Road Use Worldwide: A Financial Times Management Report*, London: Financial Times

Harris, N. and Tewdwr-Jones, M. (1995) 'The implications for planning of local government reorganisation in Wales: purpose, process, and practice', *Environment and Planning C: Government and Policy* 13: 47–66

Harris, R. (1992a) 'The Environmental Protection Act 1990: Penalising the polluter', *Journal of Planning and Environment Law* 1992: 515–24

Harris, R. (1992b) 'Integrated pollution control in practice', *Journal of Planning and Environment Law* 1992: 611–23

Harrison, A. (1992) 'What shall we do with the contaminated site?', *Journal of Planning and Environment Law* 1992: 809–16

Harrison, J. (1994) 'Who is my neighbour?' *Journal of Planning and Environment Law* 1994: 219–23

Harrison, M. (1992) 'A presumption in favour of planning permission?', *Journal of Planning and Environment Law* 1992: 121–9

Harrison, M. L. and Mordey, R. (eds) (1987) *Planning Control: Philosophies, Prospects and Practice*, London: Croom Helm

Harrison, P. (1983) *Inside the Inner City: Life under the Cutting Edge*, Harmondsworth, Middx: Penguin

Harrison, R.T. and Hart, M. (eds) (1992) *Spatial Policy in a Divided Nation*, London: Jessica Kingsley

Harrison, T. (1988) *Access to Information in Local Government*, London: Sweet & Maxwell

Hart, T. (1992) 'Transport, the urban pattern and regional change, 1960–2010', *Urban Studies* 29: 483–503

Hart, T. (1993) 'Transport investment and disadvantaged regions: UK and European policies since the 1950s', *Urban Studies* 30: 417–36

Hart, T., Haughton, G., and Peck, J. (1996) 'Accountability and the non-elected local state: calling training and enterprise councils to local account', *Regional Studies* 30: 429–41

Harte, J. D. C. (1989) 'The scope of protection resulting from the designation of Sites of Special Scientific Interest', *Journal of Environmental Law* 1: 245–54

Harte, J. D. C. (1990) 'The scheduling of ancient monuments and the role of interested members of the public in environmental law', *Journal of Environmental Law* 2: 224–49

Hasegawa, J. (1992) *Replanning the Blitzed City Centre*, Buckingham: Open University Press

Hass-Klau, C. (1990) *The Pedestrian and City Traffic*, London: Belhaven

Hass-Klau, C., Nold, I., Bocker, G., and Crampton, G. (1992) *Civilised Streets: A Guide to Traffic Calming*, Brighton: Environmental and Transport Planning

Hastings, A. (1996) 'Unravelling the process of "partnership" in urban regeneration policy', *Urban Studies* 33: 253–68

Hastings, A. and McArthur, A. (1995) 'A comparative assessment of government approaches to partnership with the local community', in Hambleton, R. and Thomas, H. (eds) *Urban Policy Evaluation: Challenge and Change*, London: Paul Chapman

Hastings, A., McArthur, A., and McGregor, A. (1996) *Less than Equal? Community Organisations and Estate Regeneration Partnerships*, Bristol: Policy Press

Haughton, G. (1990) 'Targeting jobs to local people: the British urban policy experience', *Urban Studies* 27: 185–98

Haughton, G. and Hunter, C. (1994) *Sustainable Cities*, London: Jessica Kingsley

Haughton, G. and Lawless, P. (1992) *Policies and Potential: Recasting British Urban and Regional Policies*, London: Regional Studies Association

Hausner, V. A. (1986) *Urban Economic Adjustment and the Future of British Cities: Directions for Urban Policy*, Oxford: Clarendon Press

Hausner, V. A. (ed) (1987) *Critical Issues in Urban Economic Development* (2 vols), Oxford: Clarendon Press

Hausner, V. A. and Robson, B. (1986) *Changing Cities: An Introduction to the Economic and Social Research Council Inner Cities Research Programme*, Swindon: ESRC

Hawes, D. and Perez, B. (eds) (1996) *The Gypsy and the State: The Ethnic Cleansing of British Society* (second edition), Bristol: Policy Press

Hawkins, K. (1984) *Environment and Enforcement Regulation and the Social Definition of Pollution*, Oxford: Oxford University Press

Hayton, K. (1990) *Getting People into Jobs* (Department of the Environment), London: HMSO

Hayton, K. (1992) 'Scottish Enterprise: a challenge to local land use planning?', *Town Planning Review* 63: 265–78

Hayton, K. (1993) 'Scottish Enterprise: a challenge to local land use planning?', *Planning Practice and Research* 8, 2: 5–9

Hayton, K. (1996) 'Planning policy in Scotland', in Tewdwr-Jones, M. (ed) *British Planning Policy in Transition: Planning in the 1990s*, London: UCL Press

Healey, M. J. and Ibery, B. W. (1985) *The Industrialisation of the Countryside*, Norwich: Geo Books

Healey, P. (1979) *Statutory Local Plans: Their Evolution in Legislation and Administrative Interpretations*, Oxford: Department of Town Planning, Oxford Polytechnic

Healey, P. (1983) *Local Plans in British Land Use Planning*, Oxford: Pergamon

Healey, P. (1985) 'The professionalisation of planning in Britain', *Town Planning Review* 56: 492–507

Healey, P. (1986) 'The role of development plans in the British planning system: an empirical assessment', *Urban Law and Policy* 8: 1–32

Healey, P. (1988) 'The British planning system and managing the urban environment', *Town Planning Review* 59: 397–417

Healey, P. (1989) 'Directions for change in the British planning system', *Town Planning Review* 60: 125–49; 'Comments and response' (by Goodchild, R. and Marwick, A., Grant, M., Jones, A., Lyddon, D., and Robinson, D.), *Town Planning Review* 60: 319–32

Healey, P. (1990) 'Places, people and politics: plan-making in the 1990s', *Local Government Policy Making* 17, 2: 29–39

Healey, P. (1991a) 'Urban regeneration and the development industry', *Regional Studies* 25: 97–110

Healey, P. (1991b) 'The content of planning education programmes: some comments from recent British experience', *Environment and Planning B: Planning and Design* 18: 177–89

Healey, P. (1991c) 'Models of the development process', *Journal of Property Research* 8: 219–38

Healey, P. (1992a) 'Development plans and markets', *Planning Practice and Research* 7, 2: 13–20

Healey, P. (1992b) 'The reorganisation of state and market in planning', *Urban Studies* 29: 411–34

Healey, P. (1992c) 'Planning through debate: the communicative turn in planning theory', *Town Planning Review* 63: 143–62

Healey, P. (1992d) 'An institutional model of the development process', *Journal of Property Research* 9: 33–44

Healey, P. (1994a) 'Development plans: new approaches to making frameworks for land use regulation', *European Planning Studies* 2: 39–57

Healey, P. (ed.) (1994) *Trends in Development Plan-making in European Planning Systems: First Report of a Collaborative Project on Innovation in Development Plan-making in Europe*, Centre for Research in European Urban Environments, Department of Town and Country Planning, University of Newcastle upon Tyne

Healey, P. and Barrett, S. M. (1990) 'Structure and agency in land and property development processes', *Urban Studies* 37: 89–104

Healey, P. and Nabarro, R. (1990) *Land and Property Development in a Changing Context*, Aldershot, Hants.: Gower

Healey, P. and Shaw, T. (1993) 'Planners, plans and sustainable development', *Regional Studies* 27: 769–76

Healey, P. and Williams, R. H. (1993) 'European urban planning systems: diversity and convergence', *Urban Studies* 30: 699–718

Healey, P., McDougall, G., and Thomas, M. (eds) (1982) *Planning Theory: Prospects for the 1980s*, Oxford: Pergamon

Healey, P., McNamara, P., Elson, M., and Doak, A. (1988) *Land Use Planning and the Mediation of Change*, Cambridge: Cambridge University Press

Healey, P., Davoudi, S., O'Toole, M., Tavsanoglu, S., and Usher, D. (1992a) *Rebuilding the City: Property-led Urban Regeneration*, London: Spon

Healey, P., Purdue, M., and Ennis, F. (1992b) 'Rationales for planning gain', *Policy Studies* 13: 18–30

Healey, P., Purdue, M., and Ennis, F. (1993) *Gains from Planning: Dealing with the Impacts of Development*, York: Joseph Rowntree Foundation

Healey, P., Cameron, S., Davoudi, S., Graham, S. and Madani-Pour, A. (eds) (1995a) *Managing Cities; The New Urban Context*, Chichester, W. Sussex: John Wiley

Healey, P., Purdue, M., and Ennis, F. (1995b) *Negotiating Development: Rationales and Practice for Development Obligations and Planning Gain*, London: Spon

Hebbert, M. (1992) 'The British garden city: metamorphosis', in Ward, S. (ed) *The Garden City: Past, Present and Future*, London: Spon

Heim, C. (1990) 'The Treasury as developer-capitalist? British new town building in the 1950s', *Journal of Economic History* 50: 903–24

Hendry, J. (1989) 'The control of development and the origins of planning in Northern Ireland', in Bannon, M. J., Nowlan K.I., Hendry, J., and Mawhinney, K. (1989) *Planning: The Irish Experience 1920–1988*, Dublin: Wolfhound Press

Hendry, J. F. (1992) 'Plans and planning policy for Belfast: a review article', *Town Planning Review* 63, 1: 79–85

Hendry, J. F. (1993) 'Conservation in Northern Ireland' in RTPI, *The Character of Conservation Areas*, vol. 2: *Supporting Information*, London: RTPI

Hennessy, P. (1989) *Whitehall*, London: Secker & Warburg

Hennessy, P. (1992) *Never Again: Britain 1945–1951*, London: Jonathan Cape (Vintage edition 1993).

Herbert, D. T. (ed) (1995) *Heritage, Tourism and Society*, London: Mansell

Herbert-Young, N. (1995) 'Reflections on section 54A and *plan-led* decision-making', *Journal of Planning and Environment Law* 1995: 292–305

Herington, J. (1984) *The Outer City*, London: Harper & Row

Heseltine, M. (1987) *Where There's a Will*, London: Hutchinson

Hewitt, P. (1989) *A Cleaner, Faster London: Road Pricing, Transport Policy and the Environment*, London: Institute for Public Policy Research

Heycock, M., (1991) 'Public policy, need and land use planning', in Nadin, V. and Doak, J. (eds) *Town Planning Responses to City Change*, Aldershot, Hants.: Gower

Heywood, F. (1990) *Clearance: The View from the Street. A Study of Politics, Land and Housing*, Birmingham: Community Forum

Hibbs, J. (1989) *The History of the British Bus Services*, Newton Abbot, Devon: David & Charles

Higgins, J., Deakin, N., Edwards, J., and Wicks, M. (1983) *Government and Urban Poverty: Inside the Policy-making Process*, Oxford: Oxford University Press

Higman, Roger (1991) *Local Responses to 1989 Traffic Forecasts*, London: Friends of the Earth

Hill, D. M. (1970) *Participating in Local Affairs*, Harmondsworth, Middx: Penguin

Hill, D. M. (1994) *Citizens and Cities: Urban Policy in the 1990s*, Hemel Hempstead, Herts.: Harvester Wheatsheaf

Hill, L. (1991) 'Unitary development plans for the West Midlands: first stages in the statutory responses to a changing conurbation', in Nadin, V. and Doak, J. (eds) *Town Planning Responses to City Change*, Aldershot, Hants.: Gower

Hill, M. (1968) 'A goals–achievement matrix for evaluating alternative plans', *Journal of the American Institute of Planners* 34, 1: 19–29

Hill, M. (ed) (1993) *The Policy Process*, London: Harvester Wheatsheaf

Hill, M. P. (1987) 'Housing land availability: some observations on the process of housing development in contrasting urban locations', *Land Development Studies* 4: 209–19

Hill, S. and Barlow, J. (1995) 'Single regeneration budget: hope for "those inner cities"?', *Housing Review* 44: 32–5

Hillman, J. (1988) *A New Look for London* (Royal Fine Art Commission), London: HMSO

Hillman, J. (1990) *Planning for Beauty: The Case for Design Guidelines* (Royal Fine Art Commission), London: HMSO

Hillman, J. (1993) *Telelifestyles and the Flexicity: The Impact of the Electronic Home*, Dublin: European Foundation for the Improve-

ment of Living and Working Conditions/Luxembourg: Office for Official Publications of the European Communities

Hillman, M. (1989) 'More daylight, less accidents', *Traffic Engineering and Control* 30: 191–3

Hillman, M. (1992) 'Reconciling transport and environmental policy objectives: the way ahead at the end of the road', *Public Administration* 70: 225–34

Hillman, M. (1993) *Children's Helmets: The Case For and Against*, London: Policy Studies Institute

Hillman, M. (1993a) *Time for Change: Setting Clocks Forward by One Hour Throughout the Year*, London: Policy Studies Institute

Hillman, M. (ed) (1993b) *Children, Transport and the Quality of Life*, London: Policy Studies Institute

Hillman, M. and Whalley, A. (1979) *Walking IS Transport*, London: Policy Studies Institute

Hillman, M., Adams, J., and Whitelegg, J. (1991) *One False Move: A Study of Children's Independent Mobility*, London: Policy Studies Institute

Hills, P. J. (1994) 'The car versus mixed use development', in Wood, M. (ed) *Planning Icons: Myth and Practice* (Planning Law Conference, *Journal of Planning and Environment Law*), London: Sweet & Maxwell

Himsworth, C. M. G. (1994) 'Charging for inspecting the register', *Scottish Planning and Environment Law* 44: 59–60

Hirsch, D. (ed) (1994) *A Positive Role for Local Government: Lessons for Britain from Other Countries*, London: Local Government Chronicle/Joseph Rowntree Foundation

Hobbs, P. (1992) 'The economic determinants of post-war British planning', *Progress in Planning* 38, 3: 179–300

Hobhouse Report (1947) *Report of the National Parks Committee (England and Wales)*, Cmd 7121, London: HMSO

Hobley, B. (1987) 'Rescue archaeology and planning', *Planner* 73: 25–7

Hobson, J., Hockman, S., and Stinchcombe, P. (1996) 'The future of the planning system', in Bean, D. (ed) *Law Reform for All*, London: Blackstone

Hodge, I. (1989) 'Compensation for nature conservation', *Environment and Planning A* 21: 1027–36

Hodge, I. (1991) 'Incentive policies and the rural environment', *Journal of Rural Studies* 7: 373–84

Hodge, I. (1992) 'Supply control and the environment: the case for separate policies', *Farm Management* 8: 65–72

Hodge, I. (1995) *Environmental Economics: Individual Incentives and Public Choices*, London: Macmillan

Hodge, I. and Monk, S. (1991) *In Search of a Rural Economy: Patterns and Differentiation in Non-metropolitan England*, Cambridge: Department of Land Economy

Hogwood, B. W. (1995) *The Integrated Regional Offices and the Single Regeneration Budget*, London: Commission for Local Democracy

Hogwood, B. W. (1996) *Mapping the Regions: Boundaries, Coordination and Government* (Joseph Rowntree Foundation), Bristol: Policy Press

Hogwood, B. W. and Keating, M. (eds) (1982) *Regional Government in England*, Oxford: Clarendon Press

Holford, W. G. (1953) 'Design in city centres', part 3 of MHLG, *Design in Town and Village*, London: HMSO

Hollis, G., Ham, G., and Ambler, M. (eds) (1992) *The Future Role and Structure of Local Government*, Harlow, Essex: Longman

Hollox, R. E. and Biart, S. W. (1982) 'Local plan inquiries: a case study', *Journal of Planning and Environment Law* 1982: 17–23

Holmans, A. (1995) *Housing Demand and Need in England 1991 to 2011*, York: Joseph Rowntree Foundation

Home, R. (1989) *Planning Use Classes: A Guide to the 1987 Order* (second edition), Oxford: BSP Professional

Home, R. (1992) 'The evolution of the use classes order', *Town Planning Review* 63: 187–201

Home, R. H. (1982) *Inner City Regeneration*, London: Spon

Home, R. K. (1987) 'Planning decision statistics and the use classes debate', *Journal of Planning and Environment Law* 1987: 167–73

Home, R. K. (1993) 'Planning aspects of the government consultation paper on gypsies', *Journal of Planning and Environment Law* 1993: 13–18

Hooper, A. (1979) 'Land availability', *Journal of Planning and Environment Law* 1979: 752–6

Hooper, A. (1980) 'Land for private housebuilding', *Journal of Planning and Environment Law* 1980: 795–806

Hooper, A. (1982) 'Land availability in South East England', *Journal of Planning and Environment Law* 1982: 555–60

Hooper, A. (1985) 'Land availability studies and private housebuilding', in Barrett, S. and Healey, P. (eds) *Land Policy: Problems and Alternatives*, Aldershot, Hants.: Avebury

Hooper, A., Pinch, P., and Rogers, S. (1988) 'Housing land availability: circular advice, circular arguments and circular methods', *Journal of Planning and Environment Law* 1988: 225–39

Hooper, A., Pinch, P., and Rogers, S. (1989) 'Housing land availability in the South East', in Breheny, M. J. and Congdon, P. (eds) *Growth and Change in a Core Region: The Case of South East England*, London: Pion

Hoskyns, J. (1983) 'Whitehall and Westminster: an outsider's view', *Parliamentary Affairs* 36: 137–47

Hough, B. (1992) 'Standing in planning permission appeals', *Journal of Planning and Environment Law* 1992: 319–29

House Builders Federation (1987) *Private Housebuilding in the Inner Cities*, London: HBF

Housing Choice (1991) *Planning: A Citizen's Charter*, London: Housing Choice

Howard, E. (1898) *Tomorrow: A Peaceful Path to Real Reform*, London: Swan Sonnenschein

Howard, E. (1902) *Garden Cities of Tomorrow*, London: Swan Sonnenschein reprinted Faber & Faber, 1946, with an introduction by F. J. Osborn, and an introductory essay by Lewis Mumford)

Howard, E. B. and Davies, R. L. (1993) 'The impact of regional out-of-town retail centres: the case of the Metro Centre', *Progress in Planning* 40, 2: 89–165

Howe, J. (1996) 'A case of inter-agency relations: regional development in rural Wales', *Planning Practice and Research* 11: 61–72

Hughes, D. (1996) *Environmental Law* (third edition), Oxford: Butterworth

Hughes, D. J. (1995) 'Planning and conservation areas: where do we stand following PPG 15, and whatever happened to Steinberg?', *Journal of Planning and Environment Law* 1995: 679–91

Hughes, J. T. (1991) 'Evaluation of local economic development: a challenge for policy research', *Urban Studies* 28: 909–18

Hull, A., Healey, P., and Davoudi, S. (1995) *Greening the Red Rose County: Working Towards an Integrated Sub-regional Strategy*, Newcastle upon Tyne: Department of Town and Country Planning, University of Newcastle upon Tyne

Humber, J. R. (1980) 'Land availability: another view', *Journal of Planning and Environment Law* 1980: 19–23

Humber, R. (1990) 'Prospects and problems for private housebuilders', *Planner/TCPSS Proceedings* 76, 7: 15–19

Hunt Report (1969) *The Intermediate Areas*, London: HMSO

Huppes, G. and Kagan, R.A. (1989) 'Market-oriented regulation of environmental problems in the Netherlands', *Law and Policy* 11: 215–39

Hurrell, A. and Kinsbury, B. (1992) *The International Politics of the Environment: Actors, Interests and Institutions*, Oxford: Clarendon Press

Hutter, B. M. (1989) 'Variations in regulatory enforcement styles', *Law and Policy* 11: 153–74

Hutton, N. (1986) *Lay Participation in a Public Inquiry: A Sociological Case Study*, Aldershot, Hants.: Gower

Hutton, R. H. (1991) 'Local needs policy initiatives in rural areas: missing the target', *Journal of Planning and Environment Law* 1991: 303–11

Huxley Report (1947) *Conservation of Nature in England and Wales*, Cmd 7122, London: HMSO

IAURIF (Institut d'aménagement du territoire et d'urbanisme de la région d'Ile de France) (1991) *La Charte de l'Ille de France*, Paris: IAURIF

Illich, I. (1971) *Deschooling Society*, New York: Harper & Row

Illich, I. (1977) *Disabling Professions*, London: Boyars

Imrie, R. (1996) *Disability and the City*, London: Paul Chapman

Imrie, R. and Thomas, H. (1993) *British Urban Policy and the Urban Development Corporations*, London: Paul Chapman

Imrie, R., Thomas, H., and Marshall, T. (1995) 'Business organisations, local dependence and the politics of urban renewal in Britain', *Urban Studies* 32: 31–47

Ince, M. (1984) *Sizewell Report: What Happened at the Inquiry*, London: Pluto

Inland Revenue (1992) *Company Cars: Reform of Income Tax Treatment. A Consultative Document*, London: Inland Revenue

Insight Social Research (1989) *Local Attitudes to Central Advice*, London: ISR

International Energy Agency (1989) *Energy and the Environment*, Paris: OECD

Investment Property Databank (1993) *The Investment Performance of Listed Buildings*, London: RICS

Jackson, A., Morrison, N., and Royce, C. (1994) *The Supply of Land for Housing: Changing Local Authority Mechanisms*, Department of Land Economy, University of Cambridge

Jackson, A. R. W. and Jackson, J. M. (1996) *Environmental Science: The Natural Environment and Human Impact*, Harlow, Essex: Longman

Jackson, T. (1996) *Material Concerns: Pollution, Profit and Quality of Life*, London: Routledge

Jacobs, B.D. (1992) *Fractured Cities: Capitalism, Community and Empowerment in Britain and America*, London: Routledge

Jacobs, J. (1961) *The Death and Life of Great American Cities*, London: Cape/Penguin

Jacobs, J. (1985) 'Urban development grant', *Policy and Politics* 13: 191–9

Jacobs, M. (1990) *Sustainable Development: Greening the Economy*, London: Fabian Society

Jacobs, M. (1991) *The Green Economy: Environment, Sustainable Development, and the Politics of the Future*, London: Pluto Press

James, S. (1990) 'A streamlined city: the broken pattern of London government', *Public Administration* 68: 493–504

Jansen, A.J. and Hetsen, H. (1991) 'Agricultural development and spatial organization in Europe', *Journal of Rural Studies* 7: 143–51

Jarvis, R. (1996) 'Structure planning policy and strategic planning guidance in Wales', in Tewdwr-Jones, M. (ed) *British Planning Policy in Transition: Planning in the 1990s*, London: UCL Press

Jenkins, K., Oates, G., and Stott, A. (1985) *Making Things Happen: A Report on the Implementation of Government Efficiency Scrutinies. Report to the Prime Minister*, London: HMSO

Jenkins, S. (1995) *Accountable to None: The Tory Nationalisation of Britain*, London: Hamish Hamilton

Jenkins, T.N. (1990) *Future Harvest: The Economics of Farming and the Environment. Proposals for Action*, London: Council for the Protection of Rural England and WWF

Jewell, T. (1995) 'Planning regulation and environmental consciousness: some lessons from minerals?', *Journal of Planning and Environment Law* 1995: 482–98

JMP Consultants (1995a) *Travel to Food Super-stores*, London: JMP Consultants

JMP Consultants (1995b) *PPG 13: A Guide to Better Practice. Reducing the Need to Travel through Land Use and Transport Planning*, London: HMSO

John, P. (1993) *Local Government in Northern Ireland*, York: Joseph Rowntree Foundation

Johnson, D., Martin, S., Pearce, G., and Simmons, S. (1992) *The Strategic Approach to Derelict Land Reclamation* (Department of the Environment), London: HMSO

Johnson, D. A. (1995) *Planning the Great Metropolis: The 1929 Regional Plan of New York and its Environs*, London: Spon

Johnson, P. (ed) (1994) *20th Century Britain: Economic, Social and Cultural Change*, London: Methuen

Johnson, S.P. and Corcelle, G. (1989) *The Environmental Policy of the European Communities*, London: Graham & Trotman

Johnson, W. C. (1984) 'Citizen participation in local planning: a comparison of US and British experience', *Environment and Planning C: Government and Policy* 2: 1–14

Johnston, B. F. (1995) 'Commission makes case for coherence', *Planning* no. 1141, 20 October

Johnstone, R. J. (1991) *Environmental Problems: Nature, Economy and State*, Chichester, W. Sussex: Belhaven Press

Johnston, R. J. and Gardiner, V. (eds) (1991) *The Changing Geography of the UK*, London: Routledge

Jones, A. (1996) 'Local planning policy: the Newbury approach', in Tewdwr-Jones, M. (ed) *British Planning Policy in Transition: Planning in the 1990s*, London: UCL Press

Jones, C. (1996) 'Property-led local economic development policies from advance factory to English Partnerships and strategic property investment?', *Regional Studies* 30: 200–6

Jones, C. E. and Wood, C. (1995) 'The impact of environmental assessment on public inquiry decisions', *Journal of Planning and Environment Law* 1995: 890–904

Jones, M. (1996) 'TEC policy failure: evidence from the baseline follow-up studies', *Regional Studies* 30: 509–32

Jones, M. R. (1995) 'Training and enterprise councils: a continued search for local flexibility', *Regional Studies* 29: 577–80

Jones, P. (1990) *Traffic Quotes: Public Perception of Traffic Regulation in Urban Areas. Report of a Research Study* (Department of Transport), London: HMSO

Jones, P. M. (1991a) 'Gaining public support for road pricing through a package approach', *Traffic Engineering and Control* 32: 194–6

Jones, P. M. (1991b) 'UK public attitudes to urban traffic problems and possible countermeasures: a poll of polls', *Environment and Planning C: Government and Policy* 9: 245–56

Joseph Rowntree Foundation (1994) *Inquiry into Planning for Housing*, York: JRF

Journal of Planning and Environment Law (1995) 'Are off-site effects material to material changes of use?' (and case of *Forest of Dean District Council v Secretary of State for the Environment and Howells*), *Journal of Planning and Property Law* 889 and 937–43

Jowell, J. (1975) 'Development control' [review article on the Dobry Report], *Political Quarterly* 46: 340–4

Jowell, J. (1977a) 'Bargaining in development control', *Journal of Planning and Environment Law* 1977: 414–33

Jowell, J. (1977b) 'The limits of law in urban planning, *Current Legal Problems* 1977: 30: 63–83

Jowell, J. and Grant, M. (1983) 'Guidelines for planning gain?', *Journal of Planning and Environment Law* 1983: 427–31

Jowell, J. and Millichap, D. (1987) 'Enforcement: the weakest link in the planning chain', in Harrison, M. L. and Mordey, R. *Planning Control: Philosophies, Prospects and Practice*, London: Croom Helm

Jowell, J. and Oliver, D. (1994) *The Changing Constitution*, Oxford: Clarendon Press

JURUE: ECOTEC Research and Consultancy Ltd (1987) *Greening City Sites*, London: HMSO

JURUE: ECOTEC Research and Consultancy Ltd (1988) *Improving Urban Areas*, London: HMSO

Kain, J. F. and Beesley, M. E. (1965) 'Forecasting car ownership and use', *Urban Studies* 2: 163–85

Karn, V., Lucas, J. *et al.* (1996) *Home-owners and Clearance: An Evaluation of Rebuilding Grants* (Department of the Environment), London: HMSO

Kay, P. and Evans, P. (1992) *Where Motor Car is Master*, London: Council for the Protection of Rural England

Keating, M. and Jones, B. (1985) *Regions in the European Community*, Oxford: Oxford University Press

Keeble, D., Tyler, P., Broom, G., and Lewis, J. (1992) *Business Success in the Countryside: The Performance of Rural Enterprise* (Department of the Environment), London: HMSO

Keeble, L. (1969) *Principles and Practice of Town and Country Planning*, London: Estates Gazette

Keith, M. and Rogers, A. (eds) (1991) *Hollow Promises: Rhetoric and Reality in the Inner City*, London: Mansell

Kellett, J. (1990) 'The environmental impact of wind energy developments', *Town Planning Review* 61: 139–55

Kelly, A. and Marvin, S. (1995) 'Demand-side management in the electricity sector: implications for town planning in the UK', *Land Use Policy* 12: 205–21

Kemeny, J. (1994) *From Public Housing to the Social Market: Rental Policy Strategies in Comparative Perspective*, London: Routledge

Kent County Council (1995) *A Vision for EuroRegion: Towards a Policy Framework. The First Step*, Maidstone: Kent County Council

Kenyon, R. C. (1991) 'Environmental assessment: an overview on behalf of the RICS', *Journal of Planning and Environment Law* 1991: 419–22

Keogh, G. and Evans, A. W. (1992) 'The private and social costs of planning delay', *Urban Studies* 29: 687–99

Khan, M. A. (1995) 'Sustainable development: the key concepts, issues and implications', *Sustainable Development* 3: 63–9

Kidd, S. and Kumar, A. (1993) 'Development planning in the English metropolitan counties: a comparison of performance under two planning systems', *Regional Studies* 27: 65–73

King, A. D. (ed) (1980) *Buildings and Society: Essays on the Social Development of the Built Environment*, London: Routledge

King, A. D. (1990) *Global Cities: Post-imperialism and the Internationalisation of London*, London: Routledge

King, G. and Bridge, L. (1995) *Scoping Report: Directory of Coastal Planning Initiatives in England*, Swansea: National Coasts and Estuaries Group

King, J. (1990) *Regional Selective Assistance 1980–84: An Evaluation by DTI, IDS, and WOID*, London: HMSO

Kingdom, J. (1991) *Local Government and Politics in Britain*, London: Philip Allan

Kingdon, J. W. (1984) *Agendas, Alternatives, and Public Policies*, New York: HarperCollins

Kirby, A. (1985) 'Nine fallacies of local economic change', *Urban Affairs Quarterly* 21: 207–20

Kirk, G. (1980) *Urban Planning in a Capitalist Society*, London: Croom Helm

Kirkby, J., O'Keefe, P., and Timberlake, L. (1995) *The Earthscan Reader in Sustainable Development*, London: Earthscan

Kirkwood, G. and Edwards, M. (1993) 'Affordable housing policy: desirable but unlawful?' *Journal of Planning and Environment Law* 1993: 317–24

Kirwan, R. (1989) 'Finance for urban public infrastructure', *Urban Studies* 26: 285–300

Kivell, P. (1993) *Land and the City: Patterns and Processes of Urban Change*, London: Routledge

Klein, R. (1974) 'The case for elitism: public opinion and public policy', *Political Quarterly* 45: 406–17

Klosterman, R. E. (1985) 'Arguments for and against planning', *Town Planning Review* 56: 5–20

Knox, P. L. (ed) (1993) *The Restless Urban Landscape*, Englewood Cliffs, NJ: Prentice-Hall

Knox, P. L. (1994) *Urbanization: An Introduction to Urban Geography*, Englewood Cliffs, NJ: Prentice-Hall

Knox, P. L. and Cullen, J. L. (1981) 'Planners as urban managers: an exploration of the attitudes and self-image of senior British planners', *Environment and Planning A* 13: 885–98

Knox, P. L. and Taylor, P. J. (eds) (1995) *World Cities in a World-system*, Cambridge: Cambridge University Press

Koslowski, J. and Hill, G. (1993) *Towards Planning for Sustainable Development*, Aldershot, Hants.: Avebury

Kostof, S. (1991) *The City Shaped: Urban Patterns and Meaning Through History*, London: Thames & Hudson

Kostof, S. (1992) *The City Assembled: The Elements of Urban Form Through History*, London: Thames & Hudson

Kraan, D. J. and Veld, R. J. (eds) (1991) *Environmental Protection: Public or Private Choice?*, Dordrecht: Kluwer

Krämer, L. (1991) 'The implementation of Community environmental directives within member states: some implications of the direct effect doctrine', *Journal of Environmental Law* 3: 39–56

Krämer, L. (1992) *Focus on European Environmental Law*, London: Sweet & Maxwell

Krämer, L. (1994) *EC Treaty and Environmental Law*, London: Sweet & Maxwell

Kromarek, P. (1986) 'The single European Act and the environment', *European Environment Review* 1: 10–12

Kunzmann, K. R. and Wegener, (1991) 'The pattern of urbanisation in Western Europe', *Ekistics* 350 (September/October): 282–91

Labour Party (1995) *A Choice for England: A Consultation Paper on Labour's Plans for English Regional Government*, London: Labour Party

Lambert, A.J. and Wood, C.M. (1990) 'UK implementation of the European directive on EIA', *Town Planning Review* 61: 247–62

Land Capability Consultants (1990) *Cost Effective Management of Reclaimed Derelict Sites* (Department of the Environment), London: HMSO

Land Use Consultants (1986) *Channel Fixed Link: Environmental Appraisal of Alternative Proposals* (Department of Transport), London: HMSO

Land Use Consultants (1991) *Permitted Development Rights for Agriculture and Forestry* (Department of the Environment), London: HMSO

Land Use Consultants (1993) *Local Moves: The Funding and Formulation of Local Transport Policy*, London: Council for the Protection of Rural England

Land Use Consultants (1995a) *Effectiveness of Planning Policy Guidance Notes*, London: HMSO

Land Use Consultants (1995b) *Planning Controls over Agricultural and Forestry Development and Rural Building Conversions* (Department of the Environment), London: HMSO

Land Use Consultants (1996) *Reclamation of Damaged Land for Nature Conservation* (Department of the Environment), London: HMSO

Landry, C., Montgomery, J., and Worpole, K. (1989) *The Last Resort: Tourism, Tourist Employment and 'PostTourism' in the South East*, Bournes Green, Glos.: Comedia

Lane, J. and Vaughan, S. (1992) *An Evaluation of the Impact of PPG 16 on Archaeology and Planning*, London: Pagoda Associates

Lane, P. and Peto, M. (1995) *Blackstone's Guide to the Environment Act 1995*, London: Blackstone

Lang, P. (1994) *LETS Work: Rebuilding the Local Economy*, Swanley, Kent: Grower Books

Larkham, P. J. (1990) 'The use and measurement of development pressure', *Town Planning Review* 61: 171–83

Larkham, P. J. (1995) *Patterns in the Designation of Conservation Areas*, Birmingham: School of Planning, University of Central England in Birmingham

Larkham, P. J. (1996) *Conservation and the City*, London: Routledge

Larkham, P. J. and Chapman, D. W. (1996) 'Article 4 Directions and development control: planning myths, present uses, and future possibilities', *Journal of Environmental Planning and Management* 39: 5–19

Larkham, P. J. and Jones, A. (1993) 'Conservation and conservation areas in the UK: a growing problem', *Planning Practice and Research* 8, 2: 19–29

Larkham, P. J., Jones, A., and Daniels, R. (1992) *The Character of Conservation Areas*, London: RTPI

Lash, S., Szerszynski, B., and Wynne, B. (1996) *Risk, Environment and Modernity: Towards a New Ecology*, London: Sage

Lasok, D. (1994) *Law and Institutions of the European Communities* (sixth edition), London: Butterworth

Lavers, A. and Webster, B. (1994) 'Participation in the plan-making process: financial interests and professional representation', *Journal of Property Research* 11: 131–44

Law, C. M. (1992) 'Urban tourism and its contribution to economic regeneration', *Urban Studies* 29: 599–619

Law Commission (1993) *Administrative Law: Judicial Review and Statutory Appeals*, Law Commission Consultation Paper 126, London: HMSO

Lawless, P. (1986) *The Evolution of Spatial Policy: A Case Study of Inner Urban Policy in the United Kingdom 1968–1981*, London: Pion

Lawless, P. (1989) *Britain's Inner Cities* (second edition), London: Paul Chapman

Lawless, P. (1991) 'Urban policy in the Thatcher decade: English inner-city policy, 1979–90', *Environment and Planning C; Government and Policy* 9: 15–30

Lawless, P. (1994) 'Partnership in urban regeneration in the UK: the Sheffield central area study', *Urban Studies* 31: 1301–24

Lawless, P. (1996) 'The inner cities: towards a new agenda', *Town Planning Review* 67: 21–43

Lawless, P. and Dabinett, G. (1995) 'Urban regeneration and transport investment: a research agenda', *Environment and Planning A* 27: 1029–48

Laws, F. G. (1991) *Guide to the Local Government Ombudsman Service*, Harlow, Essex: Longman

Lawton, R. and Pooley, C. G. (1992) *The New Geography of London*, London: Edward Arnold

Laxton (1996) *Laxton's Guide to Single Regeneration Budgets*, Oxford: Heinemann

Layfield, F. (1992) 'The Environmental Protection Act 1990: the system of integrated pollution control', *Journal of Planning and Environment Law* 1992: 3–13

Leach, R. (1994) 'The missing dimension to the local government review', *Regional Studies* 28, 8: 797–802

Leach, S. and Game, C. (1991) 'English metropolitan government since abolition: an evaluation of the abolition of the English metropolitan councils', *Public Administration* 69: 141–70

Leach, S. and Stewart, M. (1992) *Local Government: Its Role and Function*, York: Joseph Rowntree Foundation

Leach, S., Davis, H., Game, C., and Skelcher, C. (1991) *After Abolition: The Operation of the Post-1986 Metropolitan Government System*, Birmingham: Institute of Local Government Studies, University of Birmingham

Leach, S., Stewart, J., Spencer, K., Walsh, K., and Gibson, J. (1992) *The Heseltine Review of Local Government: A New Vision or Opportunities Missed?*, Birmingham: Institute of Local Government Studies, University of Birmingham

Leach, S., Stewart, J., and Walsh, K. (1994) *The Changing Organisation and Management of Local Government*, London: Macmillan

Lederman, P. B. and Librizzi, W. (1995) 'Brownfields remediation: available technologies', *Journal of Urban Technology* 2: 21–29

Ledgerwood, G. (1985) *Urban Innovation: The Transformation of the London Docklands 1968–84*, Aldershot, Hants.: Gower

Lees, A. (1993) *Enquiry into the Planning System in North Cornwall District*, London: HMSO

Legates, R. and Stout, F. (1996) *The City Reader*, London: Routledge

Leitch Report (1977) *Report of the Advisory Committee on Trunk Road Assessment*, London: HMSO

Leung, H. L. (1979) *Redistribution of Land Values: A Re-examination of the 1947 Scheme*, Cambridge: Department of Land Economy, University of Cambridge

Leung, H. L. (1987) 'Developer behaviour and development control', *Land Development Studies* 4: 17–34

Lever, W. F. (1991) 'Deindustrialisation and the reality of the post-industrial city', *Urban Studies* 28: 983–99

Lever, W. F. (1992) 'Local authority responses to economic change in West Central Scotland', *Urban Studies* 29: 935–48

Levett, R. (1993) *Agenda 21: A Guide for Local Authorities in the UK*, Luton, Beds.: Local Government Management Board

Levin, P. H. (1976) *Government and the Planning Process: An Analysis and Appraisal of Government Decision-making Processes with Special Reference to the Launching of New Towns and Town Development Schemes*, London: Allen & Unwin

Levin, P. H. (1979) 'Highway inquiries: a study in government responsiveness', *Public Administration* 57: 21–49

Levy, F., Meltsner, A.J., and Wildavsky, A.B. (1974) *Urban Outcomes: Schools, Streets, and Libraries*, Berkeley, CA: University of California Press

Lewis, J. and Townsend, A. (1989) *The North–South Divide: Regional Change in Britain in the 1980s*, London: Paul Chapman

Lewis, N. (1992) *Inner City Regeneration: The Demise of Regional and Local Government*, Buckingham: Open University Press

Lewis, N. C. (1994) *Road Pricing: Theory and Practice*, London: Thomas Telford

Lewis, R. (1995) 'Contaminated land: the new regime of the Environment Act 1995', *Journal of Planning and Environment Law* 1995: 1087–96

Lewis, R. P. (1992) 'The Environmental Protection Act 1990: waste management in the 1990s. Waste regulation and disposal', *Journal of Planning and Environment Law* 1992: 303–12

Lichfield, N. (1956) *Economics of Planned Development*, London: Estate Gazette

Lichfield, N. (1989) 'From planning gain to community benefit', *Journal of Planning and Environment Law* 1989: 68–81

Lichfield, N. (1992) 'From planning gain to community benefit', *Journal of Planning and Environment Law* 1992: 1103–18

Lichfield, N. (1995) *Community Impact Evaluation: Principles and Practice*, London: UCL Press

Lichfield, N. and Darin-Drabkin, H. (1980) *Land Policy in Planning*, London: Allen & Unwin

Lindblom, C. E. (1959) 'The science of "muddling through"', *Public Administration Review*, Spring; (reprinted in Faludi, A. 1973, *A Reader in Planning Theory*, Oxford: Pergamon, 1973)

Liniado, M. (1996) *Car Culture and Countryside Change*, Cirencester, Glos.: National Trust

Little, J. (1994) *Gender, Planning and the Policy Process*, Oxford: Pergamon

Little, J. (1995) 'Women's initiatives in town planning in England', *Town Planning Review* 65: 261–76

Little, J., Peake, L., and Richardson, P. (eds) (1988) *Women in Cities*, New York: New York University Press

Littlewood, S. and Whitney, D. (1996) 'Re-forming urban regeneration? The practice and potential of English Partnerships', *Local Economy* 11: 39–49

Llewelyn-Davies (1994a) *Providing More Homes in Urban Areas*, Bristol: School for Advanced Urban Studies, University of Bristol

Llewelyn-Davies (1994b) *The Quality of London's Residential Environment*, London: London Planning Advisory Committee

Lloyd, M. G. (1990) 'Simplified planning zones in Scotland: government failure or the failure of government?', *Planning Outlook* 33: 128–32

Lloyd, M. G. (1992) 'Simplified planning zones, land development, and planning policy in Scotland', *Land Use Policy* 9: 249–58

Lloyd, M. G. and Livingstone, L. H. (1991) 'Marine fish farming in Scotland: proprietorial behaviour and the public interest', *Journal of Rural Studies* 7: 253–63

Lloyd, M. G. and Rowan-Robinson, J. (1992) 'Review of strategic planning guidance in Scotland', *Journal of Environmental Planning and Management* 35: 93–9

Local Government Management Board (1992) *Citizens and Local Democracy: Charting a New Relationship*, Luton, Beds.: LGMB

Local Government Management Board (1993a) *Local Agenda 21 UK: A Framework for Local Sustainability*, Luton, Beds.: LGMB

Local Government Management Board (1993b) *Local Agenda 21: Principles and Process. A Step by Step Guide*, Luton, Beds.: LGMB

Local Government Management Board (1995a) *Sustainable Settlements: A Guide for Planners, Designers and Developers* (Barton, H., Davis, G., and Guise, R.), Luton, Beds.: LGMB

Local Government Management Board (1995b) *Indicators for Local Agenda 21*, Luton, Beds.: LGMB

Local Government Management Board and RTPI (1993) *Planning Staffs Survey 1992*, Luton, Beds.: LGMB

Lock, D. (1989) *Riding the Tiger: Planning the South of England*, London: Town and Country Planning Association

Lock, D. (1994) 'Keynote address', in Wood, M. (ed.) *Planning Icons: Myth and Practice* (Planning Law Conference), London: Sweet & Maxwell

Loftman, P. and Nevin, B. (1992) *Urban Regeneration and Social Equity: A Case Study of Birmingham 1986–1992*, Birmingham: Faculty of the Built Environment Research Paper, University of Central England

Loftman, P. and Nevin, B. (1995) 'Prestige projects and urban regeneration in the 1980s and 1990s: a review of the benefits and limitations', *Planning Practice and Research*, 10: 299–315

Lomas, J. (1994) 'The role of management agreements in rural environmental conservation', *Land Use Policy* 11: 119–23

London and South East Regional Planning Conference: *see* SERPLAN

London Boroughs Association (1990) *Road Pricing for London*, London: LBA

London Boroughs Association (1992) *Out of Order: The 1987 Use Classes Order. Problems and Proposals*, London: LBA

London Boroughs' Disability Resource Team (1991) *Towards Integration: The Participation of Disabled People in Planning*, London: Disability Resource Team

London Economics (1992) *The Potential Role of Market Mechanisms in the Control of Acid Rain* (Department of the Environment), London: HMSO

London Planning Advisory Committee (1991) *London: World City Moving into the 21st Century*, London: HMSO

London Research Centre (1991) *Much Ado About Nothing: An Examination of the Potential for the Planning System to Secure Affordable Housing with Special Reference to Planning Agreements*, London: The Centre

Longley, P., Batty, M., Shepherd, J., and Sadler, G. (1992) 'Do green belts change the shape of urban areas? A preliminary analysis of the settlement geography of South East England', *Regional Studies* 26: 437–52

Loughlin, M. (1994) *The Constitutional Status of Local Government*, Research Report No. 3, London: Commission for Local Democracy

Loughlin, M., Gelfand, M.D., and Young, K. (eds) (1985) *Half a Century of Municipal Decline*, London: Allen & Unwin

Low, N. (1991) *Planning, Politics and the State: Political Foundations of Planning Thought*, London: Unwin Hyman

Lowe, P. and Flynn, A. (1989) 'Environmental politics and policy in the 1980s', in Mohan, J. (ed.) *The Political Geography of Contemporary Britain*, London: Macmillan

Lowe, P. and Goyder, J. (1983) *Environmental Groups in Politics*, London: Allen & Unwin

Lowe, P., Murdoch, J., Marsden, T., Munton, R., and Flynn, A. (1993) 'Regulating the new rural spaces: the uneven development of land', *Journal of Rural Studies* 9: 205–22

Lutyens, E. and Abercrombie, P. (1945) *A Plan for the City and County of Kingston upon Hull*, Hull: Brown

Lyddon, D. and dal Cin, A. (1996) *International Manual of Planning Practice* (third edition), The Hague: International Society of City and Regional Planners

McAllister, A. and McMaster, R. (1994) *Scottish Planning Law: An Introduction*, Edinburgh: Butterworths

McAllister, D. (ed.) (1996) *Partnership in the Regeneration of Urban Scotland* (Scottish Office), Edinburgh: HMSO

McAuslan, J. P. W. B. (1975) *Land, Law and Planning*, London: Weidenfeld & Nicolson

McAuslan, J. P. W. B. (1980) *The Ideologies of Planning Law*, Oxford: Pergamon

McAuslan, J. P. W. B. (1991) 'The role of courts and other judicial type bodies in environmental management', *Journal of Environmental Law* 3: 195–208

McBride, J. (1973) 'The urban aid programme: is it running out of cash?' *Quest* 16 March

McCaig, E., Henderson, C., and MVA Consultancy (1995) *Sustainable Development: What it Means to the General Public*, Edinburgh: Scottish Office Central Research Unit

McCarthy, J. (1995) 'The Dundee waterfront: a missed opportunity for planned regeneration', *Land Use Policy* 12: 307–19

McCarthy, P. and Harrison, T. (1995) *Attitudes to Town and Country Planning* (Department of the Environment), London: HMSO

McClintock, H. (ed.) (1992) *The Bicycle and City Traffic*, London: Belhaven Press

McConville, J. and Sheldrake, J. (eds) (1995) *Transport in Transition: Aspects of British and European Experience*, Aldershot, Hants.: Avebury

McCormick, J. (1991) *British Politics and the Environment*, London: Earthscan

McCormick, J. (1995) *The Global Environmental Movement*, Chichester, W. Sussex: John Wiley

McCrone, G. (1991) 'Urban renewal: the Scottish experience', *Urban Studies* 28: 919–38

McCrone, G. and Stephens, M. (1995) *Housing Policy in Britain and Europe*, London: UCL Press

McDougall, G. (1993) *Planning Theory: Prospects for the 1990s*, Aldershot, Hants.: Avebury

MacEwen, A. and MacEwen, M. (1982) *National Parks: Conservation or Cosmetics?*, London: Allen & Unwin

MacEwen, A. and MacEwen, M. (1987) *Greenprints for the Countryside? The Story of Britain's National Parks*, London: Allen & Unwin

MacEwen, M. (1973) *Crisis in Architecture*, London: Royal Institute of British Architects

MacEwen, M. and Sinclair, G. (1983) *New Life for the Hills*, London: Council for National Parks

McFarlane, R. (1993a) *Community Involvement in City Challenge: A Good Practice Guide*, London: National Council for Voluntary Organisations

McFarlane, R. (1993b) *Community Involvement in City Challenge: A Policy Report*, London: National Council for Voluntary Organisations

McGill, G. (1995) *Building on the Past: A Guide to the Archaeology and Development Process*, London: Spon

MacGregor, B. and Ross, A. (1995) 'Master or servant? The changing role of the development plan in the British planning system', *Town Planning Review* 66: 41–59

MacGregor, S. and Pimlott, B. (eds) (1990) *Tackling the Inner Cities: The 1980s Revisited, Prospects for the 1990s*, Oxford: Clarendon Press

McHarg, I. L. (1967) *Design with Nature*, New York: John Wiley

Mackay, D. (1995) *Scotland's Rural Land Use Agencies*, Aberdeen: Scottish Cultural Press

McKay, D. H. and Cox, A. W. (1979) *The Politics of Urban Change*, London: Croom Helm

Mackie, P. J. (1980) 'The new grant system for local transport: the first five years', *Public Administration* 59: 187–206

Mackintosh, M. (1992) 'Partnership: issues of policy and negotiation', *Local Economy* 7: 210–24

Mackintosh, S. and Leather, P. (1992) *Home Improvement under the New Regime*, Bristol: School for Advanced Urban Studies, University of Bristol

Maclennan, D. (1986) *The Demand for Housing: Economic Perspectives and Planning Practices*, Edinburgh: Scottish Development Department

Maclennan, D. (1989) 'Housing in Scotland 1977–1987', in Smith, M. E. H. (ed) *Guide to Housing; Main Changes in Housing Law* (third edition), London: Housing Centre

McLoughlin, J. B. (1969) *Urban and Regional Planning: A Systems Analysis*, Oxford: Pergamon

McMaster, R. F. (1995) *Results of RTPI Survey of Local Planning Authorities' Views on Local Government Reorganisation: A Mixed Picture*, Edinburgh: Royal Town Planning Institute in Scotland

Macnaghten, P., Grove-White, R., Jacobs, M., and Wynne, B. (1995) *Public Perceptions and Sustainability in Lancashire* (Report by the Centre for the Study of Environmental Change, Lancaster University), Preston: Lancashire County Council

Macpherson, M. (1987) 'Local ombudsman or the courts?', *Journal of Planning and Environment Law* 1987: 92–101

McQuail, P. (1994) *Origins of the Department of the Environment*, London: DoE

McQuail, P. (1995) *A View from the Bridge*, London: DoE

Macrory, R. (1992) *Campaigners' Guide to Using EC Environmental Law*, London: Council for the Protection of Rural England

Madden, M. (1987) 'Planning and ethnic minorities: an elusive literature', *Planning Practice and Research* no. 2 (March): 29–32

Maddison, D., Pearce, D. *et al.* (1996) *The True Costs of Road Transport*, London: Earthscan

Maloney, W. A. and Richardson, J. J. (1995) *Managing Policy Change in Britain: The Politics of Water*, Edinburgh: Edinburgh University Press

Malpass, P. (1994) 'Policy making and local governance: how Bristol failed to secure City Challenge funding (twice)', *Policy and Politics* 22: 301–12

Malpass, P. (1996) 'The unravelling of housing policy in Britain', *Housing Studies*, 11, 3: 459–70

Mandelker, D. R. (1962) *Green Belts and Urban Growth*, Madison, WI: University of Wisconsin Press

Manley, J. (1987) 'Archaeology and planning: a Welsh perspective', *Journal of Planning and Environment Law* 1987: 466–84 and 552–63

Manning, P.K. (1989) 'Managing risk: managing uncertainty in the British nuclear installations inspectorate', *Law and Policy* 11: 350–69

Marriott, O. (1967) *The Property Boom*, London: Hamilton

Marris, P. (1982) *Community Planning and Conceptions of Change*, London: Routledge & Kegan Paul

Marsden, T., Murdoch, J., Lowe, P., Munton, R. J. C., and Flynn, A. (1993) *Constructing the Countryside: An Approach to Rural Development*, London: UCL Press

Marshall, R. (1988) 'Agricultural policy development in Britain', *Town Planning Review* 59: 419–35

Marshall, T. (1991) *Regional Planning in England and Germany*, Oxford: School of Planning, Oxford Polytechnic

Marshall, T. (1995) *Clearing an Industrial Area: Collingdon Road and the Cardiff Bay Development Corporation 1988–1994*, Oxford: School of Planning, Oxford Polytechnic

Marte, L. (1994) *Ecology and Society: An Introduction*, London: Policy Press

Martin, L. R. G. (1989) 'The important published literature of British planners', *Town Planning Review* 60: 441–57

Martin, S. (1989) 'New jobs in the inner city: the employment impacts of projects assisted under the urban development grant programme', *Urban Studies* 26: 627–38

Martin, S. (1995) 'Partnerships for local environment action: observations of the first two years of *Rural Action for the Environment*', *Journal of Environmental Planning and Management* 38: 149–65

Martin, S. and Pearce, G. (1992) 'The internationalisation of local authority economic development strategies: Birmingham in the 1980s', *Regional Studies* 26: 499–509

Martin, S.J., Ticker, M.J., and Bovaird, A.G. (1990) 'Rural development programmes in theory and practice', *Regional Studies* 24: 268–76

Marvin, S. and Slater, S. (1996) *'Holes in the Road': Roads and Utilities in the 1990s*, Newcastle upon Tyne: Department of Town and Country Planning, University of Newcastle

Marwick, A. (1964) 'Middle opinion in the thirties: planning, progress and political agreement', *English Historical Review* 79: 258–98

Marwick, A. (1970) *Britain in the Century of Total War: War, Peace and Social Change 1900–1967*, Harmondsworth, Middx: Penguin

Masser, I. (1983) *Evaluating Urban Planning Efforts*, Aldershot, Hants.: Gower

Matarosso, F. (1995) *Spirit of Place: Redundant Churches as Urban Resources*, Bournes Green, Glos.: Comedia

Mather, A. S. (1991) 'The changing role of planning in rural land use: the example of afforestation in Scotland', *Journal of Rural Studies* 7: 299–309

Mather, G. (1988) *Paying for Planning*, London: Institute of Economic Affairs

Maunder, W. J. (1994) *Dictionary of Global Climate Change* (second edition), London: UCL Press

Mawson, J. (ed) (1995) *The Single Regeneration Budget: The Stocktake*, Birmingham: Centre for Urban and Regional Studies, University of Birmingham

May, A. D. (1986) 'Traffic restraint: a review of the alternatives', *Transportation Research A* 20A: 109–21

Meadows, D. H., Meadows, D. L., and Randers, J. (1992) *Beyond the Limits: Global Collapse or a Sustainable Future?*, London: Earthscan

Meadows, P. (ed) (1996) *Work Out – or Work In? Contributions to the Debate on the Future of Work*, York: Joseph Rowntree Foundation

Meyerson, M. (1956) 'Building the middle-range bridge for comprehensive planning', *Journal of the American Institute of Planners* 22, 2 (reprinted in Faludi, A., *A Reader in Planning Theory*, Oxford: Pergamon, 1973)

Meynell, A. (1959) 'Location of industry', *Public Administration* 37: 9

Midwinter, A. (1995) *Local Government in Scotland: Reform or Decline?* Basingstoke, Hants.: Macmillan

Millichap, D. (1995a) 'Law, myth and community: a reinterpretation of planning's justification and rationale', *Planning Perspectives* 10: 279–93

Millichap, D. (1995b) *The Effective Enforcement of Planning Controls* (second edition), London: Butterworths

Mills, E. S. and Hamilton, B. W. (1994) *Urban Economics* (fifth edition), New York: HarperCollins

Milner Holland Report (1965) *Report of the Committee on Housing in Greater London*, Cmnd 2605, London: HMSO

Milton, K. (1991) 'Interpreting environmental policy: a social scientific approach', in Churchill, R., Warren, L. M., and Gibson, J. (eds) *Law, Policy and the Environment*, Oxford: Blackwell

Minay, C. L. W. (ed.) (1992) 'Developing regional planning guidance in England and Wales: a review symposium', *Town Planning Review* 63: 415–34

Minhinnick, R. (1993) *A Postcard Home*, Llandysul, Dyfed: Gomer Press

Mishan, E. J. (1976) *Elements of Cost-benefit Analysis*, London: Allen & Unwin

Mitchell, C. (1995) *Renewable Energy in the UK: Financing Options for the Future*, London: Council for the Protection of Rural England

Mitchell Report (1991) *Railway Noise and the Insulation of Dwellings: Report of the Committee to Recommend a National Noise Insulation Standard for New Railway Lines*, London: HMSO

Mogridge, M.J.H. (1990) *Travel in Towns*, London: Macmillan

Mohan, J. (ed) (1989) *The Political Geography of Contemporary Britain*, London: Macmillan

Mole, D. (1996) 'Planning gain after the *Tesco* case', *Journal of Planning and Environment Law* 1996: 183–93

Möltke, K. von (1988) 'The *vorsorgerprinzip* in West German environmental policy', (printed as an appendix to Royal Commission on Environmental Pollution, *Twelfth Report: Best Practicable Environmental Option* (Cm 310), London: HMSO)

Monk, S., Pearce, B. J., and Whitehead, C. M. E. (1996) 'Land-use planning, land supply, and house prices', *Environment and Planning A* 28: 495–511

Montanari, A. and Williams, A. M. (1995) *European Tourism: Regions, Spaces and Restructuring*, Chichester, W. Sussex: John Wiley

Montgomery, J. and Thornley, A. (1988) 'Phoenix ascending: radical planning initiatives for the 1990s', *Planning Practice and Research* no. 4, Spring: 3–7

Montgomery, J. and Thornley, A. (eds) (1990) *Radical Planning Initiatives: New Directions for Urban Planning in the 1990s*, Aldershot, Hants.: Gower

Moore, B. and Townroe, P. (1990) *Urban Labour Markets: Reviews of Urban Research* (Department of the Environment), London: HMSO

Moore, B., Rhodes, J., and Tyler, P. (1986) *The Effects of Government Regional Economic Policy* (Department of Trade and Industry), London: HMSO

Moore, V. (1995) *A Practical Approach to Planning Law* (fifth edition), London: Blackstone

Morgan, K. (1994) *The Fallible Servant: Making Sense of the Welsh Development Agency*, Cardiff: Department of City and Regional Planning, University of Cardiff

Morgan, K. (1995) 'Reviving the Valleys? Urban renewal and governance structures in Wales', in R. Hambleton and H. Thomas (eds), *Urban Policy Evaluation: Challenge and Change*, London: Paul Chapman

Morgan, P. and Nott, S. (1995) *Development Control: Law, Policy and Practice* (second edition), London: Butterworths

Morphet, J. (1993a) *Greening Local Authorities: Sustainable Development at the Local Level*, Harlow, Essex: Longman

Morphet, J. (1993b) *Towards Sustainability: A Guide for Local Authorities*, Luton, Beds.: Local Government Management Board

Morphet, J. and Hams, T. (1994) 'Responding to Rio: the local authority approach', *Journal of Environmental Planning and Management* 37: 479–86

Morris, A. E. J. (1994) *History of Urban Form: Before the Industrial Revolution* (third edition), Harlow, Essex: Longman

Morris, P. and Therivel, R. (eds) (1995) *Methods of Environmental Impact Assessment*, London: UCL Press

Morton, D. (1991) 'Conservation areas: has saturation point been reached?', *Planner* 7, 17: 5–8

Mowat, C. L. (1955) *Britain Between the Wars 1918–1940*, London: Methuen

Moxen, J., McCulloch, A., Williams, D., and Baxter, S. (1995) *Accessing Environmental Information in Scotland* (Scottish Office Central Research Unit), Edinburgh: HMSO

Muchnick, D. M. (1970) *Urban Renewal in Liverpool*, Occasional Papers on Social Administration 33, London: Bell

Murdoch, J. and Marsden, T. (1994) *Reconstructing Rurality: The Changing Countryside in an Urban Context*, London: UCL Press

Murray, M. (1991) *The Politics and Pragmatism of Urban Containment: Belfast since 1940*, Aldershot, Hants.: Avebury

Murray, M. R. and Greer, J. V. (1993) *Rural Planning and Development in Northern Ireland*, Aldershot, Hants.: Avebury

MVA Consultancy (1995) *The London Congestion Charging Research Programme: Principal Findings* (Government Office for London), London: HMSO

Mynors, C. (1992) *Planning Control and the Display of Advertisements*, London: Sweet & Maxwell

Mynors, C. (1993) 'The extent of listing', *Journal of Planning and Environment Law* 1993: 99–111

Mynors, C. (1995) *Listed Buildings and Conservation Areas* (second edition), London: FT Law and Tax/Pearson Professional

Nadin, V. (1992a) 'Consultation by consultants', *Town and Country Planning* 61: 272–4

Nadin, V. (1992b) 'Local planning: progress and prospects', *Planning Practice and Research* 7: 27–32

Nadin, V. and Daniels, R. (1992) 'Consultants and development plans', *Planner* 78, 15: 10–12

Nadin, V. and Doak, J. (eds) (1991) *Town Planning Responses to City Change*, Aldershot, Hants.: Gower

Nadin, V. and Jones, S. (1990) ' A profile of the profession', *Planner* 76, 3: 13–24

National Audit Office: *see* appendix on *Official Publications*

National Committee for Commonwealth Immigrants (1967) *Areas of Special Housing Need*, London: NCCI

National Farmers' Union (1994) *Real Choices: Report by the Long Term Strategy Group*, London: NFU

National Farmers' Union (1995a) *Taking Real Choices Forward*, London: NFU

National Farmers' Union (1995b) *The Rural White Paper: Submission of the National Farmers' Union of England and Wales*, London: NFU

National Housing Forum (1996) *More than Somewhere to Live*, London: National Housing Forum

National Rivers Authority: *see* appendix on *Official Publications*

National Society for Clean Air and Environmental Protection (1995) *1995 Pollution Handbook*, Brighton: NSCA

Neill, W. J. V., Fitzsimons, D.S., and Murtagh, B. (1995) *Reimaging the Pariah City: Urban Development in Belfast and Detroit*, Aldershot, Hants.: Avebury

Nettlefold, J.S. (1914) *Practical Town Planning*, London: St Catherine's Press

Nevin, B. and Shiner, P. (1995) 'Community regeneration and empowerment: a new approach to partnership', *Local Economy* 9: 308–22

Nevin, M. and Abbie, L., (1993) 'What price roads? Practical issues in the introduction of road-user charges in historic cities in the UK', *Transport Policy* 1: 68–73

Newberry, D. (1990) 'Pricing and congestion: economic principles relevant to pricing roads' *Oxford Review of Economic Policy* 5, 2: 22–38

Newby, H. (1985) *Green and Pleasant Land: Social Change in Rural England* (second edition), London: Wildwood House

Newby, H. (1986) 'Locality and rurality: the restructuring of rural social relations', *Regional Studies* 20: 209–16

Newby, H. (1990) 'Ecology, amenity, and society', *Town Planning Review* 61: 3–20

Newby, H. (1991) *'The future of rural society'*, Swindon, Wilts.: Economic and Social Research Council (unpublished mimeo)

Newman, P. (1995) 'The politics of urban redevelopment in London and Paris', *Planning Practice and Research* 10: 15–23

Newman, P. and Thornley, A. (1996) *Urban Planning in Europe: International Competition, National Systems, and Planning Projects*, London: Routledge

Newman, P. W. G. and Kenworthy, J. R. (1989) *Cities and Automobile Dependence: A Sourcebook*, Aldershot, Hants.: Gower

Newman, P. W. G. and Kenworthy, J. R. (1996) 'The land use–transport connection', *Land Use Policy* 13: 1–22

Newson, M. (ed) (1992) *Managing the Human Impact on the Natural Environment: Patterns and Processes*, Chichester, W. Sussex: Belhaven

Nijkamp, P. (1993) 'Towards a network of regions: the United States of Europe'. *European Planning Studies* 1: 149–68

Noise Review Working Party (1990) *Report of the Noise Review Working Party*, London: HMSO

Nolan Report (1995) *First Report of the Committee on Standards in Public Life* (Cm 2850), London: HMSO

Northcott, J. (1991) *Britain in Europe in 2010*, London: Policy Studies Institute

Norton, D. M. (1993) 'Conservation areas in an era of plan led planning', *Journal of Planning and Environment Law* 1993: 211–13

Nuffield Report (1986) *Town and Country Planning*, London: The Foundation

Nugent, N. (1989) *The Government and Politics of the European Community*, London: Macmillan

Oatley, N. (1995) 'Competitive urban policy and the regeneration game', *Town Planning Review* 66: 1–14

Oatley, N. and Lambert, C. (1995) 'Evaluating competitive urban policy: the city challenge inititive', in Hambleton, R. and Thomas, H. (eds) *Urban Policy Evaluation: Challenge and Change*, London: Paul Chapman

Observer, The (1995) *The Observer Blueprint for a National Travel Plan to take Britain's Transport System into the 21st Century*, London: *The Observer*

Oc, T. and Tiesdell, S. (1991) 'The London Docklands Development Corporation 1981–1991: a perspective on the management of urban regeneration', *Town Planning Review* 62: 311–30

Ofori, S. (1994) 'Urban policy and environmental regeneration in two Scottish peripheral estates', *Environmentalist* 14: 283–96

Ogden, P. (1992) *Update: London Docklands. The Challenge of Development*, Cambridge: Cambridge University Press

Oliver, D. and Waite, A. (1989) 'Controlling neighbourhood noise: a new approach', *Journal of Environment Law* 1: 173–91

Organisation for Economic Cooperation and Development: *see* appendix on *Official Publications*

O'Riordan, T. (1992) 'The environment', in Cloke, P. (ed) *Policy and Change in Thatcher's Britain*, Oxford: Pergamon

O'Riordan, T. and Cameron, J. (eds) (1994) *Interpreting the Precautionary Principle*, London: Earthscan

O'Riordan, T. and D'Arge, R.C. (1979) *Progress in Resource Management and Environmental Planning*, New York: John Wiley

O'Riordan, T. and Weale, A. (1989) 'Administrative reorganization and policy change: the case of Her Majesty's Inspectorate of Pollution', *Public Administration* 67: 277–94

Osborn, F.J. (1969) *Green Belt Cities*, London: Evelyn, Adams & Mackay

Osborne, D. and Gaebler, T. (1992) *Reinventing Government: How the Entrepreneurial Spirit is Transforming the Public Sector*, New York: Addison-Wesley

Osmotherly, E. B. C. (1995) *Guide to the Local Ombudsman Service* (looseleaf; with updating service), London: Pitman

O'Toole, M. (1996) *Regulation Theory and the New British State: The Impact of Locality on the Urban Development Corporation 1987–1994*, Aldershot, Hants.: Avebury

Owen, S. (1991) *Planning Settlements Naturally*, Chichester, W. Sussex: Packard

Owen, S. (1995a) 'Local distinctiveness in villages', *Town Planning Review* 66: 143–61

Owen, S. (1995b) 'Land, limits and sustainability: a conceptual framework and some dilemmas for the planning system', *Transactions of the Institute of British Geographers* 19: 439–56

Owen, S. (1996) 'Sustainability and rural settlement planning', *Planning Practice and Research* 11: 37–47

Owens, S. (1984) 'Energy and spatial structure: a rural example', *Environment and Planning A* 16: 1319–37

Owens, S. (1985) 'Potential energy planning conflicts in the UK', *Energy Policy* 13: 546–58

Owens, S. (1986a) 'Strategic planning and energy conservation', *Town Planning Review* 57: 69–86

Owens, S. (1986b) *Energy, Planning and Urban Form*, London: Pion

Owens, S. (1989) 'Integrated pollution control in the United Kingdom: prospects and problems', *Environment and Planning C: Government and Policy* 7: 81–91

Owens, S. (1990a) 'Land use for energy efficiency', in Cullingworth, J. B. (ed.) *Energy, Land, and Public Policy*, New Brunswick, NJ: Transaction Publishers

Owens, S. (1990b) 'The unified pollution inspectorate and best practicable environmental option in the United Kingdom', in Haigh, N. and Irwin, F. (eds) *Integrated Pollution Control in Europe and North America*, Bonn: Institute for European Environmental Policy, and Washington, DC: Conservation Foundation

Owens, S. (1991a) *Energy Conscious Planning*, London: Council for the Protection of Rural England

Owens, S. (1994a) 'Land, limits and sustainability: a conceptual framework and some dilemmas for the planning system', *Transactions of the Institute of British Geographers* NS 19: 439–56

Owens, S. (1994b) 'Can land use planning produce the ecological city?', *Town and Country Planning* 63: 170–3

Owens, S. and Cope, D. (1992) *Land Use Planning Policy and Climate Change* (Department of the Environment), London: HMSO

Owens, S. and Cowell, R. (1996) *Rocks and Hard Places: Mineral Resource Planning and Sustainability*, London: Council for the Protection of Rural England

Oxford Retail Group (1995) *The Implementation of PPG 6*, Oxford: Oxford Institute of Retail Management, Templeton College

PA Cambridge Economic Consultants (1987) *An Evaluation of the Enterprise Zone Experiment* (Department of the Environment), London: HMSO

PA Cambridge Economic Consultants (1990a) *Indicators of Comparative Regional–Local Economic Performance and Prospects* (Department of the Environment), London: HMSO

PA Cambridge Economic Consultants (1990b) *An Evaluation of Garden Festivals* (Department of the Environment), London: HMSO

PA Cambridge Economic Consultants (1992) *An Evaluation of the Belfast Action Team Initiative*, Belfast: DoENI

PA Cambridge Economic Consultants (1995) *Final Evaluation of Enterprise Zones* (Department of the Evironment), London: HMSO

Pacione, M. (1990) 'Development pressure in the metropolitan fringe', *Land Development Studies* 7: 69–82

Pacione, M. (1991) 'Development pressure and the production of the built environment in the urban fringe', *Scottish Geographical Magazine* 107: 162–9

Pacione, M. (1995) *Glasgow: The Socio-spatial Development of the City*, Chichester, W. Sussex: John Wiley

Pack, C. and Glyptis, S. (1989) *Developing Sport and Leisure* (Department of the Environment), London: HMSO

Paddison, R., Money, J., and Lever, B. (eds) (1993) *International Perspectives in Urban Studies*, London: Jessica Kingsley

Paddison, R., Money, J., and Lever, B. (eds) (1994) *International Perspectives in Urban Studies 2*, London: Jessica Kingsley

Paddison, R., Money, J., and Lever, B. (eds) (1995) *International Perspectives in Urban Studies 3*, London: Jessica Kingsley.

Paddison, R., Money, J., and Lever, B. (eds) (1996) *International Perspectives in Urban Studies 4*, London: Jessica Kingsley.

Page, S. (1995) *Urban Tourism*, London: Routledge

Pahl, R. E. (1970) *Whose City? and Other Essays in Sociology and Planning*, Harlow, Essex: Longmans

Paris, C. (ed) (1982) *Critical Readings in Planning Theory*, Oxford: Pergamon

Parker, H. R. (1954) 'The financial aspects of planning legislation', *Economic Journal* 64: 82–6

Parkhurst, G. (1995) 'Park and ride: could it lead to an increase in car traffic?', *Transport Policy* 2: 15–23

Parkinson, M. (1985) *Liverpool on the Brink*, Hermitage, Berks.: Policy Journals

Parkinson, M. (1989) 'The Thatcher government's urban policy 1979–1989', *Town Planning Review* 60: 421–40

Parry, M. and Duncan, R. (eds) (1995) *The Economic Implications of Climate Change in Britain*, London: Earthscan

Parsons, M. L. (1995) *Global Warming: The Truth behind the Myth*, New York: Insight/Plenum

Patten, C. (1989) 'Planning and local choice', *Municipal Journal*, 13 October 18–19

Paxton, A. (1994) *The Food Miles Report: The Dangers of Long Distance Food Transport*, London: SAFE Alliance

Peake, S. (1994) *Transport in Transition: Lessons from the History of Energy*, London: Royal Institute of International Affairs/Earthscan

Peake, S. and Hope, C. (1994) 'Sustainable mobility in context: three transport scenarios for the UK', *Transport Policy* 1: 195–207

Pearce, B. J. (1992) 'The effectiveness of the British land use planning system', *Town Planning Review* 63: 13–28

Pearce, D. (ed) (1991) *Blueprint 2: Greening the World Economy*, London: Earthscan (1994 edition)

Pearce, D. (ed) (1993) *Blueprint 3: Measuring Sustainable Development*, London: Earthscan

Pearce, D. and Turner, R. K. (1992) 'Packaging waste and the polluter pays principle: a taxation solution', *Journal of Environmental Planning and Management* 35: 5–15

Pearce, D., Edwards, L., and Beuret, G. (1979) *Decision-making for Energy Futures: A Case Study of the Windscale Inquiry*, London: Macmillan

Pearce, D., Markandya, A., and Barbier, E. B. (1989) *Blueprint for* a Green Economy: Report for the UK Department of the Environment, London: Earthscan

Pearce, G., Hems, L., and Hennessy, B. (1990) The Conservation Areas of England, London: Historic Buildings and Monuments Commission (English Heritage)

Pearlman, J. J. (1995) 'Modification orders made under Wildlife and Countryside Act 1981: an update', Journal of Planning and Environment Law 1995: 1106–13

Penn, C. N. (1995) Noise Control: The Law and its Enforcement (second edition), Crayford, Kent: Shaw

Pennington, M. (1996) Conservation and the Countryside: By Quango or Market?, London: Institute of Economic Affairs

Pepper, D. (1993) Eco-socialism: From Deep Ecology to Social Justice, London: Routledge

Pepper, D. (1996) Modern Environmentalism: An Introduction, London: Routledge

Percy-Smith, J. (1994) Submissions to the Commission on Aspects of Local Democracy, London: Commission on Local Democracy

Perveen, F. (ed) (1994) Urban Environment: An Annotated Bibliography, Manchester: British Council

Pezzey, J. (1989) Economic Analysis of Sustainable Growth and Sustainable Development, Washington, DC: World Bank

Pezzey, J. (1992) 'Sustainability', Environmental Values 4: 321–62

Pharoah, T. (1993) 'Traffic calming in west Europe', Planning Practice and Research 8: 20–8

Pharoah, T. (1996) 'Reducing the need to travel: a new planning objective in the UK?', Land Use Policy 13: 23–36

Pharoah, T. and Apel, D. (1995) Transport Concepts in European Countries, Aldershot, Hants.: Avebury

Phelps, R. (1995) 'Structure plans: the conduct and conventions of examinations in public', Journal of Planning and Environment Law 1995: 95–101

Pickvance, C. (1982) 'Physical forces and market forces in urban development', in C. Paris (ed.), Critical Readings in Planning Theory, Oxford: Pergamon Press

PIEDA (1990) Five Year Review of the Bolton, Middlesbrough and Nottingham Programme Authorities (Department of the Environment), London: HMSO

PIEDA (1992) Evaluating the Effectiveness of Land Use Planning (Department of the Environment), London: HMSO

PIEDA (1995) Involving Communities in Urban and Rural Regeneration (Department of the Environment), London: DoE

Plan Local (1992) The Character of Conservation Areas, London: Royal Town Planning Institute

Planning Advisory Group (PAG) (1965) The Future of Development Plans, London: HMSO.

Planning Aid for Scotland (1996) Access to Planning Information, Edinburgh: Planning Aid for Scotland

Planning Exchange (1989) Evaluation of the Use and Effectiveness of Planning Publications, Edinburgh: Scottish Development Department

Planning Inspectorate (annual) Annual Report

Planning Inspectorate (annual) Business and Corporate Plan, Bristol: PI

Planning Inspectorate (1992 and annually) Statistical Report, Bristol: PI

Planning Inspectorate (1996a) Development Plan Inquiries: Guidance for Local Authorities, Bristol: PI

Planning Inspectorate (1996b) Service Agreement between the Planning Inspectorate Executive Agency and the Council, Bristol: PI

Plowden Report (1967) Children and Their Primary Schools, London: HMSO

Plowden, S. and Buchan, K. (1995) A New Framework for Freight Transport, London: Civic Trust

Plowden, S. and Hillman, M. (1984) Danger on the Road: The Needless Scourge. A Study of Obstacles to Progress in Road Safety, London: Policy Studies Institute

Plowden, S. and Hillman, M. (1996) Speed Control and Transport Policy, London: Policy Studies Institute

Plowden, W. (1971) The Motor Car and Politics 1896–1970, London: Bodley Head

Postle, M. (1993) Development of Environmental Economics for the NRA (National Rivers Authority), London: HMSO

Potter, S. (1992) 'New town legacies', Town and Country Planning 61: 298–302

Pressman, J. L. and Wildavsky, A. B. (1984) Implementation: How Great Expectations in Washington are Dashed in Oakland; or, Why It's Amazing that Federal Programs Work At All, This Being a Saga of the Economic Development Administration as Told by Two Sympathetic Observers who Seek to Build Morals on a Foundation of Ruined Hopes, 3rd edn Berkeley: University of California Press

Pretty, J. N. (1995) Regenerating Agriculture: Policies and Practice for Sustainability and Self-reliance, London: Earthscan

Price Waterhouse (1993) Evaluation of Urban Development Grant, Urban Regeneration Grant, and City Grant (Department of the Environment), London: HMSO

Property Advisory Group (1980) Structure and Activity of the Development Industry, London: HMSO

Property Advisory Group (1981) Planning Gain, London: HMSO

Property Advisory Group (1983) The Climate for Public and Private Partnerships in Property Development, London: HMSO

Property Advisory Group (1985) Report on Town and Country Planning (Use Classes) Order 1972, London: Department of the Environment

Public Sector Management Research Centre (1992) Parish Councils in England (Department of the Environment), London: HMSO.

Public Sector Management Research Unit (1988a) An Evaluation of the Urban Development Grant Programme (Department of the Environment), London: HMSO

Public Sector Management Research Unit (1988b) Improving Inner City Shopping Centres: An Evaluation of Urban Programme Funded Schemes in the West Midlands (Department of the Environment), London: HMSO

Pugh, C. (ed.) (1996) Sustainability, the Environment and Urbanization, London: Earthscan

Pugh-Smith, J. (1992) 'The local authority as a regulator of pollution in the 1990s', Journal of Planning and Environment Law 1992: 103–9

Pugh-Smith, J. and Samuels, J. (1993) 'PPG 16: two years on', Journal of Planning and Environment Law 1993: 203–10

Pugh-Smith, J. and Samuels, J. (1996a) Archaeology in Law, London: Sweet & Maxwell

Pugh-Smith, J. and Samuels, J. (1996b) 'Archaeology and planning: recent trends and potential conflicts', Journal of Planning and Environment Law 1996: 707–24

Punter, J. V. (1985) Office Development in the Borough of Reading 1951–1984: A Case Study of the Role of Aesthetic Control, Reading: Department of Land Management, University of Reading

Punter, J. V. (1986–7) 'A history of aesthetic control: the control of the external appearance of development in England and

Wales' (parts 1 and 2), *Town Planning Review* 57: 351–81, and 58: 29–62

Punter, J. V. (1990) *Design Control in Bristol, 1940–1990: The Impact of Planning in the Design of Office Development in the City Centre*, Bristol: Redcliffe

Punter, J. V. (1992) 'Design control and the regeneration of docklands: The example of Bristol', *Journal of Property Research* 9: 49–78

Punter, J. V. (ed) (1994) 'Design control in Europe', special issue of *Built Environment* 20, 2

Punter, J. V., Carmons, M. C., and Platts, A. (1994) 'The design content of development plans', *Planning Practice and Research* 9: 199–220

Purdom, C. B. (1913) *The Garden City: A Study in the Development of a Modern New Town*, Letchworth, Herts.: Temple Press

Purdom, C. B. (1925) *The Building of Satellite Towns*, London: Dent

Purdue, M. (1986) 'The flexibility of North American zoning as an instrument of land use planning', *Journal of Planning and Environment Law* 1986: 84–91

Purdue, M. (1989) 'Material considerations: an ever expanding concept?', *Journal of Planning and Environment Law* 1989: 156–61

Purdue, M. (1991) 'Green belts and the presumption in favour of development', *Journal of Environmental Law* 3: 93–121

Purdue, M. (1995) 'When a regulation becomes a taking of land: a look at two recent decisions of the United States Supreme Court', *Journal of Planning and Environment Law* 1995: 279–91

Purdue, M. and Kemp, R. (1985) 'A case for funding objectors at public inquiries? A comparison of the position in Canada as opposed to the United Kingdom', *Journal of Planning and Environment Law* 1985: 675–85

Purdue, M., Healey, P., and Ennis, F. (1992) 'Planning gain and the grant of planning permission: is the United States' test of the *rational nexus* the appropriate solution?', *Journal of Planning and Environment Law* 1992: 1012–24

Quinn, M. J. (1996) 'Central government planning policy', in Tewdwr-Jones, M. (ed.) *British Planning Policy in Transition: Planning in the 1990s*, London: UCL Press

Rabe, B. G. (1994) *Beyond Nimby: Hazardous Waste Siting in Canada and the United States*, Washington, DC: Brookings Institution

Radcliffe, J. (1991) *The Reorganisation of British Central Government*, Aldershot, Hants.: Dartmouth

Raemaekers, J. (1995) 'Scots have a way to go on strategic waste planning', *Planning* no. 1106 (17 February): 24–5

Raemaekers, J., Prior, A., and Boyack, S. (1994) *Planning Guidance for Scotland: A Review of the Emerging New Scottish National Planning Policy Guidelines*, Edinburgh: Royal Town Planning Institute in Scotland

Ramblers Association (1993) *Open Door Planning: Access to Planning Application Documents*, London: Green Balance

Ramsay Report (1945) *Report of the Scottish National Parks Survey Committee* (Cmd 6631), Edinburgh: HMSO

Ramsay Report (1947) *National Parks and the Conservation of Nature in Scotland* (Cmd 7235), Edinburgh: HMSO

Ranson, S., Jones, G., and Walsh, K. (eds) (1985) *Between Centre and Locality: The Politics of Public Policy*, London: Allen & Unwin

Rao, N. (1990) *The Changing Face of Housing Authorities*, London: Policy Studies Institute

Ratcliffe, D. A. (1994) *Conservation in Europe: Will Britain Make the Grade? The Status of Nature Resources in Britain and the Implementation of the EC Habitats and Species Directive*, London: Friends of the Earth

Ravaioli, C. (1995) *Economists and the Environment*, London: ZED Books

Ravetz, A. (1980) *Remaking Cities*, London: Croom Helm

Ravetz, A. (1986) *The Government of Space: Town Planning in Modern Society*, London: Faber & Faber

Ravetz, A. with Turkington, R. (1995) *The Place of Home: English Domestic Environments 1914–2000*, London: Spon

Rawcliffe, P. (1995) 'Making inroads: transport policy and the British environmental movement', *Environment* 37, 3: 6–20, 29–36

Read, L. and Wood, M. (1994) 'Policy, law and practice', in Wood, M. (ed.) *Planning Icons: Myth and Practice* (Planning Law Conference, *Journal of Planning and Environment Law*), London: Sweet & Maxwell

Reade, E. J. (1982) 'If planning isn't everything . . . ' *Town Planning Review* 53: 65–78

Reade, E. J. (1983) 'If planning is anything, maybe it can be identified', *Urban Studies* 20: 159–71

Reade, E. J. (1985) 'Planning's usurpation of political choice', *Town and Country Planning* 54: 184–6

Reade, E. J. (1987) *British Town and Country Planning*, Milton Keynes, Bucks.: Open University Press

Reade, E. J. (1992) 'The little world of Upper Bangor', part 1: 'How many conservation areas are slums?'; part 2: 'Professionally prestigious projects or routine public administration?'; part 3: 'What is planning for anyway?', *Town and Country Planning* 61: 11–12, 25–7, 44–7

Redclift, M. (1987) *Sustainable Development: Exploring the Contradictions*, London: Methuen

Redclift, M. and Sage, M. (1994) *Strategies for Sustainable Development*, Chichester, W. Sussex: John Wiley

Redman, M. (1990) 'Archaeology and development', *Journal of Planning and Environment Law* 1990: 87–98

Redman, M. (1991) 'Planning gain and obligations', *Journal of Planning and Environment Law* 1991: 203–18

Redundant Churches Fund (1990) *Churches in Retirement: A Gazetteer*, London: HMSO

Reed, M. (1990) *The Landscape of Britain*, London: Routledge

Rees, W. E. (1990) 'Sustainable development as capitalism with a green face: a review article' [Review of Pearce, D., Markandya, A., and Barbier, E. B., *Blueprint for a Green Economy: Report for the UK Department of the Environment*, London: Earthscan, 1989], *Town Planning Review* 61: 91–4

Reeves, D. (1995) 'Developing effective public consultation: a review of Sheffield's UDP process', *Planning Practice and Research* 10: 199–213

Regional Policy Commission (1996) *Renewing the Regions: Strategies for Regional Economic Development*, Sheffield: Sheffield Hallam University

Regional Studies Association (1990) *Beyond Green Belts*, London: Jessica Kingsley

Reid, C. (1994) *Nature Conservation Law*, Edinburgh: W. Green

Reid, D. (1995) *Sustainable Development: An Introductory Guide*, London: Earthscan

Reith Reports (1946) *Interim Report of the New Towns Committee* (Cmd 6759); *Second Interim Report* (Cmd 6794); and *Final Report* (Cmd 6876), London: HMSO

Rendel Geotechnics (1993) *Coastal Planning and Management: A Review* (Department of the Environment), London: HMSO

Rendel Geotechnics (1995) *Coastal Planning and Management: A Review of Earth Science Information Needs* (Department of the Environment), London: HMSO

Renton, J. (1992) 'The ombudsman and planning', in *Scottish Planning Law and Practice Conference 1992*, Glasgow: Planning Exchange

Richardson, G., Ogus, A., and Burrows, A. (1982) *Policing pollution: A Study of Regulation and Enforcement*, Oxford: Oxford University Press

Ridley, F. (1986) 'Liverpool is different: Political struggles in context', *Political Quarterly* 57: 125–36

Ridley, N. (1991) *The Local Right*, London: Centre for Policy Studies

Rietveld, P. and Wissen, L. van (1991) 'Transport policies and the environment: regulation and taxation', in Kraan, D. J. and Veld, R. J. (eds) *Environmental Protection: Public or Private Choice?*, Dordrecht: Kluwer

Rittel, H. W. J. and Webber, M. M. (1973) 'Dilemmas in a general theory of planning', *Policy Sciences* 4: 155–69

Rivlin, A. M. (1971) *Systematic Thinking for Social Action*, Washington, DC: Brookings Institution

Roberts, J., Cleary, J., Hamilton, K., and Hanna, J. (1992) *Travel Sickness: The Need for a Sustainable Transport Policy for Britain*, London: Lawrence & Wishart

Roberts, P. (1991) 'Environmental priorities and the challenges of environmental management', *Town Planning Review* 62: 447–69

Roberts, P. (1995) *Environmentally Sustainable Business: A Local and Regional Perspective*, London: Paul Chapman

Roberts, P. (1996) 'Regional planning guidance in England and Wales: back to the future?', *Town Planning Review* 67: 97–109

Robertson, G. (1989) *Freedom, the Individual and the Law* (sixth edition), Harmondsworth, Middx: Penguin

Robins, N. (1991) *A European Environment Charter*, London: Fabian Society

Robinson, G. M. (1994) 'The greening of agricultural policy: Scotland's Environmentally Sensitive Areas', *Journal of Environmental Planning and Management* 37: 215–25

Robinson, M. (1992) *The Greening of British Party Politics*, Manchester: Manchester University Press

Robinson, P. C. (1990) 'Tree preservation orders: felling a dangerous tree', *Journal of Planning and Environment Law* 1990: 720–3

Robson, B. (1988) *Those Inner Cities: Reconciling the Social and Economic Aims of Urban Policy*, Oxford: Clarendon Press

Robson, B., Bradford, M. B., Deas, I., Hall, E., Harrison, E., Parkinson, M., Evans, R., Garside, P. and Harding, A. (1994) *Assessing the Impact of Urban Policy* (Department of the Environment), London: HMSO

Roche, F. L. (1986) 'New communities for a new generation', *Town and Country Planning* 55: 312–13

Rodriguez-Bachiller, A., Thomas, M., and Walker, S. (1992) 'The English planning lottery: some insights from a more regulated system', *Town Planning Review* 63: 387–402

Rogers, A. (1985) 'Local claims on rural housing', *Town Planning Review* 56: 367–80

Roome, N. J. (1986) 'New directions in rural policy: recent administrative changes in response to conflict between rural policies', *Town Planning Review* 57: 253–63

Rose, C. (1990) *The Dirty Man of Europe: The Great British Pollution Scandal*, London: Simon & Schuster

Rosenbloom, S. (1992) 'Why working families need a car', in Wachs, M. and Crawford, M. (eds) *The Car and the City*, Ann Arbor, MI: University of Michigan Press

Ross, A., Rowan-Robinson, J., and Walton, W. (1995) 'Sustainable development in Scotland: the role of Scottish Natural Heritage', *Land Use Policy* 12: 237–52

Ross, M. (1991) *Planning and the Heritage*, London: Spon

Ross, M. (1996) *Planning and the Heritage* (second edition), London: Spon

Roth, G. (1996) *Roads in a Market Economy*, Aldershot, Hants.: Avebury

Rowan-Robinson, J. and Durman, R. (1992a) *Section 50 Agreements*, Central Research Unit Papers, Edinburgh: Scottish Office

Rowan-Robinson, J. and Durman, R. (1992b) 'Conditions or agreements', *Journal of Planning and Environment Law* 1992: 1003–11

Rowan-Robinson, J. and Durman, R. (1992c) 'Planning agreements and the spirit of enterprise', *Scottish Geographical Magazine* 108: 157–63

Rowan-Robinson, J. and Durman, R. (1993) 'Planning policy and planning agreements', *Land Use Policy* 10: 197–204

Rowan-Robinson, J. and Lloyd, M. G. (1988) *Land Development and the Infrastructure Lottery*, Edinburgh: T. & T. Clark

Rowan-Robinson, J. and Lloyd, M. G. (1991) 'National planning guidelines: a strategic opportunity wasting', *Planning Practice and Research* 6, 3: 16–19

Rowan-Robinson, J. and Ross, A. (1994) 'Enforcement of environmental regulation in Britain: strengthening the link', *Journal of Planning and Environment Law* 1994: 200–18

Rowan-Robinson, J. and Young, E. (1987) 'Enforcement: the weakest link in the Scottish planning control system', *Urban Law and Policy* 8: 255–88

Rowan-Robinson, J. and Young, E. (1989) *Planning by Agreement in Scotland*, Glasgow: Planning Exchange/Edinburgh: Green

Rowan-Robinson, J., Ross, A., and Walton, W. (1995) 'Sustainable development and the development control process', *Town Planning Review* 66: 269–86

Rowan-Robinson, J., Ross, A., Walton, W., and Rothnie, J. (1996) 'Public access to environmental information: a means to what end?', *Journal of Environmental Law* 8: 19–42

Rowell, T. A. (1991) *SSSIs: A Health Check*, London: Wildlife Link

Royal Commission on Environmental Pollution: *see* appendix on *Official Publications*

Royal Institute of British Architects (1995) *Quality in Town and Country: A Response to the Secretary of State by a Special Working Party*, London: RIBA

Royal Institution of Chartered Surveyors (1991) *Britain's Environmental Strategy: A Response by the RICS to the White Paper 'This Common Inheritance'*, London: RICS

Royal Institution of Chartered Surveyors (1994) *The Effects of*

Lasting Peace on Property and Construction in Northern Ireland, Belfast: RICS Northern Ireland Branch

Royal Institution of Chartered Surveyors (1995a) *The Private Finance Initiative: The Essential Guide*, London: RICS

Royal Institution of Chartered Surveyors (1995b) *Considering Private Finance for Public Sector Works: How They Do It Over There*, London: RICS

Royal Society for the Protection of Birds (RSPB) (1990) *RSPB Planscan: A Study of Development Plans in England and Wales*, Bain, C., Dodd, A. M. and Pritchard, D. E., Sandy, Beds.: RSPB

Royal Society for the Protection of Birds (RSPB) (1992) *RSBP Planscan: A Study of Development Plans in Scotland*, Davies, C., Pritchard, D.E. and Austin, L. W., Sandy, Beds.: RSPB

Royal Society for the Protection of Birds (RSPB) (1993a) *A Shore Future: RSPB Vision for the Coast*, Sandy, Beds.: RSPB

Royal Society for the Protection of Birds (RSPB) (1993b) *RSPB Planscan: A Study of Development Plans in Northern Ireland*, Dodd, A. M. and Pritchard, D. E., Sandy, Beds.: RSPB

Royal Society for the Protection of Birds (RSPB) (1993c) *Strategies for Wildlife: A Study of Local Authority Nature Conservation Strategies in the United Kingdom*, Sandy, Beds.: RSPB

Royal Society for the Protection of Birds (RSPB) (1996) *A Step by Step Guide to Environmental Appraisal*, Sandy, Beds.: RSPB

Royal Town Planning Institute (RTPI) (1976) *Planning and the Future*, London: RTPI

Royal Town Planning Institute (RTPI) (1982) *The Public and Planning: Means to Better Participation*, London: RTPI

Royal Town Planning Institute (RTPI) with Commission for Racial Equality (1983) *Planning for a Multi-racial Britain*, London: RTPI

Royal Town Planning Institute (RTPI) (1985) *Planners and Environmental Education*, London: RTPI

Royal Town Planning Institute (RTPI) (1987) *Report and Recommendations of the working Party on Women in Planning*, London: RTPI

Royal Town Planning Institute (RTPI) (1988–95) *Practice Advice Notes*, London: RTPI
  Development Control: Handling Planning Applications (1988)
  The Appointment of Consultants by Public Authorities (1988)
  Access for Disabled People (1988)
  Chartered Town Planners at Inquiries (1989)
  Consultancy by Current and Former Public Sector Employees (1989)
  Enforcement of Planning Control (1989; revised 1996)
  Professional Practice and Maladministration (1989)
  Development Briefs (1990)
  Development Control: Handling Appeals (1991; revised 1995)
  Continuing Professional Development (1992)
  Personal Safety at Meetings and Site Visits (1992)
  Planning for Women (1995)
  Environmental Assessment (1995)

Royal Town Planning Institute (RTPI) (1988a) *Managing Equality: The Role of Senior Planners*, London: RTPI

Royal Town Planning Institute (RTPI) (1988b) *Planning for Choice and Opportunity*, London: RTPI

Royal Town Planning Institute (RTPI) (1989) *Choice and Opportunity in Planning*, London: RTPI

Royal Town Planning Institute (RTPI) (1990a) *The Impact of the Channel Tunnel on Regions* by D. Simmonds, London: RTPI

Royal Town Planning Institute (RTPI) (1990b) *Caring for Cities – Town Planning's Role*, by P. Fleming, London: RTPI

Royal Town Planning Institute (RTPI) (1991a) *Traffic Growth and Planning Policy*, by D. Hutchinson, London: RTPI

Royal Town Planning Institute (RTPI) (1991b) *Planning: Is It a Service and How Can It Be Effective?*, by A. Gunne-Jones, London: RTPI

Royal Town Planning Institute (RTPI) (1992a) *Planning Policy and Social Housing*, London: RTPI

Royal Town Planning Institute (RTPI) (1992b) *The Regional Planning Process*, London: RTPI

Royal Town Planning Institute (RTPI) (1993a) *Ethnic Minorities and the Planning System*, by H. Thomas and V. Krishnarayan, London: RTPI

Royal Town Planning Institute (RTPI) (1993b) *The Character of Conservation Areas*, by R. Daniels, A. Jones and P. Larkham, London: RTPI

Royal Town Planning Institute (RTPI) (1993c) *Access Policies for Local Plans*, London: RTPI and Access Committee for England

Royal Town Planning Institute (RTPI) (1994) *The Impact of the European Community on Land Use Planning in the UK*, by H. W. E. Davies, London: RTPI

Royal Town Planning Institute (RTPI) (1996a) *The Local Delivery of Planning Services*, by Spawforth Planning Associates, London: RTPI

Royal Town Planning Institute (RTPI) (1996b) *The Role of Elected Members in Plan Making and Development Control*, by R. Darke and R. Manson, London: RTPI

Rural Development Commission: *see* appendix on *Official Publications*

Rural Voice (1990) *Employment of the Land*, Cirencester, Glos.: Rural Voice

Rush, M. (1990) *Parliament and Pressure Politics*, Oxford: Clarendon Press

Rutherford, L. A. and Peart, J. D. (1989) 'Opencast guidance: opportunities for green policies', *Journal of Planning and Environment Law* 1989: 402–10

Ryan, J. C. (1991) 'Impact fees: a new funding source for local growth', *Journal of Planning Literature* 5: 401–7

Ryder, A. A. (1987) 'The Dounreay inquiry: public participation in practice', *Scottish Geographical Magazine* 103: 54–7

Rydin, Y. (1984) 'The struggle for housing land: a case of confused interests', *Policy and Politics* 12: 431–46

Rydin, Y. (1986) *Housing Land Policy*, Aldershot, Hants.: Gower

Rydin, Y. (1992) 'Environmental dimensions of residential development and the implications for local planning practice', *Journal of Environmental Planning and Management* 35: 43–61

Rydin, Y. (1993) *The British Planning System: An Introduction*, London: Macmillan

Rydin, Y., Home, R., and Taylor, K. (1990) *Making the Most of the Planning Appeals System: Report to the Association of District Councils*, London: Association of District Councils

Salter, J. R. (1992a) 'Environmental assessment: the challenge from Brussels', *Journal of Planning and Environment Law* 1992: 14–20

Salter, J. R. (1992b) 'Environmental assessment: the need for transparency', *Journal of Planning and Environment Law* 1992: 214–21

Salter, J. R. (1992c) 'Environmental assessment: the question of implementation', *Journal of Planning and Environment Law* 1992: 313–18

Salter, M. and Newman, P. (1992) 'Minding their own business in the planning department', *Municipal Journal* 50 (11–17 December): 28–9

Sandbach, F. (1980) *Environment, Ideology and Public Policy*, Oxford: Blackwell

Sandercock, L. and Forsyth, A. (1992) 'A gender agenda: new directions for planning theory', *Journal of the American Planning Association* 58, 1: 49–58

Sanders, A.-M. and Rothnie, J. (1996) 'Planning registers: their role in promoting public participation', *Journal of Planning and Environment Law* 1996: 539–46

Sandford Report (1974) *Report of the National Park Policies Review Committee*, London: HMSO

Sassen, S. (1991) *The Global City: New York, London, Tokyo*, Princeton, NJ: Princeton University Press

Saunders, P. (1986) *Social Theory and the Urban Question* (second edition), London: Routledge

Scanlon, K., Edge, A., and Willmott, T. (1994) *The Economics of Listed Buildings*, Cambridge: Department of Land Economy, University of Cambridge

Scargill, D. I. and Scargill, K. E. (1994) *Containing the City: The Role of Oxford's Green Belt*, Oxford: School of Geography, University of Oxford

Scarman Report (1981) *The Brixton Disorders 10–12 April 1981* (Cmnd 8427), London: HMSO

Schackleton, J. R. (1992) *Training Too Much? A Sceptical Look at the Economics of Skill Provision in the UK*, London: Institute of Economic Affairs

Schaffer, F. (1970) *The New Town Story*, London: MacGibbon & Kee

Schmidt-Eichstaedt, G. (1996) *Land Use Planning and Building Permission in the European Union*, Cologne: Deutscher Gemeindeverlag and Verlag W. Kölhammer

Schofield, J. (1987) *Cost-benefit Analysis in Urban and Regional Planning*, London: Allen & Unwin

Scholefield, G. P. (1990) 'Transport and society: the Rees Jeffreys discussion papers', *Town Planning Review* 61: 487–93

Schon, D. A. (1971) *Beyond the Stable State*, London: Temple Smith

Schubert, D. and Sutcliffe, A. (1996) 'The "Haussmanization" of London? The planning and construction of Kingsway-Aldwych, 1889–1935', *Planning Perspectives* 11: 115–44

Schuster Report (1950) *Report of the Committee on the Qualifications of Planners* (Cmd 8059), London: HMSO

Scott Report (1942) *Report of the Committee on Land Utilisation in Rural Areas* (Cmd 6378), London: HMSO

Scottish Homes (1991) *Planning Agreements and Low Cost Housing in Scotland's Rural Areas*, Edinburgh: Scottish Homes

Scrase, T. (1991) 'Archaeology and planning: a case for full integration', *Journal of Planning and Environment Law* 1991: 1103–12

Seebohm Report (1968) *Report of the Committee on Local Authority and Allied Personal Social Services* (Cmnd 3703), London: HMSO

Segal Quince Wicksteed (1992) *Evaluation of Regional Enterprise Grants: Third Stage*, London: Department of Trade and Industry

Segal Quince Wicksteed (1996) *The Impact of Tourism on Rural Settlements*, Salisbury, Wilts.: Rural Development Commission

Self, P. (1982) *Planning the Urban Region: A Comparative Study of Policies and Organisations*, London: Allen & Unwin

Self, P. (1993) *Government by the Market? The Politics of Public Choice*, London: Macmillan

Sellgren, J. (1990) 'Development control data for planning research: the use of aggregated development control records', *Environment and Planning B: Planning and Design* 17: 23–7

Selman, P. (1988) *Countryside Planning in Practice: The Scottish Experience*, Stirling: Stirling University Press

Selman, P. (1992) *Environmental Planning: The Conservation and Development of Biophysical Resources*, London: Paul Chapman

Selman, P. (1995a) 'Theories for rural–environmental planning', *Planning Practice and Research* 10: 5–13

Selman, P. (1995b) 'Local sustainability: can the planning system help get us from here to there?', *Town Planning Review* 66: 287–302

Selman, P. (1996) *Local Sustainability: Managing and Planning Ecologically Sound Places*, London: Paul Chapman

SERPLAN (1992) *SERPLAN: Thirty Years of Regional Planning 1962–1992*, London: London and South East Regional Planning Conference

Sharman, F. A. (1985) 'Public attendance at planning inquiries', *Journal of Planning and Environment Law* 1985: 152–8

Sharp, E. (1969) *The Ministry of Housing and Local Government*, London: Allen & Unwin

Sharp, T. (1947) *Exeter Phoenix*, London: Architectural Press

Shaw, D., Nadin, V., and Westlake, T. (1995) 'The compendium of spatial planning systems and policies', *European Planning Studies* 3: 390–5

Shaw, D., Nadin, V., and Westlake, T. (1996) 'Towards a supra-national spatial development perspective: experience in Europe', *Journal of Planning Education and Research*, 15: 135–42

Shaw, K. (1995) 'Assessing the performance of urban development corporations: how reliable are the official government output measures?', *Planning Practice and Research*, 10: 287–97

Shaw, T. (1992) 'Regional planning guidance for the North East: advice to the Secretary of State for the Environment', in Minay, C. L. W., 'Developing regional guidance in England and Wales: a review symposium', *Town Planning Review* 63: 415–34

Sheail, J. (1976) *Nature in Trust: The History of Nature Conservation in Britain*, Glasgow: Blackie

Sheail, J. (1983) 'Deserts of the moon: the Mineral Workings Act and the restoration of ironstone workings in Northamptonshire', *Town Planning Review* 54: 405–24

Sheail, J. (1992) 'The *amenity* clause: an insight into half a century of environmental protection in the United Kingdom', *Transactions of the Institute of British Geographers* NS 17: 152–65

Sheail, J. (1995) 'John Dower, national parks, and town and country planning in Britain', *Planning Perspectives* 10: 1–16

Shelbourn, C. (1996) 'Protecting the "familiar and cherished scene"', *Journal of Planning and Environment Law* 1996: 463–69

Shelton, A. (1991) 'The well informed optimist's view', in Nadin, V. and Doak, J. (eds) *Town Planning Responses to City Change*, Aldershot, Hants.: Gower

Shepherd, J. and Abakuks, A. (1992) *The National Survey of Vacant Land in Urban Areas of England 1990* (Department of the Environment), London: HMSO

Shere, M. E. (1995) 'The myth of meaningful environmental risk assessment', *Harvard Environmental Law Review* 19: 409–92

Sherlock, H. (1991) *Cities Are Good for Us*, London: Paladin

Shiva, V. (1992) 'Recovering the real meaning of sustainability', in Cooper, D. E. and Palmer, J. A. (eds) *The Environment in Question*, London: Routledge

Shoard, M. (1980) *The Theft of the Countryside*, London: Maurice Temple Smith

Shoard, M. (1987) *This Land is Our Land*, London: Paladin

Short, J. R., Fleming, S., and Witt, S. (1986) *Housebuilding, Planning and Community Action*, London: Routledge & Kegan Paul

Shucksmith, M. (1983) 'Second homes: a framework for policy', *Town Planning Review* 54: 174–93

Shucksmith, M. (1988) 'Policy aspects of housebuilding on farmland in Britain', *Land Development Studies* 5: 129–38

Shucksmith, M. and Watkins, L. (1991) 'Housebuilding on farmland: the distributional effects in rural areas', *Journal of Rural Studies* 7: 153–68

Shucksmith, M., Henderson, M., Raybould, S., Coombes, M., and Wong, C. (1995) *A Classification of Rural Housing Markets in England*, (Department of the Environment), London: HMSO

Sidaway, R. (1994) *Recreation and the Natural Heritage: A Research Review*, Perth: Scottish Natural Heritage

Sillince, J. (1986) *A Theory of Planning*, Aldershot, Hants.: Gower

Sim, P. A. (1994) 'Mixed use development', in Wood, M. (ed.) *Planning Icons: Myth and Practice* (Planning Law Conference, *Journal of Planning and Environment Law*), London: Sweet & Maxwell

Simmie, J. (1981) *Power, Property and Corporatism*, London: Macmillan

Simmie, J. (1993) *Planning at the Crossroads*, London: UCL Press

Simmie, J. (ed.) (1994) *Planning London*, London: UCL Press

Simmie, J. and French, S. (1989) 'Corporatism, participation and planning: the case of London', *Progress in Planning* 31

Simmie, J. and King, R. (eds) (1990) *The State in Action: Public Policy and Politics*, London: Frances Pinter

Simmonds, D. (1990) *The Impact of the Channel Tunnel on the Regions*, London: RTPI

Simmons, I. G. (1993) *Interpreting Nature: Cultural Constructions of the Environment*, London: Routledge

Simpson, I. (1987) 'Planning gain: an aid to positive planning', in Harrison, M. L. and Mordey, R. (eds) *Planning Control: Philosophies, Prospects and Practice*, London: Croom Helm

Sinclair, G. (1992) *The Lost Land: Land Use Change in England 1945–1990*, London: Council for the Protection of Rural England

Sinfield, A. (1973) 'Poverty rediscovered', in Cullingworth, J. B. (ed.) *Planning for Change; vol. 3 of Problems of an Urban Society*, London: Allen & Unwin

Skeffington Report (1969) *Report of the Committee on Public Participation in Planning*, London: HMSO

Skelcher, C. (1985) 'Transportation', in Ranson, S., Jones, G., and Walsh, K. (eds) *Between Centre and Locality: The Politics of Public Policy*, London: Allen & Unwin

Skelcher, C., McCabe, A., Lowndes, V., and Nanton, P. (1996) *Community Networks in Urban Regeneration: It All Depends Who You Know*, Bristol: Policy Press

Slater, S., Marvin, S., and Newson, M. (1994) 'Land use planning and the water sector', *Town Planning Review* 65: 375–97

Smallbone, D. (1991) 'Partnership in economic development: the case of UK local enterprise agencies', *Policy Studies Review* 10: 87–98

Smart, G. and Anderson, M. (1990) *Planning and Management of Areas of Outstanding Natural Beauty*, Cheltenham, Glos.: Countryside Commission

Smeed Report (1964) *Road Pricing: The Economic and Technical Possibilities*, London: HMSO

Smith, A. G., Williams, G., and Houlder, M. (1986) 'Community influence on local planning policy', *Progress in Planning* 25: 1–82

Smith, D. (ed.) (1993) *Business and the Environment: Implications of the New Environmentalism*, London: Paul Chapman

Smith, G. (1994) 'Vitality and viability of town centres', in Wood, M. (ed.) *Planning Icons: Myth and Practice* (Planning Law Conference, *Journal of Planning and Environment Law*), London: Sweet & Maxwell

Smith, M. E. H. (ed.) (1989) *Guide to Housing: Main Changes in Housing Law* (third edition), London: Housing Centre

Smith, M. E. H. (ed.) (1990) *The Developing Housing Scene 1989–1990* (first supplement to the third edition of the *Guide to Housing*), London: Housing Centre

Smith, M. E. H. (ed.) (1995) *Housing – Today and Tomorrow* (second supplement to the third edition of the *Guide to Housing*), London: Housing Centre

Smith, N. (1996) *The New Urban Frontier: Gentrification and the Revanchist City*, London: Routledge

Smith, R. and Wannop, U. (1985) *Strategic Planning in Action: The Impact of the Clyde Regional Plan 1946–82*, Aldershot, Hants.: Gower

Solesbury, W. (1976) 'The environmental agenda: an illustration of how situations may become political issues and issues may demand responses from government: or how they may not', *Public Administration* 54: 379–97

Solesbury, W. (1981) 'Strategic planning: metaphor or method?', *Policy and Politics* 9: 419–37

Solesbury, W. (1986) 'The dilemmas of inner city policy', *Public Administration* 64: 389–400

Solesbury, W. (1993) 'Reframing urban policy', *Policy and Politics* 21: 31–8

Sorafu, F. J. (1957) 'The public interest reconsidered', *Journal of Politics* 19: 616–39

Sorensen, A. D. and Day, R. A. (1981) 'Libertarian planning', *Town Planning Review* 52: 390–402

Southgate, M. (1994) 'The added value test of merger' [of English Nature and the Countryside Commission], *Town and Country Planning*, May 134–5

Southgate, M. (1995) 'Nature conservation and planning', *Report for the Natural and Built Environment Professions*, Issue 7, August 34–7

Sparks, L. (1987) 'Retailing in enterprise zones: the example of Swansea', *Regional Studies* 21: 37–42

Standing Advisory Committee on Trunk Road Assessment: *see* Department of Transport, in appendix on *Official Publications*

Starkie, D. N. M. (1982) *The Motorway Age: Road and Traffic Policies in Postwar Britain*, Oxford: Pergamon

Steel, J., Nadin, V., Daniels, R., and Westlake, T. (1995) *The Efficiency and Effectiveness of Local Plan Inquiries*, London: HMSO

Steele, J. (1995) *Public Access to Information: An Evaluation of the Local Government (Access to Information) Act 1985*, London: Policy Studies Institute

Steer Davies Gleave (1992) *Financing Public Transport: How Does Britain Compare?*, London: Transport 2000

Steer Davies Gleave (1994) *Promoting Rail Investment*, London: Transport 2000

Steer Davies Gleave (1995) *Alternatives to Traffic Growth: The Role of Public Transport and the Future of Freight*, London: Transport 2000

Stein, J. M. (ed.) (1995) *Classic Readings in Urban Planning*, New York: McGraw-Hill

Stevens Report (1976) *Report of the Committee on Planning Control over Mineral Working*, London: HMSO

Stewart, J. D. and Stoker, G. (eds) (1991) *The Future of Local Government*, London: Macmillan

Stewart, J. D. and Stoker, G. (eds) (1995) *Local Government in the 1990s*, London: Macmillan

Stewart, J. D., Walsh, K., and Prior, D. (1995) *Citizenship: Rights, Community, and Participation*, London: Pitman

Stewart, M. (1994) 'Value for money in urban public expenditure', *Public Money and Management* October–December: 55–61

Stewart, M. and Taylor, M. (1995) *Empowerment and Estate Regeneration*, Bristol: Policy Press

Stoker, G. (1991) *The Politics of Local Government*, London: Macmillan

Stokes, G., Goodwin, P., and Kenny, F. (1992) *Trends in Transport and the Countryside*, Cheltenham, Glos.: Countryside Commission

Stone, P. A. (1973) *The Structure, Size and Costs of Urban Settlements* (National Institute of Economic and Social Research), Cambridge: Cambridge University Press

Storey, D. J. (1990) 'Evaluation of policies and measures to create local employment', *Urban Studies* 27: 669–84

Stubbs, M. (1994) 'Planning appeals by informal hearing: an appraisal of the views of consultants', *Journal of Planning and Environment Law* 1994: 710–14

Strathclyde Regional Council (1995) *Sustainability Indicators*, Glasgow: The Council

Suddards, R. W. and Hargreaves, R. (1995) *Listed Buildings*, London: Sweet & Maxwell

Suddards, R. W. and Morton, D. M. (1991) 'The character of conservation areas', *Journal of Planning and Environment Law* 1991: 1011–13

Sustrans (1996) *The National Cycle Network: Guidelines and Practical Details*, Bristol: Sustrans

Sutcliffe, A. (1981a) *Towards the Planned City: Germany, Britain, the United States and France 1780–1914*, Oxford: Blackwell

Sutcliffe, A. (ed.) (1981b) *British Town Planning: The Formative Years*, Leicester: Leicester University Press

Sutcliffe, A. (ed.) (1984) *Metropolis 1890–1940*, London: Mansell

Tant, A. P. (1990) 'The campaign for freedom of information: a participatory challenge to elitist British government', *Public Administration* 68: 477–91

Tate, J. (1994) 'Sustainability: a case of back to basics?', *Planning Practice and Research* 9: 367–79

Taussik, J. (1992) 'Pre-application enquiries', *Journal of Planning and Environment Law* 1992: 414–19

Taylor, A. (1984) 'The planning implications of new technology in retailing and distribution', *Town Planning Review* 55: 161–76

Taylor, A. (1992) *Choosing our Future: A Practical Politics of the Environment*, London: Routledge

Taylor, M. (1995) *Unleashing the Potential: Bringing Residents to the Centre of Regeneration*, York: Joseph Rowntree Foundation

Taylor, N. (1992) 'Professional ethics in town planning: what is a code of conduct for?', *Town Planning Review* 63: 227–41

Teague, P. (1994) 'Governance structures and economic perfor-mance: the case of Northern Ireland', *International Journal of Urban and Regional Research* 18: 275–92

TEST (1984) *The Company Car Factor*, London: TEST

TEST (1991) *Wrong Side of the Tracks? Impacts of Road and Rail Transport on the Environment: A Basis for Discussion*, London: TEST

TEST (1992) *An Environmental Approach to Transport and Planning in Cardiff*, London: TEST

Tewdwr-Jones, M. (1994a) 'Policy implications of the plan-led system', *Journal of Planning and Environment Law* 1994: 584–93

Tewdwr-Jones, M. (1994b) 'The development plan in policy implementation', *Environment and Planning C; Government and Policy* 12: 145–63

Tewdwr-Jones, M. (ed.) (1996) *British Planning Policy in Transition: Planning in the 1990s*, London: UCL Press

Thake, S. and Stauerbach, R. (1993) *Investing in People: Rescuing Communities from the Margin*, York: Joseph Rowntree Foundation

Therivel, R. (1995) 'Environmental appraisal of development plans: current status', *Planning Practice and Research* 10: 223–34

Therivel, R. and Partidario, M. R. (1996) *The Practice of Strategic Environmental Assessment*, London: Earthscan

Thomas, D. (1995) *Community Development at Work: A Case of Obscurity in Accomplishment*, London: Community Development Foundation

Thomas, H. (1992) 'Disability, politics and the built environment', *Planning Practice and Research* 7: 22–6

Thomas, H. (1996) 'Public participation in planning', in Tewdwr-Jones, M. (ed.) *British Planning Policy in Transition: Planning in the 1990s*, London: UCL Press

Thomas, H. and Healey, P. (1991) *Dilemmas of Planning Practice: Ethics, Legitimacy, and the Validation of Knowledge*, Aldershot, Hants.: Avebury

Thomas, H. and Krishnarayan, V. (1993) 'Race, equality and planning', *Planner* 79: 17–20

Thomas, H. and Krishnarayan, V. (1994) 'Race, disadvantage and policy processes in British planning', *Environment and Planning A* 26: 1891–1910

Thomas, K. (1990a) *Planning for Shops*, London: Estates Gazette

Thomas, K. (1990b) 'Planning students and careers: the enrolment of planning students, and career destinations of new planning graduates, 1977 to 1989', *Planner* 76, 36: 13–16

Thomas, K. (1990c) 'Enrolment of students and output of UK planning schools, 1977 to 1989, *Education for Planning Association Newsletter* February: 17–49

Thomas, S. and Watkin, T. G. (1995) 'Oh, noisy bells, be dumb: church bells. Statutory nuisance and ecclesiastical duties', *Journal of Planning and Environment Law* 1995: 1097–1105

Thompson, B. (1985) 'Neighbour notification: recent developments in Scotland and Northern Ireland', *Journal of Planning and Environment Law* 1985: 530–5

Thompson, H. (1992) 'Contaminated land: the implications for property transactions and the property market', *National Westminster Bank Quarterly Review* November 20–33

Thompson, P. B. (1995) *The Spirit of the Soil: Agriculture and Environmental Ethics*, London: Routledge

Thomson, J. M. (1969) *Motorways in London*, London: Duckworth

Thornley, A. (1991) *Urban Planning under Thatcherism: The Challenge of the Market*, London: Routledge

Thornley, A. (1993) *Urban Planning under Thatcherism: The Challenge of the Market* (second edition), London: Routledge

Thornley, A. and Newman, P. (1996) *Replanning European Cities*, London: Routledge

Tibbs, N. (1991) 'The objectives and implementation of the Bristol Inner City Project', in V. Nadin and J. Doak (eds), *Town Planning Responses to City Change*, Aldershot, Hants.: Gower

Tiesdell, S. A., Oc, T., and Heath, T. (1996) *Revitalising Historic Urban Quarters*, Oxford: Butterworth-Heinemann

Tietenberg, T. (1990) 'Economic instruments for environmental regulation', *Oxford Review of Economic Policy* 6: 17–33

Till, J. E. (1995) 'Building credibility in public studies', *American Scientist* 83: 468–73

Tinch, R. (1996) *The Valuation of Environmental Externalities* (Department of the Environment), London: HMSO

Titmuss, R. M. (1958) 'War and social policy', in his *Essays on 'The Welfare State'*, London: Allen & Unwin

Tolba, M. (1992) *Saving our Planet: Challenges and Hopes*, London: Chapman & Hall

Tolley, R. (1990a) *Calming Traffic in Residential Areas*, Wales: Brefi Press

Tolley, R. (ed.) (1990b) *The Greening of Urban Transport: Planning for Walking and Cycling in Western Cities*, London: Belhaven

Towers, G. (1995) *Building Democracy: Community Architecture in the Inner Cities*, London: UCL Press

Town and Country Planning Association (1989) *Bridging the North–South Divide*, London: TCPA

Town and Country Planning Association (1992) *New Settlements: Planning Policy Guidance* [spoof PPG], London: TCPA

Town and Country Planning Association (1993) *Strategic Planning for Regional Development*, London: TCPA

Town and Country Planning Association (1996) *The People: Where Will They Go?* (ed. Breheny, M. and Hall, P.), London: TCPA

Townroe, P. and Martin, R. (1992) *Regional Development in the 1990s: The British Isles in Transition*, London: Jessica Kingsley

Townsend, P. (1976) 'Area deprivation policies', *New Statesman*, 6 August 186–71

Transport 2000 (1995) *Moving Together: Policies to Cut Car Commuting*, London: Transport 2000 Trust

Travers, T., Biggs, S., and Jones, G. (1995) *Joint Working Between Local Authorities: Experience for the Metropolitan Areas*, London: Local Government Chronicle and Joseph Rowntree Foundation

Travers, T., Jones, G., Hebbert, M., and Burnham, J. (1991) *The Government of London*, York: Joseph Rowntree Foundation

Treisman, M. (1995) *Traffic and Planning in Oxford: Commercial Development and the Environment*, Oxford: School of Planning, Oxford Brookes University

Trench, S. and Oc, T. (1990) *Current Issues in Planning*, Aldershot, Hants.: Avebury

Trench, S. and Oc, T. (1995) *Current Issues in Planning*, vol. 2, Aldershot, Hants.: Avebury

Tromans, S. (1991) 'Roads to prosperity or roads to ruin? Transport and the environment in England and Wales', *Journal of Environmental Law* 3: 1–37

Tromans, S. and Clarkson, M. (1991) 'The Environmental Protection Act 1990: its relevance to planning controls', *Journal of Planning and Environment Law* 1991: 507–15

Tromans, S. and Turrall-Clarke, R. (1994 with 1996 Supplement) *Contaminated Land*, London: Sweet & Maxwell

Tromans, S., Grant, M., and Nash, M. (eds) (1997) *Encyclopaedia of Environmental Law* (looseleaf; updated regularly), London: Sweet & Maxwell

Truelove, P. (1992) *Decision Making in Transport Planning*, Harlow, Essex: Longman

Tubbs, C. (1994) 'The New Forest: one step backwards', *Ecos* 15, 2: 37–42

Tugnutt, A. (1991) 'Design: the wider aspects of townscapes', *Town and Country Planning Summer School 1991: Report of Proceedings*: 19–22

Turner, T. (1992) 'Open space planning in London: from standards per 1000 to green strategy', *Town Planning Review* 63: 365–86

Turner, T. (1996) *City as Landscape: A Post-postmodern View of Design and Planning*, London: Spon

Turok, I. (1988) 'The limits of financial assistance: an evaluation of local authority aid to industry', *Local Economy* 2: 286–97

Turok, I. (1989) 'Evaluation and understanding in local economic policy', *Urban Studies* 26: 587–606

Turok, I. (1990a) 'Public investment and privatisation in the new towns: a financial assessment of Bracknell', *Environment and Planning A* 22: 1323–36

Turok, I. (1990b) *Targeting Urban Employment Initiatives* (Department of the Environment), London: HMSO

Turok, I. (1992) 'Property-led urban regeneration: panacea or placebo?', *Environment and Planning A* 24: 361–79

Turok, I. and Shutt, J. (eds) (1994) *Urban Policy into the 21st Century*: special issue of *Local Economy*, 9, 3: 211–304

Turok, I. and Wannop, U. (1990) *Targeting Urban Employment Initiatives* (Department of the Environment), London: HMSO

Tyldesely, David, & Associates (1996) *Wildlife Impact: The Treatment of Nature Conservation in Environmental Assessment*, Sandy, Beds.: Royal Society for the Protection of Birds

Tym, R. H. (1994) 'Planning in a rapidly changing economy', in Wood, M. (ed.) *Planning Icons: Myth and Practice* (Planning Law Conference, *Journal of Planning and Environment Law*), London: Sweet & Maxwell

Tym, Roger, & Partners (1982–4) *Monitoring Enterprise Zones: Year One Report* (1982); *Year Two Report* (1983); and *Year Three Report* (1984), London: Roger Tym & Partners

Tym, Roger, & Partners (1984) *Land Supply for Housing in Urban Areas*, London: Housing Research Foundation

Tym, Roger, & Partners (in association with Land Use Consultants) (1987) *Evaluation of Derelict Land Grant Schemes* (Department of the Environment), London: HMSO

Tym, Roger, & Partners (1988) *An Evaluation of the Stockbridge Village Trust Initiative* (Department of the Environment), London: HMSO

Tym, Roger, & Partners (1989a) *The Effect on Small Firms of Refusal of Planning Permission* (Department of the Environment), London: HMSO

Tym, Roger, & Partners (1989b) *The Incidence and Effects of Planning Conditions* (Department of the Environment), London: HMSO

Tym, Roger, & Partners (1990) *Development Control Performance*, London: National Planning Forum

Tym, Roger, & Partners (1991) *Housing Land Availability* (Department of the Environment), London: HMSO

Tym, Roger, & Partners (1995a) *The Use of Article 4 Directions* (Department of the Environment), London: HMSO

Tym, Roger, & Partners (1995b) *Review of the Implementation of PPG 16, Archaeology and Planning*, London: English Heritage

Tym, Roger, & Partners (1996a) *Enterprise Zones Monitoring 1994–95*, London: HMSO

Tym, Roger, & Partners (in association with Oscar Faber TPA) (1996b) *The Gyle Impact Study* [retail impact], Edinburgh: HMSO

Tyme, J. (1978) *Motorways versus Democracy*, London: Macmillan

Uman, M. F. (ed.) (1993) *Keeping Pace with Science and Engineering: Case Studies in Environmental Regulation*, Washington, DC: National Academy Press

Underwood, J. (1981a) 'Development control: a case study of discretion in action', in Barrett, S. and Fudge, C. (eds) *Policy and Action: Essays on the Implementation of Public Policy*, London: Methuen

Underwood, J. (1981b) 'Development control: a review of research and current issues', *Progress in Planning* 16, 3: 175–242

United Nations (1992) *Environmental Accounting: Current Issues, Abstracts and Bibliography*, New York: Department of Economic and Social Development, United Nations

University of Liverpool, Environmental Advisory Unit (1986) *Transforming our Waste Land: The Way Forward* (Department of the Environment), London: HMSO

Unwin, R. (1909) *Town Planning in Practice: An Introduction to the Art of Designing Cities and Suburbs*, London: T. Fisher Unwin

Upton, W. and Harwood, R. (1996) 'The stunning powers of environmental inspectors', *Journal of Planning and Environment Law* 1996: 623–32

Urban Initiatives (1995) *Hertfordshire Dwelling Provision through Planning Regeneration*, Hertford: Hertfordshire County Environment Department

Urban Villages Group (1992) *Urban Villages: A Concept for Creating Mixed-use Urban Developments on a Sustainable Scale*, London: Urban Villages Group

Urban Villages Group (1993) *Economics of Urban Villages*, London: Urban Villages Forum

Uthwatt Report (1942) *Final Report of the Expert Committee on Compensation and Betterment* (Cmd 6386), London: HMSO

Vale, B. (1995) *Prefabs: A History of the UK Temporary Housing Programme*, London: Spon

Varner, G. E. (1987) 'Do species have standing?', *Environmental Ethics* 9: 57–72

Veld, R. J. (1991) 'Road pricing: a logical failure', in Kraan, D. J. and Veld, R. J. (eds) *Environmental Protection: Public or Private Choice?*, Dordrecht: Kluwer

Vergara, C. J. (1995) *The New American Ghetto*, New Brunswick, NJ: Rutgers University Press

Verney Report (1976) *The Way Ahead: Report of the Advisory Committee on Aggregates*, London: HMSO

Vickerman, R. W. (1991) *Infrastructure and Regional Development*, London: Pion

Vickers, G. (1995) *The Art of Judgment: A Study of Policy-making* (Centenary Edition), London: Sage

Vilagrasa, J. and Larkham, P. J. (1995) 'Post-war redevelopment and conservation in Britain: ideal and reality in the historic core of Worcester', *Planning Perspectives* 10: 149–72

Viner, D. and Hulme, M. (1994) *The Climate Impacts LINK Project*, Norwich: Climatic Research Unit, University of East Anglia

Vogel, D. (1986) *National Styles of Regulation: Environmental Policy in Great Britain and the United States*, Ithaca, NY: Cornell University Press

Wachs, M. (ed.) (1985) *Ethics in Planning*, New Brunswick, NJ: Center for Urban Policy Research

Wachs, M. and Crawford, M. (1992) *The Car and the City*, Ann Arbor, MI: University of Michigan Press

Wade, E. C. S. and Bradley, A. W. (1993) *Constitutional and Administrative Law* (eleventh edition by Bradley, A. W. and Ewing, K. D.), London: Longman

Wägenbaur, R. (1991) 'The European Community's policy on implementation of environmental directives', *Fordham International Law Journal* 14: 455–77

Waite, R. (1995) *Household Waste Recycling*, London: Earthscan

Wakeford, R. (1990) *American Development Control: Parallels and Paradoxes from an English Perspective*, London: HMSO

Wakeford, R. (1993) 'Planning policy guidance: what's the use?' *Housing and Planning Review* April/May 14–18

Waldegrave, W., Byng, J., Paterson, T., and Pye, G. (1986) *Distant Views of William Waldegrave's Speech*, London: Centre for Policy Studies

Walker, M. and Reynard, M. V. (1990) *Costs in Planning Proceedings*, London: Longman

Walmsley, D. A. and Perrett, K. E. (1992) *The Effects of Rapid Transport on Public Transport and Urban Development*, Transport Research Laboratory, State of the Art Review 6, London: HMSO

Wannop, U. (1990) 'The Glasgow Eastern Area Renewal (GEAR) project: a perspective on the management of urban regeneration', *Town Planning Review* 61: 455–74

Wannop, U. (1994) *Regional Planning and Governance in Britain in the 1990s*, Glasgow: University of Strathclyde, Centre for Planning, Strathclyde Papers on Planning no. 27

Wannop, U. (1995) *The Regional Imperative: Regional Planning and Governance in Britain, Europe and the United States*, London: Regional Studies Association and Jessica Kingsley.

Ward, C. and Hardy, D. (1990) *Goodnight Campers! The History of the British Holiday Camp*, London: Spon

Ward, S. V. (1988) *The Geography of Interwar Britain: The State and Uneven Development*, London: Routledge

Ward, S. V. (ed.) (1992) *The Garden City: Past, Present and Future*, London: Spon

Ward, S. V. (1994) *Planning and Urban Change*, London: Spon

Wardroper, J. (1981) *Juggernaut*, London: Temple Smith

Warren, H. and Davidge, W. R. (eds) (1930) *Decentralisation of Population and Industry: A New Principle in Town Planning*, London: King

Waters, G. R. (1994) 'Government policies for the countryside', *Land Use Policy* 11: 88–93

Wates, N. (1977) *The Battle for Tolmers Square*, London: Routledge & Kegan Paul

Watt, P. (1992) 'Publicity for planning applications', *Scottish Planning Law and Practice* 38: 15–16.

Weale, A. (1992) *The New Politics of Pollution*, Manchester: Manchester University Press

Weale, A., O'Riordan, T., and Kramme, L. (1991) *Controlling Pollution in the Round: Change and Choice in Environmental Regulation in Britain and West Germany*, London: Anglo-German Foundation for the Study of Industrial Society

Weber, M. M. (1968–69) 'Planning in an environment of change', *Town Planning Review* 39: 179–95 and 277–95 (reprinted in Cullingworth, J. B. (ed.) 1973 *Planning for Change; vol. 3 of Problems of an Urban Society*, London: Allen & Unwin, 1973)

Webster, B. and Lavers, A. (1991) 'The effectiveness of public local inquiries as a vehicle for public participation in the plan-making process: a case study of the Barnet unitary development plan inquiry', *Journal of Planning and Environment Law* 1991: 803–13

Welford, R. (1995) *Environmental Strategy and Sustainable Development: The Corporate Challenge for the 21st Century*, London: Routledge

Wells, H. G. (1905) *A Modern Utopia*, London: Collins

Wenban-Smith, A. and Meeston, J. (1990) *Negotiating with Planning Authorities*, London: Estates Gazette

Westwood, S. and Williams, J. (1996) *Imagining Cities: Scripts, Signs and Memories*, London: Routledge

Whatmore, K. and Boucher, S. (1993) 'Bargaining with nature: the discourse and practice of environmental planning gains', *Transactions of the Institute of British Geographers* NS 18: 166–78

Whatmore, S., Munton, R., and Marsden, T. (1990) 'The rural restructuring process: emerging divisions of agricultural property rights', *Regional Studies* 24: 235–45

Whitaker, S. (1995) *First Steps: Local Agenda 21 in Practice. Municipal Strategies for Sustainability as Presented to Global Forum, Manchester, 1994*, London: HMSO

Whitbread, M. and Marsay, A. (1992) *Coastal Superquarries to Supply South-East England Aggregate Requirements* (Department of the Environment), London: HMSO

Whitbread, M., Mayne, D., and Wickens, D. (1991) *Tackling Vacant Land: An Evaluation of Policy Instruments for Tackling Land Vacancy* (Department of the Environment), London: HMSO

White, P. (1986) 'Land availability, land banking and the price of land for housing: a review of recent debates', *Land Development Studies* 3: 101–11

White, P. (1995) *Public Transport: Its Planning, Management and Operation*, London: UCL Press

Whitehand, J. W. R. (1989) 'Development pressure, development control, and suburban townscape change', *Town Planning Review* 60: 403–20

Whitehand, J. W. R. and Larkham, P. J. (1991) 'Suburban cramming and development control', *Journal of Property Research* 8: 147–59

Whitehand, J. W. R. and Larkham, P. J. (1992) *Urban Landscapes: International Perspectives*, London: Routledge

Whitelegg, J. (1993) *Transport for a Sustainable Future: The Case for Europe*, London: Belhaven

Whitelegg, J. (1994) *Roads, Jobs and the Economy*, London: Greenpeace

Whitney, D. and Haughton, G. (1990) 'Structures for development partnerships in the 1990s: practice in West Yorkshire', *Planner* 76, 21: 15–19

Widdicombe Report (1986) *The Conduct of Local Authority Business: Report* (Cmnd 9797); *Research Volumes* (Cmnd 9798, 9799, 9800, and 9801); London: HMSO. *Government Response to the Report* (Cm 433), London: HMSO, 1988

Wilcox, S. (1995) *Housing Finance Review 1995/96*, York: Joseph Rowntree Foundation

Wildavsky, A. B. (1973) 'If planning is everything, maybe it's nothing', *Policy Sciences* 4: 127–53

Wildavsky, A. B. (1987) *Speaking Truth to Power: The Art and Craft of Policy Analysis* (second edition), London: Transaction Publishers

Wilding, R. (1990) *The Care of Redundant Churches: A Review of the Operation and Financing of the Redundant Churches Fund*, London: HMSO

Wilheim, J. (1996) *Fax Messages from the Future*, London: Earthscan

Wilkes, S. and Peter, N. (1995) 'Think globally, act locally: implementing Agenda 21 in Britain', *Policy Studies* 16: 37–44

Wilkinson, D. (1992) 'Maastricht and the environment: the implications for the EC's environmental policy of the Treaty on European Union', *Journal of Environmental Law* 4: 221–39

Wilkinson, D. and Waterton, J. (1991) *Public Attitudes to the Environment in Scotland*, Edinburgh: Scottish Office

Williams, G., Bell, P., and Russell, L. (1991) *Evaluating the Low Cost Rural Housing Initiative* (Department of the Environment), London: HMSO

Williams, G., Strange, I., Bintley, M., and Bristow, R. (1992) *Metropolitan Planning in the 1990s: The Role of Unitary Development Plans*, Manchester: Department of Planning and Landscape, University of Manchester

Williams, R. and Birch, N. (1994) 'The longer-term implications of national traffic forecasts and international network plans for local roads policy: the case of Oxfordshire', *Transport Policy* 1: 95–99

Williams, R. and Wood, B. (1994) *Urban Land and Property Markets in the UK*, London: UCL Press

Williams, R. H. (1996a) *The European Union Committee of the Regions, its UK Membership and Spatial Planning*, Newcastle upon Tyne: Department of Town and Country Planning, University of Newcastle

Williams, R. H. (1996b) *European Union Spatial Policy and Planning*, London: Paul Chapman

Willis, K. and Garrod, G. (1993) *The Value of Waterside Properties*, Countryside Change Unit, University of Newcastle upon Tyne

Willis, K. G. (1995) 'Judging development control decisions', *Urban Studies* 32: 1065–79

Wilmott, P. (1994) *Urban Trends 2: A Decade in Britain's Deprived Urban Areas*, London: Policy Studies Institute.

Wilson, D. and Game, C. (1994) *Local Government in the United Kingdom*, London: Macmillan

Wilson, E. (1993) *Strategic Environmental Assessment*, London: Earthscan

Wilson, G. K. (1983) 'Planning lessons from the ports', *Public Administration* 61: 265–81

Wilson Report (1963) *Committee on the Problems of Noise: Final Report* (Cmnd 2056), London: HMSO

Winpenny, J. T. (1991) *Values for the Environment*, London: HMSO

Winter, M. (1996) *Rural Politics: Policies for Agriculture, Forestry and the Environment*, London: Routledge

Wolman, H. and Goldsmith, M. (1992) *Urban Politics and Policy: A Comparative Approach*, Oxford: Blackwell

Wolman, H. L., Cookford, C., and Hill, E. (1994) 'Evaluating the success of urban success stories', *Urban Studies* 31: 835–50

Womack, J. P. (1994) 'The real EV [electric vehicle] challenge: reinventing an industry', *Transport Policy* 1: 266–70

Wood, C. (1989) *Planning Pollution Prevention*, London: Heinemann

Wood, C. (1991) 'Urban renewal: the British experience', in R. Alterman and G. Cars, *Neighbourhood Regeneration: An International Evaluation*, London: Mansell

Wood, C. (1994) 'Local urban regeneration initiatives: Birmingham Heartlands', *Cities* 11: 48–58

Wood, C. (1995) *Environmental Impact Assessment: A Comparative Review*, Harlow, Essex: Longman

Wood, C. (1996) 'Private sector housing renewal in the UK: progress under the 'new regime' in Birmingham', paper presented to the ACSP–AESOP Joint International Congress, Toronto, Canada

Wood, C. and Jones, C. (1991) *Monitoring Environmental Assessment and Planning* (Department of the Environment), London: HMSO

Wood, C. and Jones, C. (1992) 'The impact of environmental assessment on local planning authorities', *Journal of Environmental Planning and Management* 35: 115–28

Wood, M. (ed.) (1994) *Planning Icons: Myth and Practice* (Planning Law Conference, *Journal of Planning and Environment Law*), London: Sweet & Maxwell

Wood, M. (1996) 'Local plans and UDPs: is there a better way?', *Journal of Planning and Environment Law* 1996: 807–15

Wood, W. (1949) *Planning and the Law*, London: Marshall

Woodward, G. D. and Larkham. P. J. (1994) *Ideal and Reality in Planning: Post-war Development in Worcester*, Birmingham: School of Planning, University of Central England

World Commission on Environment and Development: Brundland Report (1987) *Our Common Future*, Oxford: Oxford University Press

Worskett, R. (1969) *The Character of Towns*, London: Architectural Press

Wraith, R. E. and Hutchesson, P. G. (1973) *Administrative Tribunals*, London: Allen & Unwin

Wraith, R. E. and Lamb, G. B. (1971) *Public Inquiries as an Instrument of Government*, London: Allen & Unwin

Wrigley, N. and Lowe, M. (eds) (1996) *Retailing, Consumption and Capital*, Harlow, Essex: Longman

Wye College (1991) *Rural Society: Issues for the Nineties*, Wye, Kent: Wye College

Yardley, D. (1995) *Introduction to Constitutional and Administrative Law*, London: Butterworths

Yelling, J. A. (1992) *Slums and Redevelopment: Policy and Practice in England 1918–1945*, London: UCL Press

Yeomans, D. (1994) 'Rehabilitation and historic preservation: a comparison of American and British approaches', *Town Planning Review* 65: 159–78

Yiftachel, O. (1989) 'Towards a new typology of urban planning theories', *Environment and Planning B: Planning and Design* 16: 23–9

Young, E. (1991) *Scottish Planning Appeals*, Edinburgh: Green/Sweet & Maxwell

Young, E. and Rowan-Robinson, J. (1985) *Scottish Planning Law and Practice*, Glasgow: Hodge

Young, K. (1984) 'Metropolitan government: the development of the concept of reality', in Leach, S. (ed.) *The Future of Metropolitan Government*, Birmingham: Institute of Local Government Studies, University of Birmingham

Young, K. (1986) 'Metropolis, R.I.P.' *Political Quarterly* 57: 36–46

Young, K. (1990) *The Politics of Local Government since Widdicombe*, York: Joseph Rowntree Foundation

Young, K., Gosschalk, B., and Hatter, W. (1996) *In Search of Community Identity* (Joseph Rowntree Foundation), York: York Publishing Services

SELECTED OFFICIAL PUBLICATIONS

This list excludes agencies with only a small number of selected publications: these are included in the main bibliography.

A NOTE ON OFFICIAL PUBLICATIONS

The student of town and country planning now has a very rich library of official publications to consult. This has grown considerably over the last decade, partly because of the large research programme sponsored by the DoE. This appendix, though not comprehensive, lists the more important relevant publications.

A few words of explanation about the mysteries of government publications may be helpful. *White Papers* are technically *Command Papers*, i.e. they are 'presented to Parliament by Command of Her Majesty'. The reference to colour was at one time meaningful: a White Paper was a slim paper (often a policy statement or report) which had a white cover. It was therefore easily distinguishable from the more substantial Blue Books (which had blue covers). The Victorian Royal Commission reports were perhaps the best known of these.

Command papers are numbered sequentially and have a short prefix which varies, but which (so far) is always an abbreviation of the word 'command'. The earliest papers were prefixed C, but this was changed to Cd at the turn of the century. This prefix lasted until 1918–19 when, on approaching the number 10,000, it was changed to Cmd. This served its purpose until 1956, in which year the prefix was changed again – and for the same reason – to Cmnd. The latest (but presumably not the final) change was made in 1986, when the prefix became Cm. At the time of writing, the last number to be used was Cm 2207 – for the DoE *Annual Report* (of which more anon). Why the Queen's Printer is unable to go beyond 9,999 in its numbering is not clear. What *is* clear is that authors and publishers alike have difficulty in getting the prefix correct every time.

Though still used to denote a statement of government policy, the term 'White Paper' now has no precise meaning. Coloured covers are used extensively and, in recent years, the amount of 'art work' has increased. A White Paper no longer necessarily looks like one. Colourful and graphic presentation has become common, and some White Papers bear a strong resemblance to company reports.

Until relatively recently, it was reasonable to assume that, if a White Paper was a policy document, it represented the government's view, or its fairly firm proposals. This distinguished it from a 'Green Paper' which was of the nature of a preliminary draft White Paper. Such papers began to appear in the late 1960s, and were popular for a time.

The distinction between white and green is now blurred and it is not uncommon to hear that a White Paper has a 'green tinge' (i.e. some of the proposals are still open for discussion). Similarly, Green Papers may have a 'white tinge' (i.e. certain issues have been firmly decided and are not open for further debate). Of course, circumstances change, and so do the minds of governments. (The 1989 White Paper on *The Future of Development Plans* announced the abolition of structure plans, but in fact they have been retained.)

Green Papers have now largely been superseded by Consultation Papers which, ironically, are often not published by HMSO, but are 'available' from the department concerned. The one advantage of this is that they are free. In recent years, there have been large numbers of these. Those concerned with planning are frequently printed in the monthly update to Grant's *Encyclopaedia*.

Another recent innovation is the publication of annual reports of departments. Again these are Command Papers (with a range of colours). But they are both more and less than annual reports. They are more, in the sense that they present details of recent and planned expenditure on the services which are covered by the budget of the department(s) concerned. In this, they are parts of *The Government's Expenditure Plans* (which is their subtitle). But they are less than a full annual report in that they do not enter into discussion of the policies to which the figures relate. They merely state what these policies are (which is useful, but limited). As increasing sections of the executive parts of departments are hived off to *Next Steps* Agencies, the reports become of an even more summary nature.

Most government publications are published by HMSO, but until 1996 the market for Scottish and Welsh publications apparently did not justify the same treatment. As a consequence, many of the planning documents issued by the Scottish Office and the Welsh Office up to that year have to be obtained direct (as is the case with most DoE Consultation Papers). Many planning publications can now be obtained from HMSO Edinburgh and HMSO Cardiff. Some publications of a number of English departments (such as Agriculture, Industry and Transport) are also obtainable only from the departments concerned. In the following lists, the publisher is HMSO unless otherwise indicated. Northern Ireland publications are available either from HMSO Belfast or, more usually, from the Planning Service.

Note: The HMSO has now been privatised and renamed Stationery Office.

COMMAND PAPERS

1940 *Report of the Royal Commission on the Distribution of the Industrial Population* (Barlow Report), Cmd 6153

1942 *Report of the Committee on Land Utilisation in Rural Areas* (Scott Report), Cmd 6378

1942 *Final Report of the Expert Committee on Compensation and Betterment* (Uthwatt Report), Cmd 6386

1944 *The Control of Land Use*, Cmd 6537

1946 *Interim Report of the New Towns Committee*, Cmd 6759; *Second Interim Report*, Cmd 6794; *Final Report*, Cmd 6876 (Reith Report)

1947 *Town and Country Planning Bill 1947: Explanatory Memorandum*, Cmd 7006

1947 *Report of the National Parks Committee (England and Wales)* (Hobhouse Report), Cmd 7121; *Report of the Special Committee on Footpaths and Access to the Countryside* (Hobhouse Subcommittee Report), Cmd 7207

1950   *Report of the Committee on the Qualifications of Planners* (Schuster Report), Cmd 8059

1951   *Town and Country Planning 1943–1951: Progress Report*, Cmd 8204

1952   *Town and Country Planning Act, 1947: Amendment of Financial Provisions*, Cmd 8699

1957   *Report of the Committee on Administrative Tribunals and Enquiries* (Franks Report), Cmnd 218

1958   *Town and Country Planning Bill: Explanatory Memorandum*, Cmnd 562

1963   *London – Employment: Housing: Land*, Cmnd 1952

1963   *Central Scotland: A Programme for Development and Growth*, Cmnd 2288

1965   *Report of the Committee on Housing in Greater London* (Milner Holland Report), Cmnd 2605

1965   *The Land Commission*, Cmnd 2771

1966   *Leisure in the Countryside*, Cmnd 2928

1966   *Transport Policy*, Cmnd 3057

1967   *Town and Country Planning*, Cmnd 3333

1967   *Public Transport and Traffic*, Cmnd 3481

1968   *The Older Houses in Scotland: A Plan for Action*, Cmnd 3598

1968   *Old Houses into New Homes*, Cmnd 3602

1968   *Transport in London*, Cmnd 3686

1968   *Report of the Committee on Local Authority and Allied Personal Social Services* (Beveridge Report), Cmnd 3703

1969   *Modifications in Betterment Levy*, Cmnd 4001

1969   *Information and the Public Interest*, Cmnd 4089

1970   *The Protection of the Environment: The Fight Against Pollution*, Cmnd 4373

1970   *The Reorganisation of Central Government*, Cmnd 4506

1972   *Industrial and Regional Development*, Cmnd 4942

1972   *Development and Compensation: Putting People First*, Cmnd 5124

1973   *Land Resource Use in Scotland: The Government's Observations on the Report of the Select Committee on Scottish Affairs*, Cmnd 5248

1973   *Homes for People: Scottish Housing Policy in the 1970s*, Cmnd 5272

1973   *Widening the Choice: The Next Steps in Housing*, Cmnd 5280

1973   *Towards Better Homes: Proposals for Dealing with Scotland's Older Housing*, Cmnd 5338

1973   *Better Homes: The Next Priorities*, Cmnd 5339

1974   *Land*, Cmnd 5730

1975   *Development Land Tax*, Cmnd 6195

1975   *Sport and Recreation*, Cmnd 6200

1977   *Nuclear Power and the Environment*, Cmnd 6820

1977   *Transport Policy*, Cmnd 6836

1977   *Statement on the Non-statutory Inquiry by the Baroness Sharp into the Continued Use of Dartmoor for Military Training*, Cmnd 6837

1977   *Policy for the Inner Cities*, Cmnd 6845

1978   *Planning Procedures: The Government's Response to the Eighth Report from the Expenditure Committee, Session 1976–77*, Cmnd 7056

1978   *Policy for Roads: England*, Cmnd 7132

1978   *Report on the Review of Highway Inquiry Procedures*, Cmnd 7133

1978   *The Challenge of North Sea Oil*, Cmnd 7143

1979   *A National Heritage Fund*, Cmnd 7428

1979   *Organic Change in Local Government*, Cmnd 7457

1979   *Farming and the Nation*, Cmnd 7458
1979   *Central Government Controls over Local Authorities*, Cmnd 7634
1981   *Committee of Inquiry into Local Government in Scotland* (Stodart Report), Cmnd 8115
1981   *Lorries, People and the Environment*, Cmnd 8439
1982   *Policy for Roads: England 1981*, Cmnd 8496
1982   *Radioactive Waste Management*, Cmnd 8607
1982   *Public Transport Subsidy in Cities*, Cmnd 8735
1983   *Coal and the Environment: The Government's Response to the Commission on Energy and the Environment's Report*, Cmnd 8877
1983   *Public Transport in London*, Cmnd 9004
1983   *Rates: Proposals for Rate Limitation and Reform of the Rating System*, Cmnd 9008
1983   *Roads in Scotland: Report for 1982*, Cmnd 9010
1983   *Policy for Roads in England: 1983*, Cmnd 9059
1983   *Streamlining the Cities: Government Proposals for Reorganising Local Government in Greater London and the Metropolitan Counties*, Cmnd 9063
1984   *Report of the Committee of Inquiry into the Functions and Powers of the Islands Councils of Scotland* (Montgomery Report), Cmnd 9216
1984   *Progress in Financial Management in Government Departments*, Cmnd 9297
1984   *Buses*, Cmnd 9300
1985   Home Improvement: *A New Approach: Government Proposals for Encouraging the Repair and Improvement of Private Sector Housing in England and Wales*, Cmnd 9513
1985   *Airports Policy*, Cmnd 9542
1985   *Lifting the Burden*, Cmnd 9571
1985   *Home Improvement in Scotland: A New Approach. Government Proposals for Encouraging the Repair and Improvement of Private Sector Housing in Scotland*, Cmnd 9677
1986   *Paying for Local Government*, Cmnd 9714
1986   *Privatisation of the Water Authorities in England and Wales*, Cmnd 9734
1986   *The Channel Fixed Link*, Cmnd 9735
1986   *Building Businesses . . . Not Barriers*, Cmnd 9794
1986   *The Conduct of Local Authority Business: Report of the Committee of Inquiry* (Widdicombe Report), Cmnd 9797; *Research Volumes*, Cmnd 9798, 9799, 9800, and 9801
1986   *Planning Appeals, Call-in and Major Public Inquiries: The Government's Response to the Fifth Report from the Environment Committee, Session 1985–86*, Cm 43
1987   *Annual Review of Agriculture 1987*, Cm 67
1987   *Policy for Roads in England: 1987*, Cm 125
1988   *DTI – The Department for Enterprise*, Cm 278
1988   *Training for Employment*, Cm 316
1988   *Releasing Enterprise*, Cm 512
1988   *Civil Service Management Reform: The Next Steps* (Government Reply to the Report from the Treasury and Civil Service Committee), Cm 524
1988   *Scottish Enterprise: A New Approach to Training and Enterprise Creation*, Cm 534
1988   *Employment in the 1990s*, Cm 540
1989   *The Future of Development Plans*, Cm 569
1989   *Roads for Prosperity*, Cm 693
1989   *New Roads by New Means – Bringing in Private Finance: A Consultation Paper*, Cm 698

1989   *The Scottish New Towns: The Way Ahead*, Cm 711

1990   *This Common Inheritance: Britain's Environmental Strategy*, Cm 1200

1990   *Improving Management in Government: The Next Steps Agencies. Review 1990*, Cm 1261

1991   *The Citizen's Charter: Raising the Standard*, Cm 1599

1991   *Competing for Quality: Buying Better Public Services*, Cm 1730

1991   *Improving Management in Government: The Next Steps Agencies: Review 1991*, Cm 1760

1992   *People, Jobs and Opportunity*, Cm 1810

1992   *The Citizens' Charter: First Report*, Cm 2101

1993   *Local Government in Wales: A Charter for the Future*, Cm 2155

1993   *The Government's Response to the 3rd Report of the Welsh Affairs Committee, Session 1992–93: Rural Housing*, Cm 2375

1994   *Sustainable Development: The UK Strategy*, Cm 2426

1994   *Climate Change: The UK Programme*, Cm 2427

1994   *Biodiversity: The UK Action Plan*, Cm 2428

1994   *Sustainable Forestry: The UK Programme*, Cm 2429

1994   *Safer Ships: Cleaner Seas: Report of Lord Donaldson's Inquiry into the Prevention of Pollution from Merchant Shipping*, Cm 2560

1994   *Our Forests – the Way Ahead*, Cm 2644

1994   *Forestry and Woodlands: The Government's Response to the Select Committee on Welsh Affairs*, Cm 2645

1995   *Safer Ships: Cleaner Seas. Government Response to the Report of Lord Donaldson's Inquiry into the Prevention of Pollution from Merchant Shipping*, Cm 2766

1995   *First Report of the Committee on Standards in Public Life* (Nolan Report), Cm 2850

1995   *Prospects for Nuclear Power*, Cm 2860

1995   *Competitiveness: Forging Ahead*, Cm 2867

1995   *The Citizen's Charter: The Facts and Figures: A Report to Mark Four Years of the Charter Programme*, Cm 2970

1995   *Rural England: A Nation Committed to a Living Countryside*, Cm 3016

1995   *Government Response to the Lords Select Committee Report on Sustainable Development*, Cm 3018

1995   *Government Response to the 3rd Report of the Transport Select Committee on Urban Road Pricing*, Cm 3019

1995   *Making Waste Work: A Strategy for Sustainable Waste Management in England and Wales*, Cm 3040

1995   *Rural Scotland: People, Prosperity and Partnership*, Cm 3041

1996   *Next Steps Agencies in Government, Review 1995*, Cm 3164

1996   *Government Response to the Environment Committee First Report (Session 1995–96) into the Single Regeneration Budget*, Cm 3178

1996   *A Working Countryside for Wales*, Cm 3180

1996   *This Common Inheritance: 1996 UK Annual Report*, Cm 3188

1996   *Department of the Environment Annual Report 1996*, Cm 3207

1996   *Transport: The Way Forward*, Cm 3234

DEPARTMENT OF THE ENVIRONMENT

An *Index of Planning Guidance* (HMSO, 1995) lists all DoE circulars, planning policy guidance notes, etc.; it also provides a subject index. A more comprehensive subject index, but confined to PPGs, is published by the CPRE and the weekly magazine *Planning*.

DoE Circulars

This selected list includes significant extant circulars, together with those cancelled circulars which are of historical importance.

42/55   *Green Belts*

50/57   *Green Belts*

48/59   *Town and Country Planning Act, 1959, with Explanatory Memorandum*

56/71   *Historic Towns and Roads*

12/72   *The Planning of the Undeveloped Coast*

63/73   *Local Government Act 1972: Administration of National Parks*

 4/76   *Report of the National Parks Policies Review Committee*

16/76   *National Land Use Classification*

73/77   *Guidelines for Regional Recreational Strategies*

36/78   *Trees and Forestry*

50/78   *Report on the Advisory Committee on Aggregates*

58/78   *Report of the Committee on Planning Control over Mineral Working*

68/78   *Inner Urban Areas Act 1978*

22/80   *Development Control: Policy and Practice*

 2/81   *Development Control Functions: Act of 1980*

 8/81   *Local Government, Planning and Land Act 1980: Various Provisions*

13/83   *Purchase Notices*

20/83   *Publication by Local Authorities of Information about the Handling of Planning Applications*

14/84   *Green Belts*

15/84   *Land for Housing*

18/84   *Crown Land and Crown Development*

22/84   *Memorandum on Structure Plans*

 1/85   *The Use of Conditions in Planning Permissions*

 2/85   *Planning Control over Oil and Gas Operations*

20/85   *Enforcement Appeals and Advertisement Appeals*

28/85   *Reclamation and Re-use of Derelict Land*

 6/86   *Local Government (Access to Information) Act 1985*

18/86   *Planning Appeals Decided by Written Representations*

19/86   *Housing and Planning Act 1986: Planning Provisions*

 8/87   *Historic Buildings and Conservation Areas: Policy and Procedures*

 3/88   *Local Government Act 1985: Unitary Development Plans*

10/88   *Town and Country Planning (Inquiries Procedure) Rules 1988, Town and Country Planning Appeals (Determination by Inspectors) (Inquiries Procedure) Rules 1988*

15/88   *Town and Country Planning (Assessment of Environmental Effects) Regulations*

24/88   *Environmental Assessment of Projects in Simplified Planning Zones and Enterprise Zones*

18/89   *Publication of Information about Unused and Underused Land*

 1/90   *Compulsory Purchase by Non-ministerial Acquiring Authorities (Inquiries Procedure) Rules 1990 (SI 1990 no. 512)*

 7/91   *Planning and Affordable Housing*

12/91   *Redundant Hospital Sites in Green Belts: Planning Guidelines*

14/91   *Planning and Compensation Act 1991*

Development Control Policy Notes

This series has been superseded by the PPGs. Only one Note is still operative.

Planning Policy Guidance Notes (England)

PPG 1   *General Policy and Principles*, 1992 (revised 1997)
PPG 2   *Green Belts*, 1995
PPG 3   *Housing*, 1992
PPG 4   *Industrial and Commercial Development and Small Firms*, 1992
PPG 5   *Simplified Planning Zones*, 1992
PPG 6   *Town Centres and Retail Development*, 1993
PPG 7   *The Countryside and the Rural Economy*, 1992 (revised 1997)
PPG 8   *Telecommunications*, 1992
PPG 9   *Nature Conservation*, 1994
PPG 12   *Development Plans and Regional Planning Guidance*, 1992
PPG 13   *Transport*, 1994
PPG 14   *Development on Unstable Land*, 1990
PPG 15   *Planning and the Historic Environment*, 1994
PPG 16   *Archaeology and Planning*, 1990
PPG 17   *Sport and Recreation*, 1991
PPG 18   *Enforcing Planning Control*, 1991
PPG 19   *Outdoor Advertising Control*, 1992
PPG 20   *Coastal Planning*, 1992
PPG 21   *Tourism*, 1992
PPG 22   *Renewable Energy*, 1993
PPG 23   *Planning and Pollution Control*, 1994
PPG 24   *Planning and Noise*, 1994

Minerals Policy Guidance Notes

MPG 1   *General Considerations and the Development Plan System*, 1996
MPG 2   *Applications, Permissions and Conditions*, 1988
MPG 3   *Coal Mining and Colliery Spoil Disposal*, 1994
MPG 4   *The Review of Mineral Working Sites*, 1988
MPG 5   *Minerals Planning and the General Development Order*, 1988
MPG 6   *Guidelines for Aggregates Provision in England*, 1994
MPG 7   *The Reclamation of Mineral Workings*, 1989
MPG 8   *Planning and Compensation Act 1991: Interim Development Order Permissions (IDOS). Statutory Provisions and Proceedings*, 1991
MPG 9   *Planning and Compensation Act 1991: Interim Development Orders*, 1992
MPG 10   *Provision of Raw Material for the Cement Industry*, 1991
MPG 11   *The Control of Noise at Surface Mineral Workings*, 1993
MPG 12   *Treatment of Disused Mine Openings and Availability of Information on Mined Ground*, 1994
MPG 13   *Guidelines for Peat Provision in England (including the Place of Alternative Materials)*, 1995
MPG 14   *Environment Act 1995: Review of Mineral Planning Permissions*, 1995

Regional Planning Guidance Notes

RPG 1   *Strategic Guidance for Tyne & Wear*, 1989
RPG 2   *Strategic Guidance for West Yorkshire*, 1989
RPG 3   *Strategic Guidance for London Planning Authorities*, 1996
RPG 3 (Annex)   *Supplementary Guidance for London on the Protection of Strategic Views*, 1991
RPG 4   *Strategic Guidance for Greater Manchester*, 1989
RPG 5   *Strategic Guidance for South Yorkshire*, 1989
RPG 6   *Regional Planning Guidance for East Anglia*, 1991
RPG 7   *Regional Planning Guidance for the Northern Region*, 1993
RPG 8   *Regional Planning Guidance for the East Midlands Region*, 1994
RPG 9   *Regional Guidance for the South East*, 1994
RPG 9a   *The Thames Gateway Planning Framework*, 1995
RPG 10   *Strategic Guidance for the South West*, 1994
RPG 11   *Regional Planning Guidance for the West Midlands Region*, 1995
RPG 12   *Regional Planning Guidance for Yorkshire and Humberside*, 1996
RPG 13   *Regional Planning Guidance for the North West*, 1996

Derelict Land Grant Advice Notes

DLGA 1   *Derelict Land Grant Policy*, 1991

Case Studies of Good Practice in Urban Regeneration

Community Businesses (Land and Urban Analysis Ltd), 1990
Creating Development Trusts (Warburton, D.), 1988
Developing Businesses (Johnstone, D.), 1988
Getting People into Jobs (Hayton, K.), 1990
Greening City Sites (JURUE: ECOTEC Research and Consultancy Ltd), 1987
Improving Urban Areas (JURUE: ECOTEC Research and Consultancy Ltd), 1988
Managing Workspaces (Jackson, A., Mair, D., and Nabarro, R.), 1987
Re-using Redundant Buildings (Urban and Economic Development Ltd), 1987

DoE Research Reports

(HMSO unless otherwise stated)

Analysis of Land Use Change Statistics (SERRL/Birkbeck), 1994, SERRL/Birkbeck, 7–15 Gresse Street, London W1P 1PA
Assessing the Impact of Urban Policy (Robson, B. *et al.*), 1994
Assessment of the Effectiveness of Derelict Land Grant in Reclaiming Land for Development, 1994
Attitudes to Town and Country Planning (McCarthy, P. and Harrison, T.), 1995
Barnsbury Environmental Study (Ministry of Housing and Local Government), London: MHLG, 1968
Business Success in the Countryside: The Performance of Rural Enterprise (Keeble, D., Tyler, P., Broom, G., and Lewis, J.), 1992
Climate Change and the Demand for Water (Herrington, P.), 1996

Coastal Planning and Management: A Review (Rendel Geotechnics), 1993

Coastal Planning and Management: A Review of Earth Science Information Needs (Rendel Geotechnics), 1995

Coastal Superquarries to Supply South-East England Aggregate Requirements (Whitbread, M. and Marsay, A.), 1992

Coastal Superquarries: Options for Wharf Facilities on the Lower Thames (Arup Economics and Planning), 1995

Community Involvement in Planning and Development Processes, 1995

Cost Effective Management of Reclaimed Derelict Sites (Land Capability Consultants), 1989

Costs of Determining Planning Applications and the Development Control Service (Price Waterhouse), 1994

Deeplish Study: Improvement Possibilities in a District of Rochdale (Ministry of Housing and Local Government), 1966

Derelict Land Prevention and the Planning System (Arup Economics and Planning), 1995

Developing Indicators to Assess the Potential for Urban Regeneration (Coombes, M., Raybould, S., and Wong, C.), 1992

Dynamics of the Rural Economy (ECOTEC Research and Consulting Ltd), London: DoE, 1990

Economic Instruments and Recovery of Resources from Waste (Environmental Resources Ltd), 1992

Effectiveness of Planning Policy Guidance Notes (Land Use Consultants), London: DoE, 1995

Efficiency and Effectiveness of Local Plan Inquiries, 1995

Enterprise Zones Monitoring 1994–95 (Roger Tym & Partners), 1996

Evaluating the Effectiveness of Land Use Planning (PIEDA in association with CUDEM, Leeds Polytechnic, and Professor Derek Diamond), 1992

Evaluation of Derelict Land Schemes (Roger Tym & Partners in association with Land Use Consultants), 1987

Evaluation of Environmental Projects Funded under the Urban Programme (JURUE), 1986

Evaluation of Garden Festivals (PA Cambridge Economic Consultants), 1990

Evaluation of Industrial and Commercial Improvement Areas (JURUE), 1986

Evaluation of Planning Enforcement Provisions (Arup Economic and Planning, and Linklaters & Paines), 1995

Evaluation of the Enterprise Zone Experiment (PA Cambridge Economic Consultants), 1987

Evaluation of the Stockbridge Village Trust Initiative (Roger Tym & Partners), 1988

Evaluation of the Urban Development Grant Programme (Public Sector Management Research Unit, Aston University), 1988

Examination of the Effects of the Use Classes Order 1987 and the General Development Order 1988 (Wootton Jeffreys Consultants and Bernard Thorpe), 1991

Feasibility Study for Deriving Information about Land Use Stock, (Harrison, A. R.), Bristol: Department of Geography, University of Bristol, 1994

Final Evaluation of Enterprise Zones (PA Cambridge Consultants), 1995

Five Year Review of the Bolton, Middlesbrough and Nottingham Programme Authorities (PIEDA), 1990

Home-owners and Clearance: An Evaluation of Rebuilding Grants (Karn, V., Lucas, J. *et al.*) 1996

Housing Land Availability (Roger Tym & Partners), 1991

Housing Land Availability: The Analysis of PS3 Statistics on Land with Outstanding Planning Permission (Bibby, P. and Shepherd, J.), 1993

Impact of Environmental Improvements on Urban Regeneration (PIEDA Planning and Economic Consultants), 1995

Improving Inner City Shopping Centres: An Evaluation of Urban Programme Funded Schemes in the West Midlands (Public Sector Management Research Centre, Aston University), 1988

Indicators of Comparative Regional–Local Economic Performance and Prospects (PA Cambridge Economic Consultants), 1990

Integrated Planning and Granting of Permits in the EC (GMA Planning in association with P-E International and Jacques & Lewis), 1993

Involving Communities in Urban and Regional Regeneration: A Guide for Practitioners, London: DoE, 1995

Land Use Planning and Indicators of Housing Demand (Coopers & Lybrand), London: Coopers & Lybrand, 1987

Land Use Planning and the Housing Market (Coopers & Lybrand), London: Coopers & Lybrand, 1985

Land Use Planning Policy and Climate Change (Owens, S. and Cope, D.), 1992

Managing Demolition and Construction Wastes: Report of the Study on the Recyling of Demolition and Construction Wastes in the UK (Howard Humphreys & Partners), 1994

Managing Urban Change: A Report on the Management Training Needs of Urban Programme Project Managers (Urban and Economic Development Ltd), 1988

Mineral Policies in Development Plans (Arup Economic Consultants), 1990

Monitoring Environmental Assessment and Planning (Wood, C. and Jones, C.), 1991

National Survey of Vacant Land in Urban Areas of England 1990 (Shepherd, J. and Abakuks, A.), 1992

Patterns and Processes of Urban Change in the United Kingdom (Fielding, T. and Holford, S., University of Sussex Centre for Urban and Regional Research), 1990

People, Parks and Cities: A Guide to Current Good Practice in Urban Parks and Environmental Spaces (Greenhalgh, L.,Worpole, K., and Grove-White, R.), 1996

Permitted Development Rights for Agriculture and Forestry (Land Use Consultants), 1991

Planning Controls over Agricultural and Forestry Development and Rural Building Conversions (Land Use Consultants), 1995

Planning for Affordable Housing (Barlow, J., Cocks, R., and Parker, M.), 1994

Planning for Rural Diversification (Elson, M., Macdonald, R., Steenberg, C., and Brown, G.), 1995

Planning for Rural Diversification: A Good Practice Guide (Elson, M., Steenberg, C., and Wilkinson, J.), 1995

Planning, Pollution and Waste Management (Environmental Resources Ltd in association with Oxford Polytechnic School of Planning), 1992

Potential Effects of Climate Change in the United Kingdom (UK Climate Change Impacts Review Group, First Report), 1991

Potential Role of Market Mechanisms in the Control of Acid Rain (London Economics), 1992

PPG 13: A Guide to Better Practice. Reducing the Need to Travel through Land Use and Transport Planning (JMP Consultants), 1995

Process of Local Plan Adoption and Inspectors' Recommendations on Local Plans: Final Report to the DoE, vol. 1: Main Findings; vol. 2: Case Studies (Crispin, G., Fidler, P., and Nadin, V.), Coventry: Coventry Polytechnic, 1985

Rates of Urbanization in England 1981–2001 (Bibby, P. R. and Shepherd, J. W.), 1990

Relationship between House Prices and Land Supply (Gerald Eve with the Department of Land Economy, University of Cambridge), 1992

Review of Data Sources for Urban Policy (ECOTEC Research and Consulting Ltd), 1987

Review of UK Environmental Expenditure: A Final Report to the Department of the Environment (ECOTEC Research and Consulting Ltd), 1993

Simplified Planning Zones: Progress and Procedures (Arup Economic Consultants), 1991

Slate Waste Tips and Workings in Britain (Richards Moorehead and Laing), 1995

Socio-demographic Change and the Inner City (Boddy, M., Bridge, G., Burton, P., and Gordon, D.), 1995

Stockbridge Village Trust: Building a Community (Roger Tym & Partners), 1988

Strategic Approach to Derelict Land Reclamation (Public Sector Management Research Centre, Aston University), 1992

Tackling Vacant Land: An Evaluation of Policy Instruments for Tackling Urban Land Vacancy (Whitbread, M., Mayne, D., and Wickens, D., Arup Economic Consultants), 1991

Targeting Urban Employment Initiatives (Turok, I. and Wannop, U.), 1990

Tourism and the Inner City: An Evaluation of the Impact of Grant Assisted Tourism Projects (Polytechnic of Central London School of Planning, Leisureworks, and DRV Research), 1990

Transforming Our Waste Land: The Way Forward (University of Liverpool, Environmental Advisory Unit), 1986

Urban Industrial Change: The Causes of the Urban–Rural Contrast in Manufacturing Employment Trends (Fothergill, S., Kitson, M., and Monk, S.), 1985

Urban Labour Markets: Reviews of Urban Research (Moore, B. and Townroe, P.), 1990

Urban Land Markets in the UK (Chubb, R. N.), 1988

Urban Programme and the Young Unemployed (Whitting, C.), 1986

US Experience in Evaluating Urban Regeneration: Reviews of Urban Research, (Barnekov, T., Hart, D., and Benfer, W.), 1990

Use of Planning Agreements (Grimley J.R. Eve incorporating Vigers in association with Thames Polytechnic School of Land and Construction Management and Alsop Wilkinson), 1992

Vacant Urban Land: A Literature Review (Cameron, G. E., Monk, S., and Pearce, B. J.), London: DoE, 1988

Other DoE Publications

(HMSO unless otherwise stated)

Air Quality: Meeting the Challenge. The Government's Strategic Policies for Air Quality Management, London: DoE, 1995

Analysis to Responses to Quality in Town and Country, 1996

Biodiversity: The UK Action Plan, Cm 2428, 1994

Biodiversity Steering Group Report (2 vols), 1995

British Government Panel on Sustainable Development: First Report, London: DoE, 1995

Climate Change: Our National Programme for Carbon Dioxide Emissions, London: DoE, 1992

Climate Change: Report on United Kingdom National Programme for Limiting Carbon Dioxide Emissions, London: DoE, 1992

Climate Change: The UK Programme (United Kingdom's Report under the Framework Convention on Climate Change), Cm 2427, 1994

Commercial and Industrial Floorspace Statistics 1995, 1995

Community Projects Review (Elliott, S., Lomas, G., and Riddell, A.), London: DoE, 1984

Cost Effective Management of Reclaimed Derelict Sites, 1989

Countryside Survey 1990: Main Report, London: DoE, 1993

The Countryside: Environmental Quality and Economic Development, London: DoE, 1996

Development Below Low Water Mark (Discussion Paper), London: DoE, 1993

Development Control Statistics: England, London: DoE, (annual)

Development Plans: A Good Practice Guide, 1992

Development Plans: What You Need to Know, London: DoE, 1992

Digest of Data for the Construction Industry, (annual)

Digest of Environmental Protection and Water Statistics, (annual)

Ecclesiastical Exemption: What it is and How it Works, London: DoE and Cardiff: Cadw, 1994

Eco-management and Audit Scheme for UK Local Government, 1993

Environmental Assessment: A Guide to Procedures (revised edition), 1994

Environmental Facts: A Guide to Using Public Registers of Environmental Information, London: DoE, 1995

Enterprise Zone Information 1981–1994, 1996

Estate Action: Annual Report 1991–92, London: DoE, 1992

Estate Action: New Life for Local Authority Estates: Guidelines for Local Authorities on Estate Action and Housing Action Trusts, and Links with Related Programmes, London: DoE, 1992

Evaluation of Six Early Estate Action Schemes, 1996

Evaluation of Urban Development Grant, Urban Regeneration Grant, and City Grant (Price Waterhouse), 1993

Functions of Local Authorities in England (Local Government Review), 1992

Global Climate Change, 1st edition 1989, 2nd edition 1991, 3rd edition 1994

Green Rights and Responsibilities: A Citizen's Guide to the Environment, London: DoE, 1992

Guidance on Safeguarding the Quality of Public Water Supplies, 1989

Guide to Risk Assessment and Risk Management for Environmental Protection, 1995

Housing and Construction Statistics, annual

Improving Environmental Quality: The Government's Proposals for a New Independent Environment Agency, London: DoE, 1991

Improving the Local Plan Process: Consultation Paper, London: DoE, 1994

Indicators of Sustainable Development for the United Kingdom, 1996

Integrated Pollution Control: A Practical Guide, (revised edition) London: DoE, 1996

Land Use Change in England no. 7 (Statistical Bulletin 92/4), London: DoE, 1992

Land Use Change in England no. 8 (Statistical Bulletin 93/1), London: DoE, 1993

Land Use Change in England no. 9 (Statistical Bulletin 94/1), London: DoE, 1994

Land Use Change in England no. 10 (Statistical Bulletin), London: DoE, 1995

Landslide Investigation and Management in Great Britain: A Guide for Planners and Developers, London: DoE, 1996

Making Waste Work: A Strategy for Sustainable Waste Management in England and Wales, Cm 3040, 1995

Managing the Coast (discussion paper), London: DoE 1993

Mediation: Benefits and Practice. Information for Those Considering Mediation as a Way of Resolving Neighbour Disputes, London: DoE, 1994

Minerals Planning Policy and Supply Practices in Europe, 1996

OECD Environmental Performance Review: United Kingdom, London: DoE, 1995

Origins of the Department of the Environment (McQuail, P.), London: DoE, 1994

Ozone (Report of the Expert Panel on Air Quality Standards), 1994

Ozone Layer, London: DoE, 1995

People. Parks and Cities: A Guide to Current Good Practice in Urban Parks, London: DoE, 1996

Policy Appraisal and the Environment: A Guide for Government Departments, 1991

Policy Guidelines for the Coast, London: DoE, 1995

Potential Effects of Climate Change in the United Kingdom (First Report of the Climate Change Impacts Review Group, 1991

Preparation of Environmental Statements for Planning Projects that Require Environmental Assessment: A Good Practice Guide, 1995

Priority Estates Project 1981: Improving Problem Council Estates, 1981

Projections of Households in England to 2016, 1995

Quality in Town and Country (Discussion Paper), 1994

Quality in Town and Country: Urban Design Guidelines, 1995
Reclamation of Damaged Land for Nature Conservation (Land Use Consultants), 1996
Reclamation of Mineral Workings to Agriculture, 1996
Report of the Noise Review Working Party (Batho Report), 1990
Review of the Department of the Environment 1995: Report of the Departmental Task Force, London: DoE
Risk Assessment and Risk Management for Environmental Protection, 1995
Sea Change 1995: A Review of the Prospects for the UK Commercial Property Market, London: DoE
Slate Waste Tips and Workings in Britain, 1995
Speeding Planning Appeals: A Review of the Handling of Transferred Written Representation Planning Appeals. Report of an Efficiency Scrutiny (Wakeford, R. and Heywood, R.), 1986
Speeding Planning Appeals: The Handling of Inquiries Planning Appeals: Action Plan and Review, 1987
Structure of Local Government in England: Consultation Paper (Local Government Review), London: DoE, 1991
Survey of Derelict Land in England 1993, 1995
Survey of Mineral Workings in England, 1996
Sustainable Development: The UK Strategy, Cm 2426, 1994
Transport and the Environment Study (Joint Memorandum by the Departments of Environment and Transport to the Royal Commission on Environmental Pollution), London: DoE, 1992
United Kingdom National Air Quality Strategy: Consultation Draft, London: DoE, 1996
UK Environment, 1992
Urban Air Quality in the United Kingdom (First Report of the Quality of Urban Air Review Group), London: DoE, 1993
Urban Programme: Annual Programme Guidance 1993/94, London: DoE, 1993
Urban Programme 1985: A Report on its Operation and Achievements in England, London: DoE, 1986
View from the Bridge (McQuail, P.), London: DoE 1995
Waste Management: The Duty of Care. A Code of Practice: Consultation Draft, London: DoE, 1995
Waste Management: The Duty of Care: A Code of Practice, 1996
Waste Management Planning: Principles and Practice: A Guide on Best Practice for Waste Regulators, 1995
Water Conservation: Government Action, London: DoE, 1995

GOVERNMENT OFFICE FOR LONDON

London: Facts and Figures 1995 Edition, 1995
The London Congestion Charging Research Programme: Principal Findings (MVA Consultancy), 1995
London's Urban Environment: Planning for Quality (Building Design Partnership), 1996
Thames Strategy: A Study of the Thames, Prepared for the Government Office for London (Ove Arup Partnership), 1995
A Transport Strategy for London, 1996

DEPARTMENT OF TRANSPORT

White Papers and Road Programmes

1977 *Transport Policy*, Cmnd 6836
1983 *Public Transport in London*, Cmnd 9004
1987 *Policy for Roads in England: 1987*, Cm 125

1989   *Roads for Prosperity*, Cm 693
1990   *Trunk Roads, England: Into the 1990s* (not a White Paper)
1994   *Trunk Roads in England: 1994 Review* (not a White Paper)
1995   *Managing the Trunk Road Programme* (not a White Paper)
1996   *Transport: The Way Forward*, Cm 3234

Reports of the Standing Advisory Committee on Trunk Road Assessment

1979   *Trunk Road Proposals: A Comprehensive Framework for Appraisal*
1986   *Urban Road Appraisal*
1992   *Assessing the Environmental Impact of Road Schemes*
1994   *Trunk Roads and the Generation of Traffic*

Other DoT Publications

(HMSO unless otherwise stated)

Best Environmental Practice: A Manager's Guide for Transport Distribution Centres (Wordford, F.), 1994
Better Places Through Bypasses: Report of the Bypass Demonstration Project, 1995
Channel Fixed Link: Environmental Appraisal of Alterative Proposals (Land Use Consultants), 1986
Cycling in Great Britain (Transport Statistics Report) 1996
Design Manual for Roads and Bridges, vol. 11: *Environmental Assessment*, 1979–94 and continuing
Developers' Contributions to Highway Works: Consultation Document, London: DoT, 1992
Developers' Contributions to Highway Works: Efficiency Scrutiny Report and the Department's Response, London: DoT, 1992
Development in the Vicinity of Trunk Roads, DoT Circular Roads 6/91, London: DoT, 1991
Disability Unit Annual Report, London: DoT, annual
Emissions from Heavy Duty Diesel Engined Vehicles: The Government's Response to the Fifteenth Report of the Royal Commission on Environmental Pollution 1992
Guidance on Decriminalised Parking Enforcement Outside London, DoT Local Authority Circular 1/95, 1995
Guidance on Induced Traffic, DoT Guidance Note 1/95, London: DoT, 1995
Interdepartmental Review of Road Safety Policy: Report by the Department of Transport, London: DoT, 1987
Keep Buses Moving: A Guide to Traffic Management to Assist Buses in Urban Areas, Local Transport Note 1/91, 1991
London Congestion Charging Research Programme: Principal Findings (MVA Consultancy), Government Office for London, 1995
London Heliport Study, London: DoT, 1995
Managing the Trunk Road Programme, 1995
National Cycling Strategy, London: DoT, 1996
National Road Traffic Forecasts (Great Britain) 1989, 1989
National Travel Survey 1989/91, 1993
National Travel Survey 1992/94, 1995
New Roads by New Means: Bringing in Private Finance: A Consultation Paper, Cm 698, 1989
Paying for Better Motorways, Cm 2200, 1993

Railway Noise and the Insulation of Dwellings: Report of the Committee to Recommend a National Noise Insulation Standard for New Railway Lines (Mitchell Report) 1991

Report of a Field Study of Aircraft Noise and Sleep Disturbance: A Study Commissioned by the Department of Transport from the Civil Aviation Authority (Ollerhead, J. B.), 1994

Road Traffic Act 1991, DoT Local Authority Circular 4/91, 1991

Road Traffic Statistics, Great Britain, London: DoT, annual

Role of Investment Appraisal in Road and Rail Transport, London: DoT, 1992

Section 56 Grant for Public Transport, DoT Circular 3/89, 1989

Traffic in London: Traffic Management and Parking Guidance, DoT Local Authority Circular 5/92, 1992

Traffic Quotes: Public Perception of Traffic Regulation in Urban Areas: Report of a Research Study (Jones, P.), 1990

Transport: A Guide to the Department, London: DoT, 1989

Transport and the Environment, London: DoT, 1988

Transport and the Environment Study: Joint Memorandum by the Departments of the Environment and Transport (Prepared for the Royal Commission on Environmental Pollution), London: DoE/DoT, 1992

Transport Policies and Programme Submissions, DoT Local Authority Circular, annual

Transport Statistics, Great Britain, annual

Transport Statistics for London, annual

Transport Statistics for Metropolitan Areas, annual

Travel by London Men and Women: Why Do Women Travel Less Than Men?, London: DoT, 1995

Valuation of Environmental Externalities (Tinch, R.), 1996

SCOTTISH OFFICE

(Published by the Scottish Office unless otherwise stated)

Scottish Development Department Circulars

19/77   *National Planning Guidelines*
32/83   *Structure and Local Plans*
 6/85   *Code of Practice for the Examination in Public of Structure Plans*
 7/85   *Code of Practice for Local Plan Inquiries and Hearings*
17/85   *Development Control Priorities and Procedures*
24/85   *Development in the Countryside and Green Belts*
18/86   *The Use of Conditions in Planning Permissions*
37/86   *Housing and Planning Act 1986*
38/86   *Location of Major Retail Developments*
29/88   *Notification of Applications*
 2/89   *Administration of Planning Appeals*
 6/89   *The Town and Country Planning (Use Classes) (Scotland) Regulations 1989*
16/89   *Housing Plans*
 8/90   *Changes to the General Development Order*
 3/91   *Electricity Generating Stations and Overhead Lines Permitted Development for Electricity Undertakings*
13/91   *Indicative Forestry Strategies*
21/91   *Planning and Compensation Act 1991: Land Compensation and Compulsory Purchase*
22/91   *Planning and Compensation Act 1991*

Scottish Office Environment Department Circulars

2/92   *Planning and Compensation Act 1991: Mineral Provisions*
5/92   *Town and Country Planning (General Permitted Development) (Scotland) Order 1992*
6/92   *Town and Country Planning (General Development Procedure) (Scotland) Order 1992*
8/92   *Enforcing Planning Control*
9/92   *Planning and Compensation Act 1991: Enforcement of Tree Preservation Orders*
31/92   *Control over Advertisements and Fish Farming*
36/92   *Lawful Development and Enforcement*
4/93   *Part I of the Natural Heritage (Scotland) Act 0991: Scottish Natural Heritage*
5/93   *Planning Controls for Hazardous Substances*
27/93   *Memorandum of Guidance on Listed Buildings and Conservation Areas* (new edition)
2/94   *Housing and Crime Prevention*
3/94   *The Planning System and Development Plan Departures* (introducing NPPG1)
20/94   *Town and Country Planning (General Permitted Development) (Scotland) Order 1992: Amendments to Permitted Development Rights for Mineral Working and Exploration*
25/94   *Town and Country Planning (Notification of Applications) (Scotland) Amendment Direction 1994*
26/94   *The Environmental Assessment (Scotland) Amendment Regulation 1994*
6/95   *Nature Conservation: Implementation in Scotland of EC Directives on the Conservation of Natural Habitats and of Wild Flora and Fauna, and the Conservation of Wild Birds: The Conservation (Natural Habitats etc) Regulations 1994*
15/95   *The Town and Country Planning (Demolition Which is Not Development) (Scotland) Direction 1995*
18/95   *Planning and Compensation Act 1991: Simplified Planning Zones*
18/96   *Closed Circuit Television Cameras*
30/96   *Consultation with the Royal Fine Art Commission*
32/96   *Town and Country Planning: Code of Practice for Local Plan Inquiries*
34/96   *Environment Act 1995, Section 96: Guidance on the Statutory Provisions and Procedures*
42/96   *Town and Country Planning (General Permitted Development Order) (Scotland) Amendment (No. 2) Order 1996: Water and Sewerage Authorities and Liquified Petroleum Gas*

Planning Advice Notes

PAN 30   *Local Planning*, 1994
PAN 31   *Simplified Planning Zones*, 1987
PAN 32   *Development Opportunities and Local Plans*, 1988
PAN 33   *Development of Contaminated Land*, 1988
PAN 34   *Local Plan Presentation*, 1989
PAN 35   *Town Centre Improvement*, 1989
PAN 36   *Siting and Design of New Housing in the Countryside*, 1991
PAN 37   *Structure Planning*, 1992
PAN 38   *Structure Plans: Housing Land Requirements*, 1993
PAN 39   *Farm and Forestry Buildings*, 1993
PAN 40   *Development Control*, 1993
PAN 41   *Development Plan Departures*, 1994
PAN 42   *Archaeology: The Planning Process and Scheduled Monuments Procedures*, 1994

PAN 43   *Golf Courses and Associated Developments*, 1994
PAN 44   *Fitting New Housing Development into the Landscape*, 1994
PAN 45   *Renewable Energy Technologies*, 1994
PAN 46   *Planning for Crime Prevention*, 1994
PAN 47   *Developing the Relationship between Local Authorities and Community Councils*, 1996
PAN 48   *Planning Application Forms*, 1996
PAN 49   *Community Councils and Planning*, 1996
PAN 50   *Controlling the Effects of Surface Mineral Workings*, 1996

National Planning Guidelines

1   *North Sea Oil and Gas: Coastal Planning Guidelines 1974*
2   *National Planning Guidelines for Aggregate Working 1977*
3   *National Planning Guidelines 1981: Priorities for Development Planning; Land for Housing; Land for Large Industry; Land for Petrochemical Development; Rural Planning Priorities; National Scenic Areas; Nature Conservation; Forestry*
4   *National Planning Guidelines 1984: Skiing Developments*
5   *National Planning Guidelines 1985: High Technology; Individual High Amenity Sites*
6   *National Planning Guidelines 1986: Location of Major Retail Development*
7   *National Planning Guidelines 1987: Agricultural Land*

National Planning Policy Guidelines

NPPG 1   *The Planning System*, 1994
NPPG 2   *Business and Industry*, 1993
NPPG 3   *Land for Housing*, 1993
NPPG 4   *Land for Mineral Working*, 1994
NPPG 5   *Archaeology and Planning*, 1994
NPPG 6   *Renewable Energy*, 1994
NPPG 7   *Planning and Flooding*, 1995
NPPG 8   *Retailing*, 1996
NPPG 9   *The Provision of Roadside Facilities on Motorways and Other Trunk Roads in Scotland*, 1996
NPPG 10   *Planning and Waste Management*, 1996
NPPG 11   *Sport, Physical Recreation and Open Space*, 1996

Scottish Natural Heritage

An Agenda for Investment in Scotland's Natural Heritage, 1992
Agriculture and Scotland's Natural Heritage, 1994
Annual Report
An Area Sustainability Study of Ettrick and Lauderdale, 1994
Enjoying the Outdoors: A Programme for Action: An SNH Policy Paper, 1994
Plans and Progress 1995–96
Public Access to the Countryside: A Guide to the Law, Practice and Procedure in Scotland, 1993

Sustainable Development and the Natural Heritage, 1993
The Environment: Who Cares? (Crofts, R.), 1995

Other Scottish Office Publications

Accessing Environmental Information in Scotland (Moxen, J. *et al.*), 1995
Assessing Environmental Information in Scotland (Robert Gordon University, Aberdeen), 1995
Cairngorms Partnership: A Statement of Intent by the Secretary of State for Scotland Following the Advice of the Cairngorms Working Party, 1994
Code of Practice of Local Planning Inquiries, 1996
Counting Travellers in Scotland: The 1992 Picture (Gentleman, H.), 1993
Demand for Housing: Economic Perspectives and Planning Practices (MacLennan, D.), 1986
Derelict Land Survey 1993 (1995)
Development Control Performance in Scotland (MVA Consultancy), 1985
Enforcement of Planning Control in Scotland (Consultants' Report by Rowan-Robinson, J., Young, E., and McLarty, I.) (Scottish Development Department), 1984
Environmental Education: Is The Message Getting Through? (MORI), Scottish Office Central Research Unit, 1995
Environmentally Sensitive Areas in Scotland (Department of Agriculture and Fisheries for Scotland), 1989
Evaluation of the Use and Effectiveness of Planning Publications (Planning Exchange) (Scottish Development Department), 1989
Fitting Roads: A Balanced Approach to Rural Road Design: A Review for Discussion, 1995
Football Stadia: Policy Guidance (Scottish Office Education Department, Circular 10/91), 1991
Future Management Orders for Regional and Country Parks (Peter Gibson Associates and Scottish Agricultural College), 1995
Guide to Measures Available to Control the Recreational Use of Water (Cobham Resources Consultants), 1995
Gyle Impact Study (Roger Tym and Partners in association with Oscar Faber TPA), Edinburgh: HMSO, 1996
Historic Buildings and Monuments: Guide for Grant Applicants, 1988
House Price Monitoring Systems and Housing Planning in Scotland (Report by the Centre for Housing Research, University of Glasgow), 1989
Implementation of Natural Heritage Areas: Statement of Government Position Following Consultation, 1992
Improving Scotland's Environment: Consultation Paper on the Scottish Environmental Protection Agency (SOEnD), 1992
Interim Evaluation of the Castlemilk Partnership (O'Toole, M., Snape, D., and Stewart, M.), 1995
Interim Evaluation of the Ferguslie Partnership (Gaster, L.), 1995
Interim Evaluation of the Wester Hailes Partnership (McGregor, A.), 1995
Interim Report of the Whitfield Partnership (Kintrea, K.), 1995
Land Supply and House Prices in Scotland (PIEDA), 1987
Listed Buildings and Conservation Areas: Memorandum of Guidance, 1987; Amended 1988
Natural Heritage Areas: Consultation Paper, 1991
New Life for Urban Scotland, 1988
Partnership in the Regeneration of Urban Scotland (McAllister, D.), 1996
Programme for Partnership: Announcement of the Outcome of the Scottish Office Review of Urban Regeneration Policy, 1995
Progress in Partnership: A Consultation Paper on the Future of Urban Regeneration Policy in Scotland, 1993

Public Attitudes to the Environment in Scotland (Wilkinson, D. and Waterton, J.), 1991

Retail Impact Methodology (Drivers Jonas), 1992

Review of Information on Rural Issues (Burbridge, V.), 1990

Review of Neighbour Notification (School of Planning and Housing, Edinburgh College of Art, Heriot-Watt University, and P.P.C. Allan Chartered Town Planning Consultants Ltd), 1995

Review of the Town and Country Planning System in Scotland: Consultation Paper, 1994

Review of the Town and Country Planning System in Scotland: Digest of Responses to Consultation, 1995

Review of the Town and Country Planning System in Scotland: The Way Ahead, 1995

Review of the Use Classes Order (Strathclyde University Centre for Planning in association with James Barr and Son), 1994

Roads, Bridges and Traffic in the Countryside, 1992

Roads, Bridges and Traffic in the Countryside: Environmental Policies and Practices on Rural Roads, in Villages and Historic/Conservation Areas (Draft), 1995

Rural Framework, 1992

Rural Road Hierarchy and Lorry Routeing: A Review for Discussion, 1995

Safer Cities Programme in Scotland: Overview Report (Carnie, J. K.), 1996

Scotland's Travelling People (Gentleman, H. and Swift, S.), Edinburgh: HMSO, 1971

Scottish Bathing Waters: Progress Towards Compliance 1995, Edinburgh: HMSO, 1995

Scottish Environment Protection Agency: A Consultation Paper, 1992

Scottish Rural Life: A Socio-economic Profile of Rural Scotland, 1992

Section 50 Agreements (Rowan-Robinson, J. and Durman, R.), 1992

Simplified Planning Zones: Streamlining of Procedures, 1990

Sustainable Development: What it Means to the General Public (McCaig, E. and Henderson, C., and MVA Consultancy), 1995

Town and Country Planning (Inquiries Procedures) (Scotland) Rules 1980: Consultation Paper, 1995

Urban Scotland into the 90s: New Life: Two Years On, 1990

Urban Scotland into the 90s: Report of the Conference held in Glasgow, May 1990, 1990

Use and Effectiveness of Scottish Development Department Planning Publications (The Planning Exchange, Glasgow), 1990

1992–based Household Projections for Scotland, Scottish Office Statistical Bulletin, Housing Series, HSG/1995/3

WELSH OFFICE

(Published by the Welsh Office; not available from HMSO, unless otherwise stated. Obtainable from the Map Library, Welsh Office, Crown Building, Cathays Park, Cardiff CF1 3NQ.)

Planning Policy Guidance (Wales)

Old Series

PPG 3   (Wales) *Land for Housing in Wales*, 1992

PPG 12   (Wales) *Development Plans and Strategic Guidance in Wales*, 1992

PPG 16   (Wales) *Archaeology and Planning*, 1991

New Series

Planning Guidance Wales – Unitary Development Plans, Cardiff: HMSO, 1996

Planning Guidance Wales – Planning Policy, Cardiff: HMSO, 1996

Technical Advice Notes (Wales)

TAN (W) 1   *Joint Housing Land Availability Studies*
TAN (W) 2   *Planning and Affordable Housing*
TAN (W) 3   *Simplified Planning Zones*
TAN (W) 4   *Town Centres and Retail Development*
TAN (W) 5   *Historic Buildings and Conservation Areas*
TAN (W) 6   *Archaeology and Planning*
TAN (W) 7   *Outdoor Advertisement Control*
TAN (W) 8   *Renewable Energy*
TAN (W) 9   *Nature Conservation and Planning (Consultation Draft)*
TAN (W) 10   *Transport (Consultation Draft)*

Other Welsh Office Publications

Circular 61/81   *Historic Buildings and Conservation Areas: Policy and Procedure*
Circular 47/84   *Land for Housing in Wales*
Circular 53/88   *The Welsh Language: Development Plans and Planning Control*
Circular 31/91   *Planning and Affordable Housing*

The Environmental Agenda for Wales, 1995
Environmentally Sensitive Areas Wales: Socio-economic Aspects of Designation (Hughes, G. O. and Sherwood, A. M.), 1992
Programme for the Valleys: Building on Success, 1993
The Role of Community and Town Councils in Wales (Consultation paper), 1992
The Welsh Environment: A Guide to Your Rights and Responsibilities, 1992

NORTHERN IRELAND OFFICE

(Available from the Planning Service, Clarence Court, 10–18 Adelaide Street, Belfast, BT2 8GB.)

Development Control Advice Notes

1   *Amusement Centres*
2   *Multiple Occupancy*
3   *Bookmaking Offices*
4   *Hot Food Bars*
5   *Taxi Offices*
6   *Restaurants and Cafés*
7   *Public Houses*
8   *Small Unit Housing in Existing Residential Areas*
9   *Residential and Nursing Homes*
10   *Environmental Impact Assessment*
11   *Access for People with Disabilities*
11a   *Nature Conservation and Planning*

12   *Hazardous Substances*
13   *Crèches, Day Nurseries and Pre-school Playgroups*

Proposed Planning Policy Statements

1   *The Northern Ireland Planning System*
2   *Planning and Nature Conservation*
3   *Planning and Roads Considerations*
4   *Industrial Development*
5   *Retailing and Town Centres*

Other Northern Ireland Office Publications

Belfast City Region: Towards and Beyond the Millennium, 1996
Community Economic Regeneration Scheme and Community Regeneration and Improvement Programme, NI Audit
    Office, HC 439 (1994–95), Belfast: HMSO, 1995
Criteria and Standards for Listing, DoENI, Memorandum 163, 1991
Department of the Environment: Town and Country Planning Service, NI Audit Office, HC 3 (1995–96), Belfast:
    HMSO, 1995
Design Guide for Rural Northern Ireland, DoENI, 1994
Enterprise Zones Annual Report 1990, DoENI, 1991
Evaluation of the Enterprise Zone Experiment in Northern Ireland, Belfast: HMSO, 1988
Evaluation of the Belfast Action Team Initiative (PA Cambridge Economic Consultants), 1992
Finding and Minding: A Report on the Archeological Work of the Department of the Environment Northern Ireland,
    1993
Growing a Green Economy: Strategy for the Environment and the Economy in Northern Ireland, DoENI and
    Department of Economic Development, 1993
Houses in Harmony with the Countryside, DoENI, 1988
Making Belfast Work: Strategy Proposals 1994–95 to 1996–97, 1994
New Look at the Northern Ireland Countryside (Jean Balfour), 1983
Planning Strategy for Rural Northern Ireland, Belfast: HMSO, 1993
Rural Housing Policy: The Way Ahead. A Policy Statement, Belfast: Northern Ireland Housing Executive, 1991
Tourism in Northern Ireland: An Indicative Plan, Belfast: Northern Ireland Tourist Board, 1990

AUDIT COMMISSION FOR LOCAL AUTHORITIES AND THE NATIONAL HEALTH SERVICE IN ENGLAND AND WALES

(HMSO unless otherwise stated)

Annual Report, annual
Building in Quality: A Study of Development Control, 1992
Citizen's Charter Performance Indicators, London: Audit Commission, 1992
Putting Quality on the Map: Measuring and Appraising Quality in the Public Service, 1993
Urban Regeneration and Economic Development: The Local Government Dimension, 1989
Urban Regeneration and Economic Development: The European Community Dimension, 1991

COUNTRYSIDE COMMISSION

(All published by the Countryside Commission)

Note: The Commission publishes a *Catalogue of Publications* and a series of *Bibliographies*, e.g. on national parks, agricultural landscapes, and the uplands. These are available, free of charge, from Countryside Commission Postal Sales, PO Box 124, Walgrave, Northampton NN6 9TL.

Annual Report (annual)
Areas of Outstanding Natural Beauty: A Policy Statement (CCP 356), 1991
City Links with National Parks: Developing Partnerships (CCP 401), 1992
Climate Change, Acidification and Ozone (CCP 458), 1995
Countryside and Nature Conservation Issues in District Local Plans (CCP 317), 1990
Countryside Planning File (CCP 452), 1995
Countryside Stewardship: An Outline (CCP 455), 1994
Countryside Stewardship Handbook (CCP 453), 1995
Design in the Countryside (CCP 418), 1994
Down to Earth: Environmental Problems Associated with Soil Degradation in the English Landscape, 1994
Enjoying the Countryside, 1994
Environmental Assessment: The Treatment of Landscape and Countryside Recreation Issues (CCP 326), 1991
Fit for the Future: Report of the National Parks Review Panel (CCP 334), 1991 [An *Executive Summary* is also available: CCP 335, 1991]
Fit for the Future: The Countryside Commission's Response to the National Parks Review Panel (CCP 337), 1991
Forests for the Community (CCP 340), 1991
Grants and Payment Schemes (CCP 422), 1995
Green Capital: Planning for London's Greenspace (CCP 344), 1991
Groundwork: The First Decade (CCP 417), 1993
Heritage Coasts in England and Wales (CCP 252), 1993
National Forest: The Strategy (CCP 468), 1994
National Park Authority: Purposes, Powers and Administration (CCP 230), 1993
National Target for Rights of Way: A Guide to the Milestone Approach (CCP 435), 1993
Paying for a Beautiful Countryside: Securing Environmental Benefits and Value for Money from Incentive Schemes (CCP 413), 1993
Protected Landscapes in the United Kingdom (CCP 362), 1992
Quality of Countryside: Quality of Life, The Countryside Commission's Prospects into the Next Century (CCP 470), 1995
Rights of Way Condition Survey (CCP 426), 1993
Role of the Countryside Commission in the Town and Country Planning System (CCP 415), 1993
Sustainable Rural Tourism (CCP 483), 1995
Trends in Transport and the Countryside: Technical Report (CCP 382), 1992
Wind Energy Development and the Landscape (CCP 357), 1991

COUNTRYSIDE COUNCIL FOR WALES

(Published by Countryside Council for Wales/Cyngor Cefn Gwlad Cymru, Plas Penrhos, Fford Penrhos, Bangor, Gwynedd LL57 2LQ.)

Annual Report (annual)
Countryside Council for Wales: An Introduction, 1991
Tir Cymen: A Farmland Stewardship Scheme, 1992
Welsh Estuaries Review, 1993

ENGLISH HERITAGE (Historic Buildings and Monuments Commission for England)

(Published by English Heritage, Fortress House, 23 Savile Row, London W1X 1AB.)

Buildings at Risk: A Sample Survey, 1992
Colliery Landscapes, 1994
Conservation Areas of England, 1990
Conservation Area Management, 1993
Conservation Bulletin (three issues a year)
Conservation in London: A Study of Strategic Planning Policy in London, 1995
Conservation Issues in Local Plans, 1996
Conservation Issues in Strategic Plans, 1993
Development Plan Policies for Archaeology, 1992
Ecclesiastical Exemption: What It Is and How It Works, 1994
Future for Our Past? An Introduction to Heritage Studies, 1996
In the Public Interest: London's Civic Architecture at Risk, 1995
Local Government Reorganization: Guidance to Local Authorities on Conservation of the Historic Environment, 1995
Managing England's Heritage: Setting our Priorities for the 1990s, 1992
Register of Buildings at Risk in Greater London, 1994
Something Worth Keeping: Post-war Architecture in England, 1996

ENGLISH NATURE

(Published by English Nature, Northminster House, Peterborough PE1 1UA.)

Annual Report
Conserving England's Marine Heritage: A Strategy, 1993
England's National Nature Reserves (Marren, P. R.), London: Poyser Academic Press, 1994
Important Areas for Marine Wildlife around England, 1994
Nature Conservation in Environmental Assessment, 1994
People, Economics and Nature Conservation (Burgess, J.), 1994
Planning for Wildlife in Towns and Cities (David Tyldesley and Associates), 1994
Rebuilding the English Countryside: Habitat Fragmentation and Wildlife Corridors in Practical Conservation, 1995
Roads and Nature Conservation: Guidance of Impacts, Mitigation and Enhancement, 1994

Strategic Planning and Sustainable Development, 1992
Strategy for the 1990s, 1993
Targets for Coastal Habitat Re-creation (Pye, K. And French, P. W.), 1993

NATIONAL AUDIT OFFICE

(Published by HMSO)

Achievements of the Second and Third Generation Urban Development Corporations, HC 54 (1993–94), 1993
Arrangements for Regional Industrial Incentives, HC 346 (1987–88), 1988
Coastal Defences in England, HC 9 (1992–93), 1992
Control and Monitoring of Pollution, HC 637 (1990–91), 1991
Derelict Land Grant, HC 689 (1987–88), 1988
Enterprise Zones, HC 209 (1985–86), 1986
Environmental Factors in Road Planning and Design, HC 389 (1993–94), 1994
Expenditure on Motorways and Trunk Roads, HC 571 (1984–85), 1985
Investment Activities of the Scottish Development Agency, Welsh Development Agency, and the Highlands and Islands Development Board, HC 230 (1984–85), 1985
Protecting and Managing England's Heritage Property, HC 132 (1992–93), 1992
Protecting and Presenting Scotland's Heritage Properties, HC 430 (1994–95), 1995
Protecting and Managing Sites of Special Scientific Interest in England, HC 379 (1993–94)
Regenerating the Inner Cities, HC 169 (1989–90), 1990
Regulation of Heavy Lorries, HC 92 (1987–88), 1987
Review of Forestry Commission Objectives and Achievements, HC 75 (1985–86), 1986
River Pollution from Farms in England, HC 235 (1994–95), 1995
Road Planning, HC 688 (1988–89), 1988
Trunk Roads, HC 571 (1984–85), 1985
Urban Development Corporations, HC 492 (1987–88), 1988
Urban Programme, HC 513 (1984–85), 1985

NATIONAL RIVERS AUTHORITY (now Environment Agency)

(Published by HMSO unless otherwise noted)

Abandoned Mines and the Water Environment, 1994
Bathing Water Quality in England and Wales (annual)
Contaminants Entering the Sea, 1995
Contaminated Land and the Water Environment, 1994
Demand for Irrigation Water, 1994
Development of Environmental Economics for the NRA (Postle, P.), 1993
Discharge Consents and Compliance: The NRA's Approach to Control of Discharges to Water, 1994
Guidance Notes for Local Planning Authorities on the Methods of Protecting the Water Environment through Development Plans, Bristol: NRA, 1994
Guide to Groundwater Protection Zones in England and Wales, 1995

Guide to Groundwater Vulnerability Mapping in England and Wales, 1995

Implications of Climate Change for the NRA (Arnell, N. W., Jenkins, A., and George, D. G.), 1994

Measures to Safeguard Public Water Supplies: Second Report, Bristol: NRA, 1995

Policy and Practice for the Protection of Groundwater, 1992

Water: Nature's Precious Resource. An Environmentally Sustainable Water Resources Development Strategy for England and Wales, 1994

Water Pollution Incidents in England and Wales (annual)

Water Resources Development Strategy: A Discussion Document, Bristol: NRA, 1992

ORGANISATION FOR ECONOMIC COOPERATION AND DEVELOPMENT

(Published by OECD, Paris; available in UK from HMSO)

Cities and New Technologies, 1992

Economic Appraisal of Environmental Projects and Policies, 1995

Environment and Taxation: The Cases of the Netherlands, Sweden and the United States, 1994

Environmental Indicators, 1994

Environmental Performance Indicators: United Kingdom, 1994

Environmental Policy: How to Apply Economic Instruments, 1991

Environmental Taxes in OECD Countries, 1995

Motor Vehicle Pollution: Reduction Strategies Beyond 2010, 1995

OECD Environmental Performance Reviews: United Kingdom, 1994

OECD Societies in Transition: The Future of Work and Leisure, 1994

Planning for Sustainable Development: Country Experiences, 1995

Strategies for Housing and Social Integration in Cities, 1996

Territorial Development and Structural Change: A New Perspective on Adjustment and Reform, 1993

Urban Travel and Sustainable Development, 1995

Women in the City: Housing, Services and the Urban Environment, 1995

PARLIAMENTARY INQUIRIES

Charging for the Use of Motorways, HC Transport Committee, 5th Report, Session 1993–94, HC 376 (1994). Government Observations, HC Transport Committee, 1st Special Report, Session 1994–95, HC 71 (1994)

Coastal Zone Protection and Planning, HC Environment Committee, 2nd Report, Session 1991–92, HC 17 (1992)

Coastal Zone Protection and Planning: The Government's Response, Cm 2011 (1992)

Consequences of Bus Deregulation, HC Transport Committee. 1st Report, Session 1995–96, HC 54 (1995)

Contaminated Land, HC Environment Committee, 1st Report, Session 1989–90, HC 170 (1990)

Contaminated Land: The Government's Response to the 1st Report from the House of Commons Select Committee on the Environment, Cm 1161 (1990)

Control and Monitoring of Pollution: Review of the Pollution Inspectorate, HC Public Accounts Committee, 16th Report, Session 1991–92, HC 50, (1992)

Cost of Nuclear Power, HC Energy Committee, 4th Report, Session 1989–90, HC 205 (1990)

Cycling, HC Transport Committee, Minutes of Evidence, 8 May 1991, Session 1990–91, HC 423–I (1991)

The Defence Estate, HC Defence Committee (1994) 1st Report, Session 1994–95, HC 67 (1994)

EC Draft Directive on the Landfill of Waste, HC Environment Committee, 7th Report, Session 1990–91, HC 263 (1991)

EC Draft Directive on the Landfill of Waste: The Government's Reply to the Seventh Report from the House of Commons Select Committee on the Environment, Cm 1821 (1992)

Energy and the Environment, HL Select Committee on the European Communities, 13th Report, Session 1990–91, HL 62 (1991)

Environment White Paper: This Common Inheritance, HC Environment Committee, Session 1990–91, Minutes of Evidence, Department of the Environment, HC 48–I (1991)

Environmental Aspects of the Reform of the Common Agricultural Policy, Select Committee on the European Communities, 14th Report, Session 1992–93 (1992)

Environmental Impact of Leisure Activities, HC Environment Committee, 4th Report, Session 1994–95, HC 246 (1995); *Government Response*, HC 761 (1995)

Environmental Issues in Northern Ireland, HC Environment Committee, 1st Report, Session 1990–91, HC 39 (1990)

European Community Environmental Policy, HC Environment Committee, 2nd Report, Session 1989–90, HC 372 (1990)

Fifth Environmental Action Programme: Integration of Community Policies, HL Select Committee on the European Communities, 8th Report, Session 1992–93, HL Paper 27 (1992)

Forestry and the Environment, HC Environment Committee, 1st Report, Session 1992–93, HC 257 (1993)

Forestry and Woodlands, HC Select Committee on Welsh Affairs, 1st Report, Session 1993–94, HC 35 (1994). *Government's Response*, Cm 2645 (1994)

The Government's Proposals for an Environment Agency, HC Environment Committee, 1st Report, Session 1991–92, HC 55 (1992)

Government's Proposals for the Deregulation of Buses in London, HC Transport Committee, 4th Report, Session 1992–93, HC 623 (1993)

Historic Buildings and Ancient Monuments, HC Environment Committee, 1st Report, Session 1986–87, HC 146 (1987)

Housing Need, HC Environment Committee, 2nd Report, Session 1995–96, HC 22 (1996)

Implementation and Enforcement of Environmental Legislation, HL Select Committee on the European Communities, 9th Report, Session 1991–92, HL Paper 53 (1992)

Implementation of the Reform of the Common Agricultural Policy, Select Committee on the European Communities, 9th Report, Session 1992–93, HL Paper 28 (1992)

Indoor Pollution, HC Environment Committee, 6th Report, Session 1990–91, HC 61 (1991)

Indoor Pollution: The Government's Response to the Sixth Report from the House of Commons Select Committee on the Environment, Cm 1633 (1991)

Industrial Change: Retraining and Redeployment, HC Employment Committee, 2nd Report, Session 1991–92, HC 71 (1992)

Industry and the Environment, HL Select Committee on the European Communities, 18th Report, Session 1992–93, HL 73 (1993)

Nature Conservancy Council, HL Select Committee on Science and Technology, 2nd Report, Session 1989–90, HL 33 (1990)

Nature Conservancy Council: Government Response, HL Select Committee on Science and Technology, 6th Report, Session 1989–90, HL 60 (1990)

Operation of the Enterprise Agencies and the LECs, HC Scottish Affairs Committee, First Report, Session 1994–95, HC 339 (1995)

Our Heritage: Preserving It, Prospering from It, HC National Heritage Committee, Third Report, Session 1993–94, HC 139 (1994)

Planning System in Northern Ireland, HC Northern Ireland Affairs Committee, 1st Report, Session 1995–96, HC 53 (1996)

Preservation of Historic Buildings and Ancient Monuments, HC Welsh Affairs Committee, 2nd Report, Session 1992–93, HC 403 (1993)

Protecting and Managing England's Heritage Property, HC Public Accounts Committee, Session 1992–93, Minutes of Evidence, 2 November 1992, HC 252–I (1992)

Protecting and Managing Sites of Special Scientific Interest in England, HC Committee of Public Accounts, 11th Report, Session 1994–95, HC 375 (1995)

Protection of Wild Birds, HL Select Committee on the European Communities, 11th Report, Session 1993–94, HL 70 (1994)

Recycling, HC Environment Committee, 2nd Report, Session 1993–94, HC 63 (1994)

Regenerating the Inner Cities, HC Public Accounts Committee, 33rd Report, Session 1989–90, HC 216 (1990)

Remedying Environmental Damage, HL Select Committee on the European Communities, 3rd Report, Session 1993–94, HL 10 (1993)

Rural England: The Rural White Paper, HC Environment Committee, 3rd Report, Session 1995–96, HC 163 (1996)

Rural Housing, HC Welsh Affairs Committee, 3rd Report, Session 1992–93, HC 621 (1993); *Government Response*, Cm 2375 (1993)

Single Regeneration Budget, HC Environment Committee, 1st Report, Session 1995–96, HC 26 (1995)

Sustainable Development, HL Select Committee on Sustainable Development Report, Session 1994–95, HL 72 (1995); *Government Response*, Cm 3018 (1995)

Tourism and the Community, House of Lords Select Committee on the European Communities, Session 1995–96, HL 39 (1996)

Urban Public Transport: The Light Rail Option, HC Transport Committee, 4th Report, Session 1990–91, HC 14 (1991)

Urban Road Pricing, HC Transport Committee, 3rd Report, Session 1994–95, HC 104 (1995); *Government Response*, Cm 3019 (1995)

Welsh Development Agency, HC Committee of Public Accounts, 29th Report, 1994–95, HC 376 (1995)

Wind Energy, HC Welsh Affairs Committee, 2nd Report, Session 1994, HC 336 (1994)

World Trade and the Environment, HC Environment Committee, 4th Report, Session 1995–96, HC 149 (1996)

ROYAL COMMISSION ON ENVIRONMENTAL POLLUTION

First Report, Cmnd 4585, 1972

Second Report: Three Issues in Industrial Pollution, Cmnd 4894, 1972

Third Report: Pollution in Some British Estuaries and Coastal Waters, Cmnd 5054, 1972

Fourth Report: Pollution Control – Progress and Problems, Cmnd 5780, 1974

Fifth Report: Air Pollution Control – An Integrated Approach, Cmnd 6371, 1976

Sixth Report: Nuclear Power and the Environment, Cmnd 6618, 1976
Seventh Report: Agriculture and Pollution, Cmnd 7644, 1979
Eighth Report: Oil Pollution of the Sea, Cmnd 8358, 1981
Ninth Report: Lead in the Environment, Cmnd 8852, 1983
Tenth Report: Tackling Pollution – Experience and Prospects, Cmnd 9149, 1984
Eleventh Report: Managing Waste: The Duty of Care, Cmnd 9675, 1985
Twelfth Report: Best Practicable Environmental Option, Cm 310, 1988
Thirteenth Report: The Release of Genetically Engineered Organisms to the Environment, Cm 720, 1989
Fourteenth Report: GENHAZ: A System for the Critical Appraisal of Proposals to Release Genetically Modified Organisms into the Environment, Cm 1557, 1991
Fifteenth Report: Emissions from Heavy Duty Diesel Vehicles, Cm 1631, 1991
Sixteenth Report: Freshwater Quality, Cm 1966, 1992
Seventeenth Report: Incineration of Waste, Cm 2181, 1993
Eighteenth Report: Transport and the Environment, Cm 2674, 1994
Nineteenth Report: Sustainable Use of Soil, Cm 3165, 1996

RURAL DEVELOPMENT COMMISSION

(Published by the Rural Development Commission, 141 Castle Street, Salisbury, Wilts. SP1 3TP.)

The Economy and Rural England (Tarling, R., Rhodes, J., North, J., and Broom, G.), 1993
English Rural Communities: An Assessment and Prospect for the 1990s (Rogers, A.), 1993
European Experience of Rural Development (Clout, H.), 1993
An Evaluation of the Rural Development Programme Process (Aston Business School), 1990
Fair Shares for Rural Areas: An Assessment of Public Resource Allocation Systems, 1996
Homelessness in Rural Areas (Lambert, C., Jeffers, S., Burton, P., and Bramley, G.), 1992 [*Statistical update to 1992/93*, 1994]
Impact of Tourism on Rural Settlements (Segal Quince Wicksteed), 1996
Lifestyles in Rural England (Cloke, P., Milbourne, P., and Thomas, C.), 1994
Planning for People and Prosperity, 1995
Rural Development and Statutory Planning (ARUP Economics and Planning), 1993
Rural Economic Activity: Background Paper for the Rural White Paper, 1995
Rural Incomes and Housing Affordability, 1995
Rural Sustainable Development (Lowe, P. and Murdoch, J.), 1993
Rural Transport Problems and Needs (Huntley, P. and Taylor, J.), 1993
Small Business in Rural Areas (North, D. and Smallbone, D.), 1993
Submission on the Rural White Paper, 1995
Survey of Rural Services, 1994
Telecommunications in Rural England (Economic and Transport Planning Group), 1989
Tourism in the Countryside, 1995
Village Shops: A Report on Community Action (Woollett, S.), 1993
1994 Survey of Rural Services (BMRB International), 1994

EUROPEAN UNION

This is necessarily a selective list, given the very large number of official publications from the European Union. Publications are available from the national agents. In the UK this is HMSO, although many publications (particularly free summaries) can be obtained from the Commission Information Office, Jean Monnet House, 8 Storey's Gate, London SW1P 3AT, or from DGXVI: rue de la Loi 200, B-1049 Brussels (Fax: +32 2 296 60 03). New publications from the EU are publicised in *EUR-OP News* which is published four times a year by the Office for Official Publications of the European Communities, and can be obtained free from EUR-OP Office 172, 2 rue Mercier, L-2985 Luxembourg. Lists of recent publications of the EU such as Green Papers, White Papers (whose colours denote the same purposes as UK government papers) and reports are available via the Internet and can be accessed via the Europa site at http://europa.eu.int.

The principal publication of the European Union is the *Official Journal (OJ)*, which is published five days a week and comprises two main parts (the L, series which contains all legislation, and the C series, which carries other information such as proposals for legislation, and opinions of the consultative bodies). COM documents are official communications of the EU and will have appeared in the *OJ*. DGXVI publishes a regular newsletter called *Inforegio News* which provides summaries of regional policy news. The monthly bulletin of the *European Information Service* provided by the Local Government International Bureau notes publications also. The following is a very selective list of the many relevant publications available from the EU.

Papers

Green Paper on the Urban Environment, COM (90) 218
The Impact of Transport on the Environment: A Community Strategy for Sustainable Mobility, Green Paper, COM (92) 46
The Future Development of the Common Transport Policy, COM (92) 494
Proposal for a Council Decision on the Creation of Trans-European Road Network and on the Creation of a European Inland Waterway Network, COM (92) 321 Final
Remedying Environmental Damage, Green Paper, COM (93) 47
The Citizen's Network: Fulfilling the Potential of Public Passenger Transport in Europe, Green Paper, COM (95) 691, 1995
Growth, Competitiveness and Employment: The Challenge and Ways Forward into the 21st Century, White Paper, COM (93) 700
Towards Fair and Efficient Policy in Transport, Green Paper, COM (95) 691
A Strategy for Revitalising the Community's Railways, White Paper, COM (96)

Official Reports

Report on the Regional Problems of the Enlarged Community, COM (73) 550
Europe 2000: Outlook for the Development of the Community's Territory, CX-71–91–518–EN-C 1991
Europe 2000+: Cooperation for European Territorial Development, CX-85–94–486-C, 1994
Competitiveness and Cohesion: Trends in the Regions – Fifth Periodic Report on the Social and Economic Situation and Development of the Regions of the Community CX-85–94–147-C
Fifth Action Programme on the Environment 1992–2000: Towards Sustainability, COM (92) 23

Proposal for a Council Decision on the Creation of a Trans-European Road Network and on the Creation of a European Inland Waterway Network, COM (92) 321

The Trans-European Transport Network, 1995

European Commission Expert Group on the Urban Environment Sustainable Cities Project: European Sustainable Cities: Final Report, 1996

Guides

(Available from the UK Commission Information Office.)

Community Structural Funds 1994–99: Regulations and Commentary, 1993

Guide to Community Initiatives 1994–99 CM-84–94–056-C, EN, 1994

Guide to Innovative Actions for Regional Development (ERDF Article 10) 1995–99, CX-90–95–720-C, EN, 1995

Serving the European Community: A Citizen's Guide to the Institutions of the European Union, 1996

Sponsored Studies

4 *Urbanisation and the Function of Cities in the European Community*, CX-75–92–259-C, 1992

8 *Study of Prospects in the Atlantic Regions: Europe 2000*, CX-79–93–485-C, 1993

18 *The Prospective Development of the Northern Seaboard*, CX-85–94–535-C, 1994

21 *The Regional Impact of the Channel Tunnel throughout the Community*, CX-85–94–559-C, 1994

22 *The Prospective Development of the Central and Capital Cities and Regions* CX-85–95–470-C, 1995

COUNCIL OF EUROPE

European Planning Charter (The Torremolinos Charter), 1983

The European Urban Charter, 1983

European Regional Planning Strategy, (prepared by CEMAT) 1992

THE INTERNET

An increasing number of sources of information are available on the Internet, the global computer network that links more than a million host computers worldwide. The World-Wide Web protocol has made access to the Internet very easy with hypertext links between hundreds of thousands of documents. It is expanding very rapidly and there are many sites that will be of value to planners. Some of the main ones, together with their addresses (or uniform resource locator – URL), are listed here. All the URLs start with *http://*. Users should be aware that web pages and their addresses are constantly updated. If in doubt use one of the search engines.

EUROPE

The European Union Europa site   www.europa.eu.int
     This is an excellent site which provides access to comprehensive information about the Union, its institutions, publications and some news releases

European Centre for Nature Conservation   www.ecnc.nl

European Environment Agency   www.eea.dk/

CENTRAL GOVERNMENT

Central Office of Information   www.coi.gov.uk/coi
     Provides access to the press releases of government departments

Commission for New Towns   www.cnt.org.uk

Department of the Environment   www.open.gov.uk/doe/

Department of Transport   www.open.gov.uk/dot/dothome.htm

Environment Agency of England and Wales   www.environment-agency.gov.uk
     Provides press releases and list of publications

Resline DoE research database for construction and planning   www.open.gov.uk/doe/resline.html

Rural Development Commission   www.open.gov.uk/rdc/rdchome.htm

Scottish Office

UK Government Information Service
    Provides a full index of central government
    departments and other government organisations
    with pages on the WWW

Welsh Department Agency

Welsh Office

www.open.gov.uk/scotoff/scofhom.htm

www.open.gov.uk/index.htm

www.networks.co.uk/wda/home.htm

www.open.gov.uk/woffice/whome.htm

LOCAL GOVERNMENT

Local authority associations (via the Association of
County Councils)

Local government directory of sites
    More than 60 authorities had Internet sites in 1996

Local government information service
    Subscription required

London Government Information Unit

London Research Centre
    and access to the Accompline and Urbaline
    databases by subscription

www.demon.co.uk/logon.acc/acc-home.html

www.tagish.co.uk/tagish.links.localgov.htm
or
www.cityscape.co.uk/users/cw68/index.html

www.demon.co.uk/bcclgis/

www.lgiu.uk/

www.london-research.gov.uk

PLANNING EDUCATION

ACSP (The Association of Collegiate Schools of
Planning (USA))

AESOP (The Association of European Schools of
Planning)

APSA (Asia Planning Schools Association)
    Includes the newsletter and reports together with
    links to the sites of member schools
Planning schools
    Many schools have their own home page with access
    to information on courses and sometimes planning
    information in the region.

For example:

Bristol, West of England

Cardiff (UWCC)
    (with links to other planning related pages)

Coventry

Dundee

Edinburgh, Heriot Watt

Liverpool John Moores

Liverpool
    (with links to other planning pages)

www.gis2.arch.gatech.edu/acsp

www.regplan.kth.se/aesop/

www.hku.hk/cupem/apsa

www.uwe.ac.uk.facults/fbe.html

www.cf.ac.uk/uwcc/cplan/page_16.html

www.coventry.ac.uk/publicat/fullt96/sect7/sbe/
    home.htm

www.hp.dundee.ac.uk

nameserve.hw.ac.uk/ecawww/planning/plan.htm

www.livjm.ac.uk/university/academic/blt/blt.html

www.liv.ac.uk/pjbbrown/cirdes.html

London, South Bank	www.pices.sbu.ac.uk/be/sudp/home.html
London, University College	www.doric.bart.ucl.ac.uk
Manchester	www.man.ac.uk.arts/planning_and_landscape
Newcastle upon Tyne	www.ncl.ac.uk/undergraduate/subject/town.html
(Internet planning directory)	www.ncl.ac.uk/"n270556/urp.html
Nottingham	www.geog.nottingham.ac.uk/urbplan/home.htm
Oxford Brookes	www.cs3.brookes.ac.uk
Reading	www.reading.ac.uk/lm/lm
Sheffield Hallam	www.shu.ac.uk/courses/schools/ursinfo.htm
Sheffield	www.shef.ac.uk/uni/academic/R-Z/trp
Strathclyde	www.strath.ac.uk/department/architecture/index.html

OTHER

Environmental Organisations Web Directory	www.webdirectory.com
Friends of the Earth	www.foe.co.uk
Greenpeace	www.greenpeace.org
International Society of City and Regional Planners (ISOCARP)	www.soc.titech.ac.ip/titsoc/higuchi-lab/isocarp/index/html
Local Sustainability – European Good Practice Information Service	www.cities21.com/europractice
On-line planners forum A US-based forum with a library of Web content on planning, directory and bulletin board	www.asu.edu/caed/onlineplanner
Planning and Architecture Internet Resource Centre (USA)	www.arch.buffalo.edu/pairc
Planning and Internet Resource Centre Links to 4,000 related sites	www.arch.buffalo.edu/pairc/
Resource for Urban Design Information	www.rudi.herts.ac.uk
Royal Town Planning Institute	www.rtpi.org.uk
UK Development Plans Directory	www.apr.co.uk

INDEX OF STATUTES

GENERAL INDEX